T0400060

Advanced Materials and Technologies for Wastewater Treatment

Emerging Materials and Technologies
Series Editor: Boris I. Kharissov

Recycled Ceramics in Sustainable Concrete: Properties and Performance
Kwok Wei Shah and Ghasan Fahim Huseien

Photo-Electrochemical Ammonia Synthesis: Nanocatalyst Discovery, Reactor Design, and Advanced Spectroscopy
Mohammadreza Nazemi and Mostafa A. El-Sayed

Fire-Resistant Paper: Materials, Technologies, and Applications
Ying-Jie Zhu

Sensors for Stretchable Electronics in Nanotechnology
Kaushik Pal

Polymer-Based Composites: Design, Manufacturing, and Applications
V. Arumugaprabu, R. Deepak Joel Johnson, M. Uthayakumar, and P. Sivaranjana

Nanomaterials in Bionanotechnology: Fundamentals and Applications
Ravindra Pratap Singh and Kshitij RB Singh

Biomaterials and Materials for Medicine: Innovations in Research, Devices, and Applications
Jingan Li

Advanced Materials and Technologies for Wastewater Treatment
Sreedevi Upadhyayula and Amita Chaudhary

Advanced Materials and Technologies for Wastewater Treatment

Edited by

Sreedevi Upadhyayula and Amita Chaudhary

CRC Press is an imprint of the
Taylor & Francis Group, an informa business

First edition published 2022
by CRC Press
6000 Broken Sound Parkway NW, Suite 300, Boca Raton, FL 33487-2742

and by CRC Press
2 Park Square, Milton Park, Abingdon, Oxon, OX14 4RN

CRC Press is an imprint of Taylor and Francis Group, LLC

Library of Congress Cataloging-in-Publication Data
A catalog record for this title has been requested

ISBN: 978-0-367-68616-1 (hbk)
ISBN: 978-0-367-68617-8 (pbk)
ISBN: 978-1-003-13830-3 (ebk)

DOI: 10.1201/9781003138303

Typeset in Times
by KnowledgeWorks Global Ltd.

Contents

Preface

This is a handbook on advanced materials and treatment techniques for wastewater for undergraduate and postgraduate students, researchers, and professionals involved in water/wastewater treatment. We found that engineering students (especially chemical and civil), biologists, chemists, environmental scientists, hydrologists, and public-service technical officials have been introduced to wastewater treatment materials and techniques that are currently practiced. However, there are many drawbacks in most of these techniques, and they fall short of giving full-fledged solutions to the new load of pollutants that are being continuously added to water bodies due to the industrial and anthropological activities of the exponentially growing population of the world and its rapid urbanization. Hence, in the authors' view, deeper insights into advanced water treatment materials and technologies are the need of the hour that will allow the world to keep pace with the ever-increasing water pollution problem. This book is an attempt in this direction; it compiles the research and knowledge base of many scientists across the continents, starting with an introduction to waste water and its analysis, followed by methods and technologies combining physical, chemical, biological, and thermo-catalytic treatment techniques, including description of advanced treatment materials and broad areas of treatment techniques with resource recovery and finally, concluding with future challenges and viable solutions.

This book is divided into 15 chapters. It attempts to provide as wide a coverage as possible of all those materials and technologies and their advancements applicable to both potable water and wastewater from industrial and municipal sources.

Chapter 1 deals with characteristics of water, various pollutants affecting water quality, and introduction to conventional and advanced water treatments and challenges. The importance of the circular economy and resource recovery is equally important in today's context and is also discussed briefly.

Chapter 2 covers water quality assessment for a variety of regular domestic and industrial uses and certain essential physicochemical parameters and their role from the point of view of qualitative and quantitative analysis of water.

Chapter 3 describes the role of advanced materials, their functional properties and technologies for sustainable development, their effect on the economy and society, and their significance in preventing environmental degradation and depletion of natural resources.

In Chapter 4, the physical processes involved in wastewater treatment are covered along with advanced methods used in the physical treatment of water, such as membrane filtration and nanotechnology methodology. Also, development of novel adsorbents using natural or recycled materials is explained.

Chapter 5 describes the inorganic and organic polymers used for coagulation in the wastewater treatment industry, with a main focus on the more recent technology of grafted polymers.

Chapter 6 explains pollutant degradation by using ultrasonic treatment and hybrid treatment, in which the ultrasonic method is combined with other techniques such as adsorption and oxidation.

Chapter 7 covers the application of chemistry, chemical reactions, and reagents in almost every step of wastewater treatment.

Chapter 8 covers advanced oxidation processes (AOPs), details the advantages and disadvantages of different AOP technologies, and provides a perspective on future direction needed in the development of AOP technologies for wastewater treatment.

Chapter 9 presents a general review of the catalytic ozonation process and its application to treat potable and wastewater. It comprises eight sections covering fundamentals; homogeneous catalytic ozonation; heterogeneous catalytic ozonation; photocatalytic ozonation; influence factors in the catalytic ozonation system; reactor configuration, catalysts, and their utilization in water treatment; applications of catalytic ozonation in combination processes; and lastly, economic aspects.

Chapter 10 discusses the biological characteristics of water and the use of traditional biological/biochemical methods for treating pollutants, including biological/biochemical technologies in wastewater treatment.

Chapter 11 describes the role of biochar in the removal of contamination due to heavy metals, pesticides, antibiotics, dyes, inorganic and organic compounds in water through π-π surface complexation, electrostatic interaction, co-precipitation, ion exchange, pore filling, diffusion, hydrophobic interaction, and hydrogen bonding mechanisms.

Chapter 12 discusses removal mechanisms of pharmaceutical compounds in microalgae-based systems including sorption, biodegradation, and photodegradation, depending on the physicochemical characteristics of compounds and the environmental conditions.

Chapter 13 focuses on processes for the treatment of industrial wastewater after reviewing current technologies that are being used by different industries to treat various types of wastewater effectively.

In Chapter 14, sustainable technologies for wastewater treatment are reviewed, starting from the most important conventional and emerging wastewater treatment technologies and focusing on their sustainability while comparing their advantages and disadvantages.

Finally, in Chapter 15, challenges, innovations, and future prospects in transforming the future wastewater treatment plants into resource recovery facilities are discussed at length. This chapter brings the readers' attention to the contribution of wastewater treatment plants to the overall nutrient enrichment in our rivers and oceans, advocating the future transformation of conventional wastewater treatment plants into resource recovery factories as the only sustainable solution to reducing both carbon and environmental footprints.

Furthermore, there is a glossary of several hundred terms at the end of this book. This will prove useful to you not only when reading through the chapters but as a general resource reference. Overall, this volume can be used as a comprehensive text on water treatment technologies by students, researchers, and industry professionals equally.

To effectively communicate to a large reader community, the material has been presented in plain, simple English in a concise format with pictorial depictions and summary tables where necessary.

Finally, our heartfelt thanks are extended to all the chapter contributors for their timely submissions and patience in formatting and improving the chapters. Our deep sense of gratitude is extended to the publishers of this book for considering this very useful topic in their book series and giving us the opportunity to participate in this book project and edit this volume.

Sreedevi Upadhyayula
Amita Chaudhary
Indian Institute of Technology Delhi, India
March 1, 2021

Editors

Sreedevi Upadhyayula received her B.Tech in Chemical Engineering at Andhra University in 1991. She completed her Master's degree at the Indian Institute of Technology Kharagpur (IITKGP) and worked in the engineering industry for five years at BHPV Ltd. and M/s APL, India, before becoming a senior research fellow at the National Chemical Laboratory, Pune, in 1999. She worked there until 2001, when she submitted her doctoral thesis at IITKGP. She then joined the Department of Chemical Engineering at the University of Notre Dame as a post-doctoral fellow in 2001. She joined the Department of Chemical Engineering at IITKGP as an assistant professor in 2004. In 2006, she moved to the Indian Institute of Technology Delhi (IIT D). She is currently a chair professor in the Department of Chemical Engineering at IIT Delhi. Her research program focuses on heterogeneous catalysis and its applications to wastewater treatment, environmental pollutant mitigation, reaction kinetics, and modeling of heterogeneous reactions in chemical industry.

Amita Chaudhary received her Bachelor of Science in Chemistry at Chaudhary Charan Singh University, India, in 2000. She completed her Master's degree in Organic Chemistry from Chaudhary Charan Singh University, India. She completed her Bachelor of Education in Chemistry from Guru Gobind Singh Indraprastha University, India, before joining the Department of Chemical Engineering, IIT Delhi, India, as a Junior Research Fellow in 2011 under the supervision of Professor Ashok N. Bhaskarwar. She worked as a senior research fellow in the Department of Chemical Engineering, IIT Delhi, since June 2015 and completed her PhD in January 2019. Currently she is working as an assistant professor at Nirma University. Her research areas include waste management, global warming, and gas-liquid reactions.

Contributors

Gul Afreen
Department of Chemical Engineering
Indian Institutes of Technology Delhi
New Delhi, India

Parveena Bano
Division of Entomology
Sher-e-Kashmir University of Agricultural Sciences & Technology of Kashmir
Srinagar, India

Sai Harsha Bhamidipati
Indian Institute of Petroleum and Energy
Visakhapatnam, India

Amita Chaudhary
Department of Chemical Engineering
Indian Institutes of Technology Delhi
New Delhi, India

Institute of Technology
Nirma University
Ahmedabad, India

Gunjan Deshmukh
School of Chemistry and Chemical Engineering
Queen's University
Belfast, UK

Jennyfer Diaz-Angulo
Research Group in Development of Materials and Products
GIDEMP, CDT ASTIN SENA, Tecnoparque
Cali, Colombia

Escuela de Ingeniería Agroindustrial
Instituto Universitario de la Paz
Barrancabermeja, Colombia

Ankur H. Dwivedi
Institute of Technology
Nirma University
Ahmedabad, India

Eliana M. Jiménez-Bambague
Escuela de Recursos Naturales y Medio Ambiente
Universidad del Valle
Cali, Colombia

Neeraj Kumari
School of Basic and Applied Science
K.R. Mangalam University
Gurugran, India

Jose A. Lara-Ramos
Escuela de Ingeniería Química
Universidad del Valle
Ciudad Universitaria Meléndez
Cali, Colombia

Fiderman Machuca-Martinez
Escuela de Ingeniería Química
Universidad del Valle
Cali, Colombia

Carlos A. Madera-Parra
Escuela de Recursos Naturales y Medio Ambiente
Universidad del Valle
Cali, Colombia

Haresh Manyar
School of Chemistry and Chemical Engineering
Queen's University
Belfast, UK

Sudheshna Moka
Shriram Institute for Industrial Research
New Delhi, India

Miguel A. Mueses
Department of Chemical Engineering
Universidad de Cartagena
Cartagena, Colombia

Krishna Kumar Nedunuri
Environmental Engineering Program
International Center for Water Resources Management
Wilberforce, Ohio

Pablo Ortiz
Department of Chemical Engineering
Universidad de los Andes
Bogotá, Colombia

Aura C. Ortiz-Escobar
Escuela de Recursos Naturales y Medio Ambiente
Universidad del Valle
Cali, Colombia

Chandan Kumar Pal
Department of Chemistry
Scottish Church College
University of Calcutta
Kolkata, India

Firdaus Parveen
Department of Chemical Engineering
Indian Institutes of Technology Delhi
New Delhi, India

Department of Chemistry
Imperial College London
White City London, UK

Shibu G. Pillai
Department of Chemical Engineering
School of Engineering
Institute of Technology
Nirma University
Ahmedabad, India

C.R. Venkateswara Rao
Department of Humanities and Basic Science
Gokaraju Rangaraju Institute of Engineering and Technology
Hyderabad, India

Mushtaq Ahmad Rather
Department of Chemical Engineering
National Institute of Technology Hazratbal
Srinagar, India

Biswajit Saha
Department of Chemical Engineering
Indian Institutes of Technology Kharagpur
Kharagpur, India

Department of Chemical and Biological Engineering
University of Saskatchewan
Saskatoon, Canada

Mamta Saiyad
Department of Chemical Engineering
School of Engineering
Institute of Technology
Nirma University
Ahmedabad, India

Sonali Sengupta
Department of Chemical Engineering
Indian Institutes of Technology Kharagpur
Kharagpur, India

Nimish Shah
Department of Chemical Engineering
School of Engineering
Institute of Technology
Nirma University
Ahmedabad, India

Sreedevi Upadhyayula
Department of Chemical Engineering
Indian Institutes of Technology Delhi
New Delhi, India

Dharani Prasad Vadlamudi
Indian Institute of Petroleum and Energy
Visakhapatnam, India

Ning Zhang
International Center for Water Resources Management
Wilberforce, Ohio

Sushma
Department of Industrial Waste Management
Central University of Haryana
Mahendergarh, India

1 Introduction to Wastewater

Mamta Saiyad
Chemical Engineering Department, School of Engineering, Institute of Technology Nirma University, Ahmedabad, India

CONTENTS

DOI: 10.1201/9781003138303-1

1.1 INTRODUCTION

Water is a necessity for the survival of all living beings, and it is the lifeline of the environment, plants, animals, and humans. Everything possible must be done to maintain its quality for today and the future. Water naturally exists in various sources. It is in the air, underground, in the rivers and springs, and in the ocean. Water is the lifeblood for living beings. It is needed for daily consumption, irrigation, aquatic life, recreational uses, fisheries, and reuse and recycling of sewage and industrial waste. Water circulation systems are critical for supplying safe water. Due to the increase in the worlds' population, water consumption has increased hugely, and the resources are fewer; hence the water demand is growing day by day. Fresh liquid water must be accessible for human use such as drinking, preparing food, washing clothes, and other necessary functions. In dry areas, groundwater is used to supply drinking water, support farming, and agriculture. As long as the water is not withdrawn faster than nature can refill it, groundwater may be considered renewable, but in many regions the water does not renew by itself or it renews very slowly. Provision of good-quality water, usually in terms of its suitability for a specific purpose, is of prime importance. Acceptable values for chemical, physical, and biological parameters depend on the use and not on the source of the water. Each specific use has its quality criteria that must meet the standards for the application. Therefore, water suitable for one use may be not suitable for another. For instance, drinking water can be used for industry, but the water used for the industry or agriculture may not be appropriate for drinking purpose.[1]

From a technical aspect, quality is determined by the types and amounts of substances suspended and dissolved in the water and their effects on the inhabitants of the ecosystem. Physical and chemical criteria are fixed to the maximum amount of pollutants, acceptable ranges of physical parameters, and the minimum amount of desirable parameters, such as dissolved oxygen. Biological criteria describe the presence of species and classes of organisms such as viruses, bacteria, plants, and other entities, reasonable for public health and established values.

To protect the quality of water, administrations have fixed standards for concentration limits and issued scientifically determined guidelines that can be acceptable for particular water use such as drinking, irrigation, or recreation such as swimming. These standards are also applicable to the selection of raw water sources and the choice of treatment processes.

A process to remove impurities from collected wastewater to convert it into an effluent that can be recycled with an acceptable effect on the environment and reused for various purposes is called wastewater treatment. It is also known as water recovery.

1.2 WASTEWATER

Any water that has been polluted by living beings is called wastewater. It can be defined as "water used from sources like domestic, industrial, commercial and agricultural activities, surface runoff, stormwater, sewer inflow/infiltration." Therefore, wastewater is a by-product of these activities.

Human activities involve drainage components that can seriously affect surface water pollution. Very fine substances that are dispersed in the surface water are responsible for turbidity. These substances cannot be separated by sedimentation or filtration of the water. Water also contains soluble and insoluble pollutants. Groundwater, surface runoff, and stormwater can be further contaminated by the addition of man-made products such as petroleum, oil, road dust, and a variety of chemicals that get into it and make it unsafe and unfit for human use (Fig. 1.1).

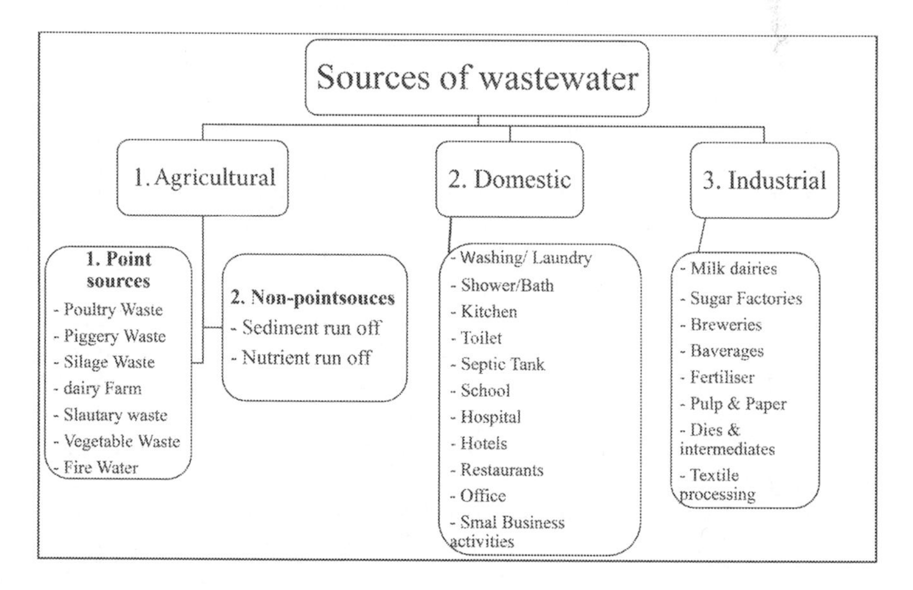

FIGURE 1.1 Sources of wastewater.

Apart from the main sources of water pollution such as industrial waste, sewage, agricultural waste, and other wastewater, mining activities, marine dumping, unplanned oil leakage, burning of fossil fuels, chemical fertilizers, pesticides, and leakage of sewer lines, etc. are also sources of water pollution as these add an excess of different contaminants, toxic waste, petroleum, and disease-causing microorganisms to water bodies.

1.3 POLLUTANTS

Following are the main groups of pollutants that affect surface water:

- **Organic material:** Wastewater contains organic materials such as protein, carbohydrate, fat, oil, grease, urea, and various synthetic organic molecules in large numbers (e.g., surfactants, volatile organic compounds, agricultural pesticides). Organic material can be found in surface waters either in a suspended form or as a dissolved compound. It can be determined by analyzing the biological oxygen demand (BOD) and chemical oxygen demand (COD) of the wastewater.
- **Phosphorus:** Phosphorus is found in surface water as phosphate and organic compounds. It can originate from distributed and point sources. The main source is rainwater overflow from the ground, including agricultural soil. The source of phosphorus pollution is from human waste, urban and rural sewage, industrial wastewater, and wastewater from animal husbandry.
- **Heavy metals:** Heavy metals such as mercury, lead, chromium, and cadmium can be found mainly in industrial discharges, rainwater overflows, city streets, mining activities, and landfills leachate, and agriculture can also contribute to heavy metal pollution that is highly contained with pesticides and fertilizers. Most of these heavy metals are generally hazardous for underwater life, ecosystems, and human health as well. The polluting ability of heavy metals depends on their concentration in water and in the form in which they are present.
- **Detergents and surfactants:** Detergents and surfactants have a good amount of biodegradability and reasonable toxicity for aquatic living beings; they can affect the oxygen demand of water and slow down the sedimentation process, thus delaying water clarification. They are common contaminants of water due to their heavy use in washing and cleaning processes. These synthetic compounds are characterized by a negatively charged polar head and a nonpolar tail consisting of a carbon-based linear or branched chain. The group constituting the polar head is responsible for the water solubility of detergents, which can be anionic, cationic, or polar. The alkylbenzene sulfonate salts are generally used in everyday washing products.
- **Traces of synthetic organic compounds:** Many traces of synthetic organic compounds can be found as contaminants in surface water. Even very low concentrations can lead to a long-term risk due to the accumulation either in food particles or in water residue. The source of these organic compounds

can be agricultural activities, such as fertilizer and pesticide use and industrial discharges. The type of pollutant changes according to the type of processed material. Examples of hazardous contaminants present in industrial activities are petroleum products, solvents, polychlorinated biphenyls, polychlorinated phenols, polyaromatic hydrocarbons, dioxins, and many more. The toxicity of synthetic organic compounds is due to many factors, such as their molecular structure, water solubility, biodegradability, photo-degradability, and volatility. Insoluble or fatty soluble substances are more dangerous than soluble chemicals as they accumulate in organisms.

- **Biodegradable organic chemicals:** Biodegradable organic chemicals such as detergents and surfactants have low toxicity risk as they are decomposed by microorganisms present in water and therefore are less susceptible to the accumulation process. Volatile substances reduce contamination potential for surface water as they are likely to leave the liquid phase. Their presence in water favors the aeration process. Organic compounds, which are susceptible to light or photo-degradation, generally limit their polluting effect, unless the decomposition process produces toxic substances.

1.4 CHARACTERISTICS OF WASTEWATER

The characteristics of wastewater mainly depend on the sources from which it is collected. Types of wastewater include domestic (from households), municipal (from communities, such as sewage) industrial, and agricultural activities.[2]

Wastewater is classified mainly according to the following three characteristics (Fig. 1.2):

- Physical characteristics
- chemical characteristics due to the chemical impurities
- biological characteristics due to the contaminants

Analysis, standards, and regulation of physical, chemical, and biological characteristics of wastewater are necessary and have a critical role in the process.

1.4.1 Physical Characteristics of Wastewater

1.4.1.1 Turbidity

Suspended solids in wastewater, such as mud, sludge, finely dispersed organic and inorganic matter, colored organic compounds, and microorganisms, impart turbidity or unclean opacity to water. Hence, turbid water has a muddy, dirty, and cloudy appearance that looks unattractive.

The higher acidic pH value has a lower range of turbidity, and the more basic value of pH has a higher range of turbidity values. Hence, the more basic the water is, the cloudier it will be. Settling and decanting is the method to reduce turbidity by leaving the water to settle for 2–24 hours so that the particulates settle down to the

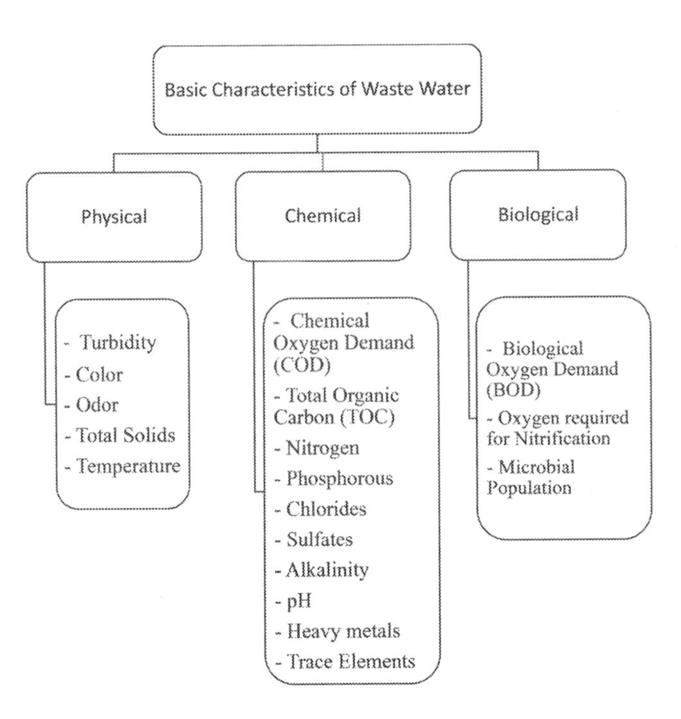

FIGURE 1.2 Basic characteristics of wastewater.

bottom of the tank. Then the clear water is decanted from the top into another tank. An increase in turbidity in streams affects the growth rate of algae and aquatic and micro-aquatic plants and decreases the amount of light for photosynthesis. Turbidity can also increase water temperature because suspended particles absorb more heat.

1.4.1.2 Color

Water can get color from many natural dissolved organics along with synthetic chemical dyes. To supply acceptable water for consumers and industrial processes, the color must be removed. To remove color, inorganic coagulants such as alum, lime, ferric or ferrous sulfate, and polyaluminum chloride are generally used. In most cases, these coagulants prove to be effective in removing color from wastewaters containing dissolvable solids. Sewage is a light brownish-gray color when it is fresh. In two to six hours, sewage will change from fresh to stale at temperatures above 20°C. Due to anaerobic activities, older sewage converts to dark gray and black color, called stale or septic color.

1.4.1.3 Odor

Due to anaerobic digestion in the wastewater, hydrogen sulfide (H_2S) is generated as a natural by-product that gives off a strong, unpleasant smell. It has low solubility in wastewater; hence it is released into the atmosphere, producing an aggressive odor. Amines and mercaptans also cause odor at treatment plants. These gases release a very strong odor that directly affects operators and people who live or work nearby.

Increasing the oxygen supply will help to control odors as anaerobic digestion will slow down. The aerobic bacteria use this oxygen for digestion of wastewater. Also the circulation of wastewater can create nearly equal dissolved oxygen throughout. This in turn will facilitate odor-free aerobic digestion.

1.4.1.4 Temperature

Normally sewage temperature varies between 30°C and 35°C. Wastewater has a higher temperature because of more biological activity. Wastewater treatment greatly affects operating temperatures. However, it is desirable to keep the temperature of wastewater between 35°C and 40°C to avoid the growth of microorganisms.

1.4.1.5 Total Solids

Matter that is suspended, dissolved, and settleable in water or wastewater with particle size more than 2 μ (micron), is called "total solids." It is related to specific conductance and turbidity. Analytically, total solids can also be referred to as the residue left in a container after evaporation and drying of a water sample at 103°C–105°C. Total solids, or residue upon evaporation, can be classified as either suspended or filterable solids by passing a known volume of liquid through a filter with a pore size of 1μ (micron). The matter with a significant vapor pressure at this temperature may evaporate and is not defined as a solid.

1.4.1.6 Total Suspended Solids

Suspended solids have a diameter of more than 1 μ, which remains in the suspension of water. They may be removed by settling if their density is greater than that of the water. They may be removed by sedimentation and filtration of water. Filterable solid elements consist of colloidal and dissolved solids. The colloidal fraction is particulate matter with an approximate diameter range from 1 mμ to 1μ. They do not settle and cause turbidity. Dissolved solids consist of organic and inorganic molecules and ions that are present in water. Generally, biological oxidation or coagulation, followed by sedimentation, is required to remove these particles from suspension.[4] Suspended solids restrict effective drinking water treatment as high sediment loads restrict coagulation, filtration, and disinfection. A high load of total solids will make drinking water unpleasant and can have an adverse effect on vulnerable people. Total solids also affect water clarity. Higher solids increase the turbidity and decrease the amount of light for the photosynthesis process of aquatic plants.

The solids content of wastewater may be classified approximately as shown in Fig. 1.3.

1.4.1.7 Total Dissolved Solids

Total dissolved solids (TDS) are the substances present in water that can pass through a filter. They represent the concentration of dissolved substances in water. TDS are generally a form of inorganic salts as well as a small amount of organic matter. Common salts that can be found in water include calcium, magnesium, potassium, carbonate, bicarbonate, chloride, sulfate, phosphate, nitrate, sodium, organic ions, and other ions. Among them, calcium, magnesium, potassium, and sodium are all cations; and carbonates, nitrates, bicarbonates, chlorides, and sulfates are all anions.

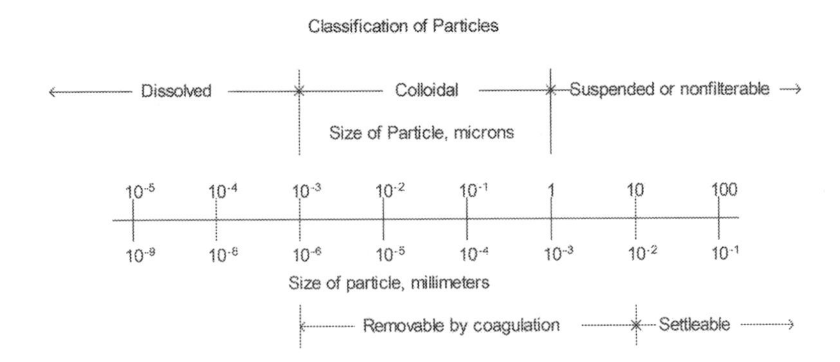

FIGURE 1.3 Classification and size of particles found in wastewater.

Cations are positively charged ions, and anions are negatively charged ions. Thus dissolved solids refer to any minerals, salts, metals, cations, or anions dissolved in water. According to the Bureau of Indian Standards (BIS), water is not acceptable for drinking if the upper limit of the TDS level in water is 500 ppm. According to the World Health Organization (WHO), however, the limit is 300 ppm.

1.4.2 Chemical Characteristics of Wastewater

Wastewater contains chemicals such as organic materials, inorganic materials, gases, and dissolved oxygen as well as others that may affect its composition and pH rating. Acidity or alkalinity in wastewater is the main reason for industrial waste and requires pre-treatment before discharge to the sewer or spring water. Wastewater contains various chemicals in different forms as mentioned below.

1.4.2.1 Organic Materials

Organic compounds are mainly composed of carbon, hydrogen, oxygen, and nitrogen. Other elements such as sulfur, phosphorus, and iron, may also be present. The principal groups of organic substances found in wastewater are protein, carbohydrates, fats and oils. Urea also contributes to wastewater as it is a main constituent of urine. Biodegradable organic substance in the water resource reduces the oxygen of the water. Non-biodegradable matter makes the wastewater treatment processes complicated being difficult to degrade. This non-degradable element imparts some of the harmful effects such as the odor and color, to the water resource, imparts toxic effects on aquatic life, and increases the additional cost for treatment.

1.4.2.2 Inorganic Materials

Common inorganic substances present in normal wastewater are chloride, alkaline compounds, hydrogen ions, nitrogen, sulfur, phosphorus, and heavy metals. Even a fractional presence of these substances can significantly affect organisms due to their growth limiting characteristics. Normally in natural water, the presence of phosphorus is significantly low in concentrations than nitrogen or carbon.

However, even extremely low concentrations of phosphorus may sustain the algae growth. Chloride, even in a very low concentration, can pollute a large amount of water.

1.4.2.3 Gases

In raw wastewater commonly found gases are oxygen, nitrogen, carbon dioxide, ammonia, hydrogen sulfide, and methane. All these gases should be considered while designing the treatment processes; however, more attention should be given to oxygen, methane, and hydrogen sulfide concentrations in the treated wastewater. Dissolved oxygen is essential for all aerobic life, during the treatment and in the water resource. In the absence of aerobic conditions, oxidation is brought about by reducing inorganic salts such as sulfates or by reducing the activity of methane forming bacteria. Generally, wastewater end-products are very noxious. To eliminate such conditions in the wastewater treatment and the natural water resource receiving the effluent, an anaerobic state must be maintained.

1.4.2.4 Chemical Oxygen Demand

Chemical oxygen demand (COD) is a significant parameter of water and wastewater quality. It is the oxygen required to oxidize the organic materials in the wastewater. The COD test is a measurement to monitor the efficiency of water treatment plants. The basis of this test is that a strong oxidizing agent, under acidic conditions, can fully oxidize almost any organic compound to carbon dioxide. High COD levels in stormwater runoff are caused by sources like residual food and beverage waste, antifreeze, emulsified oils from industrial food processing, and agricultural activities. As most sources of COD are water-soluble, this pollutant spreads easily into streams. Higher COD levels indicate a higher amount of oxidizable organic material in the sample, which can reduce dissolved oxygen levels. A reduction in dissolved oxygen levels can lead to anaerobic conditions, which is harmful to aquatic life.

1.4.2.5 pH

pH is the measurement of the hydrogen ion concentration in a solution. It is an important parameter in the characterization of water and wastewater. It is a very important factor in biological and chemical wastewater treatment. Water and wastewater can be classified as neutral, alkaline, or acidic according to the following ranges: $pH = 7$ neutral, $pH > 7$ alkaline, $pH < 7$ acidic.

1.4.2.6 Total Organic Carbon

Total organic carbon (TOC) is the number of carbon atoms present in organic compounds of a water sample. It is an indicator of water quality because pure water does not contain carbon. It does not identify specific carbon-containing compounds; it only specifies the presence of undesirable organic compounds in pure water. TOC levels in water can be increased upon exposure to air, so in the case of bottled water, the baseline TOC levels will increase after the containers are opened. Hence for protection against contamination, the water from any high-purity water system should be used immediately.

1.4.2.7 Heavy Metals

The adverse effects and allowable concentrations of heavy metals in waste water (Table 1.1):

TABLE 1.1
The Adverse Effects and Allowable Concentrations of Heavy Metals

Heavy Metal	Toxicity Effects	MCL (mg/L)
Mercury	Rheumatoid arthritis and diseases of the kidneys, circulatory system, and nervous system	0.00003
Arsenic	Skin manifestations, visceral cancers, vascular disease	0.050
Lead	Damage to the fetal brain and diseases of the kidneys, circulatory system, and nervous system	0.006
Zinc	Depression, lethargy, neurological signs, and increased thirst	0.80
Cadmium	Kidney damage, renal disorder, human carcinogen	0.01
Copper	Liver damage, Wilson disease, insomnia	0.25
Nickel	Dermatitis, nausea, chronic asthma, coughing, human carcinogen	0.20
Chromium	Headache, diarrhea, nausea, vomiting, carcinogenic	0.05

Methods such as chemical precipitation, ion exchange, adsorption, membrane filtration, reverse osmosis, solvent extraction, and electrochemical treatment have been used to remove heavy metals from contaminated water. Many heavy metals are toxic; however, some heavy metals are essential. Metals such as bismuth have low toxicity. Mostly the term "toxic metals" includes cadmium, manganese, lead, mercury, thallium, and radioactive metals. The adsorption method is used to remove heavy metals from wastewater. In this process, a liquid solute — an adsorbate — accumulates on the surface of a solid adsorbent and forms a molecular or atomic film.

1.4.2.8 Others

- **Phosphates (PO_4^-):** Phosphates are chemicals containing phosphorous. They cause excessive algae growth which unbalances the ecosystem and thus affects the water quality by producing harmful toxins.
- **Chlorides (Cl^-):** Chloride in water is a tough problem to solve. A very small amount can pollute water permanently. At high concentrations, chloride can harm fish and plant life. Chlorides can contaminate freshwater streams and lakes. Chlorides increase the electrical conductivity of water and thus corrode metals. It affects the taste of food products. Once it enters the water, there is no easy way to remove it.
- **Sulfates (SO_4^{-2}):** Sulfate compounds are found in nature and are naturally present in water. A high level of sulfate in water causes a bitter taste. Sulfates are also found in minerals, soil, rocks, plants, and food. At very high concentrations its toxicity is harmful to cattle. It may contribute to the acidification of surface water and soil. Sulfates and hydrogen sulfide are common troublesome contaminants. Although there is not any significant health hazard due to sulfates, they can have a temporary laxative

effect on humans. Sulfates also may clog plumbing and stain clothing and make it difficult to clean clothes. Sulfates are also harmful to the skin and hair. Reverse osmosis, distillation, or ion exchange treatment systems will remove sulfates from drinking water. Carbon filters, water softeners, and sediment filters will not remove sulfates.

1.4.3 Biological Characteristics of Wastewater

1.4.3.1 Biochemical Oxygen Demand

Biochemical oxygen demand (BOD) is the amount of oxygen required by microorganisms to decompose organic matter under aerobic conditions at the specified temperature. High BOD conditions make it difficult for aquatic life to survive. This can occur from sources such as sewage water, septic tank leakages, agricultural runoffs, landslides, erosion, etc. It can be removed by supplying oxygen during treatment.

1.4.3.2 Nitrogenous Oxygen Demand

Nitrogenous oxygen demand (NOD) is the amount of dissolved oxygen required for the biological oxidation of nitrogenous materials. Nitrogen in ammonia and organic nitrogen in wastewater may increase NOD.

1.4.3.3 Microbial Life in Wastewater

Wastewater contains the following microbes:

1. Bacteria
2. Protozoa
3. Fungi
4. Viruses
5. Algae
6. Rotifers
7. Nematodes

These can be removed by floc formation, harvesting of bacteria, and removal of suspended material.

1.4.3.4 Oil and Grease

Oil and grease originate from food waste and petroleum products.

Domestic, municipal, and industrial wastewaters are the major sources of oils and grease. Apart from that, marine activities, oil leakages, etc. may also contaminate water. Microbes such as bacillus are used to treat microorganisms, fats, oils, and grease in the wastewater treatment plant.

1.5 WASTEWATER TREATMENTS — AN OVERVIEW

Wastewater treatment is a process to remove contaminants from wastewater or sewage and convert it into effluent that can be returned to the water supply with an acceptable impact on the environment and can be reused for suitable purposes. Effluent can be defined as an outflow of treated water to the receiving water body. The contaminants from the collected water can be removed to a large extent by separation processes such as sedimentation, flotation, and filtration to produce high-quality water. Direct filtration of flocs can be used for raw water with low suspended solids. The primary objective of the treatment is to remove the particulate substances and dissolved inorganic and organic substances. These particulate substances must be removed by coagulation and flocculation using precipitation, coprecipitation, or adsorption. Phosphates, heavy metals, and dissolved organic substances of natural or anthropogenic origin are also removed by this treatment.[3] See Table 1.2 for standards to be maintained for Common Effluent Treatment Plants by Government of India.

TABLE 1.2
Standards Set by the Ministry of Environment and Forests, Government of India for Common Effluent Treatment Plants per Environment Protection Rules, 1986

A. Primary Treatment

Parameter for Inlet Effluent Quality of CETP	Standards (Concentration in mg/l)
pH	5.5–9.0
Temperature (°C)	45
Oil and grease	20
Phenolic compounds (as C6H5OH)	5
Ammonical nitrogen (as N)	50
Cynide (as N)	2
Chromium hexavalent (as Cr+6)	2
Chromium (total) (as Cr)	2
Copper (as Cu)	3
Lead (as Pb)	1
Nickel (as Ni)	3
Zinc (as Zn)	15
Arsenic (as As)	0.2
Mercury (as Hg)	0.01
Cadmiun (as Cd)	1
Selenium (as Se)	0.05
Fluoride (as F)	15
Boron (as B)	2
Radioactive Materials:	
Alpha emitters, Hc/mL	10^{-7}
Beta emitters, He/ml	10^{-8}

Notes: 1. These standards apply to the small-scale industries (i.e., total discharge up to 25KL/Day).
2. For each CETP and its constituent units, the state board will prescribe standards according to local needs and conditions; these can be more stringent than those prescribed above. However, in case of clusters of units, the state board, with the concurrence of CPCB in writing, may prescribe suitable limits.

B. Treated Effluent Quality of Common Effluent Treatment Plant

(Concentration in mg/l except pH and Temperature)

Parameters	Into Inland Surface Waters	On Land for Irrigation	Into Marine Coastal areas
pH	5.5–9.0	5.5–9.0	5.5–9.0
BOD (3 days at 27°C)	30	100	100
Oil and grease	10	10	20
Temperature	Shall not exceed 40°C in any section of the stream within 15 meters down steam		
From the effluent outlet	–	45°C at the point of discharge.	
Suspended solids	100	200	a. For process waste water 100 b. For cooling water effluent 10 percent above total suspended matter of effluent cooling water
Dissolved solids (inorganic)	2100	2100	–
Total residue chlorine	1.0	-	1.0
Ammonical nitrogen(As N) 50	-	50	
Total Kjeldahl nitrogen (as N)	100		–100
Chemical oxygen Demand	250	-	250
Arsenic (as As)	0.2	0.2	0.2
Mercury (as Hg)	0.01	-	0.01
Lead (as Pb)	0.1	-1.0	
Cadmium (as Cd)	1.0	-	2.0
Total cadmium (as Cr)	2.0	-	2.0
Copper (as Cu)	3.0	-	3.0
Zinc (as Zn)	5.0	-	15
Selenium (as Se)	0.05	-	0.05
Nickel (as Ni)	3.0	-	5.0
Boron (as B)	2.0	2.0	-
Percent sodium	-	60	-
Cynide (as CN)	0.2	0.2	0.2
Chloride (as Cl)	1000	600	-
Fluoride (as F)	2.0	-	15
Sulfate (as SO4)	1000	1000	-
Sulfide (as S)	2.8	-	5.0
Pesticides	Absent	Absent	Absent
Phenolic compounds (as C6H5OH)	1.0	-	5.0

Note: All efforts should be made to remove color and unpleasant odor as far as possible.

1.5.1 Stabilization

Stabilization reduces organic content and removes pathogens from wastewater. It can be done biologically and chemically. Biological sludge stabilization can prevent anaerobic breakdown, and chemical adsorption is done by using polymers in which Van der Waals forces present in the particles adsorb by the electrostatic charges on the adsorbent surface or by forces of hydration. Colloidal and suspended particles can be stabilized by aggregation.

1.5.2 Destabilization

In adsorption, stable particles are extracted by destabilizing through neutralization of their negative surface charges using adsorption of positively charged coagulants. In hydrolysis, the metal salts transform into metal hydroxides. The aggregated particles are separated by either sedimentation or filtration.

1.5.3 Precipitation

Precipitates are formed by insoluble substances from dissolved matter and the chemicals that are present. In the precipitation process, salts that can precipitate as hydroxides in the neutral pH region can be removed from the water using coagulants such as Al^{3+} and Fe^{3-}.

1.5.4 Coprecipitation

In the coprecipitation process, dissolved and particulate substances are precipitated using Al^{3+} and Fe^{3-} hydroxide to aggregate the particles; they are removed by coagulation mechanisms and then flocculated using amorphous, gelatinous hydroxide flocculants for particle collection.

1.5.5 Coagulation

Coagulation is the process of removing particles in colloidal or suspended forms. Wastewater colloidal particles less than 1 μ in size undergo Brownian motion and usually do not settle down by suspension. Coagulation combines small particles into larger aggregates that are removed by using the solid/liquid separation process. In this process, the electrical repulsion between individual particles is reduced, inverted, or neutralized. This will cause the individual particles to agglomerate at one point as floc. Some common coagulants are aluminum sulfate, aluminum chloride, polyaluminum chloride, ferric chloride, ferrous sulfate, lime, calcium chloride, and magnesium chloride. Coagulants are chemicals used for destabilization in the initial phase of floc formation. Nowadays, organic polymers are used instead of conventional Al^{+3} and Fe^{-3} coagulants. Technically, coagulation is preferable for the floc formation after chemical mixing and destabilization and before dosing of flocculants.

1.5.6 Flocculation

Flocculation is the process by which the coagulated particulates are clumped together in a floc, which will generally float to the top of the liquid or settle down at the bottom of the liquid to be filtered out. Flocculants are substances that promote agglomeration of fine particles present in water, creating a floc, which either floats or settles down and can be easily removed from the liquid. Flocculants counterbalance the charge of the particles, adsorb on the particles, and cause destabilization by bridging and/or charge neutralization. In water treatment, flocculation is the phase of formation of large flocs by the addition of a secondary chemical (after the addition of the coagulant) called a flocculent. Nowadays polymeric flocculants are widely used. The polymeric material forms bridges between individual particles. Bridging occurs when segments of a polymer chain adsorb on different particles and help the particles agglomerate into larger particles. This can largely flocculate the particles with a very small amount of flocculant.

1.6 NOVEL TREATMENTS AND MATERIALS FOR WASTEWATER PURIFICATION

1.6.1 Adsorbents and Absorbents

Adsorption is a wastewater purification technique used to remove a wide range of compounds from industrial wastewater. This process uses adsorbents that have a very large surface area which allows molecules to adsorb on the surface of a solid substance. Industries such as textiles, leather, paper, cosmetics, plastics, food, etc. follow a dyeing process that uses organic dyes, intermediates, and water. These are highly colored organic compounds with low biodegradability. Their disposal into the water resources causes a major water pollution.[6] It is essential to remove these chemicals from wastewaters before discharge into water resources. For this purpose, coagulation, flocculation, biodegradation, adsorption, ion-exchange, and advanced oxidation are the most commonly used techniques.[7] Among them, adsorption is accounted as the most effective and economical one.[8] Adsorbents such as fly ash, coal, mulch, sawdust, lignite, and wood are widely used, mainly because of their availability, low cost, and substantial surface area. Graphene-based materials, including graphene oxide[9] and reduced graphene oxide,[10] are also gaining attention for adsorption of cationic and anionic dyes and intermediates. Polymers are another class of adsorbents compared to conventional adsorbents such as activated carbon and clay. Advantages of polymers are their favorable physicochemical stability, high selectivity and structural diversity, eco-friendliness, and regeneration abilities.[11–13] Accordingly, they have found various applications in the removal of cationic and anionic dyes.[14–16]

1.6.2 Coagulants

As mentioned in Section 1.5, normal coagulants are used because of the low cost. They are available in various commercial forms and purities. Nowadays soluble mixed coagulants are widely used for wastewater treatment. They easily mix in water and form hexaaqua complexes. Hexaaqua complexes vary from one metal to another, with six water molecules attached to the central ion by a coordinate bond. The metal ions from $Me(H_2O)^{3+}$ (hexaaqua complexes) are acidic species that give protons and aggregates colloidal particles. Cationic polymers with high charges are also used as coagulants, as an alternative to common aluminum and ferrous salts. See Table 1.3 for coagulants commonly used in industries.

TABLE 1.3
Coagulants

Type	Formula
Alum	$Al_2 (SO_4)_3 \cdot 14 H_2O$
Polyaluminum chloride	$Al (OH)_{1.5} (SO_4)_{.125} \cdot Cl_{1.25}$
Sodium aluminate	$NaAlO_2$
Iron (III) chloride	$FeCl_3$
Iron (III) sulfate	$Fe_2(SO_4)_3$
Iron (III) sulfate Chloride	$FeClSO_4$
Iron (II) sulfate	$FeSO_4 \cdot 7H_2O$
Sodium silicate (Activated silica)	Na_2SiO_3

1.6.3 Flocculants

Flocculants are high molar mass water-soluble salts. Nowadays water-soluble polymers are also widely used as flocculants. Selected chemicals and polymers are used to increase the formation of larger flocs. Inorganic flocculants form insoluble hydroxide precipitates in water. Organic flocculants are linear, water-soluble polymers that can be synthetic, natural, or modified natural polymers. See Table 1.4 for flocculants commonly used in industries.

1.6.3.1 Polymeric Flocculants

Nonionic acrylamide, anionic acrylic acid, cationic N, N (dimethyl aminopropyl) methacrylate, and cationic ethyleneamine monomers are copolymerized to use as polymer flocculants. Depending on copolymer composition, the charge density is very high; most commercial polymers are copolymers. Synthetic organic flocculants are water-soluble polymers; they can be classified into nonionic polymers and ionic polymers and can be further subdivided into anionic, cationic, and amphoteric types. Examples of flocculants from each of these groups are given in the table.

In addition to synthetic polymers, there are natural organic flocculants[18] which include water-soluble starch, guar gum, alginates, and products based on chitin, glue, and gelatin. These flocculants have relatively low molecular masses with moderate efficiency. Despite the very low cost of some of these natural products, the cost-to-performance ratio of synthetic flocculants is considerably more favorable due to

TABLE 1.4
Organic Synthetic Flocculants

TYPE	FORMULA
Polyacrylamide	$-[CH_2CH(CONH_2)]_n-$
Poly (ethylene oxide)	$-[CH_2CH_2O]_n-$
Poly(sodium acrylate)	$-[CH_2CH(COO^-Na^+)]_n-$
Poly[2-(N, N, N – trimethylamino)-ethyl acrylate] (chloride salt)	$-[CH_2CH(COOCH_2CH_2N^+(CH_3)_3\ Cl^-)]_n-$
Polyethyleimine	$-[NCH_2CH_2NHCH_2CH_2(CH_2CH_2NHCH_2CH_2NH_2)]_n-$
Poly[N-(dimethylamino-methyl)acrylamide]	$-[CH_2CH(CONHCH_2N(CH_3)_2)]_n-$

their very high relative molecular masses. The natural polymers are less effective, unstable, and may support bacterial growth compared to synthetic polymers. But in another way, they are safe for aquatic life.

1.6.3.2 Synthetic Biopolymer Flocculants

The term "synthetic biopolymer" is applied to polymers prepared from the comonomers of synthetic and same or similar monomer units of natural polymers. These flocculants are soluble synthetic polymers that display biological activity.

1.6.4 Nanomaterials for Wastewater Treatment

The application of nanomaterials in water and wastewater treatment is increasing day by day. At present, the nanomaterials for water and wastewater treatment under consideration are zero-valent metal nanoparticles, metal oxides nanoparticles, carbon nanotubes (CNTs), and nanocomposites. In addition to this, the graphene-coated ultrathin nanofilter is the most effective filter, which could be commercialized for water purification. These membranes can be used in various forms such as free, surface modified, and graphene cast in membranes in the range of micro, nano, or ultrafilters.

Nanotechnology in water purification uses nanoscopic materials like carbon nanotubes and alumina fibers for nanofiltration. For example, carbon nanotube membranes can remove almost all kinds of water contaminants, including turbidity, oil, bacteria, viruses, and organic contaminants.

1.6.5 Membrane Filtration Technique

The membrane-processing technique permits concentration and separation without using heat. Particles are separated by their molecular size and shape using pressure and specially designed semi-permeable membranes. In this technique, the membrane acts as a filter that allows water to flow through; during the flow it catches suspended solids and other substances. There are four basic pressure-driven membrane filtration processes for liquid separations. These are, in ascending order of the size of particles that can be separated: reverse osmosis, nanofiltration, ultrafiltration, and microfiltration.

Membrane filters are constructed out of a wide range of synthetic materials, including cellulose acetate, cellulose nitrate (collodion), polyamide (nylon), polycarbonate, polypropylene, and polytetrafluoroethylene (Teflon).

1.6.6 Eco-Friendly Techniques for the Remediation of Pollutants

Physicochemical methods used to remove a variety of pollutants from industrial wastewater are costly and environmentally harmful and may create secondary pollution and ultimately damage environmental quality. To overcome these problems, various eco-friendly technologies are emerging and becoming popular for the removal of pollutants from industrial wastewaters.[1]

Natural resources such as wetlands, stream buffers, and vegetated land cover can also naturally filter out pollutants such as metals, pesticides, sediment, and overabundant nutrients that may affect water quality.

1.6.7 Photocatalysis as Oxidation Processes

Photocatalysis is the process by which a light source interacts with the surface of semiconductor materials. During this process, two simultaneous reactions occur—oxidation from photo-generated holes and reduction from photo-generated electrons. The photocatalytic oxidation process combines UV irradiation with a catalyst, which results in a reaction that changes malicious contaminants into water, carbon dioxide, and waste. The produces harmless water molecules, carbon dioxide, and waste.

1.6.8 Smart Materials

Smart materials are used to develop economical and high-performance treatment systems. They can also instantly and continuously monitor water quality. Smart materials can be helpful to reuse, recycle, desalinize, and to detect biological and chemical contamination from municipal, industrial, or man-made water sources. In the research and development of water treatment technology, smart materials are now extensively used for treatment, remedy, and pollution prevention. They can maintain water quality for the long term. A large number of researchers are working on concepts, approaches, and applications of smart materials in the area of water treatment techniques like nanofiltration, ultrafiltration, reverse osmosis, adsorption, and nanoreactive membranes, etc.

The focus of the development of smart materials is on various carbon nanomaterials for wastewater treatment, synthetic nanomaterials for pollutants removal, bio-polymeric nanomaterials, natural polymers in composite and nanocomposite material for wastewater treatment.[20]

1.6.9 Ultrasonic Wastewater Treatment

The ultrasonic method is an innovative technology that may be used for water and wastewater treatment for pollution removal. This is an advanced oxidation process. The application of this technology is to decompose many complex organic compounds into much simpler compounds during the cavitation process[21] and to contribute to eliminating various types of contaminants. Ultrasound waves and high pressure combine to create thousands of tiny cavities in the water. They decay spontaneously, generating enough energy to break down the impurities. In this method, known as cavitation, the treatment steps are screening and pumping, grit removal, primary settling, aeration/activated sludge, secondary settling, filtration, disinfection, and oxygen uptake.

1.7 CHALLENGES AND PROSPECTS

Today India faces major environmental challenges associated with waste generation, collection, transport, treatment, and disposal. Current systems in India have limitations to cope with the volumes of waste generated by an increase in urban population, and this directly impacts the environment and public health. Of course, the challenges and constraints are significant, but so are the opportunities.[22] Wastewater treatment is essential to protect one of the most valuable resources. The Pollution Control Board has estimated that only about one-fifth of the sewage that urban India generates is treated. There are no estimates of the sewage from rural India, where it is estimated that about 33 percent communities have septic tanks that collect sewage among which only a few villages have any arrangements to handle wastewater.[23]

India needs comprehensive practices for wastewater management. Limited and inadequate resources must be used wisely. Managing and treating wastewater is a huge part of this management. In this scenario, it is necessary to quantify the challenge, and one can only guess and estimate the wastewater generated from urban and rural areas. Besides, there is no uniformity in the collection and handling of wastewater, with each area adopting a different system of waste management. It is observed that proper solutions for sludge disposal are generally lacking both in rural and urban areas. Towns and cities have relied excessively on centralized sewage systems or informal mechanical and manual septic tank pit emptiers. Rural India relies predominantly on manual pit emptiers, typically manual scavengers. Sludge management is usually limited to a desludging service that is provided by municipal agencies or the private sector. The recent proliferation of mini-sewage treatment plants in urban housing complexes comes with its dangers. There have been many reported deaths caused by poorly equipped pit emptiers employed by unskilled plant operators. There is a similar danger associated with fecal sludge treatment plants as well. This is the other aspect of the unregulated market for maintaining privately owned and operated sewage treatment units. To deal with the existing wastewater management challenges there is a need to establish standardized practices. Consulting the users of wastewater or sewage is necessary. While developing standards, it is necessary to balance the concerns of the users with the costs and consequently provide an appropriate solution. Farmers use wastewater from municipalities under contracts in several cities in some regions. These cities have no sewage treatment plants but have sewage or drain networks. The water is transmitted to sewage farms, where it is naturally treated. However, there are problems with this system as it can be unhygienic for agriculture. However, it has also been recognized that sewage and sludge can be reused in agriculture. Water from sewage is recycled for industrial and non-potable urban use. For instance, thermal power plants have been mandated to use treated sewage water.

The growing population, urbanization, and industrialization have created an almost unmanageable wastewater problem in India. This is leading to toxic contamination of our water sources. To work toward achieving sustainable development goals, it is essential to plan holistically for urban and rural areas, looking at the water resources on which they draw from and to which they return. To influence planning at the levels of local government, civil society organizations (CSOs) working in the water and sanitation sector need to be well informed of the interconnectedness between water resources, water supply, and wastewater.

1.8 CIRCULAR ECONOMY IN WASTEWATER TREATMENTS

Linear economy (LE) is based on the “take-make-use-dispose” system, in which waste is found at the end of the product life cycle. In the concept of the circular economy (CE), raw materials and products remain in the economy as individual products themselves at each stage of the life cycle for as long as possible. In this concept, waste should be treated as secondary raw materials that can be recycled and reused. This is how CE and LE differ. CE encourages the sustainable organization of materials and energy by reducing the quantity of waste generation and their reuse. The limited availability of raw materials, the dependence of the import of raw materials, high prices, market volatility, and decreasing competitiveness of the global economies, provide encouragement to implement a CE effectively.

1.8.1 Resource Recovery at the Wastewater Treatment Plant

Nutrient recovery: Nutrient recycling helps to recover nutrients from raw wastewater, and semi-treated wastewater and to recover biosolids from sewage sludge. The tradition of using biosolids is the oldest method, which generates natural fertilizer and reduces their spread or penetration into the soil. Before their utilization, biosolids can be treated by anaerobic or aerobic digestion, composting, drying, and chemical treatment processes.[24] Bio solid recovery has a positive influence on the environment because it reduces the demand for conventional natural fossil-based fertilizers and consequently reduces the consumption of water and energy.

Water reuse: The reuse of treated wastewater is the key element to cope with the demand for agriculture, irrigation, industrial applications, groundwater refilling, toilet flushing, and other domestic use. Improvement in the effluent quality can also increase the quality of resource waters used as a source of drinking water. The use of treated wastewater for irrigation in agriculture has been known for many years, and it can replace the agricultural demand and consequently reduces the stress on local water resources. Effluent from secondary treatment is recommended to use in irrigation of non-food crops, and effluent from tertiary treatment for irrigation of food crops.[25]

Energy recovery: Energy recovery at wastewater treatment plants is very important for sustainability. Energy can be recovered by producing biogas, establishing heat pumps in the effluent treatment plants, and energy recovery from various high-temperature streams by heat exchangers.[26,27]

This is the preliminary overview of the specific aspects of water quality in terms of chemical, physical, and biological contents, water treatments, and novel techniques and materials used in water treatments. It will be elaborated in the next chapters.

REFERENCES

1. Tyagi, S., et al. (2013). Water quality assessment in terms of water quality index. *American Journal of Water Resources,* 1(3), 34–38. West Water Engg. Treatment Disposal by Metcalf and Eddy, Inc.
2. *Encyclopedia of Chemical Engg.* Ulmmann's Vol. 28.
3. https://www.whitman.edu/chemistry/edusolns_software/TSSBackground.pdf, 24 August 2020
4. Metcalf, L., & Eddy H. P. *Sewerage and sewage disposal: A textbook.* McGraw-Hill Book Company, Incorporated, 1922.
5. Saravanan, R., Karthikeyan, N., Gupta, V. K., Thirumal, E., Thangadurai, P., Narayanan, V., & Stephen, A. J. M. S. (2013). ZnO/Ag nanocomposite: an efficient catalyst for degradation studies of textile effluents under visible light. *Materials Science and Engineering: C, 33*(4), 2235–2244.
6. Janaki, V., Oh, B. T., Shanthi, K., Lee, K. J., Ramasamy, A. K., and Kamala-Kannan, S. (2012). Polyaniline/chitosan composite: An eco-friendly polymer for enhanced removal of dyes from aqueous solution. *Synthetic Metals, 162*(11-12), 974–980.
7. Gupta, V. K. (2009). Application of low-cost adsorbents for dye removal: A review. *Journal of Environmental Management, 90*(8), 2313–2342.

8. Scalese, S., Nicotera, I., D'Angelo, D., Filice, S., Libertino, S., Simari, C., ... & Privitera, V. (2016). Cationic and anionic azo-dye removal from water by sulfonated graphene oxide nanosheets in Nafion membranes. *New Journal of Chemistry, 40*(4), 3654–3663.
9. Xiao, J., Lv, W., Xie, Z., Tan, Y., Song, Y., & Zheng, Q. (2016). Environmentally friendly reduced graphene oxide as a broad-spectrum adsorbent for anionic and cationic dyes via π–π interactions. *Journal of Materials Chemistry A, 4*(31), 12126–12135.
10. Umoren, S. A., Etim, U. J., & Israel, A. U. (2013). Adsorption of methylene blue from industrial effluent using poly (vinyl alcohol). *Journal of Materials and Environmental Science, 4*(1), 75–86.
11. Valderrama, C., Gamisans, X., De las Heras, F. X., Cortina, J. L., & Farran, A. (2007). Kinetics of polycyclic aromatic hydrocarbons removal using hyper-cross-linked polymeric sorbents Macronet Hypersol MN200. *Reactive and Functional Polymers, 67*(12), 1515–1529.
12. Gezici, O., Küçükosmanoğlu, M., & Ayar, A. (2006). The adsorption behavior of crystal violet in functionalized sporopollenin-mediated column arrangements. *Journal of Colloid and Interface Science, 304*(2), 307–316.
13. Malana, M. A., Ijaz, S., & Ashiq, M. N. (2010). Removal of various dyes from aqueous media onto polymeric gels by adsorption process: Their kinetics and thermodynamics. *Desalination, 263*(1-3), 249–257.
14. Crini, G. (2008). Kinetic and equilibrium studies on the removal of cationic dyes from aqueous solution by adsorption onto a cyclodextrin polymer. *Dyes and Pigments, 77*(2), 415–426.
15. Dhodapkar, R., Rao, N. N., Pande, S. P., & Kaul, S. N. (2006). Removal of basic dyes from aqueous medium using a novel polymer: Jalshakti. *Bioresource Technology, 97*(7), 877–885.
16. Encyclopedia of Chemical Engg. Ulmmann's Vol. 11.
17. Houghton, J. I., & Quarmby, J. (1999). Biopolymers in wastewater treatment. *Current Opinion in Biotechnology, 10*(3), 259–262.
18. Saxena, G., Goutam, S. P., Mishra, A., Mulla, S. I., & Bharagava, R. N. (2020). Emerging and eco-friendly technologies for the removal of organic and inorganic pollutants from industrial wastewaters. In *Bioremediation of Industrial Waste for Environmental Safety* (pp. 113–126). Springer, Singapore.
19. Mishra, A. K. (2016). *Smart materials for wastewater applications.* John Wiley & Sons.
20. Mahvi, A. H. (2009). Application of ultrasonic technology for water and wastewater treatment. *Iranian Journal of Public Health*, 1–17.
21. Kumar, S., Smith, S. R., Fowler, G., Velis, C., Kumar, S. J., Arya, S., ... & Cheeseman, C. (2017). Challenges and opportunities associated with waste management in India. *Royal Society Open Science, 4*(3), 160764.
22. https://www.ircwash.org/sites/default/files/insights_10_wastewater_challenges_solutions_ver_fin.pdf, 24 August 2020
23. Zhang, Q., Hu, J., Lee, D. J., Chang, Y., & Lee, Y. J. (2017). Sludge treatment: Current research trends. *Bioresource Technology, 243*, 1159–1172.
24. Becerra-Castro, C., Lopes, A. R., Vaz-Moreira, I., Silva, E. F., Manaia, C. M., & Nunes, O. C. (2015). Wastewater reuse in irrigation: A microbiological perspective on implications in soil fertility and human and environmental health. *Environment International, 75*, 117–135.
25. Bertanza, G., Canato, M., & Laera, G. (2018). Towards energy self-sufficiency and integral material recovery in wastewater treatment plants: Assessment of upgrading options. *Journal of Cleaner Production, 170*, 1206–1218.
26. Stillwell, A. S., Hoppock, D. C., & Webber, M. E. (2010). Energy recovery from wastewater treatment plants in the United States: A case study of the energy-water nexus. *Sustainability, 2*(4), 945–962.
27. Neczaj, E., & Grosser, A. (2018). Circular economy in wastewater treatment plant–challenges and barriers. *Multidisciplinary Digital Publishing Institute Proceedings, 2*(11), 614.

2 Qualitative and Quantitative Analysis of Water

C R Venkateswara Rao
Department of Humanities and Basic Science,
GRIET, Hyderabad, India

CONTENTS

2.1 INTRODUCTION

Historically, treatment of water for drinking dates back nearly 4000 years according to inscriptions on ancient Egyptian tombs. Entire civilizations thrived near mighty rivers like Nile and Sindh. As the water was being used for everything from drinking and cleaning to irrigation and transportation, it became contaminated and polluted. People realized the necessity of clean drinking water. The London Metropolitan Water Act of 1852 made water-quality maintenance mandatory. The Federal Water Pollution Control Act of 1948 in the United States emphasized the importance of water quality and health. When we are searching for extraterrestrial life, we look for the signature of water on other planets, because it is vital for life and organic life does not exist.

Pure water, with a chemical formula of H_2O, is a colorless, odorless, and tasteless liquid. It has wonderful physical and chemical properties. It exists in all three phases in the normal temperature range on Earth, having a melting point of 0°C and boiling point of 100°C. It shows anomalous expansion (solid water expands and

DOI: 10.1201/9781003138303-2

lowers density instead of contraction), allowing aquatic life to survive in winters as floating ice forms a shield, preventing further freezing of water body. Water is a polar molecule composed of hydrogen and oxygen atoms with partial positive charge and partial negative charge on them, respectively. It does not exist in pure state in nature due to the property of universal solvency. It can dissolve anything that comes in its contact, making itself impure. In its pure form, water is the most essential natural resource. Gases in the air, such as oxygen and carbon dioxide, minerals in the soil and rocks, such as chlorides, carbonates of alkali, and alkaline earth metals, naturally contaminate water. Use, misuse, and overuse by humans make water more contaminated and polluted, making it an environmental concern.

Though available in huge quantity on Earth's surface, only less than 1 percent water is usable. All life processes take place in water as the medium. Water is used for drinking, cleaning, agriculture, fisheries, industries, and recreation purposes. Each of these uses has definite physical, chemical, and biological standards to be met with. Dissolved, colloidal, and suspended impurities contaminate water, precluding us from using it directly. Microbiological and organic impurities are another set of impurities in water bodies.

Insufficient availability of pure surface water places ground water resources under pressure. Many municipalities supply drinking water by treating bore-well water. Standards for drinking water are different from those of water used for agriculture and fisheries. The main criteria considered are differences in action of impurities present in water used in the field. Similarly recreation purposes impose different standards (WHO 2017; BIS 2019).

Until recently, water analysis was required to be done in regional and local laboratories on a regular basis. A simple sampling process along with portable testing equipment is used in remote areas. Nowadays, application of advanced technology makes routine analysis of water easy and fast. Microbiological indicators associated with fecal pollution are one of the primary testing targets. *Escherischia coli* and *Streptococcus* are known to predominantly contaminate water and are tested by membrane filtration and multiple-tube methods. However, chemical contaminants play an important role in the properties of water in daily usage in many areas of human life. It is the qualitative and quantitative chemical analysis that is of primary focus in this chapter. There are a number of physicochemical parameters, from color to conductivity, that play very important role in use of water for domestic and industrial purposes. Hardness, alkalinity, pH, conductivity, total dissolved solids (TDS), biological oxygen demand (BOD), chemical oxygen demand (COD), turbidity, etc., are primarily discussed in this chapter.

2.2 HARDNESS

Dissolved calcium and magnesium bicarbonates, chlorides, and sulfates cause hardness of water (Diskant 1952; Ramya 2015). Hardness can be demonstrated by soap-consuming capacity by the formation of a white, curdy precipitate of Ca or Mg salt of fatty acid such as Ca-stearate. The more soap required to produce lather, the harder the water is. It has high mineral content. Hardness is not good for domestic as well as industrial uses. Hard water produces a deposit of precipitate in containers. In

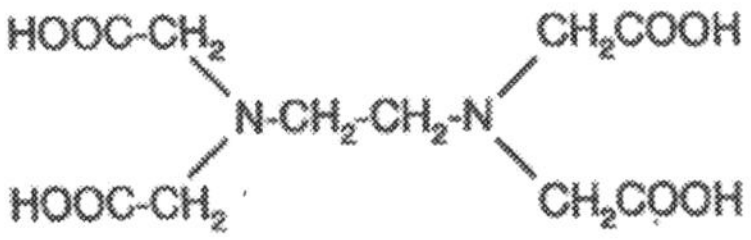

FIGURE 2.1 Structure of EDTA.

industries, the problems are even more pronounced and serious. Three hardness parameters are defined.

i. **Temporary hardness**
   It is caused by bicarbonates, and it can be removed by boiling the water.
ii. **Permanent hardness**
   It is caused by sulfates, chlorides, and nitrates, and it can be removed by chemical precipitation using Na_2CO_3.
iii. **Total hardness (TH)**
   Total hardness is the sum total of both these parameters. Total concentration of metal ions is expressed in terms of mg/L of equivalents $CaCO_3$.

Hardness is often used as sum of the two most important cations, namely Ca and Mg, to an approximation. In fact, TH is the sum total of concentrations of all divalent ions Ca, Mg, Fe, Mn, Al, etc. There are different hardness scales in use. Parts per million (ppm), mg/L, degree Clarke, degree French, etc., are some of the popularly used units.

The complexometric method of estimating hardness of water is a volumetric titration method that involves the use of a complexing agent ethylene diamine tetraacetic acid (EDTA) and a blue dye indicator Eriochrome Black-T (EBT), buffered to a pH of 9 to 10. The principle of EDTA titration involves EBT forming a complex with Ca^{2+} and Mg^{2+}, having a wine-red color that would turn blue by EDTA due to abstraction of Ca and Mg. EDTA is a hexadentate ligand forming a stable complex. Figs. 2.1 and 2.2 represent the structures of EDTA and EBT, respectively.

EBT is an indicator dye, having blue color, that combines with Ca^{2+} and Mg^{2+} in hard water to form an unstable wine-red complex at a pH between 9 and 10.

$$Ca^{2+} + EBT \rightarrow \left[Ca^{2+} - EBT\right]$$

$$Mg^{2+} + EBT \rightarrow \left[Mg^{2+} - EBT\right]$$

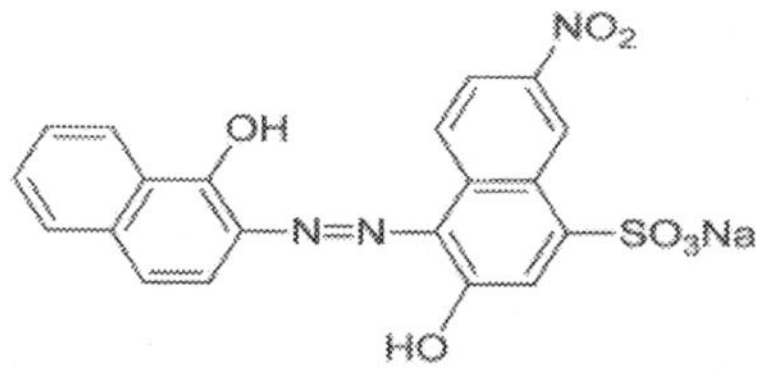

FIGURE 2.2 Structure of Eriochrome Black T.

TABLE 2.1
Levels of Hardness of Water in Normal Use

S. No.	Hardness Value (mg/L)	Degree of Hardness
1.	60	Soft
2.	60–120	Moderately hard
3.	120–180	Hard
4.	>180	Very hard

EDTA pulls out Ca^{2+} and Mg^{2+} from the above unstable complex and forms a colorless, stable complex with regeneration of blue-colored EBT.

$$\left[Ca^{2+} - EBT\right] + EDTA \rightarrow \left[Ca^{2+} - EDTA\right] + EBT$$

$$\left[Mg^{2+} - EBT\right] + EDTA \rightarrow \left[Mg^{2+} - EDTA\right] + EBT$$

The amount of EDTA used corresponds to TH of water. Temporary hardness and permanent hardness values are calculated from TH:

$$\text{Permanent hardness} = \text{Total hardness} - \text{Temporary hardness}$$

Temporary hardness is removed as carbonates simply by boiling water and filtering off $CaCO_3$ precipitate.

On the basis of dissolved concentration of calcium carbonate equivalent salts, the levels of water hardness can be tabulated as shown in Table 2.1.

2.3 ALKALINITY

Alkalinity is the capacity to neutralize acid. It is due to carbonates, bicarbonates, and hydroxides of Ca, Mg, Na, and K in water. Carbon dioxide is dissolved in water to form carbonic acid, which will be in equilibrium with bicarbonate and carbonate ions.

Alkalinity is generally determined by titrimetric analysis against a standard acid, using phenolphthalein and methyl orange indicators successively.

During titration with acid, hydroxide ions are neutralized to water as indicated by phenolphthalein indicator at the first end point as shown in the chemical equation:

$$OH^- + H^+ \rightarrow H_2O$$

Carbonate ions are neutralized to bicarbonate ions:

$$CO_3^{2-} + H^+ \rightarrow HCO_3^-$$

and bicarbonate ions are neutralized to water and carbon dioxide.

$$HCO_3^- + H^+ \rightarrow H_2O + CO_2$$

This is indicated by methyl orange indicator.

2.4 ACIDITY (pH)

pH is a measure of the acidity or basicity and is mathematically defined as negative logarithm of H^+ ion concentration. Danish chemist Sorensen introduced the concept of pH. It has values between 1 and 14. Each unit change in pH represents a 10-fold change in H^+ ion concentration. When the number of H^+ ions equals the number of OH^- ions, water is neutral and has a pH of 7. Carbon dioxide forms carbonic acid and reduces pH of water, whereas lime (calcium hydroxide), with a formula of $Ca(OH)_2$, raises pH of water.

$$pH = -\log\left[H^+\right] \quad \text{or} \quad \left[H^+\right] = 10^{-pH}$$

pH is measured using pH meter which works by the principle of potentiometer connected to a glass electrode and calomel electrode.

Approximate values of pH can be measured by dipping pH paper that has dye smeared on its surface in test solution. It is an important parameter by which we can know the presence of certain metals and nutrients in water.

The pH range of drinking water is between 6.5 and 8.5. Water in the lower pH range tends to be corrosive and in the higher pH range it tends to be scale-forming. These pH changes affect the performance of water treatment plants. Control of pH is necessary for minimizing corrosion of water distribution systems. pH plays an important role in biological systems. pH of water makes certain metals and minerals more or less available to the body because it determines the solubility and biological availability of nutrients and minerals. Heavy metals in low pH tend to be more toxic as they are more available to the body. A safer range is between 6.5 and 8.5 (Naik et al. 1987).

The cause of variation in pH is the composition of bedrock, soil, or other things that water comes in contact with. Alkalinity is due to flow of water through limestone rocks where carbonates, bicarbonates, and hydroxides dissolve in water raising its pH. Dissolved CO_2 forming carbonic acid reduces the pH of pure water. Decomposing plants also may cause pH reduction.

In an aqueous solution, the following equilibrium takes place:

$$H_2O_l \rightleftharpoons H^+_{aq} + OH^-_{aq}$$

In pure water, $[H^+] = [OH^-] = 1 \text{ X } 10^{-7}$ M

$[H^+]\,[OH^-] = k_w = 1 \text{ X } 10^{-14}$, k_w is ionic product of water.

pH of pure water is 7. Solutions where $[H^+] > [OH^-]$ are acidic, and solutions where $[H^+] < [OH^-]$ are basic.

The measurement of pH of a solution is generally carried out by the following two methods depending on the accuracy needed.

TABLE 2.2
Indicators vs. Color Change

S. No.	Indicator	pH Range	Color Change
1.	Thymol blue	1.2–2.8	Red to yellow
2.	Methyl orange	3.1–4.4	Red to orange to yellow
3.	Bromocresol green	3.8–5.4	Yellow to blue
4.	Methyl red	4.4–6.2	Red to yellow
5.	Thymol blue	8.0–9.6	Yellow to blue
6.	Indigo carmine	12.0–13.0	Blue to yellow

The first method employs chemical dyes called indicators. These substances are weak acids or weak bases. They can exist in two different forms, exhibiting different colors based on whether the dye molecule is protonated or deprotonated:

$$HI_n \rightleftharpoons H^+ + I_n^-$$

where I_n is indicator. Indicators change colors over a short range of pH. A given indicator is useful for determining pH only in the region in which it changes color. There are indicators that cover all pH ranges.

By comparing the color of the unknown solution with the color of the indicator, the pH of the unknown can be determined as shown in Table 2.2.

The second method, based on electrochemistry principles, uses a digital pH meter and glass electrode coupled with a reference electrode. A glass electrode is a type of membrane electrode having a bulb made up of special glass. A gel layer is present both inside and outside of the glass bulb. Depending on the pH of the solution in which we dip the electrode, the H^+ ions either diffuse out of the gel layer or diffuse into the gel layer.

In case of an alkaline solution, the H^+ ions diffuse out and a negative charge is established on the outside of the gel layer. Potential at inner surface is maintained constant as it is exposed to an internal buffer with constant pH.

Total membrane potential is the difference between inner and outer charge:

$$E = E^0 - S(pH_a - pH_i)$$

The reference electrode used is a saturated calomel electrode. This electrode and a glass electrode are immersed in the solution whose pH is to be determined. A digital electronic circuit takes the measurement and displays the digital reading directly.

2.5 ELECTRICAL CONDUCTIVITY

Electrical conductivity is a measure of a substance's ability to allow electrical current to pass through. It can be viewed as the reciprocal of resistance. In ionic solutions, current is carried by ions. Pure water will show very low conductivity, whereas solutions having ions dissolved in them display high conductivity (Jones 2002).

Electrical conductivity measurement has widespread applications in water pollution and industrial quality control (Bozkurt et al. 2009). High sensitivity, reliability, speed, and low cost of a conductivity-measuring instrument make it a good choice. It is measured using a conductivity cell that has platinum plates of unit area placed at a constant distance of 1 cm and passing AC current.

Cations in the electrically conducting solution migrate to negative electrode and anions migrate to positive electrode.

Resistance of the solution R can be measured using Ohm's law:

$$V = IR,\ R = V/I$$

where, V= voltage, I = current, R = resistance

Conductance is defined as the reciprocal of resistance:

$$G = 1/R$$

Conductivity $\kappa = GC$

C is the cell constant defined as the ratio of distance between the electrodes to the area of electrodes, i.e., $C = \ell/A$

Conductivity measurement is possible for aqueous solutions having electrolytes dissolved in them. These are the solutions of ionic salts or compounds that, when ionized in water, give ions upon dissociation. These ions are responsible for carrying electrical current. Acids, bases, and salts can be strong electrolytes or weak electrolytes, depending on their degree of dissociation.

2.6 TOTAL DISSOLVED SOLIDS

Total Dissolved Solids (TDS) is a measure of the combined total of inorganic and organic substances dissolved in water in the ionic and molecular form (Singh & Kalra 1975). The most common dissolved inorganic ions of metals, including calcium, magnesium, sodium, potassium, iron, manganese along with the anions such as bicarbonates, chlorides, sulfates, nitrates, and carbonates, can pass through a filter with a pore size of around 2 microns. Suspended particles include silt and clay, plankton, fine organic debris, and other particulate matter. These particles will not pass through 2-micron filters. Small quantities of chloromethanes are formed by reaction of chlorine (used to disinfect water) with humic and fulvic acids from soil. They do not get filtered off through normal filtration.

High concentration of TDS will make drinking water unpalatable. The concentration of TDS affects the water balance in the cells of aquatic organisms. An organism placed in water with high concentration of solids will shrink somewhat because the water in its cells will tend to move out.

A TDS test will only reveal that compounds are present in water, not their identity. Further tests are required to determine the exact nature of the dissolved solids.

Conductivity and gravimetry are two different methods used to measure TDS. Dissolved ionic solids in the form of ions cause conductivity. The more are the ions in water, greater is the conductivity of water. It is an easy and quick way of measuring ionic content of impurities in water. It can be readily done in the field.

A more complete and accurate way of measuring TDS is by taking a known quantity of water in a high silica porcelain dish and allowing all the water to evaporate to dryness at a temperature above 100°C. The mass of the remaining residue is weighed to give TDS by gravimetry. This method is time consuming and requires a laboratory setting.

The conversion of electrical conductivity (EC) to TDS can be done by multiplying EC with a conversion factor that varies from 0.5 to 0.8 depending on the magnitude of EC.

2.7 RESIDUAL CHLORINE

Water from all sources contains microorganisms. Some of them cause diseases. Water becomes a major source of diseases if pathogens are not killed. Chlorination is a popular disinfection method of public water supply. Started over a century ago, it is used for water treatment all over the world.

If enough chlorine is added, some will remain in the water after organic matter is destroyed. Left over chlorine is free chlorine that also will be consumed when a new contaminant enters water. The presence of residual chlorine can prevent recontamination during transportation, distribution, and storage of water. Cl_2 kills a microorganism by damaging its cell membrane. Once the cell membrane is weakened and ruptured, Cl_2 can enter the cell and disrupt cell respiration and DNA function, thereby killing the organism. When Cl_2 is added to water, it forms hypochlorous acid (HOCl) and hypochlorite ions (OCl^-) which are responsible for disinfection. Residual chlorine is available chlorine remaining in water after chlorination.

Residual chlorine is determined by iodometry. It is based on the principle that Cl_2 liberates free iodine from potassium iodide (KI) solution in an acidic medium quantitatively. Liberated I_2 is titrated against a standard sodium thiosulfate solution ($Na_2S_2O_3$). A starch indicator produces a blue color that disappears at the end point.

2.8 BIOLOGICAL OXYGEN DEMAND

Biological oxygen demand (BOD) or biochemical oxygen demand is the amount of oxygen consumed by microorganisms while they decompose organic matter under aerobic conditions at a fixed temperature. The presence of sufficient concentration of dissolved oxygen (DO) is essential to aquatic life for surviving in water body. Oxygen can come from atmosphere or as a by-product of photosynthesis of aquatic plants. Some environmental conditions, such as hot weather, high elevation, presence of fertilizers, affect DO levels. These factors decrease DO causing stress to aquatic life and such water is not good for drinking either. BOD is an indication of the degree of organic pollution in wastewater.

BOD is measured by comparing the DO levels of a water sample before and after five days of incubation in the dark. The BOD requirement for different samples are tabulated in Table 2.3. One liter of water sample is used for measuring the amount

TABLE 2.3
Biological Oxygen Demand at Different Pollution Levels

S. No.	Pollution Level	BOD (mg/L)
1.	Unpolluted water	<1
2.	Moderately polluted water	2–8
3.	Untreated sewage	200–600
4.	Treated sewage	20

of molecular oxygen required to convert organic molecules into carbon dioxide. Organic matter comes from dead leaves, animals, and microorganisms, sewage, etc.

BOD = initial dissolved oxygen levels – final dissolved oxygen levels after 5 days

2.9 CHEMICAL OXYGEN DEMAND

Chemical oxygen demand (COD) determines the quantity of oxygen required to oxidize organic matter in a water sample. It is based on the fact that a strong oxidizing agent, under acidic conditions, can fully oxidize almost any organic compound to carbon dioxide (Geerdink et al. 2017; Himebaugh & Smith 1979).

A known quantity of potassium dichromate ($K_2Cr_2O_7$) and sulfuric acid (H_2SO_4) are added to the water sample. Organic matter reacts with this reagent and Cr(IV) in $K_2Cr_2O_7$ is converted to Cr(III). Excess dichromate is titrated with ferrous ammonium sulfate until all the excess oxidizing agent has been reduced to Cr(III). Ferroin indicator is used and the color changes from blue-green to reddish-brown at the end point.

2.10 TURBIDITY

Turbidity in water is caused by colloidal and suspended particles such as silt, clay, finely divided organic and inorganic matter, and other microorganisms. Turbidity is a measure of water clarity. The first thing that we notice about water is how cloudy it is. Cloudiness typically comes from suspended particles in the water that we cannot see individually with naked eye. Murky water has higher turbidity and clear water has low turbidity. It is caused by particles of silt and soil formed by weathering of rocks during the flow of water. Organic matter from the decaying trees, plants, and animals also contributes to turbidity (Gillet & Marchiori 2019).

Turbidity is measured as the amount of light scattered by suspended particles in a water sample. It is commonly measured in Nephelometric Turbidity Units (NTUs) in which light scattering in water sample is compared with a standard reference. Particles in turbid water can carry disease-causing pathogens or toxic pollutants. Turbidity affects taste and odor of the water. River water with a lot of turbulence gathers mud and silt to the point that it is impossible to see through the water; on the other hand, underground water, after extensive natural filtration through sand and rocky formations, is crystal clear. When the concentration of particles increases, more particles reflect light, which increases the intensity of scattered light. Turbidity is different from the total suspended solids (TSS) which is measured by gravimetry wherein the mass of the solids suspended in a sample is measured by weighing the separated solid mass.

2.11 WATER QUALITY INDEX

Water resources are misused and mismanaged by people. Water supply systems currently in use are not sustainable. Water quality is a key parameter that needs to be seriously considered for sustainability. Water supply management must ensure a supply of good-quality water to ensure public life in healthy conditions without diseases; and for this, qualitative and quantitative analysis of water is necessary. Unless the analysis is done properly and systematically on a regular basis, the supply of good-quality water cannot be ensured.

The water quality index (WQI) integrates combined influence of various physicochemical parameters of water samples. This enables comparison of different samples for quality on the basis of a single index value for each sample. This reduces many different parameters into a single expression and helps facilitate interpretation of collected data. It is a mathematical expression used to transform large quantities of water-quality data into a single number, and it is a measure of how water-quality parameters compare to quality guidelines or objectives for a specific area. It gives managers and decision makers an idea about the seriousness of the water quality problem.

The first report of water quality quantification was given by Horton in 1965 (Hoton 1965). The National Sanitation Foundation Water Quality Index (NSFWQI) was proposed by Brown et al. in 1970, using the Delphi method, which uses certain carefully chosen physicochemical parameters in devising a formula for the index. Each parameter is given a weightage according to its importance in quality (Mercy & Babiyola 2018; Bhargava 1983; CCME 2001; Cude 2001; Garde & Jadhav 2013; Landwehr & Deininger 1976; Patil et al. 2012; Poonam et al. 2013; Rusydi 2018; Steinhart et al. 1982; Tyagi et al. 2013).

WQI is an index parameter in the form of a single number that may be calculated as follows:

1. $WQI = \sum (Wi \times qi)$ here assign weight w_i for chosen parameters on a scale of 1 to 5 according to relative importance in overall quality of water under consideration.
2. Assign relative weight $W_i = \frac{w_i}{\sum_{i=1}^{n} w_i}$ where w_i is weight of each parameter
3. Quality rating $q_i = \frac{c_i}{s_i} \times 100$ where c_i is concentration of water quality parameter and s_i represents BIS standard for every parameter of drinking water.

Measurement is done on many samples. The parameters chosen are based on their effectiveness for drinking purposes. Index values are obtained in the range of 0 to 100. This is meant for performing on a regular basis. If there is any deviation from the normal values, respective authorities would be alerted.

2.12 CONCLUSIONS

Water, being an essential resource used on daily basis in substantial quantities in all walks of life, requires constant analysis at regular periods of time to assess its purity and quality, be it for domestic human consumption, industrial application as a coolant or in a boiler, or for recreation such as swimming. A number of physico-chemical parameters such as hardness, alkalinity, pH, EC, TDS, residual chlorine, BOD, COD, and turbidity are measured by standard methods to assess the quality of water for these purposes. It is equally important to be done in all parts of the world. Qualitative as well as quantitative chemical analysis provides standard methods to perform tests both in the laboratory and in the field quickly and at a lower price without requiring sophisticated electronic equipment. WQI is a composite parameter of all these variables that enables easy comparison of quality of different water samples for continuous monitoring.

REFERENCES

Bhargava D S, 1983, Use of water quality index for river classification and zoning of Ganga river, *Environ. Pollut. Ser. B*, 6(1): 51–67.

Bozkurt A, Kurtulus C, & Endes H, 2009, Measurements of apparent electrical conductivity and water content using a resistivity meter, *Int. J. Phy. Sci.*, 4(12): 784–795.

Brown R M, McClelland N I, Deininger R A, & Tozer R A, 1970, Water quality index: Do we dare? *Water Sewage Works*, 117: 339–343.

Bureau of Indian Standards (BIS) (2019). Specifications for drinking water 2019.

Canadian Council of Ministers of the Environment (CCME), Canadian Water Quality Index 1.0: Technical report and user's manual, (2001). Canadian Environmental Quality Guidelines Water Quality Index Technical Subcommittee, Gatineau, QC, Canada.

Cude C, 2001, A tool for evaluating water quality management effectiveness, *J. Am. Water Resour. Assess.*, 37: 125–137.

Diskant E M, 1952, Stable indicator solutions for complexometric determination of total hardness in water, *Anal. Chem.*, 24(11): 1856–1857.

Garde S P & Jadhav M V, 2013, Assessment of water quality parameters: A review, *Int. J. Eng. Res. App.*, 3(6): 2029–2035.

Geerdink R R, Hurk R S, & Epema O J, 2017, Chemical oxygen demand: Historical perspectives and future challenges, *Analytica Chimica Acta*, 961: 1–11.

Gillett D & Marchiori A, 2019, A low-cost continuous turbidity monitor, *Sensors*, 19: 30–39.

Himebaugh R R & Smith M J, 1979, Semi-micro tube method for chemical oxygen demand, *Analytic. Chem.*, 51(7): 1085.

Horton R K, 1965, An index number system for rating water quality, *J. Water Pollut. Control Fed.*, 37(3): 300–306.

Jones R G, 2002, Measurements of the electrical conductivity of water, *IEE Proc. Sci.*, 149 (6): 320–322.

Landwehr J M, & Deininger R A, 1976, A comparison of several water quality indexes, *J. Water Pollut. Control Fed.*, 48(5): 954.

Medha N, Khemani L T, Momin G A, & Prakasa Rao P S, 1987, Measurement of pH and chemical analysis of rain water in a rural area of India, *Acta Met. Sinica*, 2(1): 91–100.

Mercy A C & Babiyola D, 2018, Analysing the water quality parameters from traditional to modern methods in aquaculture, *Int. J. Sci., Environ. Tech.*, 7(6): 1954–1961.

Patil P N, Sawant D V, & Deshmukh R N, 2012, Physico-chemical parameters for testing of water: A review, *Int. J. Environ. Sci.*, 3, 1194–1207.

Poonam T, Tanushree B, & Chakraborty S, 2013, Water quality indices –important tools for water quality assessment: A review, *Int. J. Adv. Chem.*, 1(1): 15–28.
Ramya P, Babu J, Tirupathi R E, & Venkateswara R L, 2015, A study on the estimation of hardness in ground water samples by edtatritrimetric method, *Int. J. Rec. Sci. Res.* 6(6): 4505–4507.
Rusydi A F, 2018, Correlation between conductivity and total dissolved solid in various type of water: A review, *IOP Conf. Series: Earth Environ. Sci.,* 118: 012019.
Singh T & Kalra Y P, 1975, Specific conductance method for in situ estimation of total dissolved solids, *J. Am, Water Works Assoc.*, 67(2): 99.
Steinhart C E, Schierow L J, & Sonzogni W C, (1982) Environmental quality index for the Great Lakes, *Water Resour. Bull.*, 18(6): 1025–1031.
Tyagi S., Sharma B., Singh P., Dobhal R, 2013, Water quality assessment in terms of water quality index. *Am. J. Water Res.*, 1(3): 34–38.
World Health Organization, 2017, Guidelines for Drinking-Water Quality, 4th edition.

3 Need for Advanced Materials and Technologies

The Sustainability Argument

Neeraj Kumari
School of Basic and Applied Science,
K. R. Mangalam University, Gurugram, India

Sushma
Department of Industrial Waste Management,
Central University of Haryana, Mahendergarh, India

Firdaus Parveen
Department of Chemical Engineering, IIT Delhi,
New Delhi, India;
Department of Chemistry, Imperial College London,
White City London, UK

CONTENTS

DOI: 10.1201/9781003138303-3

3.1 INTRODUCTION

Efforts to design materials that achieve advantageous properties and applications at the atomic/molecular scale are going on in the field of advanced materials and technology (1). Research on advanced materials varies from basic scientific studies related to their fabrication and interactions to applied engineering efforts that help in translation of such basic knowledge to advanced technological developments. Engineered materials designed at the atomic/molecular scale having dimensions in the range of 1–100 nm and enhanced properties such as tensile strength, thermal stability, and electrical and optical properties are known as advanced materials. They provide better performance with cost effectiveness, reduce energy consumption, and show less dependency on imports of critical and strategic materials (2). Over the past few decades, these advanced materials have played a key role in the advancement of technology, which is considered the present stage of human knowledge and is used to generate valuable products to fulfill all necessities by resolving all problems. These advanced materials include metal alloys, composites, carbon fiber, polymer composites, ceramics, metals, and metal oxide nano-materials and are used as lightweight materials for application in the automotive and building sectors. They are used as bio-derived materials for food packaging and agriculture by replacing petrochemical products, as graphene in touch screens and solar cells, as nano-materials for water purification and radioactive waste cleanup, and in many more applications (3–5). These novel physicochemical properties of advanced materials make them more desirable and functional for sustainable development of technologies. Sustainability means the ability to meet all necessary requirements of the present without comprising with the possibilities of upcoming generations. For sustainable development, three pillars have great importance: economic (profit), social (people), and environment (planet) (6). For sustainable development, it is necessary that these advanced materials and technologies should be used for real-life applications involving all the conditions under which a material will be used.

The main objective of this chapter is to study the role of advanced materials and technologies for sustainable development, how they are affecting the economy and society, and how they are responsible for environmental degradation and depletion of natural resources instead of having a large number of functional properties.

3.2 ADVANCED MATERIALS

Technology developers are continually discovering new and useful materials that exhibit novel or enhanced physicochemical properties such as tensile strength, hardness, thermal stability, and electrical and optical properties in comparison with conventional materials (7). These innovative and useful materials are known as advanced materials. Over the past three decades, these materials have been used continuously for different purposes (8), but in recent years they have become a legend for nano-enabled materials.

However, no protocol exists in the field of environmental safety and occupational health (ESOH) for defining these materials. In several ways, the definition of advanced materials is a great challenge, such as defining nano-materials.

There are different ways to define the advanced materials. The broader definition is materials that signify enhancements over traditional materials that have been used for several hundred years.

Alternatively, they are defined as "Materials having engineered properties formed through the development of specified processing and synthesis technology." There are mainly six advanced materials subgroups:

a. Alloys and metals, such as amorphous and shape-memory alloys, aluminum-lithium alloys, and metals that are porous and solidify rapidly;
b. Structural ceramics, such as alumina (Al_2O_3), silicon carbide (SiC),nitride (Si_3N_4), beryllia (BeO), boron nitride (BN_x), titanium carbide (TiC), and thoria (ThO_2);
c. Industrial polymers, such as polyphenylenesulphide, polyacrylate, polyetheretherketone, and anarray of polyamide-imides;
d. Advanced composites, including metal, ceramic, or polymer matrix-comprising particles, whisker and fiber reinforcements made up of different atoms/metal oxides such as carbon, boron, zirconia, aluminum silicates;
e. Materials having magnetic, electrical, and optical properties, such as gallium (Ga), indium (In), yttrium (Yt), zirconium (Zr), barium (Ba), lanthanum (La), and rare earth elements called lanthanides;
f. Medical and dental materials, including alumina (Al_2O_3) and calcium phosphate (Ca_3PO_4), glasses and carbon fiber-strengthened polylactic acid composites (8, 9);
g. Nano-materials, such as metal oxide nano-materials TiO_2, ZnO, SnO_2, ZrO_2, and Fe_3O_4 (10).

All the above materials provide the potential of reduced energy consumption, better performance with cost effectiveness, and less dependence on imports of critical and strategic materials.

3.3 TECHNOLOGY

Technology is the knowledge of science that can be applied for practical purposes or can be considered as the sum of methods, processes, techniques, and skills that are applied to produce essentials to fulfill various purposes such as scientific investigation. In simple way, it is the advancement and application of basic tools (11). Technology is the present stage of knowledge used to produce required products by combining resources, to solve undefined problems, and to fulfill all necessities by using technical methods, tools, techniques, skills, processes, and raw materials (12).

Over the past few decades, technology has advanced to nanotechnology, in which materials are manipulated to atomic, molecular, or supra-molecular scale, having at least one dimension in the range of 1–100 nm. Nanotechnology has a huge range of applications using innovative advanced materials and devices in different fields such as nano-electronics, nano-medicines, production of biomaterials, etc. (13).

When nanotechnology merges with another term, such as "medical nanotechnology," "automobile nanotechnology," or "state-of-the-art nanotechnology," it describes the state of knowledge and tools of the respective field and refers to the highest technology accessible to human beings in any field (14).

Technology has various consequences in positive and negative ways. If we talk in a positive way, due to technology, more advanced economies have been developed, resulting in the growth of a leisure class. Most of the nano-technological practices generate unwanted effects, such as pollution, that are responsible for the depletion of natural resources and damage the Earth's environment, demonstrating the negative impact of technology. Inventions of new technology always show an effect on the morals of a society, resulting in a generation of new issues in the principles of technology, such as the concept of effectiveness regarding human efficiency and the challenges of bioethics (15).

3.4 NEED FOR ADVANCED MATERIALS

Materials are at the core of our day-to-day life that ensure the functioning, stability, long-term durability, security, and environmental compatibility of all the devices and services around us. Over the past few decades, the so-called conventional/traditional materials, such as metals, concrete, and plastics, have undergone innovative developments, such as new-fangled composites, smart coatings, biomimetic materials (such as hydrogel and surface texturing), and light alloys to diminish weight and CO_2 emissions; in other words, the applications of advanced materials have spread from nanotechnology to ceramics, biomaterials related to health products, and to the newest energy materials for reducing our carbon footprint (16).

As discussed in Section 3.2, advanced materials outperform traditional materials because they contain certain properties that are far superior including physicochemical properties such as thermal stability, strength, hardness, and flexibility. They have capacity to carry these novel characteristics with ability to retain their form, or they can sense that variations have occurred in their nature and respond to all variations (17). Advanced materials and technologies are the new home for research and applications of technology-related materials, with specific efforts on advanced design, production, and incorporation of devices.

Recently, new and novel materials, also called advanced materials, are classified in terms of three distinct engineering and social demands:

1. sustainability and materials security
2. materials for energy
3. high-value markets

3.4.1 Sustainability and Materials Security

a. **Lightweight materials:** Since 1970, there have been sustained attempts by different industries to reduce weight in materials used for various purposes, such as transportation, in order to achieve advanced energy efficiency. For example, carbon and glass fiber-reinforced epoxy composites and aluminum-magnesium alloys are applied in niche vehicles and will gradually have application in high-volume car manufacturing. These light materials have great importance in automobiles and aircraft as these materials are optimized for their particular requirements, specifically for the reduction in vehicle weight, resulting in a 6 – 8 percent improvement in fuel economy (Fig. 3.1) (18, 19). On the other hand, some structural, lightweight composites also proposed new opportunities which are unavailable in monolithic materials; for instance, polymeric/epoxy-based composites are assembled to introduce the materials besides the reinforcement (elongated fibers) or matrix (epoxy-based) phase to show extra function (20).
b. **Materials with lower environmental effect throughout their life:** Materials that have harmful/hazardous effects on the environment are replaced by new materials that can be sustained for long time, thus creating various opportunities. Materials with lower environmental effects offer possibilities for cross-sector technology transfer. Globally, buildings are a main cause for degradation of the environment as they consume more than 40 percent of global energy, resulting in approximately 33 percent of carbon dioxide equivalent emissions (CO_2). The insertion of lightweight materials with effective hybrid arrangements into the building sector offers outstanding load-carrying capacity, resulting in a reduction of emission of CO_2 in the atmosphere (21). Products obtained during industrial processes such as mining, the pharmaceutical industry, and smelting, are responsible for the mobilization of heavy metals such as hexavalent

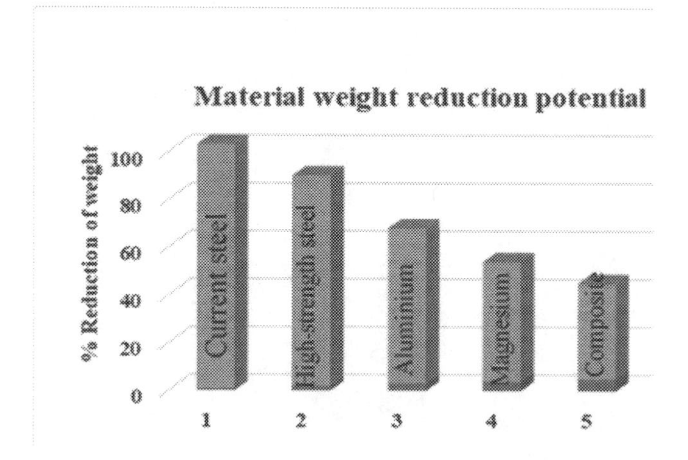

FIGURE 3.1 Comparison of lightweight materials for automotive applications.

chromium, cadmium, lead, cobalt, mercury, bismuth, and a range of different organic compounds in the environment, which is a public health concern due to the contamination of food chain as shown in Fig. 3.2 (19). For near-term trends, aluminum is replacing some metals various industrial products and is used in novel surface treatments during the production process. Medium- and longer-term developments will pave way for low-priced and accessible nanotechnology in the form of nano-materials implanted in various matrices (22).

c. **Bio-derived materials:** Materials that are not easily discarded, whose price is directly linked to natural resources, or where the complete security of resources is susceptible are being replaced with those obtained from sustainable sources or materials that easily undergo bio-degradation after their use. The main efforts have been to find to petro-chemically derived materials having strong instability in price, such as structural composites containing matrix and fibers obtained from more sustainable bio-feedstock (23). Bio-derived materials are mainly aerogel, bio plastics, polylactic acid, low-cost carbon fibers known as cellulose fiber, composite materials, barrier films, engineering wood, materials for coating and 3D printing.

Various sectors, such as automotive, construction, food packaging, agriculture, and many more, show the highest opportunities for the application of bio-based materials that offer considerable market prospects for structural purposes, especially in load-bearing structures such as load bearing wall.

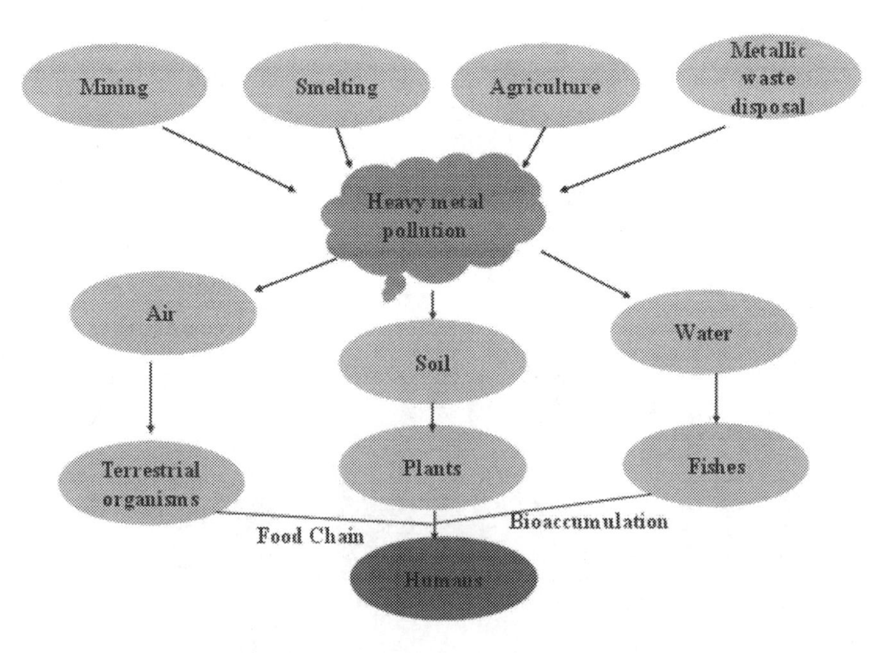

FIGURE 3.2 Role of heavy metals in various industries and their harmful effects on environment.

3.4.2 Materials for Energy Storage

Currently, the need for clear, safe, and sustainable energy is the biggest challenge facing society to improve standards of living. Advanced materials and technologies include three focus areas for energy storage: advanced battery materials, materials for storage of chemical energy (such as energy carriers based on hydrogen and CO_2, i.e., technologies such as power-to-liquid and power-to-gas), and materials for storage of thermal energy (phase-change materials) (24).

The materials required to complete this task are found in form of renewable power generation, energy storage, wind energy, fossil fuel and nuclear power production, and electricity transmission. For example, for the generation of future energy, solid oxide fuel cells play an important role. For this purpose, materials comprising rare-earth elements with superior properties must be developed, such as extended high-temperature environmental resistance-counting insulators, conductors, and functional ceramics. In case of extreme environmental conditions such as high temperatures in power generation, advanced materials such as glass fiber-reinforced plastics will also play a critical role in extending the lifetime of windows by improving their protective coating, such as thermal barrier coating and coating against corrosion and oxidation resistance, etc. (25, 26).

3.4.3 High-Market-Value Materials

Materials having high market value offer many prospects where only a few materials are considered, based on their relative imperfection but high potential value.

a. **Meta-materials:** Synthetic materials having electromagnetic (EM) properties are known as meta-materials. These properties cannot be attained with natural materials or any other way; these are the supportive technology for those complex that show manipulations in their physical structure that can be applied to observe the concepts such as nearly prefect absorbance, ultra-low observability, electrically small but efficient antenna, etc. (27, 28). Examples of meta-material are negative index meta-materials, photonic meta-materials, chiral meta-materials, and plasmonic meta-materials, etc. (Fig. 3.3). In fact, meta-materials having negative permittivity and permeability were first investigated by Veselago (1968), who called these materials media left-hand (LH), formed by electric and magnetic fields and phase propagation, also called triads of vectors (29).
   Applications of these meta-materials are found in various areas such as physics, electrical engineering, material science, and optics, where efficient meta-materials are constructed for their novel applications in radar design and antenna and invisibility of microwave cloak design.
b. **Carbon nano-materials:** When carbon atoms in the form of a flat monolayer are packed tightly into a two-dimensional (2D) honeycomb lattice, graphene formation takes place, which is the fundamental unit of graphitic materials for all dimensionalities. Each flake of a pristine monolayer of graphene has mixed properties that were not seen in a single material before.

FIGURE 3.3 Meta-materials for different applications.

Graphene is used for wide range of applications, but the main application was found in electronic field: fast transistors and efficient emitters. In optical devices, graphene is used for various applications such as touch screens, solar cells in form of thin and transparent film by replacing conductive metal oxide such as indium-tin-oxide (30).

For structural applications, graphene is used in polymers to enhance their strength and rigidity. The process through which graphite is exfoliated to form graphene has produced and furthermore, showing interest in the exfoliation of other 2D, planar materials. Most 2D materials, such as MoS_2, WS_2 and BN with several atomic layers of thickness, have already been recognized as flakes (31, 32).

c. **Biomaterials:** The past decade has seen a huge increment in implantable medical devices such as artificial hips and knees, with replacement and transplant surgery with the help of novel materials and technology in the biomedical and healthcare market becoming almost commonplace. Combinations of methods in implantable devices are further facilitated by the addition of different manufacturing techniques, such as combining bioactive tissue or bone scaffolds; 3D printing will be used for each patient, with a controlled-release active molecular therapy that is fixed in or coated over the entire structure (Fig. 3.4) (33, 34).

3.5 SUSTAINABILITY

Sustainability mainly concentrates on meeting the necessities of the present without compromising the capability of upcoming generations to meet their necessities. Sustainability is the combined perception where sustainment of prosperity can be possible for long term only by focusing on the economic, social and environmental

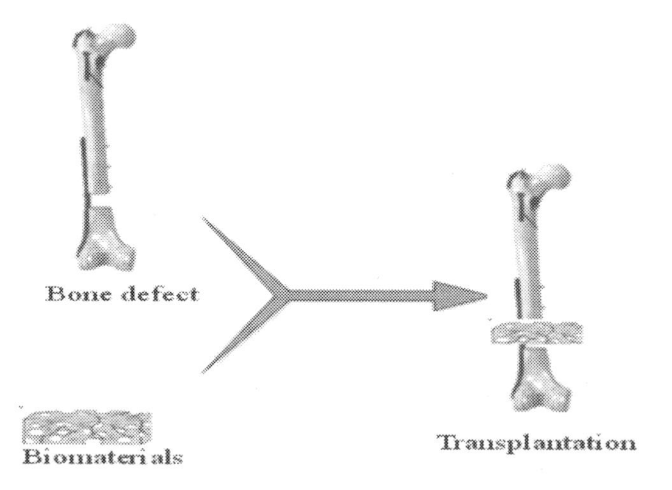

FIGURE 3.4 Transplantation surgery of defective bone using biomaterials.

impacts of determinations produced by individuals, institutes, business and nations (Fig. 3.5). The concept of sustainability includes three pillars: economic, social, and environmental, also termed informally as profits, people, and planet (35).

3.5.1 Economic Sustainability

The main requirement of economic sustainability is that a business, institution or nation consumes resources effectively for working in a sustainable manner to generate an operational profit. It also provides incentives to all organizations to adhere to rules and principles of sustainability beyond usual legislative requirements. Economic development is about giving people what they want in their life by reducing the financial burden but without compromising the product, especially in the developing world (36).

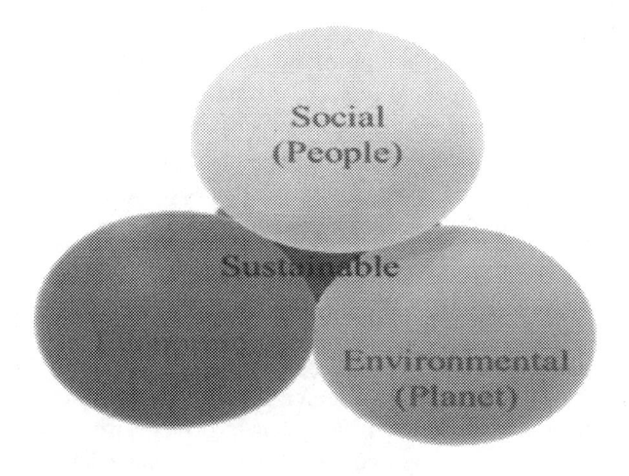

FIGURE 3.5 Three pillars of sustainable development.

3.5.2 Social Sustainability

It is the competence of a society, or any social organization, to constantly accomplish good social well-being. Accomplishing social sustainability confirms that the social well-being of a community, an organization, or a nation can be maintained for the long term. This pillar has many facets, and one of the most important is awareness and protection of people's health, which is degrading due to some harmful activities, such as pollution created by industries and organizations (37). For example, in some developed regions such as North America and Europe, strong legislation and programs have been implemented for the protection of people's health and their well-being and also to maintain access to fundamental resources without compromising the quality of life. For sustainable development, there should be a balance between individual and group needs.

3.5.3 Environmental Sustainability

Environmental sustainability often attracts most attention. We all know about the need for action to protect the environment through various efforts such as recycling and reducing power consumption by turning off electronic appliances, walking, etc. Protection of the environment is the third pillar of sustainability, the major focus of which is the future of humanity. It emphasizes the role of the individual to protect the ecosystems and air quality and to maintain sustainability of all the assets. It also focuses on the components that are responsible for putting stress on our environment (12). These days, industries are working to reduce packaging waste, water consumption, carbon footprints, etc. As a result, such industries are having a positive influence on the planet with a positive commercial effect.

Policies related to sustainable development are required for modification of economic growth rather than limit it. The policies are established on the belief that development of nation is only feasible through the possible technology that enable the individuals to offer new alternatives of resources. Therefore, all three pillars of sustainable development are designed to achieve changes and replace materials that are responsible for pollution. Although there is still much uncertainty and disagreement about the precise meaning of sustainable development, the concept of sustainable development is generally accepted and relatively simple to grasp. The main difficulty occurs when the principles of sustainable development are put into practice. Sustainability indicators have to include certain desirable characteristics: easy to calculate, helpful for decision making, and vigorous in indicating progress toward sustainability (38–40).

3.6 ADVANCED MATERIALS AND TECHNOLOGY FOR SUSTAINABLE DEVELOPMENT

Advanced materials and technology research is based on the basic principles of science (physics, chemistry, and mathematics). These principles are used to control the properties of these new materials, and these materials can be developed for real-life situations by including all the conditions under which a material will be used (41).

Advanced materials exhibit vital physicochemical properties that make them fascinating and functional for sustainable development. Advanced materials used for research into sustainable technologies are dedicated to the development of innovative materials for a sustainable pathway. These innovative materials are used for various purposes such as CO_2 emission reduction, exploration of new energy adaptation and storage techniques, and use of 3D printing for metals as direct assembling measures, thus allowing cost- and resource-effective production etc. (42).

On the other hand, a good example of innovation is technology having great potential to stimulate sustainability and enable the manipulation of individual atoms and molecules at nanoscale to generate materials, devices, and systems with enhanced properties. Advanced materials have always been technology facilitator, and they play an important role in every technology: functional ceramics used in mobile phones, high-power aluminum alloys used in aviation industry, steel girders used in skyscrapers. In the same way, advanced materials are continually playing a role in sustainable development (43). There is an immediate and precise link between material science and sustainable development involving effective use of materials. This relationship is so apparent and elementary that, paradoxically, most of the time, users cannot visualize the effects of advanced technologies (44). There are many valuable products that help to maintain sustainable development, such as lighter, stronger steel; more efficient solar panels; hydrogen fuel cells; and smaller and faster computers.

For sustainable development, the biggest challenge might be human nature. As global population increases, the crucial question is not how many persons Earth can hold, but how many persons Earth can bear sustainably with their living standards. To solve this problem, scientists, government leaders and corporate world face some critical issues regarding the supply of resources; therefore, they form some policies to solve the problem (shortages of the materials) (45).

3.7 SOCIO-ECONOMIC CONSIDERATIONS

Public support and economic feasibility are important aspects for the implementation of schemes related to sustainable development, along with technological advancement and ecological impact. This holistic approach interconnects the three pillars of sustainable development: society, economy, and the environment. The importance of economic concerns to accelerate the implementation of innovative materials and different technologies is broadly recognized, and economic drivers are considered important parameters for policy-makers to build efficient strategies. Also, the study of social awareness and approaches can offer an insight into several considerations that affect the development of public awareness on environmental actions. Social acceptability may be a factor restricting the implementation, increase, and widespread use of environmental actions. Many different strategies are known for social participation; however, the public can have involvement in a social project through a) information about continuing development (information), b) involvement in the decision-making process (planning participation), and c) financial participation in the project (45).

Over the past few decades, advanced materials have played an important role in the modern global technological revolution due to their innovative and transformative potential. The effect of advanced materials has a strategic nature, and their application is essential for sustaining the effectiveness of the economy on global markets. Advanced materials and technology will show a connection with the developments of localization, finalizing global structure with local substructures as also with ephemeralization of economy on all levels.

The main purpose of advanced materials and technology is to support endeavors and individuals to take advantage of technologies more productively with cost effectiveness and productivity gains. The use of these materials and technologies in various fields specifies the method for production of new cost-effective products and for capital collection. The value of development ratios is as significant as their size.

As we said in the beginning, advanced materials are lightweight materials, including composites, having an important role in every technology, such as fiber- and textile-based technologies and many more. They are primarily designed to acquire positive improvements; alternatively they could be used for unfavorable things (46). Further, let's have a look at the positive and negative impacts in the progress of advanced materials and technology in general to our society and economy.

Industrial productivity has increased due to the use of advanced materials, for example, the shape-memory alloy (an alloy that retains its structure and can be come back in its original shape after deformation during heating process). When the shape-memory effect is properly exploited, this material turns out to be a lightweight, solid-state substitute. Therefore, these alloys (also called smart alloys) have various applications in different sectors. They act effectively as thermal sensors and actuators in broad-range applications, like kettles, air conditioners, and vehicles. These metals are also used in artificial muscles that are electrically stimulated, such as robotic hands and surgical endoscopes. Therefore, in this way, these smart alloys are drawing growing investment from businesses and governments around the world in a sustainable way.

Among the small wonders, another advanced material is the carbon nanotubes, as strong as a diamond, having multiple times the stress strength of steel, and multiple times (almost 10 000 times) finer than human hair. These are considered better heat and electricity conductors than any other material and can even function as a semiconductor. Carbon nanotubes are used in hydrogen fuel cells as composite electrodes or membranes that help to lessen and eliminate the demand for rare and expensive catalysts. Carbon nanotubes are also used for different purposes such as in energy storage, water filters, device modeling, thin-film electronics, automotive parts, boat hulls, sporting goods, actuators coatings, and electromagnetic shields. In another way, they are designed to provide more energy-efficient substitutes and offer the key to developing next-generation products that are stronger, lighter, and more durable. When the technology is mature and the cost is reduced, it will replace traditional bulky materials (46).

Over the past few decades, advanced materials have been playing a critical role in enhancing the economy of advanced automobile technologies (mainly fuel economy) while retaining performance and safety. Advanced materials are lightweight materials that result in a 10 percent reduction in weight, which offers great potential for increasing fuel efficiency by almost 6–8 percent because less energy is required to accelerate a lighter object than a heavier one. For example, the components of conventional steel are replaced with lightweight materials such as high-strength

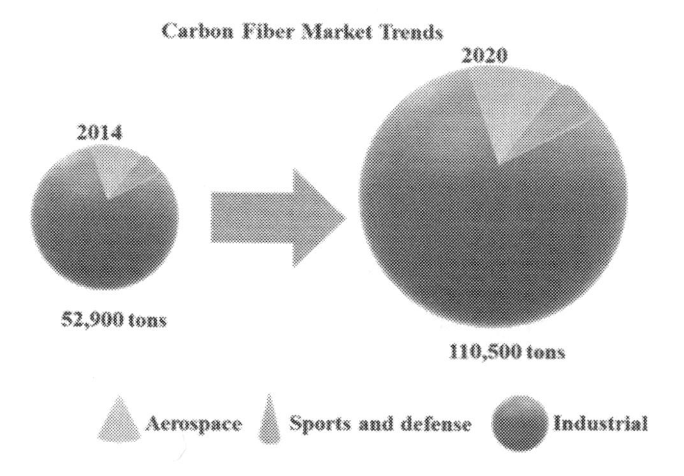

FIGURE 3.6 Carbon fiber growth in the automotive market.

steel, aluminum (Al) and magnesium (Mg) alloys, carbon fiber, and polymer composites. By using these lightweight materials, the weight of a vehicle's body and frame can be cut by up to 50 percent, resulting in a reduction of a vehicle's fuel consumption (Fig. 3.6, Table 3.1). Over the past few years, it has been observed that advanced materials ranging from Al and Mg alloys to carbon fiber composites have advanced into mass-produced passenger vehicles (47, 48).

Therefore it can be concluded from all the above points that advanced materials have an important role in the development of technology, with contributions to economic and social-cultural life.

But there are some disadvantages of these lightweight materials. They have high cost. They can be grated more as compared to less grade materials. The ductility diminishes by increasing the strength, creating issues in forming and joining of the materials. The engineers also faced some challenges in designing, component processing, and material behavior in harsh environment (49).

Actually, negative outcomes of advanced materials, technological revolution, and industrialization are critically limiting the possibilities and sources for the future generations by contaminating natural habitat, interrupting ecological balances, and consuming natural resources; this poses a threat for conflicts and hostilities among countries. In addition, it is clear that the cost to replenish rapidly consumed assets

TABLE 3.1
Comparison of Reduction of Weight of Advanced Materials w.r.t. Steel

Advanced Materials	Weight Reduction vs. Low Carbon Steel
High-strength steel	15%–25%
Aluminum	40%–50%
Magnesium	55%–60%
Carbon fiber composite	55%–60%

will be very high in terms of economic challenges for the future generations. In a report titled "The Limits of Growth," the Roman Club group reported five global trends that are of great concern for nations (3).

These trends are rapid growth in industrialization, population, widespread deficiency of nutrition/famine, rapid depletion of non-renewable resources, and degradation of environment due to pollution (3).

3.8 ENVIRONMENTAL CONSIDERATIONS

The design and development of advanced materials and technologies for enhancing various energy transformation and storage techniques as well as environmental issues have shown fascinating outcomes and a promising future. Advanced materials are getting tremendous attention and research interest because of the increasing interest in the sustainable development of energy, society, and economy, which is firmly identified with high efficiency, versatile storage, and use of energy.

The advanced materials for sustainable future technologies research line are mainly focused on the production of novel materials which are energy- and resource-efficient for reduction of CO_2 emission in a sustainable pathway and contribute significantly to the protection of environment by retaining energy, water, and raw materials. They reduce toxic waste and ozone-depleting substances to explore new methods for energy conversion and its storage. Advanced materials and technological products, processes, and applications promise certain environmental advantages and sustainability impacts.

Here are some specific examples of advanced materials and technology that benefits the environment:

i. **Recycling of batteries to be economically attractive:** Heavy metals such as mercury, lead, cadmium, and nickel are present in certain batteries. These are responsible for polluting the atmosphere and pose a possible danger to human health. If batteries are disposed of improperly it is complete waste of a potential and modest raw material. These cheap raw materials can be replaced by pure zinc oxide (ZnO) nanoparticles, which are prepared from spent Zn-MnO_2 alkaline batteries. The used cathode particles are collected from spent lithium ion batteries, recycled and regenerated in form of new material (50).

ii. **Advanced materials for cleanup of radioactive waste from water:** Over the past few years, scientists have been continuously working to find the solution for removal of radioactive waste from water bodies. Titanate nano-fibers are used as adsorbents for the removal of toxic, radioactive Ra^{2+} and Sr^{2+} ions from water bodies with subsequent safe disposal. Titanate nanotubes and nanofibers are considered superior materials for the removal of radioactive cesium and iodine ions due to their unique structural properties and ion exchange properties (51). Some artificial inorganic cation exchange materials, such as niobate molecular sieves, synthetic micas, and g-zirconium phosphate, are also found to be superior to natural materials for removal of these waste radioactive materials (52).

iii. **Nanotechnology-based solutions for oil spills:** Traditional techniques are not found to be satisfactory for cleanup of massive oil spills. Although the application of advanced materials and technology for removal of oil spills is still in its emerging phase, it offers promise for the future. A multi-shelled nanoparticle known as "Nano-Carbo-Scavenger" has been developed, having natural properties and considered a highly suitable, economical, environmentally friendly, and biocompatible nano-dispersant. The raw and distillate forms of petroleum oil were treated using NCS with substantial productivity (80 percent and 91 percent, respectively) applying sequestration and dispersion abilities in cycle with a ~10:1 (oil: NCS; w/w) loading capacity (53).

iv. **Carbon dioxide capture:** CO_2 is mixture of gases resulting from some industrial processes and combustion. Since the beginning of the Industrial Revolution, there has been a huge increment in atmospheric CO_2, which has increased from about 280 to nearly 390 ppm today. The most effective way to minimize CO_2 in the environment is to capture it during its release and its storage, commonly called CO_2 capture and storage (CCS), a process in which CO_2 is separated from other gases that are emitted from industrial processes occurring at high temperatures where different solvents are used in columns and inoculated deeply under the ground. Then it is transferred to storage tanks, thereby diminishing greenhouse gas (GHG) emissions. There are three basic techniques used to capture CO_2: pre-combustion, post-combustion, and oxy-fuel combustion (Fig. 3.7).

In recent years, various advanced materials have been used for separation and absorption of CO_2. Nano-fillers with nano-composite membranes, a hollow-fiber membrane using water-based nano-fluids, and different types

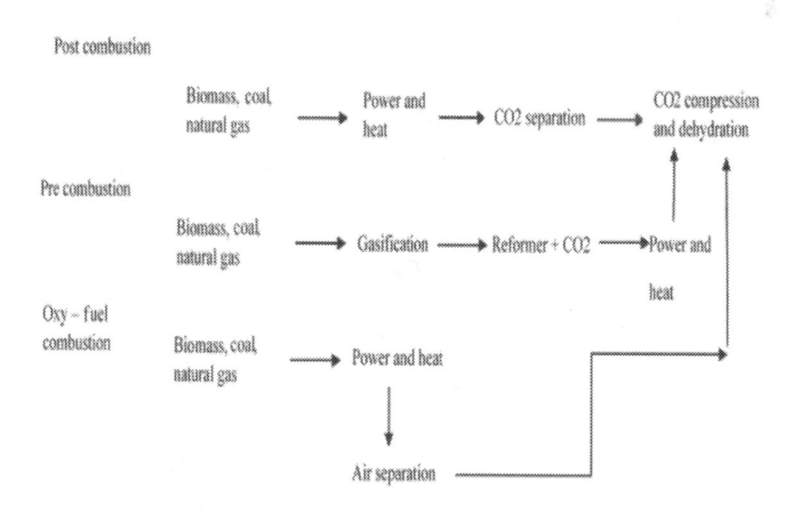

FIGURE 3.7 Various CCS techniques.

of nano-materials such as aluminum oxide and silica oxide are used for the removal/separation of CO_2(54).

v. **Hydrogen production from sunlight — artificial photosynthesis:** A chemical reaction in which water and carbon dioxide are converted into carbohydrate and oxygen results in biomimic of the photosynthesis cycle. This term is generally applied to any system such as photocatalytic water splitting system used for taking the energy from sunlight and stored it in form of the chemical bonds of a fuel (a solar fuel).

There are four stages in artificial photosynthesis similar to natural photosynthesis:

a. **Light harvesting:** Antenna molecules collect the light particles (photons) and sunlight in the form of stored energy in a reaction center.
b. **Separation of charge:** The stored energy is applied to split positive and negative charges (holes and electrons) from each other at reaction center.
c. **Splitting of water:** Holes are used to split water into hydrogen ions (protons) and oxygen after incorporated into catalytic centers.
d. **Production of fuel:** From new photons, electrons are offered more energy and consequently attached with the hydrogen ions and possibly CO_2 to generate hydrogen/carbon-based fuel.

The chemical reactions used to produce hydrogen are found to be easier than the production of carbon-based fuels. On splitting, water releases four H^+ and four electrons. These H^+ ions are then converted into two H_2 molecules after receiving energy from next four protons and this process is continuous:

$$4H^+ + 4e^- \xrightarrow{4hy} 2H_2$$

Hydrogen is found to be an attractive carbon-free energy carrier as it play a key role in future renewable energy technology. As discussed in Section 3.5, human beings are using advanced materials and technology in different fields, such as in automobiles (battery waste in electric cars, palladium and platinum nanoparticles as catalytic convertors, and fine carbon particles in tires, etc.), pharmaceuticals(silver nanoparticles as antibacterial and gold nanoparticles as therapeutics), energy resources (quantum dots as solar cells, metal oxides in solar cells and batteries), laundry (silver nanoparticles as antibacterial fabric and metal oxide as UV-proof fabric), mining (fine metal and metal oxide), and combustion (fly ash, fullerenes, soot) as shown in Fig. 3.8 (55).

With increasing demand, consumption of natural resources and deforestation results in increase in GHG emissions and global warming.

There are different ways through which advanced materials can become harmful or toxic, not only for human beings but also for the environment. For example, nanoparticles are more harmful after release into the environment because they

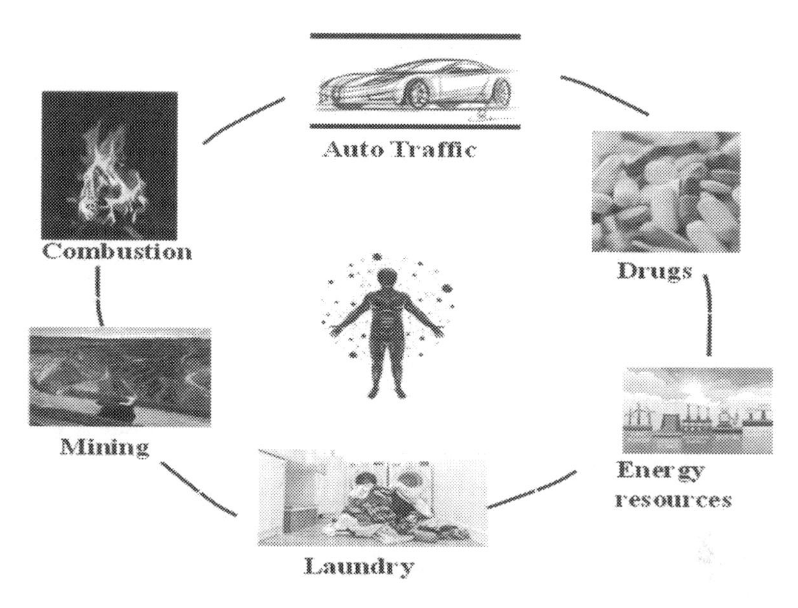

FIGURE 3.8 Applications of heavy metal/metal oxide nanoparticles in various fields.

are not visible/detectable; this creates different difficulties because they facilitate chemical reactions that can harm plankton, bacteria, and small animals.

Most metal and metal oxide nanoparticles are used as catalysts in various chemical reactions, and if they are released into the environment, they can generate different toxic chemicals by facilitating chemical reactions such as reactive oxygen species (ROS) or free radicals, as shown in Fig. 3.9.

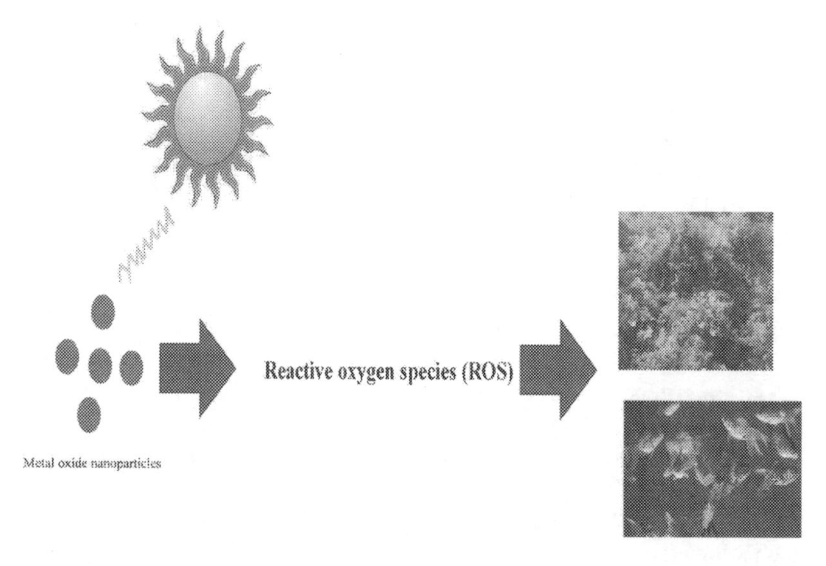

FIGURE 3.9 Mechanism of formation of reactive oxygen species (ROS) in the environment.

Nanoparticles present in the atmosphere or in water resources could be released either during their production or through a disposal process. The production of nano-materials by industries has increased from 1000 to 58 000 tons in 2020, which is a significant concern for the human being.

Like nano-materials, advanced composites such as metal composites or ceramics have drastic effects on the environment, such as air pollution, global warming, acidification, and eutrophication. When they are released into the atmosphere along with nano and ultrafine particles, they directly affect human life by emitting hazardous chemicals, toxic gases, and waste (56).

Industrial/synthetic polymers, due to their non-biodegradable nature, affect the environment in a terrible manner. During incineration, these synthetic polymers release harmful gases such as carbon dioxide, nitrogen oxide, and carbon monoxide into the environment, causing air pollution, global warming, and acidification. Some polymers containing chlorine produce HCl during combustion, resulting in corrosion in building and damage to crop production in surrounding areas (57).

There are some common but major impacts of these materials and technology on the environment:

a. Development of advanced materials, including nanomaterial and their applications, in industries, requires high amounts of energy resulting in high demand of natural resources.
b. The persistent toxic and carcinogenic nature of advanced materials is harming the environment.
c. Environmental pollution due to waste materials of these advanced materials is a resultant factor.
d. Contribution to environmental degradation is the secondary impact of growing technology.

3.9 NATURAL RESOURCES

Natural resources are materials such as air, sunlight, soil, water, and fossil fuels that exist in nature without any human activity. These resources are used for various purposes such as aesthetic, commercial, industrial, cultural, and scientific purposes (58). There are mainly two types of natural resources based on their rate of generation and replenishment compared to their consumption, as shown in Fig. 3.10:

1. **Renewable resources:** Resources that can be replenished faster than their consumption, such as solar energy, wind energy, biomass, water energy. These resources can be replenished naturally over a short period of time.
2. **Non-renewable resources:** Resources that are available in limited quantities and cannot be replenished over a short period. The rate of consumption is more than the rate of formation. Examples are coal, natural gas, petroleum, etc.

Natural resources are being depleted significantly due to the advancement of science and technology, consumption at a pace faster than replenishment, overpopulation, competing the better quality of developed countries, less awareness of environmental issues, and so on (59).

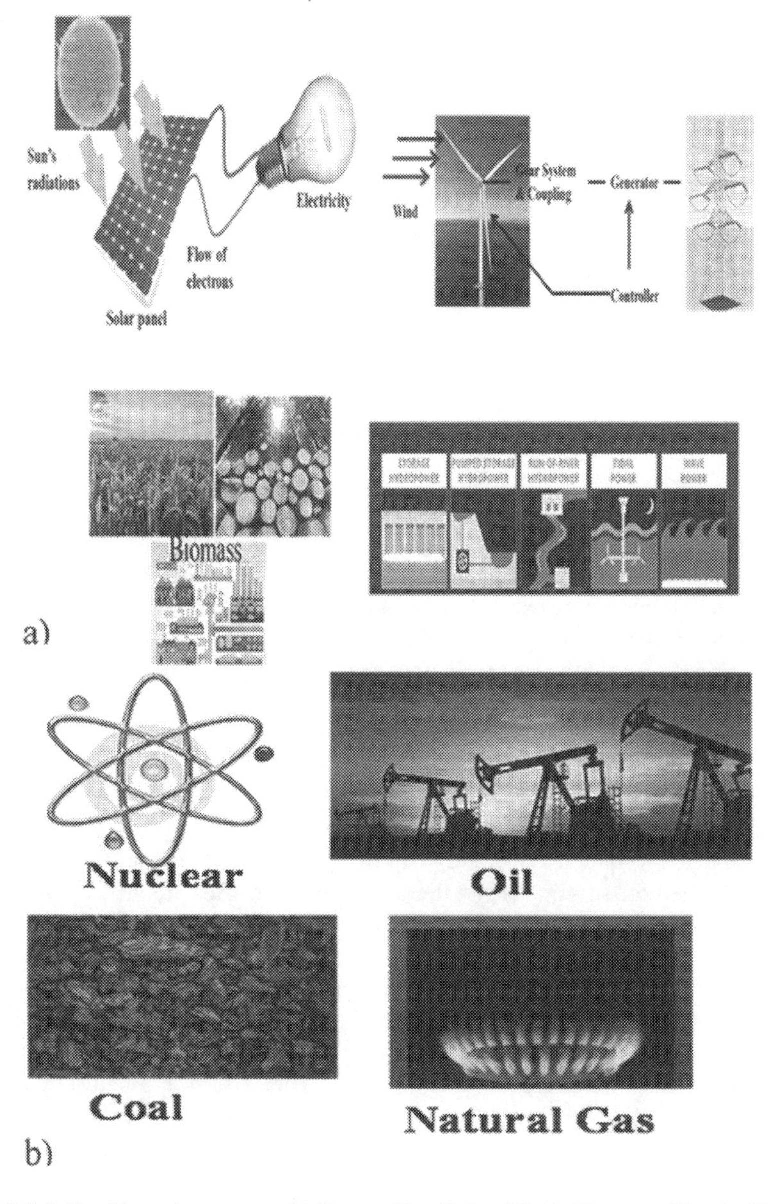

FIGURE 3.10 Natural resources (a) Renewables (Solar, Wind, Biomass, Water); (b) Non-Renewables (Nuclear, Crude Oil, Coal, Natural Gas).

3.9.1 Depletion of Natural Resources

The continuous use of natural resources beyond their replacement rate is considered to be cause of their depletion. The importance of any resource mainly depends on its availability in nature and extraction from nature. The greater the depletion rate of are source, the more its value will increase. Many types of resources are being depleted. A few of them are discussed below:

3.9.1.1 Fossil Fuels: Coal, Petroleum, and Natural Gas

Fossil fuels were formed as a consequence of natural calamities such as earthquakes, volcano eruptions and floods. During these natural calamities, animals, humans, and plants got buried deep inside the Earth's crust. The high temperature and high pressure over a period of time led to the transformation of these carbon-based materials to fossil fuels. This transformation takes thousands of years; hence, their conservation is essential.

3.9.1.2 Metallic and Non-Metallic Minerals

Metallic and non-metallic minerals are associated with igneous and sedimentary rocks respectively and yield metal and non-metal from their ores, clays, and salts. Cumulative exergy consumption (CExC) and thermo-ecological cost concepts are used to calculate the depletion of natural resources containing metallic minerals (60).

3.9.2 Depletion of Renewable Natural Resources

Renewable resources such as solar energy, wind, biomass, and water can be replenished at a faster pace, but they still present environmental concerns (61). Even exploitation of these renewable resources may cause air and water pollution. Alternative chemicals or fuel from biomass raise questions about the sustainability of the process, whether the land should be used to grow food crops or feedstock for chemical production.

3.10 IMPACTS OF DEPLETION OF NATURAL RESOURCES

The rapid increase in the population will deplete almost all resources by 2050. In the past few decades, the middle-class population has increased by a factor of two and is expected to increase further. With this growing population, more resources will be required for sustenance. At the same time industries are growing at a faster pace and developing and undeveloped nations are moving toward development and using more of the resources and contributing to their scarcity for future generations. With the depletion of natural resources other major problems will arise, such as deforestation, natural disasters, soil erosion, greenhouse gases and climate change, ozone depletion, oil depletion, forced migration, desertification, extinction of species, extreme energy shortages, water pollution. The existence of natural resources and their availability are two different aspects that depend on other various factors (62), such as:

a. **Technological factors:** Lack of technological expertise may leave resources unexplored and unextracted. These resources might be available in future with the development of expertise.
b. **Economic factors:** Price determines whether it is viable to invest in extraction of the resource.
c. **Ecological factors:** Increasing risk in extracting resources from particular geographical regions affects resource availability.
d. **Geopolitical and political factors:** Conflicts and political interference affect the availability of resources.

3.11 ECONOMIC DEVELOPMENT AT THE COST OF RESOURCE DEPLETION

There is a direct connection between economic development and resource development, leading to the debate between ecologists and economists. In developing countries, economic development is increasing at very fast pace with the exploitation of natural resources. They are exploiting their land, air, and water for economic growth without concern for environment. Also, developing countries are exporting natural resources and further putting pressure internationally and exploiting the natural resources both at national and international level (63). Efforts should be taken to manage the natural resources not only for sustainability of the environment but also for sustaining the process of extraction of resources. Management of natural resources should increase awareness about the existence of the resources and economic sustainability.

3.12 CONCLUSION

Advanced materials are at the core of daily life and provide stability, long-term durability, security, and environmental compatibility of the devices and services around us. Innovative and useful materials with enhanced physicochemical properties are known as advanced materials and have a key role in sustainable development of technology. Advanced materials are also known as engineered materials formed through the development of specified processing and synthesis technology and categorized on the basis of their properties and applications. In the 21st century, advanced materials are used in form of nano-materials or lightweight materials and applied in different sectors such as automotive, construction, pharmaceuticals, agriculture, water purification, and food Packaging, where they are used as cost-effective materials with less environmental effect. Environmental issues related to social, economic, and climate change are mainly due to human activities. Advanced materials and technology are significantly used in development and production to fulfill human wants, but if they are used in uncontrolled manner, they will obviously effect the environment in negative manner by increasing pollution, disturbing the ecological balance, and depleting the natural resources.

REFERENCES

1. I. L. Feitshans (2020). Nanotechnology law for commercialization of nano-enabled products. *Adv. Mater. Sci. Res.*, 1(2).
2. L. Yan, Y. Yang, W. Zhang, & X. Chen (2014). Advanced materials and nanotechnology for drug delivery. *Adv. Mater.*, 26 (31), 5533–5540. doi: https://doi.org/10.1002/adma.201305683
3. K. S. Lokesh & P. Prasad (2018). *Advanced materials.* Munich: GRIN Verlag. https://www.grin.com/document/452980
4. A. Varma, A. S. Rogachev, A. S. Mukasyan, & S. Hwang (1998). Combustion synthesis of advanced materials: Principles and applications. *Adv. Chem. Eng.*, 24, 79–226.
5. J. M. Karp & R. Langer (2007). Development and therapeutic applications of advanced biomaterials. *Curr. Opin. Biotech.*, 18 (5), 454–459.

6. A. D. Basiago (1998). Economic, social, and environmental sustainability in development theory and urban planning practice. *Environment*, 19, 145–161.
7. S. Palit & C. M. Hussain (2018). Green sustainability, nanotechnology and advanced materials: A critical overview and a vision for the future. *Green Sustain. Adv. Mater.*, Wiley online library, 1–3. doi: https://doi.org/10.1002/9781119528463.ch1
8. Advanced Materials Scientific Journal (2018). Overview: Aims and scope. Retrieved from https://onlinelibrary.wiley.com/page/journal/15214095/homepage/productinformation.html (August, 2020).
9. A. Kennedy, J. Brame, T. Rycroft, M. Wood, V. Zemba, & I. Linkov, (2019). A definition and categorization system for advanced materials: The Foundation for Risk-Informed Environmental Health and Safety Testing. *Risk Anal.*, 39(8), 1783–1795. https://dx.doi.org/10.1111%2Frisa.13304
10. Definition of advanced materials (2020). Retrieved from https://www.mindat.org/glossary/advanced_materials (August 2020).
11. Definition of advanced materials (2020). Retrieved from https://www.lawinsider.com/dictionary/advanced-materials (August 2020).
12. Technology, New world Encyclopaedia (2020). Retrieved from https://www.newworldencyclopedia.org/entry/Technology (August 2020).
13. K. E. Drexler (1992). *Nanosystems: Molecular machinery, manufacturing, and computation.* New York: John Wiley & Sons. ISBN 978-0-471-57547-4.
14. A. Borgmann (2006). Technology as a cultural force: For Alena and Griffin. *Canad. J. Sociol.*, 31(3), 351–360. doi:10.1353/cjs.2006.0050
15. H. K. Çaloukan (2015). Technological change and economic growth. *Procedia Soc. Behav. Sci.*, 195, 649–654.
16. M. H. Van de Voorde (2016). The importance of materials for advanced energy technologies. *High Temp. Tech.*, 1 (4), 195–200. doi:10.1080/02619180.1983.11753208
17. P. Grant (2013). New and advanced materials: Section A. Retrieved from https://assets.publishing.service.gov.uk/government/uploads/system/uploads/attachment_data/file/283886/ep10-new-and-advanced-materials.pdf
18. B. Cantor, P. S. Grant, & C. Johnston (2008). *Automotive engineering: Lightweight, functional, and novel materials.* Boca Raton: CRC Press.
19. J. E. King (2007). *The King review of low-carbon cars. Part I: the potential for CO_2 reduction.* London: HM Treasury.
20. L. Gene (2017). Lightweight materials for automotive applications. Retrieved from https://atecentral.net/downloads/1585/Lightweight_Materials_for_Automotive.pdf (August 2020).
21. G. F. Matthews, P. Edwards, T. Hirai, M. Kear, A. Lioure, & M. Way (2007). Overview of the ITER-like wall project. *Phy. Scr.*, 2007.
22. Health and Safety Executive (2013). Using nanomaterials at work. Retrieved from http://www.hse.gov.uk/pubns/books/hsg272.pdf (August, 2020).
23. A. K. Mohanty, M. Misra, & G. Hinrichsen (2000). Biofibres, biodegradable polymers and biocomposites: An overview. *Macromol. Mater. Eng.*, 276/277, 1–24.
24. A. Atkinson, S. Barnett, R. J. Gorte, J. T. S. Irvine, A. J. McEvoy, & J. Vohs (2004). Advanced anodes for high-temperature fuel cells. *Nature Mat.*, 3, 17–27.
25. G. Derrien, J. Hassoun, S. Panero, & B. Scrosati (2007). Nanostructured Sn–C composite as an advanced anode material in high-performance lithium-ion batteries. *Adv. Mat.*, 19, 2336–2340.
26. A. Magasinski, P. Dixon, B. Hertzberg, A. Kvit, J. Ayala, & G. Yushin (2010). High-performance lithium-ion anodes using a hierarchical bottom-up approach. *Nature Mat.*, 9, 353–358.
27. L. Feng, Y. L. Xu, W. S. Fegadolli, M. H. Lu, J. E. B. Oliveira, & A. Scherer (2013). Experimental demonstration of a unidirectional reflectionless paritytime metamaterial at optical frequencies. *Nature Mat.*, 12, 108–113.

28. Y. Liu & X. Zhang (2011). Metamaterials: A new frontier of science and technology. *Chem. Soc. Rev.*, 40, 2494–2507.
29. W. J. Krzysztofik & T. N. Cao (2018). Metamaterials in application to improve antenna parameters. doi: 10.5772/intechopen.80636
30. S. Bae, H. Kim, Y. Lee, X. Xu, J. S. Park, & S. Iijima (2010). Roll-to-roll production of 30-inch graphene films for transparent electrodes. *Nature Nano.*, 5, 574–578.
31. J. N. Coleman, M. Lotya, A. O'Neill, S. D. Bergin, P. J. King, & V. Nicolosi (2011). Two-dimensional nanosheets produced by liquid exfoliation of layered materials. *Science*, 331, 568–571.
32. J. C. Reichert, A. Cipitria, D. R. Epari, S. Saifzadeh, P. Krishnakanth, & D. W. Hutmacher (2012). A tissue engineering solution for segmental defect regeneration in load-bearing long bones. *Sci. Transl. Med.*, 4 (141), 141ra93. doi: 10.1126/scitranslmed.3003720
33. C. D. Taylor, B. Gully, A. N. Sanchez, E. Rode, & A. S. Agarwal (2016). Towards materials sustainability through materials stewardship. *Sustainability*, 8, 1001. doi :10.3390/su8101001
34. A. Beattie (2019). The 3 pillars of corporate sustainability. Retrieved from https://www.investopedia.com/articles/investing/100515/three-pillars-corporate-sustainability.asp#:~:text=Sustainability%20is%20most%20often%20defined,economic%2C%20environmental%2C%20and%20social (August 2020).
35. A. D. Basiago (1999). Economic, social, and environmental sustainability in development theory and urban planning practice. *Environment*, 19, 145–161.
36. P. Baruah (2020). 3 pillars of sustainability: Economic, environmental, and social. Retrieved from https://planningtank.com/environment/3-pillars-of-sustainability (August 2020).
37. M. Diallo & C. J. Brinker (2020). Nanotechnology for sustainability: Environment, water, food, minerals, and climate. *Nanotechnol. Res. Direc. Societ. Needs in 2020, Sci. Policy Rep.*, 221–259.
38. C. J. Brinker & D. Ginger (2020). Nanotechnology for sustainability: Energy conversion, storage, and conservation. *Nanotechnol. Res. Direc. Societ. Needs in 2020, Sci. Policy Rep.*, 261–303.
39. F. Kongoli (2016). Role of science and technology in sustainable development. Sustainable industrial processing summit and exhibition plenaries. Retrieved from http://www.flogen.org/pdf/sips16_524FS.pdf
40. M. S. Diallo, N. A. Former, & M. S. Jhon (2013). Nanotechnology for sustainable development: Retrospective and outlook. *J. Nanopart. Res.*, 15, 2044. doi: 10.1007/s11051-013-2044-0
41. S. T. Picraux (2020). Nanotechnology. Encyclopedia Britannica. Retrieved from https://www.britannica.com/technology/nanotechnology (August 2020).
42. P. Koltun (2010). Materials and sustainable development. *Prog. Nat. Sci. Mater. Internat.*, 20, 16–29.
43. M. M. Moulaii, M. R. Bemanian, M. Mahdavinejad, & N. Mokary (2013). How smart materials can help occupants to live in more sustainable buildings. *Asian J. App. Sci.*, 01, 16–20.
44. M. L. Green, L. Espinal, E. Traversa, & E. J. Amis (2012). Materials for sustainable development. Material Research Society. Retrieved from https://www.cambridge.org/core/terms (August 2020).
45. L. Friedman (2011). China, India could "lock" world in a high-carbon energy system, IEA warns, Scientific American. Retrieved from www.scientificamerican com/article.cfm?id=china-india-could-lock-wo (August, 2020).
46. R. Wennersten, J. Fidler, & A. Spitsyna (2008). *Nanotechnology: A new technological revolution in the 21st century. Handbook of performability engineering.* Springer, London, 943–852. https://doi.org/10.1007/978-1-84800-131-2_57
47. G. Kardys (2017). Magnesium car parts: A far reach for manufacturers? Part 1. Retrieved from ttps://insights.globalspec.com/article/7243/magnesium-car-parts-a-far-reach-for-manufacturers-part-1 (August 2020).

48. Carbon fiber growth in the automotive market (2020). Retrieved from https://www.reinforcer.com/en/category/event-detail/Carbon-Fiber-Growth-in-the-Automotive-Market/39/323/0 (August 2020).
49. J. P. Clark & M. C. Flemings (1986). Advanced materials and the economy. *Sci. Am.*, 255 (4), 51–57.
50. W. Zhang, L. Du, Z. Chen, J. Hong, & L. Yue (2016). ZnO nanocrystals as anode electrodes for Lithium-Ion batteries. *J. Nanomater.*, 2016. doi: https://doi.org/10.1155/2016/8056302
51. D. J. Yang, Z. F. Zheng, H. Y. Zhu, H. W. Liu, & X. P. Gao (2008). Titanate nanofibers as intelligent absorbents for the removal of radioactive ions from Water. *Adv. Mater.*, 20 (14), 2777–2781.
52. M. Berger (2011). New nanomaterials for radioactive waste clean-up in water. Retrieved from https://www.nanowerk.com/spotlight/spotid=22950.php (August 2020).
53. E. A. Daza, S. K. Misra, J. Scott, I. Tripathi, C. Promisel, & D. Pan (2017). Multi-shell nano-carboscavengers for petroleum spill remediation. *Sci Rep.*, 7, 41880. doi: 10.1038/srep41880.
54. R. Kumar, R. Mangalapuri, M. H. Ahmadi, N. Vo Dai-Viet, R. Solanki, & P. Kumar (2020). The role of nanotechnology on post-combustion CO_2 absorption in process industries. *Internat. J. Low-Carbon Tech.*, 15 (3), 361–367. doi: https://doi.org/10.1093/ijlct/ctaa002
55. R. Purchase, H. de Vriend, H. de Groot, & P. Harmsen (2012). Artificial photosynthesis for the conversion of sunlight to fuel. Wageningen.
56. Z. Huo, C. H. Wu, Z. Zhu, & Y. Zhao (2015). Advanced materials and nanotechnology for sustainable energy development. *J. Nanotech.*, 2015, 1. doi: http://dx.doi.org/10.1155/2015/302149
57. M. Muthukannan, A. Sankar, & C. Ganesh (2019). The environmental impact caused by the ceramic industries and assessment methodologies. *Internat. J. Quality Res.*, 13(2) 315–334.
58. X. A. Rodríguez, C. Arias, & A. Rodríguez-González (2015). Physical versus economic depletion of a nonrenewable natural resource. *Resour. Policy*, 46, 161–166. doi: 10.1016/j.resourpol.2015.09.008
59. I. R. K. G. Mittal (2006). Natural resources depletion and economic growth in present era, SOCH- Mastnath. *J. Sci. Technol.*, 122, 25–27.
60. B. Boryczko, A. Hołda, & Z. Kolenda (2014). Depletion of the non-renewable natural resource reserves in copper, zinc, lead and aluminium production. *J. Clean. Prod.*, 84(1), 313–321. doi: 10.1016/j.jclepro.2014.01.093
61. D. Watts (1988). Development and renewable resource. *Journal of Biogeography*, 15 (1), 119–126.
62. S. Petersen, A. Krätschell, N. Augustin, J. Jamieson, J. R. Hein, & M. D. Hannington (2016). News from the seabed: Geological characteristics and resource potential of deep-sea mineral resources. *Mar. Policy*, 70, 175–187. doi: 10.1016/j.marpol.2016.03.012
63. A. Khan, Y. Chenggang, J. Hussain, S. Bano, & Aa. A. Nawaz (2020). Natural resources, tourism development, and energy-growth-CO_2 emission nexus: A simultaneity modeling analysis of BRI countries. *Resour. Policy*, 68, 101751. doi: 10.1016/j.resourpol.2020.101751

4 Physical Processes in Wastewater Treatment

Ning Zhang
International Center for Water Resources Management
Wilberforce, OH

Ankur Dwivedi and Amita Chaudhary
Institute of Technology, Nirma University, Ahmedabad, India

CONTENTS

DOI: 10.1201/9781003138303-4

4.1 INTRODUCTION

Water is one of the most important and basic needs of life. Without water, the existence of life cannot be imagined. But the irony is that the total amount of water available is constant and is not sufficient to meet all the demands of everyone. Today the world is facing the problem of scarcity and the suitability of water. Despite the incredible importance of water, water sources are suffering from severe contamination, leading to severe destruction in ways beyond description. Around the world, millions of people do not have access to clean and safe water (WHO, 2017). As a result, approximately 6–8 million people die every year due to water-borne diseases. Water also plays a vital role in industries such as textile, paper and pulp, pharmaceutical, and power-generating plants. Thus, the effluent discharged by industries contains heavy metals, hazardous chemicals, and a high amount of organic and inorganic impurities. Contaminants in the water can affect the environment and human health. Therefore, water quality is the first priority in the environmental policy agenda in many parts of the world (Swenson & Baldwin, 1965). The water supplied for household purposes is commonly called domestic water. This domestic water quality has to be maintained properly as it is also used for drinking. There are various factors for checking the quality and suitability of potable or drinking water, such as taste, odor, color, and concentration of organic and inorganic substances (Ghaderpoori et al., 2018).

The main physical and chemical parameters used in analysis include color, turbidity, pH, total undissolved solids, total dissolved solids (TDS), biological oxygen demand (BOD), chemical oxygen demand (COD), and hazardous metals. These parameters are mainly responsible for degradation of the quality of the drinking water. The levels of these parameters should be less than the permissible limits set out by the World Health Organization (WHO) and other environmental organizations (WHO, n.d.). Therefore, it is mandatory to maintain these parameters under permissible limits.

4.2 SOURCES OF WATER

There are limited sources of water and those can be classified into three broad categories:

1. **Surface sources:** ponds, lakes, rivers, storage reservoirs, and oceans
2. **Groundwater sources:** springs, wells, and tube-wells
3. **Rainwater.**

Further classifications are shown in Fig. 4.1 (Mushak, 2011).

Below are the quantitative water consumption range per capita per day are tabulated in Table 4.1.

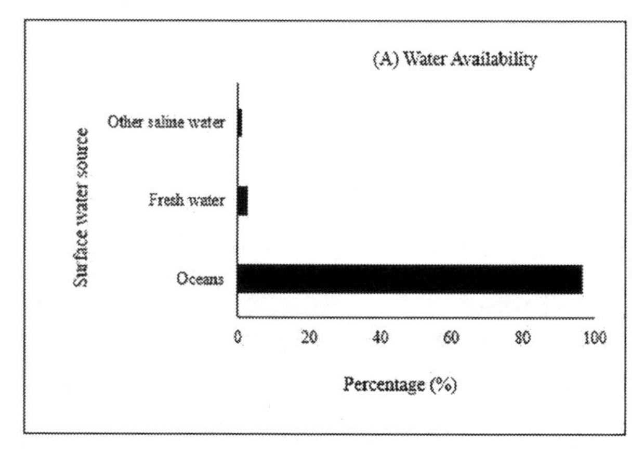

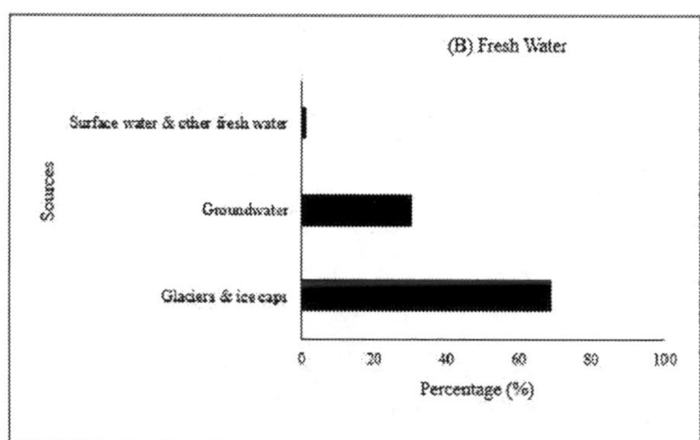

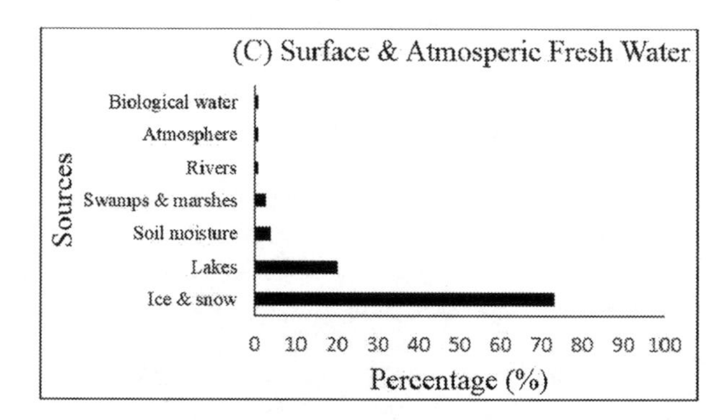

FIGURE 4.1 Scenario of water availability (a), (b), and (c).

TABLE 4.1
Water Consumption in Various Sectors

S. No.	Sectors	Normal Range (l/capita/day)	Average (l/capita/day)
1	Domestic purposes	65–300	160
2	Industry and manufacturing demands	45–450	135
3	Fire extinguisher demand	20–90	45
4	Inappropriate usage	45–150	62

4.3 TYPES OF IMPURITIES PRESENT IN WASTEWATER

The impurities present in raw water can be classified as follows:

a. **Dissolved impurities:**
   I. Inorganic and organic salts such as salts of Ca^{2+}, K^{+}, Fe^{2+}, Al^{3+}, etc.
   II. Gases such as CO_2, SO_2, NO_2, etc.
b. **Insoluble impurities:**
   I. Suspended impurities include organic and inorganic wastes.
   II. Colloidal impurities have particles of sand, clay, silica, etc.
c. **Microbial impurities:**
   This includes micro and macro pathogenic organisms.

To identify these impurities raw water is analyzed by testing its physical, chemical, and bacteriological characteristics (Swenson & Baldwin, 1965).

4.3.1 Physical Parameters

The physical parameters involved in testing the turbidity, color, taste, odor, and temperature of the raw water are tabulated in Table 4.2 (WHO, n.d.).

TABLE 4.2
Criteria for Physical Parameters

S. No.	Parameters	Acceptable Limit	Permissible Limit
1	Color	5 Hazen	15 Hazen
2	Odor	Not defined	Not defined
3	pH value	6.5–8.5	Not defined
4	Taste	Not defined	Not defined
5	Turbidity	1 NTU	5 NTU
6	Total dissolved solids	500 ppm	2000 ppm

TABLE 4.3
Criteria for the Chemical Parameters

S. No.	Chemical Parameters	Acceptable Limit (ppm)	Permissible Limit (ppm)
1	Aluminum	0.03	0.2
2	Ammonia	0.5	undesirable
3	Anionic surfactants	0.2	1.0
4	Barium	0.7	undesirable
6	Boron	0.5	1.0
7	Calcium	75	200
8	Chloramines as Cl_2	4.0	undesirable
9	Chloride	250	1000
10	Copper	0.05	1.5
11	Fluoride	1.0	1.5
12	Free residual chlorine	0.2	1
13	Iron	0.3	undesirable
14	Magnesium	30	100
15	Manganese	0.1	0.3
16	Mineral oil	0.5	undesirable
17	Nitrate	45	undesirable
18	Phenolic compounds	0.001	0.002
19	Selenium	0.01	undesirable
20	Silver	0.1	undesirable
21	Sulfate	200	400
22	Sulfide as H_2S	0.05	undesirable
23	Total alkalinity	200	600
24	Total hardness as $CaCO_3$	200	600
25	Zinc	5	15

4.3.2 Chemical Parameters

Table 4.3 shows the acceptable and permissible limits of various chemical parameters as per Bureau of Indian Standards (BIS) (Gupta et al., 2017).

4.3.3 Biological Parameters

The turbidity and the unpleasant smell in raw water are due to the dead and living microorganisms suspended in it that makes the water unfit. Some of such biological matters are tabulated in Table 4.4 (Gupta et al., 2017).

4.4 METHODS INVOLVED IN PHYSICAL TREATMENT OF WASTEWATER

Treatment of wastewater mainly involves three main processes: physical, chemical, and biological, as illustrated in Fig. 4.2. The appropriate process can be selected on the basis of the type of impurities present in the wastewater. The primary process

TABLE 4.4
Criteria for Biological Matters

S. No.	Organisms	Acceptable	Permissible
1	For drinking water:	Shall not be detectable	Undesirable
	a. *Escherichia coli* or thermotolerant coliform bacteria	in any 100 ml sample	
2	For treated water:		
	a. *E. coli* or thermotolerant coliform bacteria		
	b. Total coliform bacteria		

mainly involves preliminary treatments such as filtration, sedimentation, or settling process. Secondary treatment involves the disinfectant process, followed by tertiary treatment. Advanced techniques used in the tertiary process are nanofiltration, membrane separation, and novel adsorbent materials to get potable water. After this sequential process, the solid end product is called sludge which needs standardized procedures to dispose of, either being used as a manure for agriculture or thrown in landfill areas for further decomposition (Canziani & Spinosa, 2019). The schematic diagram of a typical wastewater treatment plant is shown in Fig. 4.3.

4.4.1 Primary Treatment Method

The primary treatment includes operations for removal of insoluble and suspended solids impurities from the wastewater. The main steps involved in physical operations are shown in Fig. 4.4.

4.4.1.1 Screening

Coarse screening and fine screening are commonly used screening processes in wastewater treatment plants. Screening is the first step in treating wastewater from any source. This process involves using strainers or screens for the removal of suspended and floating impurities, such as rags, paper, plastics, tins, and other containers (Taučer-Kapteijn et al., 2016). These screens contain parallel rods that purify the wastewater. The purification of water depends on the type of grits, screens,

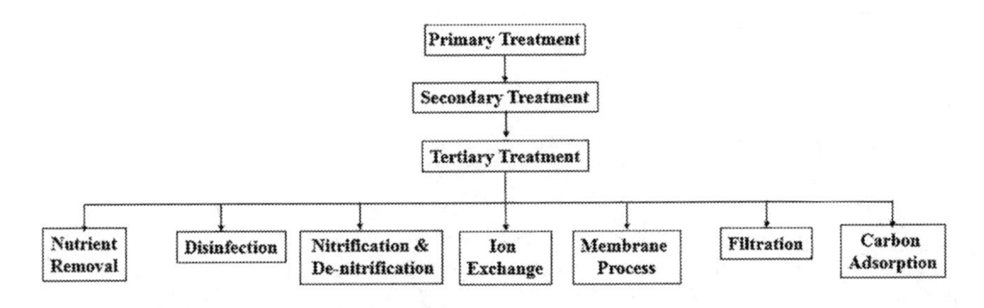

FIGURE 4.2 Processes involved in wastewater treatment.

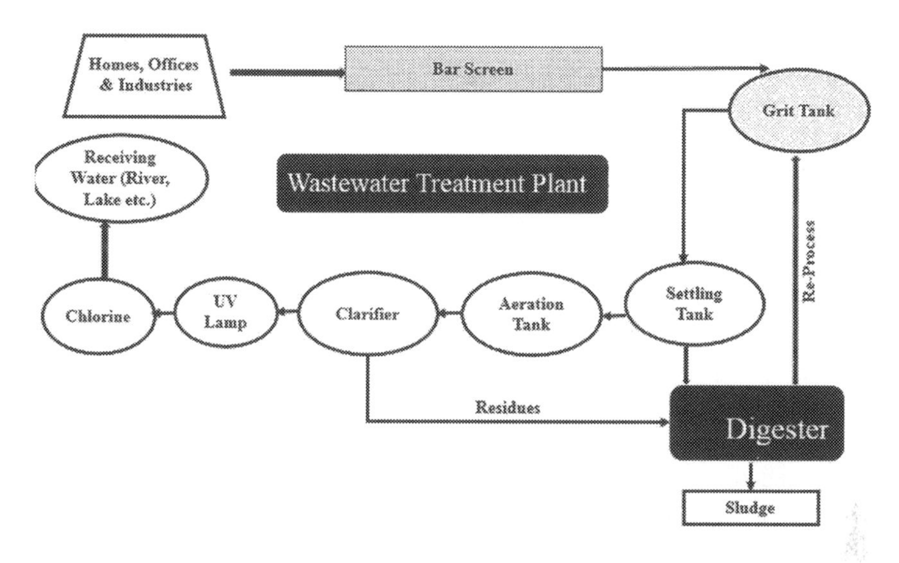

FIGURE 4.3 Schematic diagram of the wastewater treatment plant.

perforations, and meshes of different sizes used in the strainer. For separating large suspended and floating impurities, a coarse strainer (between 20 mm and 0.05 mm) is used (Nemerow, 2007). The main challenge to this process is the suspended textile fibers such as wet wipes and non-woven materials. They accumulate and clog the screens and cause significant damage to pumps and mixers.

4.4.1.2 Grit Chamber

In this chamber, all the insoluble precipitates, such as sand, clay, and other putrescible domestic wastes that may clog the pipelines and channels, are removed (Grit removal challenge for treatment plant, 2016) using fine separators. The small grit tank consists of skimmers to collect the floating fat and oil globules present on the surface of water (Englande et al., 2015).

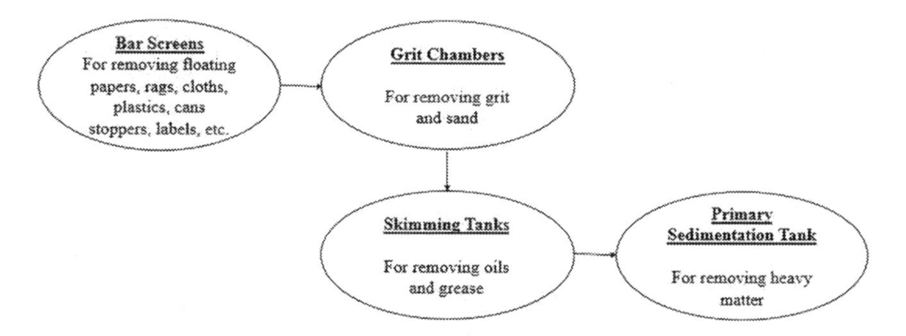

FIGURE 4.4 Steps involved in primary treatment process.

4.4.1.3 Flotation

This process is widely used to treat effluents of oil refineries, petrochemical industries, lubricant industries, and similar oil-processing industries. Flotation methods are used in the wastewater treatment to isolate oils, fats, and finely suspended solids. Particles suspended on the wastewater surface are removed using the air floatation process in which air is bubbled at a high flow rate from the bottom of the tank. This forms a foam of these oil particles, and the foam can be removed by skimmers from the surface (Ødegaard, 1995). By means of very fine gas bubbles that transport the substances to the surface, flotation separates dispersed or suspended substances. The smaller the micro-bubbles, the better they perform through collecting particles or droplets. Nowadays, wastewater technology often uses dissolved air flotation (DAF), which is economically efficient (Moursy & Abo El-Ela, 1982). In addition, auxiliary agents such as collectors, frothers, controllers, and pushers support flotation processes.

4.4.1.4 Sedimentation

A sedimentation process is used to segregate heavy or bulky impurities in water. A sedimentation tank is a large tank; at the bottom of the tank, larger particles settle down by gravity. Undissolved contaminants settle in the initial cleaning tank and form primary sludge that is eventually accumulated in the digestion tower, where it is anaerobically digested and converted into nutrient-rich manure or biogas and used for energy production (Jasim, 2020). Aerobically induced sludge is also applied to the digestion tower after it has been separated from the wastewater by sedimentation in the clarifier tank (Demirbas et al., 2017).

4.4.1.5 Filtration

Sand filters, fabric filters, and drum screens are often used in wastewater filtration systems. Filtration systems extract organic and inorganic suspended solids, sands, and dust particles from wastewater (Bali & Gueddari, 2019). This mechanical separation method is used by sewage treatment plants to drain sludge. In filtration, multi-stage processes are used for the purification of surface water to provide domestic and potable water. Another filter method used to separate very fine particles is membrane filtration.

4.4.1.6 Equalization Tank

The equalization tank is a rectangular tank equipped with air diffusers at the bottom. Aeration is carried out using coarse bubble diffusers that help in proper mixing of the chemicals to maintain the pH of the treated water. These diffusers must be properly maintained. In the water treatment plant, this tank is placed just before aeration tanks. It helps to improve the cleaning efficiency of secondary and tertiary process of wastewater treatment. The primary function of the equalization tank is to maintain the suitable pH of the treated water and collect the raw waste incoming at highly fluctuating rates and pass it on at a steady flow rate to the secondary and tertiary treatment units.

4.4.1.7 Pre-aeration

The pre-aeration process helps to separate suspended impurities prior to the primary sedimentation tank. It improves the primary treatment process by separating grease, removing grit, and making agglomerates of minute suspended particles, etc. This can be attained by increasing the detention time in the aerated chamber up to 5 minutes. The flow-rate of air in the aeration tank should be between 0.02 and 0.05 m^3/min per unit length of the tank. In the case of a separate pre-aeration tank, detention time should be kept between 10 and 40 minutes and depth should be 3–5 m. The volume of air required in relation to the volume of tank should be 0.75–3.0 m^3/m^3 (Tchobanoglous, 1991).

4.4.1.8 Flocculation

Flocculation is basically used to treat textile industries' wastewater. It is located before primary sedimentation tank in treatment plant. Industrial wastewater contains high concentrations of organic matter in colloidal form. These colloidal impurities cannot be separated out using filtration and pre-aeration. These finely divided suspended solids are so small and light that they cannot settle down by themselves. To settle them, flocculants are added to the colloidal solution to form large floc particles that can be settled easily and separated properly from wastewater as a sludge. If flocculation is required, it has to be used before the primary sedimentation tank (Muralikrishna & Manickam, 2017). Flocculation depends on the frequency of formation of large floc particles. Agitation rate play a very important role in it. If the agitation rate is very fast, it may result in the collapse of the floc particles. Therefore, the agitation rate should be moderate and maintained at 10 to 75 rotations per second in flocculators (Teh et al., 2016).

4.5 TRADITIONAL WATER TREATMENT

Traditional wastewater treatment processes are designed to remove particulates and unwanted chemicals. Treatment includes various processes:

4.5.1 Coagulation

Coagulation is the process of using chemical and/or physical techniques to promote particulate settling by reducing net electrical repulsion forces between particles. Coagulation is the process by which colloidal particles and very fine suspended solid particles present in wastewater are combined into larger agglomerates that can be separated via flocculation, sedimentation, filtration, centrifugation, or other gravity-based separation methods. Coagulation is commonly achieved by adding different types of chemicals (coagulants) to the wastewater to promote the destabilization of the colloid dispersion and agglomeration of the resulting individual colloidal particles. The main coagulants are $Al_2(SO_4)_3$, $NaAlO_2$, $FeSO_4$, and other coagulants, such as polyelectrolytes, can also be used. Coagulation requires the rapid dispersion of the coagulants throughout the wastewater through rapid agitation (Teh et al., 2016).

4.5.2 Flocculation

Flocculation is the process of agglomerating particles in water or wastewater to promote settling by using high-molecular-weight materials such as polymers, starches, and multiple charged ions.

4.5.3 Sedimentation

Sedimentation is common in wastewater treatment and water pre-treatment. It is a process that allows the flocculated or coagulated particles to settle down by gravity in a sedimentation tank. Typically, a hydraulic residence time of 4 hours is desired for settling.

4.5.4 Filtration

Filtration is a process that removes solids from water by passing water through a porous medium. The types of filters are discussed in Section 4.4.1.5. To remove accumulated solids from the filter bed, backwashing is required. Backwashing is the process of reversing the direction of flow through the filter, removing solids from the filter bed.

4.5.5 Activated Carbon Absorption

This type of physical absorption is used in tertiary treatment process to remove very small impurities. A porous activated carbon has been used for the adsorption of these contaminants due to high surface area. Activated carbon may have a surface area as great as 1500 m^2/g (7.3 million ft^2/lb) (Diaz-Elsayed et al., 2019).

4.6 MEMBRANE OXIDATIVE PROCESSES IN WATER TREATMENT

Membrane processes are an advanced separation technique widely applied in various industrial sectors including food, chemical and petrochemical, pharmaceutical, cosmetics, water desalination, and wastewater treatment. The main advantages of membrane oxidative processes are low energy consumption, eco-friendly process, and easy scale-up process. Available membranes are of different types, such as organic or inorganic, with homogeneous pores or heterogeneous pores, electrically neutral or charged. The membrane separation techniques works on the pressure difference, concentration difference, and partial pressure difference.

Most of the membranes are embedded in nanoparticles of a photocatalyst that can work in the presence of sunlight. The operating cost is very low. A few of them are discussed in subsequent sections.

4.6.1 Microfiltration

Particles, bacteria, and yeasts are separated by microfiltration. It is often used to isolate oil–water emulsions. The membrane is used in this filtration procedure to extract sand, silt, clay, algae, and bacteria. Pore sizes in membranes vary from 0.03 to 10 m.

4.6.2 Ultrafiltration

Ultrafiltration is an effective technique used to remove pollutants and produce drinkable water. It filters the particles, microorganisms, proteins, and turbidity present in wastewater. Ultrafiltration is used in the membrane activation reactor (MBR).

The membrane pore sizes range from 0.01 to 0.03 μm. Because the pores are minute, water that contain micro-sized impurities can be treated. In old and conventional filtration systems, the setup can be placed directly inside or as a separate stage after the activation tank in order to increase the efficiency of the biological wastewater treatment plant.

4.6.3 Nanofiltration

In nanofiltration, a membrane is used to remove natural organic matter and synthetic organic chemicals. The membrane has pore sizes of 0.001 μm to 1 nm (Shon et al., 2013).

4.6.4 Reverse Osmosis

Reverse osmosis can extract all types of impurities which are dissolved by applying pressure greater than the hydrostatic pressure through a semi-permeable membrane (Das et al. 2020). When the applied pressure is higher than the respective osmotic pressure, the molecules of the solvent diffuse to the side of the membrane, where the dissolved substance is already less concentrated. Reverse osmosis is capable of producing ultrapure water and thus is also known as superfiltration.

4.6.5 Electrodialysis

Electrodialysis is a desalination process. With the aid of electric current, it purifies brackish water. Dissolved salts are eliminated by electrolysis due to the movement of cations or anions of the dissolved salts through the ion-selective membrane under the influence of the electric field (Scarazzato et al. 2020).

4.7 NANOMATERIALS FOR WASTEWATER TREATMENT

Advanced oxidation processes (AOPs) have emerged as an environmentally friendly and innovative water treatment technology. AOPs are good alternatives for removal the toxic compounds from wastewater. These can be successfully used in wastewater treatment to degrade persistent organic pollutants, the oxidation process being determined by the very high oxidative potential of the OH^- radicals generated into the reaction medium by different mechanisms. AOPs can be applied to fully or partially oxidize pollutants, usually using a combination of oxidants. Photochemical and photocatalytic advanced oxidation processes, including UV/H_2O_2, UV/O_3, $UV/H_2O_2/O_3$, $UV/H_2O_2/Fe^{2+}(Fe^{3+})$, UV/TiO_2, and $UV/H_2O_2/TiO_2$, can be used for oxidative degradation of organic contaminants (Tlili & Alkanhal, 2019). AOPs are ambient pressure and temperature processes that involve in situ generation of highly reactive species such as hydroxyl free radicals (OH*) and superoxide radicals. The OH* radicals are known as powerful oxidizing agents, with an oxidation potential of 2.8 V and can act as the precursor for degradation of any organic and inorganic compound. Among AOPs, heterogeneous photocatalysis using semiconductor catalysts such as TiO_2, ZnO, Fe_2O_3, CdS, etc. (Byrne et al., 2017) has proved to be of real research interest as it is an efficient method for degrading organic contaminants. Among the various semiconductor catalysts known, titanium dioxide (TiO_2) has received immense interest in photocatalytic technology for wastewater treatment. The use of titania was first unfolded by the pioneering research of Fujishima and Honda in 1972 (Taučer-Kapteijn et al., 2016), which revealed the possibility of water splitting by a photo-electrochemical cell using a rutile Titania anode. This opened the doors for application of titania in environmental frontiers. In 1977, Frank and Bard (1977) first reported the photocatalytic oxidation of cyanide using TiO_2 powder. Later, in 1980, degradation of many harmful chemical compounds present in the air and water was reported by many research groups ("Grit Removal Challenge for Treatment Plant" 2016).

Heterogeneous photocatalysis uses ultraviolet light ($\lambda \leq 400$ nm), a photocatalyst (TiO_2), and oxygen for complete mineralization and degradation of the parent and the intermediate compound. This method has a feasible application in water treatment because:

i. It is a low-cost method
ii. There is no generation of secondary waste
iii. It uses ambient operating temperature and pressure

It was also observed that maximum quantum yield and best photocatalytic performance is always achieved using titanium dioxide nanoparticles. Degussa P-25 commercial TiO_2 photocatalyst with a mixture of anatase and rutile in an approximate ratio of 80:20 is the most active form of catalyst and is found to give very good degradation efficiencies. These particles are approximately 21 nm in size, with an approximate surface area of 50 m^2/g.

Another general agreement is about the development of highly efficient photo catalytic reactors to realize nano-sized TiO_2 intrinsic performance. It has been stressed that, due to the additional presence of light, several aspects of design, optimization, and operation should be taken into consideration during fabrication of photocatalytic reactors, which are not usually considered in conventional reactors. Thus, it is important to design an efficient photocatalytic reactor to improve the overall degradation efficiency of the wastewater treatment unit. This paper aims to give a brief overview of understanding of photocatalytic water treatment technology, especially titanium dioxide photocatalysts and photoreactor development.

4.8 NOVEL ADSORBENT FOR WASTEWATER TREATMENT

A number of technologies now are available to remove metals and organic materials in water and wastewater treatment plants. Adsorption is one of the promising technologies that has been proven effective and capable to remove a broad range of pollutants from water, such as heavy metals, inorganic anions, pharmaceuticals, and volatile organic compounds. Compared to other technologies, adsorption is considered as a process with low operational and maintenance costs; it also has a lower chemical footprint given that it generates lower residues for further handling and disposal. In recent years, more research on adsorption focused on the development of highly efficient and low cost adsorbent materials. Conventionally, activated carbons (ACs) were used as adsorbents to remove various pollutants from water and wastewater, owing to their high surface areas, abundant active surface sites with a diversity of functional groups, and high chemical stability. ACs can be prepared from several natural and industrial materials, such as coal, agricultural by-products and solid wastes, and industrial sludge and waste; however, the conventional parent material is coal. Based on the parent materials and particle size, activated carbon is prepared in micro-, meso-, or microporous varieties in powder, granular, or fibrous forms (He et al. 2016). However, due to the extreme high temperature required and complicated activation process involved, coal-based commercial ACs are expensive and complicated for regeneration (Anirudhan & Sreekumari, 2011). For example, several studies have obtained high adsorption removal of perfluoroalkyl substances (PFAS) using ACs, but they also reported difficulties in regeneration, which in return, increased the cost substantially (Senevirathna et al., 2010). Therefore, recent research interests have focused on the development of novel adsorbents using abundant and recycled materials as parent materials, such as clay minerals, agricultural products and wastes, industrial solid wastes and sludge. These materials were extensively modified and tested for their feasibility, adsorption capacities, and regeneration effectiveness as applied in adsorptive removal of different pollutants typically found in modern water treatment plants and wastewater treatment facilities. In this section, a brief introduction on adsorption mechanisms will be conducted, followed by detailed discussions on the development of novel adsorbents and their applications.

4.8.1 Adsorption Mechanism

Adsorption is a mass transfer process that involves the accumulation of substances at the interface of two phases, such as liquid–liquid, gas–liquid, gas–solid or liquid–solid interface. The substance being adsorbed is called *adsorbate* and the adsorbing material is called *adsorbent*. Specifically, according to the type of interaction occurring between the two phases, adsorption can be classified as physisorption or chemisorption. Physisorption happens when intermolecular forces, such as Van der Waals force, hydrogen bonds, and hydrophobic interactions, are observed on the interface. Physisorption is typically not stable and can be reversed through the alteration of environmental conditions. Therefore, regeneration of saturated adsorbents by physisorption can be done with some simple changes of environmental conditions, such as temperature and pH. Chemisorption occurs when chemical bonds form between adsorbent surface sites/functional groups and adsorbates. Adsorbates generally form a monolayer on the adsorbent surface through chemical bonding, and stronger interaction is usually observed. Regeneration of saturated adsorbents due to chemisorption is more difficult and usually involves a chemical solution as ion exchange or wash agents, such as acid/alkaline solutions, methyl solution, etc. Additionally, physisorption was found to be slightly specific, with minor thermal effects, whereas chemisorption is selective on surface sites, and broad thermal changes are observed.

Several environmental conditions have been reported as the crucial factors that can influence adsorption efficiency to a great extent, such as pH, temperature, contact time, and ionic strength. Many studies reported maximum adsorption taking place at slightly acidic pH values (i.e., pH between 5 and 6), such as dyes, heavy metals, perfluorinated compounds, and oxyanions (Rita et al., 2010). As pH rises to alkaline range, particularly higher than the pH for the point of zero surface charge, the adsorption efficiency is observed to decrease significantly. The pH values for high adsorption may vary from slightly alkaline to low acidic depending on the adsorbent surface net charge and the adsorbate charge characteristics. For instance, adsorption using iron-impregnated GACs was more effective in slightly acidic pH in removing selenate from aqueous solutions. Selenates formed inner sphere complexes with Fe-GAC at pH lower than 7 (Zhang et al., 2018). Exceptions have been reported by studies conducted on the adsorption of perfluorinated compounds (You et al., 2010; Kwon et al., 2012). They reported that the presence of divalent cations (e.g., Ca^{2+} and Mg^{2+}) in the aqueous solution led to an increase of the adsorption along with the rise of pH.

4.8.2 Synthesized Nanostructured Adsorbents

Among many base materials, nanostructured materials have received tremendous attention for adsorbent development, given their large surface areas, microscopic porous structure, high pore volumes, and varied functional groups that allow them to conduct efficient removal toward a great variety of pollutants. It has been a couple of decades since the first nano-sized particles were discovered in early last century, and many of their advanced characteristics have been revealed and applied to improve our knowledge in various scientific fields. Nanotechnology, which involves the development and application of a variety of nano-sized materials in different morphologies and structures, is recognized as a platform that can support many branches of science and technology to modify, endorse, or clarify existing scientific concepts. Benefits of nanotechnology applied in water and wastewater treatment have been reported in a number of studies. For instance, metallic oxide-based nanomaterials (nano-TiO_2, nano Fe_3O_4, and nano ZnO) have been used as catalysts to deconstruct and reduce persistent organic compounds in industrial wastewater treatment, and they have also been used as antimicrobial agents to inactivate pathogenic microorganism in municipal wastewater treatment. Carbonaceous nanomaterials (carbon nanotubes (CNT), graphene sheets, and fullerenes) have been reported as effective and low-cost materials for water and wastewater treatment through adsorption, membrane filtration, and ion exchange. Particularly, nano-adsorbents have been employed in industrial wastewater treatment for decolorization and removal of multiple organic dye constitutes, such as Congo red and methyl orange; they have also been found to be effective in removal of various inorganic ions, including nitrate, arsenic, selenium, etc. In addition, CNTs are widely applied in heavy metal adsorptive removal, such as Cr (VI), Cd (II), Hg (II), etc. Table 4.5 lists recent studies conducted using CNTs and their composites for heavy metal adsorption.

CNTs were first discovered in 1991 (Iijima 1991) and have been modified extensively since then to increase their surface properties and further improve the adsorption efficiency. CNTs are sp^2 allotropic carbon of graphite with cylindrical sheets structure. Microtubules graphitic carbon has an outer diameter between 0.4 and 100 nm, with length up to 1 μm. The graphite allotropes are aggregated together through Van der Waal interactions, which reduces their surface area and results in larger mesoporous volumes that can be utilized as adsorption sites to accommodate adsorbate contaminants (Poudel & Li, 2018). CNTs can be fabricated in single-walled

TABLE 4.5

Findings of Recent Studies Conducted with CNTs on Heavy Metal Adsorption

Adsorbent	Target Contaminant	Max. Capacity (mg/g)	Initial Contaminant Conc. Range (mg/L)	Contact Time (hr)	Optimal pH	Reference
Untreated carbon nanotubes (CNT)	Cd (II)	1.66	1.0	2	7	Al-Khaldi et al., 2015
Untreated CNTs	Pb (II)	17.44	2–40	4	7	Stafiej and Pyrzynska, 2007
Acidified CNTs	Cd (II)	4.35	1.0	2	7	Al-Khaldi et al., 2015
Refluxed CNTs	Cu (II)	3.49	2.0–20	4	9	Stafiej and Pyrzynska, 2007
Activated CNTs	Zn (II)	1.05	1.1	2	10	Mubarak et al., 2013
Untreated multi-walled CNTs (MWCNTs)	Cr (VI)	3.12	1.0	4	3	Ihsanullah et al., 2016
Acidified MWCNTs	Cr (VI)	1.31	1.0	4	3	
Untreated single-walled CNTs (SWCNTs)[a]	Hg (II)	214.24	NA	NA	NA	Anitha et al., 2015
Modified SWCNTs with $-COO^-$ function group[a]	Hg (II)	661.98	NA	NA	NA	Anitha et al., 2015
MWCNTs	Cr (VI)	13.2	20	0.67	2.5–4	Kumar et al., 2015
MWCNT	Pb (II)	52.17	5–65	10	6	Yu et al, 2013

Note:

[a] Molecular dynamics (MD) simulation results.

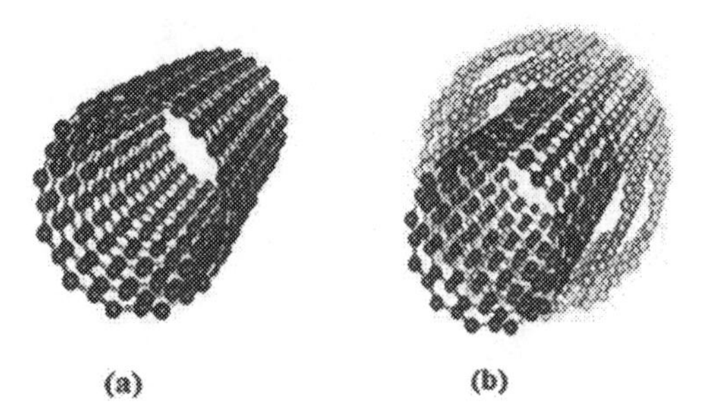

FIGURE 4.5 Cylindrical structure for single-walled carbon nanotubes (a) and multi-walled carbon nanotubes and (b) multi-walled carbon nanotubes (Dubey et al, 2017).

structures (SWCNTs) with smaller tube outer diameters or multi-walled structures (MWCNTs) with larger outer diameters. Although SWCNTs typically exhibit higher surface areas than MWCNTs, they were found with lower adsorption capacities under the same experimental conditions than MWCNTs. A picture showing the structure of SWCNTs and MWCNTs is presented in Fig. 4.5.

Various surface modifications have been conducted to enhance the surface properties of CNTs. For example, several acids were used as modification agents under different conditions to introduce a carboxyl functional group (i.e., –COOH) onto a CNT surface for heavy metal adsorptive removal. Yu et al. (2013) prepared MWCNTs with different outer tube diameters through acidification and high temperature reflux. CNTs can also be grafted with other materials such as magnetic nanoparticles, graphene oxide, surfactants, chitosan, cyclodextrin, and etc. (Kyzas & Matis, 2015) to overcome the limitations of industrial application caused by the nanoparticles aggregation and difficulties in separation from aqueous solutions. Koochaki-Mohammadpour et al. (2014) studied rare earth metal adsorption removal by oxidized MWCNTs. They reported that maximum adsorption of Dysprosium (Dy^{3+}) occurred within the pH range of 5–6.

Graphene sheets (GSs) are single-layered sp^2 allotropic carbon atoms that are arranged in a two-dimensional hexagonal honeycomb lattice structure. GSs have been employed as adsorbents in the removal of many contaminants, such as dyes, antibiotics, phenol and its derivatives, pesticide, etc. (Bai et al., 2012; Mahmoodi et al., 2019; Chen et al., 2018). Similar to the CNTs, GSs can be prepared in single layer or multiple layers. Additionally, GSs are modified to obtain graphene oxide (GO) through oxidation of graphite; and GO can be reduced to further obtain reduced graphene oxide (RGO). Both GO and RGO exhibit enhanced surface characteristics and improved adsorption capacities toward multiple contaminants in water and wastewater. For instance, oxidation of GO can add hydrophilic groups to the surface and develop high affinities to heavy metals (Bisht et al., 2016).

4.8.3 Clay-Based Adsorbents

Clays are a group of hydrous aluminosilicates occurring naturally with layered structures made of tetrahedral and octahedral sheets in various proportions. Natural clays contain varieties of minerals, such as montmorillonite, hectorite, sepiolite, laponite, saponite, kaolinite, zeolite, etc. They generally develop a highly porous structure on the surface, making them great candidates for adsorption. Additionally, natural clays are negatively charged on the surface, so through electrostatic attractions and ion exchange (Bhattacharyya and Gupta, 2008), clays can adsorb cations such as heavy metals effectively. Meanwhile, clays have multiple exchangeable ions on their surface, such as H^+, K^+, Na^+, Ca^+, Cl^-, etc., which can allow removal of anions as well. Overall, clays were found to possess strong adsorption affinities toward many pollutants, including dyes, heavy metals, organic compounds, and pharmaceuticals. Particularly, the natural clays display greater adsorption capacity toward basic dyes (e.g., methylene blue and malachite green) than acidic dyes (e.g., Congo red and methyl orange).

However, several researchers have shown that the crystal structure and negative charge of natural clays limit the natural clay's industrial applications. Rafatullah et al. (2010) investigated varied adsorption capacities between clays and activated carbon toward methylene blue. The noticeable low capacity of clays was attributed to the layered structure of the raw clays. Moreover, the colloidal characteristic of natural clays significantly affects their regeneration and recovery from aqueous solutions.

Hence, single and combined surface modification strategies have been employed to develop natural clay-based adsorbents with enhanced surface properties and improved application feasibility. Various modification methods, such as acid activation, calcination, magnetization, and surfactant or polymer activation, have been proven to be effective. The modified clays are collectively called clay composites. The most promising examples of clay composites include granulated clay composites and cross-linked clay beads. For instance, activated carbon zeolite composites were synthesized using elutrilithe and pitch through a three-step process. Three modification steps—calcination, CO_2 activation, and hydrothermal conversion—all take place at high temperature for certain amount of time. Some major constitutes (e.g., carbon, SiO_2, and Al_2O_3) in elutrilithe eventually can be transformed into an activated carbon zeolite composite. Elutrilithe is a kaolinite rich material that is composed of aluminosilicates and organic carbon, whereas pitch acts as binder and produces porous carbon. The embedding of raw clays into natural polymeric materials, such as chitosan, alginate, peanut hull, and cellulose, also attracts tremendous interest to address the challenge associated with adsorbent regeneration and recovery.

4.8.4 Agricultural By-Products and Solid Wastes

Due to the high cost of using coal as a base material in the manufacturing of activated carbon, further efforts have been made to substitute coal with inexpensive and abundant materials, such as agricultural wastes and residues. Agricultural wastes are organic substances discarded during agricultural production and primarily include plant wastes, livestock waste, and poultry manure (Dai et al. 2018). Agricultural and forestry wastes are generally rich in cellulose, hemicellulose and lignin (Salleh et al. 2011), which make them great substitutes to coals in the production of carbonaceous adsorbent. Both hemicellulose and cellulose contain multiple oxygen functional groups, such as carbonyl groups, hydroxyl groups, and ether. Lignin is an aromatic polymer consisting of several functional groups, such as carboxyl, hydroxyl, and methyl

(Dai et al. 2018). All these functional groups are found to efficiently bond with various contaminants through chelation, coordination, and hydrogen bonding. Many agricultural by-products have been investigated for activated carbon production through low temperature pyrolysis and chemical activation treatment. A number of chemicals are employed as activation agents to convert the agricultural materials to activated carbon. Examples of the activation agents include alginate, tannic acid, magnesium, zinc chloride, etc. Activation typically takes place at temperatures between 500°C and 900°C for several hours. Portinho et al. (2017) used phosphoric acid to modify grape stalks and further employed them as adsorbents to remove caffeine from aqueous solution. Results showed higher adsorption capacities for acid-treated adsorbents than for raw adsorbents developed from grape stalks. Although agriculture-based activated carbon usually exhibits lower removal capacities than commercial activated carbon, large-scale application is considered more economically attractive.

Agricultural by-products and residues, such as nutshells, rice husks, barks, switch grass, and barley straw, have proven to be suitable materials for activated carbon production. Several types of nutshells (e.g., hazelnut, walnut, almond) were studied to produce granular activated carbon (GAC) by Aygun et al. (2003). The GAC prepared by nutshells was further applied to remove methylene blue from dye wastewater, and hazelnut shells were found to be the best suitable raw material for GAC conversion. Kongsuwan et al. (2009) studied the eucalyptus bark–based activated carbon for binary adsorption of copper and lead. Wood-based activated carbon with mesoporous surface structure was found to possess greater adsorption capacities than coal-based and coconut-based activated carbon adsorbents in removing microcystin-LR from drinking water (Huang et al., 2007). Rice husk was activated by $ZnCl_2$ to obtain activated carbon to conduct cupper removal (Yahaya et al., 2010). The results show that activation temperature, activation time, and activation agent-to-parent material ratio are major factors influencing the removal efficiency.

Agricultural solid wastes, such as poultry litter and apricot stones, are also potential parent base materials for activated carbon production. Several of them have been tested for their adsorption capacity for dyes, heavy metals, cyanobacterial toxins, and pharmaceuticals. Juang et al. (2002) prepared bagasse-based activated carbon by steam at varied temperatures in the range of 750–840°C. The bagasse-activated carbon was found to have microporous structure with pore diameter averaging less than 2 nm. Poultry litter was used to manufacture activated carbon for heavy-metal removal and was found to have significantly higher adsorption capacity than commercial-grade activated carbon. Apricot stone-based activated carbons were investigated for multiple heavy-metal adsorptive removal (Kobya et al., 2005). The removal percentage was reported as high as 85% under slight acidic condition. Other agricultural solid wastes studied for activated carbon production include coconut shells, corncobs, banana peels, etc.

Biochar is a carbon-rich material that has been widely applied as an adsorbent to remove various contaminants from aqueous solutions. Biochar is prepared by slow pyrolysis of parent materials that contain high organic matter under an inert atmosphere, such as agricultural products and wastes. The pyrolysis temperature required for biochar production is lower than that for ACs, making it more economically attractive. Moreover, activation of biochar adsorbents is not necessary, whereas it is crucial for ACs (Zhang et al., 2018). It was also found that the parent materials for biochar development can greatly affect its surface properties. For instance, biochar produced from woody plants, such as pine wood and hickory, was found to have larger surface areas than that produced from herbaceous plants, such as alfalfa, cotton, and switchgrass (Abe et al., 2000; Uchimiya et al., 2010).

4.8.5 Industrial Solid Wastes for Adsorption Application

Industrial by-products and associated solid wastes are also potential substitutes for conventional and costly parent materials used in adsorbent production. Some examples of industrial solid wastes that have been investigated for adsorption application include fly ash, metal hydroxide sludge, red mud, paper mill sludge, and municipal sewage sludge. The active components found in these waste by-products are cellulose, hemicellulose, lignin, and lipids. Raw industrial solid wastes are chemically or physically modified to reduce their hazards and further increase their adsorption capacity.

Metal hydroxide sludge is an industrial by-product that contains insoluble metal hydroxides and other salts and can be used to remove varieties of aqueous pollutants such as dyes, oxyanions, and heavy metals through adsorption. The primary controlling mechanism of metal hydroxide sludge as adsorbent is ion exchange (Zhou et al., 2019). Gunes and Kaygusuz (2015) investigated Al(III) containing metal hydroxide sludge for Reactive Blue 222 adsorption. Maximum adsorption was reported to occur under slightly acidic pH range.

Fly ash, a by-product of coal-based thermal power plants, carries out adsorption through molecular attraction, which depends on the pore size and specific areas of fly ash. It has been discovered that fly ash contains adequate silica and alumina, making them efficient materials to remove dyes, organic compounds, heavy metals, and several other pollutants. Additionally, aluminum salt and iron salt contained in fly ash can complex with water to form $Al(H_2O)^{3+}$ and $Fe(H_2O)^{3+}$ once they mix with water and further produce flocs to enhance the adsorption. However, fly ash is also reported as a hazardous material owing to various heavy metal constituents present. Swarcewicz et al. (2013) developed adsorbent materials from a mixture of soil and fly ash for carbamazepine removal. The removal was reported as high as 92.8% as the fly ash contents present in the adsorbent are greater than 30%.

Red mud is a bauxite-processing residue generated during alumina production. The red mud–based adsorbents conduct adsorption primarily through ion exchange. Red mud has high adsorption capacities for several heavy metals; however, it also has limitations in regeneration and activation residue handling (Zhu et al., 2007). Cengeloglu et al. (2006) investigated both raw and activated red mud–based adsorbents for nitrate removal. Adsorption capacity was reported to increase by approximately threefold after the raw red mud adsorbent was activated with HCl. The acid activation was found to effectively induce the leaching of sodalite compounds from the raw red mud, which were originally present on surface to block the available surface binding sites. Acid activated red mud–based adsorbents were found highly effective (as high as 96.5% removal) in removing selected pharmaceuticals from aqueous solutions, such as ciprofloxacin (Balarak et al., 2017). Meanwhile, red mud–based adsorbents also exhibited high adsorption capacities for several heavy metals, such as Cd (II), Zn (II), and Cu (II).

4.8.6 Biomaterials for Adsorption Application

Adsorbent development from biomass has also received a lot of attention for removal of contaminants from aqueous solutions. Biomass such as algae, bacteria, and fungi was investigated for the potential application as adsorbents in water treatment. The cell walls of these microorganisms contain a high content of cellulosic constituents, such as hydroxyl and carboxyl groups, which have been proven to be effective binding sites to retard selective contaminants. The adsorption taking place on the cell walls of microorganisms is known as *biosorption*, which involves multiple mechanisms, such as ion exchange, surface adsorption, complexation, chelation, etc. (Crini, 2006; Sadhasivam et al., 2007; Wang and Hu, 2007). Major characteristics of the biosorption include (1) limited generation of sludge residues, (2) passive and reversible process, and (3) rapid adsorption (Rangabhashiyam et al., 2014). Non-living biomass is primarily used as adsorbent to conduct biosorption. In general, biosorption by non-living biomass adsorbents is simply based on the affinities between surface binding sites and adsorbates. Biosorption differs from bioaccumulation in that bioaccumulation primarily relies on live cells and removes contaminants through a two-step process. The first step occurs on the cell wall, and contaminant ions are inactivated through chemical or physical binding with the cell wall constituents. The second step involves the contaminant transportation into the algae cytoplasm, which is generally known as the intracellular uptake or bioaccumulation. The second step is more dependent on the biomass metabolism and may further increase the contaminants removal from aqueous solutions. Factors that enhanced the growth of *Chlorella vulgaris* belonging to algae family and its adsorption to soil contributed positively to active accumulation of Pb in a column study conducted to study the leaching potential of Pb in soils (Nedunuri et al., 1998).

Biosorbents were studied for the removal of heavy metals, dyes, persistent organic compounds, and nutrients. Among the different biosorbents studied, algae are among the most promising types of biomass owing to their high metal binding capacities. Algae-based biosorbents have multiple functional groups contained in their cell walls, for example, amino, hydroxyl, carboxyl, and sulfate (Rangabhashiyam & Vijayaraghavan, 2019), which are favorable binding sites for a variety of contaminants. Abdel-Aty al. (2013) revealed that dead microalgae are highly effective biosorbents because of their easy cultivation, high yield, and large surface area. For example, *Chlorella sp.*, *Spirulina sp.*, and *Chaetophora elegans*, were found to have significant potential toward heavy-metal biosorption removal (Nedunuri et al., 1998). Selected algae species, such as *Gracilaria verrucosa*, *Macrocystis integrifolia Bory*, and *Lessonia nigrescens Bory*, were reported to have biosorption capacities for phenolic compounds, and the primary interaction form was found to be the hydrophobic affinity at the interface (Karthik & Meenakshi, 2014). Algae biosorbents were also investigated for dye adsorption removal. Optimal adsorption of dye was found to occur under acidic condition and highly dependent on the initial dye concentration and biosorbent dosage. Microalgae biomass was reported with higher adsorption capacity for dye than marcoalgae (e.g., seaweed); this was attributed to their cell wall constituents and functional groups.

However, raw biosorbents commonly show lower adsorption capacities than conventional adsorbents, so chemical modification is desired to enhance the biosorption. Immobilizing nonliving biomass may enhance the biosoption, improve mechanical strength, and ease biosorbent recovery (Aksu & Gönen, 2004). Vinod et al. (2010) investigated both raw algae biosorbents developed from *Oedogonium hatei* and acid-treated (0.1 M HCl) algae biosorbents. The adsorption capacity for Ni (II) removal for acid treated algae biosorbent was increased by 8% compared to the untreated biosorbent under the same adsorption condition.

REFERENCES

Abdel-Aty, A.M., Ammar, N.S., Ghafar, H.H.A., & Ali, R.K. (2013) Biosorption of cadmium and lead from aqueous solution by fresh water alga *Anabaena sphaerica* biomass. *Journal of Advanced Research*, 4, 367–374.

Abe, M., Kawashima, K., Kozawa, K., Sakai, H., & Kaneko, K. (2000) Amination of activated carbon and adsorption characteristics of its animated surface. *Langmuir*, 16, 5059–5063.

Aksu, Z., & Gönen, F. (2004) Biosorption of phenol by immobilized activated sludge in a continuous packed bed: Prediction of breakthrough curves. *Process Biochemistry*. https://doi.org/10.1016/S0032-9592(03)00132-8

Al-Khaldi, F.A., Abu-Sharkh, B., Abulkibash, A.M., & Atieh, M.A. (2015) Cadmium removal by activated carbon, carbon nanotubes, carbon nanofibers, and carbon fly ash: a comparative study. *Desalination and Water Treatment*, 53, 1417–1429.

Anirudhan, T. S., & Sreekumari, S. S. (2011) Adsorptive removal of heavy metal ions from industrial effluents using activated carbon derived from waste coconut buttons. *Journal of Environmental Sciences*. https://doi.org/10.1016/S1001-0742(10)60515-3

Anitha, K., Namsani, S., & Singh, J.K. (2015) Removal of heavy metal ions using a functionalized single walled carbon nanotube: a molecular dynamics study. *Journal of Physical Chemistry*, 119, 8349–8358

Aygun, A., Yenisoy-Karakas, S., & Duman, I. (2003) Production of granular activated carbon from fruit stones and nutshells and evaluation of their physical, chemical and adsorption properties. *Microporous Mesoporous Material*, 66, 189–195.

Balarak, D., Mostafapour, F.K., & Joghataei, A. (2017) Kinetics and mechanism of red mud in adsorption of ciprofloxacin in aqueous solution. *Bioscience Biotechnology Research Communications*, 10, 241–248.

Bali, M., & Gueddari, M. (2019) Removal of phosphorus from secondary effluents using infiltration–percolation process. *Applied Water Science*, 9(3), 54. https://doi.org/10.1007/s13201-019-0945-5

Bai, L., Yuan, R., Chai, Y., Zhuo, Y., Yuan, Y. & Wang, Y. (2012) Simultaneous electrochemical detection of multiple analytes based on dual signal amplification of single walled carbon nanotubes and multi-labeled graphene sheets, *Biomaterials*, 33, 1090–1096.

Bhattacharyya, K. G., & Sen Gupta, S. (2008). Adsorption of a few heavy metals on natural and modified kaolinite and montmorillonite: a review. *Advances in Colloid and Interface Science*, 140, 114–131.

Bisht, R., Agarwal, M., & Singh, K. (2016) Heavy metal removal from wastewater using various adsorbents: a review. *Journal of Water Reuse and Desalination*, 7, 387–419.

Byrne, C., Subramanian, G., & Pillai, S. (2017) Recent advances in photocatalysis for environmental applications. *Journal of Environmental Chemical Engineering*. https://doi.org/10.1016/j.jece.2017.07.080

Canziani, R., & Spinosa, L. (2019) Sludge from wastewater treatment plants. In M. N. V. Prasad, P. J. de Campos Favas, M. Vithanage, & M. S. Mohan (Eds.) *Industrial and Municipal Sludge: Emerging Concerns and Scope for Resource Recovery* (pp. 3–30) Butterworth-Heinemann. https://doi.org/10.1016/B978-0-12-815907-1.00001-5

Cengeloglu, Y., Tor, A., Ersoz, M., & Arslan, G. (2006) Removal of nitrate from aqueous solution by using red mud. *Separation and Purification Technology*, 51, 374–378.

Chen, L., Han, Q., Li, W., Zhou, Z., Fang, Z., Xu, Z., Wang, Z. & Qian, X (2018) Three-dimensional graphene-based adsorbents in sewage disposal: a review, *Environmental Science and Pollution Research*, 25, 25840–25861.

Crini, G. (2006) Non-conventional low cost adsorbents for dye removal: a review. *Bioresources Technology*, 97, 1061–1085.

Dai, Y., Sun, Q., Wang, W., Lu, L., Liu, M., Li, J., Yang, S., Sun, Y., Zhang, K., Xu, J., Zheng, W., Hu, Z., Yang, Y., Gao, Y., Chen, Y., Zhang, X., Gao, F., & Zhang, Y. (2018) Utilizations of agricultural waste as adsorbent for the removal of contaminants: a review. *Chemosphere*, 211, 235–253.

Demirbas, A., Coban, V., Taylan, O., & Kabli, M. (2017) Aerobic digestion of sewage sludge for waste treatment. *Energy Sources, Part A: Recovery, Utilization, and Environmental Effects*, 39(10), 1056–1062. https://doi.org/10.1080/15567036.2017.1289282

Diaz-Elsayed, N., Rezaei, N., Guo, T., Mohebbi, S., & Zhang, Q. (2019) Wastewater-based resource recovery technologies across scale: a review. *Resources, Conservation and Recycling*, 145, 94–112. https://doi.org/10.1016/j.resconrec.2018.12.035

Dubey, S., Banerjee, S., Nath, S., & Chandra, Y. (2017) Application of common nano-materials for removal of selected metallic species from water and wastewaters: a critical review. *Journal of Molecular Liquids*, 240, 656–677.

Englande, A. J., Krenkel, P., & Shamas, J. (2015). Wastewater Treatment &Water Reclamation in Reference Module in Earth Systems and Environmental Sciences. Elsevier. https://doi.org/https://doi.org/10.1016/B978-0-12-409548-9.09508-7

Frank, S. N., & Bard, A. J. (1977) Heterogeneous photocatalytic oxidation of cyanide ion in aqueous solutions at titanium dioxide powder. *Journal of the American Chemical Society*, 99(1), 303–304. https://doi.org/10.1021/ja00443a081

Ghaderpoori, M., Kamarehie, B., Jafari, A., Ghaderpoury, A., & Karami, M. (2018) Heavy metals analysis and quality assessment in drinking water – Khorramabad city, Iran. *Data in Brief*, 16, 685–692. https://doi.org/10.1016/j.dib.2017.11.078

(2016) Grit removal challenge for treatment plant. *Filtration + Separation*, 53(6), 22–23. https://doi.org/10.1016/S0015-1882(16)30252-X

Gunes, E., & Kaygusuz, T. (2015) Adsorption of Reactive Blue 222 onto an industrial solid waste included Al(III) hydroxide: pH, ionic strength, isotherms, and kinetics studies. *Desalination and Water Treatment*, 53, 2510–2517.

Gupta, N., Pandey, P., & Hussain, J. (2017) Effect of physicochemical and biological parameters on the quality of river water of Narmada, Madhya Pradesh, India. *Water Science*, 31(1), 11–23. https://doi.org/10.1016/j.wsj.2017.03.002

He, X., Liu, Y., Conklin, A., Westrick, J., Weavers, L.K., Dionysiou, D.D., Lenhart, J.J., Mouser, P.J., Szlag, D., Walker, H.W. (2016) Toxic cyanobacteria and drinking water: Impacts, detection, and treatment. *Harmful Algae*, 54, 174–193.

Huang, W., Cheng, B., & Cheng, Y. (2007) Adsorption of microcystin-LR by three types of activated carbon. *Journal of Hazardous Materials*, 141, 115–122.

Ihsanullah Al-Khaldi, F.A., AbuSharkh, B., Abulkibash, A.M., Qureshi, M.I, & Laoui, T. (2016) Effect of acid modification on adsorption of hexavalent chromium (Cr(VI)) from aqueous solution by activated carbon and carbon nanotubes. *Desalination and Water Treatment*, 57(16), 7232–7244. DOI: 10.1080/19443994.2015.1021847.

Iijima S. (1991) Helical microtubules of graphitic carbon. *Nature*, 354, 56–58.

Jasim, N.A. (2020) The design for wastewater treatment plant (WWTP) with GPS X modelling. *Cogent Engineering*, 7(1), 1723782. https://doi.org/10.1080/23311916.2020.1723782

Juang, R.S., Wu, F.C., & Tseng, R.L. (2002) Characterization and use of activated carbons prepared from bagasses for liquid phase adsorption. *Colloids and Surfaces A: Physicochemcal and Engineering Aspects*, 201, 191–199.

Karthik, R., & Meenakshi, S. (2014) Removal of hexavalent chromium ions using polyaniline/silica gel composite. *Journal of Water Process Engineering*. https://doi.org/10.1016/j.jwpe.2014.03.001

Kobya, M., Demirbas, E., Senturk, E., & Ince, M. (2005) Adsorption of heavy metal ions from aqueous solutions by activated carbon prepared from apricot stone. *Bioresources Technology*, 96, 1518–1521.

Kongsuwan, A., Patnukao, P. & Pavasant, P. (2009) Binary component sorption of Cu(II) and Pb(II) with activated carbon from Eucalyptus camaldulensis Dehn bark, *Journal of Industrial and Engineering Chemistry*, 15, 465–470.

Koochaki-Mohammadpour S.M.A., M. Torab-Mostaedi, A. Talebizadeh-Rafsanjani & F. Naderi-Behdani (2014) Adsorption isotherm, kinetic, thermodynamic, and desorption studies of Lanthanum and Dysprosium on oxidized multivalled carbon nanotubes, *Journal of Dispersion Science and Technology*, 35, 244–254.

Kumar, A.S.K., Jiang, S.J., & Tseng, W.L. (2015) Effective adsorption of chromium (VI)/Cr(III) from aqueous solution using ionic liquid functionalized multiwalled carbon nanotubes as a super sorbent. *Journal of Materials Chemistry A*, 3, 7044–7057.

Kwon, Y.N., Shih, K., Tang, C., & Leckie, J.O. (2012) Adsorption of perfluorinated compounds on thin-film composite polyamide membranes. *Journal of Applied Polymer*, 124, 1042–1049.

Kyzas, G.Z., & Matis, K.A. (2015) Nanoadsorbents for pollutants removal: a review. *Journal of Molecular Liquids*, 203, 159–168.

Mahmoodi, E., Ang, W.L., Ng, C.Y., Ng, L.Y., Mohammad, A.W. & Benamor, A. (2019) Distinguishing characteristics and usability of graphene oxide based on different sources of graphite feedstock, *Journal of Colloid and Interface Science*, 542, 429–440.

Moursy, A.S., & Abo El-Ela, S.E. (1982) Treatment of oily refinery wastes using a dissolved air flotation process. *Environment International*, 7(4), 267–270. https://doi.org/10.1016/0160-4120(82)90116-7

Mubarak, N.M., Alicia, R.F., Abdullah, E.C., Sahu, J.N., Haslija, A.A., & Tan, J. (2013) Statistical optimization and kinetic studies on removal of Zn2+ using functionalized carbon nanotubes and magnetic biochar. *Journal of Environmental Chemical Engineering*, 1, 489–495.

Muralikrishna, I. V, & Manickam, V. (2017) Wastewater Treatment Technologies. In I. V Muralikrishna & V. Manickam (Eds.) *Environmental Management: Science and Engineering for Industry* (pp. 249–293) Butterworth-Heinemann. https://doi.org/10.1016/B978-0-12-811989-1.00012-9

Mushak, P. (2011) Regulation and regulatory policies for lead in water. In *Lead and Public Health* (Vol. 10, pp. 899–922) Elsevier. https://doi.org/10.1016/B978-0-444-51554-4.00028-6

Nedunuri, K. V., Erickson, L. E., & Govindaraju, R. S. (1998) Modelling the role of active biomass on the fate and transport of a heavy metal in the presence of root exudates. *Journal of Hazardous Substance Research*, 1, 1–25.

Nemerow, N. L. B. T.-I. W. T. (Ed.) (2007) Removal of organic dissolved solids (pp. 105–148) Burlington: Butterworth-Heinemann. https://doi.org/10.1016/B978-012372493-9/50043-0

Ødegaard, H. (1995) Optimization of flocculation/flotation in chemical wastewater treatment. *Water Science and Technology, 31*(3), 73–82. https://doi.org/10.1016/0273-1223(95)99878-8

Portinho, R., Zanella, O., & Féris, L.A. (2017) Grape stalk application for caffeine removal through adsorption. *Journal of Environmental Management*, 202, 178–187.

Poudel, Y. R., & Li, W (2018) Synthesis, properties and applications of carbon nanotubes filled with foreign materials: a review. *Materials Today Physics*, 7, 7–34.

Rafatullah, M., Sulaiman, O., Hashim, R., & Ahmad, A. (2010) Adsorption of methylene blue on low-cost adsorbents a review. *Journal of Hazardous Materials*, 177, 70–80.

Rangabhashiyam, S., Suganya, E., Selvaraju, N& Varghese, L.A. (2014) Significance of exploiting non-living biomaterials for the biosorption of wastewater pollutants. *World Journal of Microbiology and Biotechnology*, 30, 1669–1689.

Rangabhashiyam, S., & Vijayaraghavan, K. (2019) Biosorption of Tm(III) by free and polysulfone-immobilized *Turbinaria conoides* biomass. *Journal of Industrial and Engineering Chemistry.* https://doi.org/10.1016/j.jiec.2019.08.010

Rita, F.L.R., Sergia, M.S.M., Francisco, A.R.B., Clesia, C.N., Iara, C.C., Debora, C.M. (2010) Evaluation of the potential of microalgae Microcystis novacekii in the removal of Pb2+ from an aqueous Medium. *Journal of Hazardous Materials*, 179, 947–953.

Sadhasivam, S., Savitha, S., & Swaminathan, K. (2007) Feasibility of using Trichoderma harzianum biomass for the removal of erioglaucine from aqueous solution. *World Journal of Microbiology and Biotechnology*, 23, 1075–1081.

Salleh, M.A.M., Mahmoud, D.K., Karim, W.A.W.A., & Idris, A. (2011) Cationic and anionic dye adsorption by agricultural solid wastes: a comprehensive review. *Desalination*, 280, 1–13.

Scarazzato, T., Barros, K. S., Benvenuti, T., Rodrigues, M. A. S., Espinosa, D. C. R., Bernardes, A. M. B., … Pérez-Herranz, V. (2020). Chapter 5- Achievements in electrodialysis processes for wastewater and water treatment. In A. Basile & K. B. T.-C. T. and F. D. on (Bio-) M. Ghasemzadeh (Eds.), *Recent achievements in wastewater and water treatments* (pp. 127–160) Elsevier. https://doi.org/10.1016/B978-0-12-817378-7.00005-7

Senevirathna, S.T.M.L.D. (2010) Dissertation title: *Development of effective removal methods of PFCs (perfluorinated compounds) in water by adsorption and coagulation*, Japan School of Engineering, Kyoto University.

Shon, H.K., Phuntsho, S., Chaudhary, D.S., Vigneswaran, S., & Cho, and J. (2013) Nanofiltration for water and wastewater treatment: a mini review. *Drinking Water Engineering and Science*, 6, 47–53. https://doi.org/10.5194/dwes-6-47-2013

Stafiej, A., & Pyrzynska, K. (2007) Adsorption of heavy metal ions with carbon nanotubes. *Separation and Purification Technology*, 58, 49–52.

Swarcewicz, M.K., Sobczak, J., & Pazdzioch, W. (2013) Removal of carbamazepine from aqueous solution by adsorption on fly ash amended soil. *Water Science and Technology*, 67, 1396–1402.

Swenson, H.A., & Baldwin, H.L. (1965) *A Primer on Water Quality* (1990 reprint) General Interest Publication. https://doi.org/10.3133/7000057

Taučer-Kapteijn, M., Hoogenboezem, W., Heiliegers, L., de Bolster, D., & Medema, G. (2016) Screening municipal wastewater effluent and surface water used for drinking water production for the presence of ampicillin and vancomycin resistant enterococci. *International Journal of Hygiene and Environmental Health*, 219(4), 437–442.

Tchobanoglous, G., Burton, F.L., & Metcalf & Eddy. (1991) *Wastewater engineering: Treatment, disposal, and reuse.* New York: McGraw-Hill.

Teh, C.Y., Budiman, P.M., Shak, K.P.Y., & Wu, T.Y. (2016) Recent advancement of coagulation–flocculation and its application in wastewater treatment. *Industrial & Engineering Chemistry Research*, 55(16), 4363–4389. https://doi.org/10.1021/acs.iecr.5b04703

Tlili, I., & Alkanhal, T.A. (2019) Nanotechnology for water purification: Electrospun nanofibrous membrane in water and wastewater treatment. *Journal of Water Reuse and Desalination*, 9(3), 232–248. https://doi.org/10.2166/wrd.2019.057

Uchimiya, M., Wartelle, L.H., Lima, I.M., & Klasson, K.T. (2010) Sorption of deisopropylatrazine on broiler litter biochars. *Journal of Agricultural and Food Chemistry*, 58, 12350–12356.

Vinod, K.G., Arshi, R., & Arunima, N. (2010) Biosorption of nickel onto treated alga (*Oedogonium hatei*): application of isotherm and kinetic models. *Journal of Colloid and Interface Science*, 342, 533–539.

Wang, B.E., & Hu, Y.Y. (2007) Comparison of four supports for adsorption of reactive dyes by immobilized Aspergillus fumigatus beads. *Journal of Environmental Science*, 19, 451–457.

WHO (n.d.) *Guidelines for Drinking-Water Quality: Fourth edition incorporating first addendum* (4th ed + 1) Geneva: World Health Organization. Retrieved from https://apps.who.int/iris/handle/10665/254637

WHO (2017) *Progress on Drinking Water and Sanitation, 2017 Update and MDG Assessment.*

Yahaya, N.E.M., Latiff, M.F.P.M., Abustan, I., Bello, O.S., Ahmad, M.A. (2010) Process optimization for Zn(II) removal by activated carbon prepared from rice husk using chemical activation. *International Journal of Basic & Applied Sciences*, 10, 79–83.

You, C., Jia, C., Pan, G. (2010) Effect of salinity and sediment characteristics on the sorption and desorption of perfluorooctane sulfonate at sediment-water interface. *Environmental Pollution*, 158, 1343–1347.

Yu, F., Wu, Y., Ma, J. & Zhang, C. (2013) Adsorption of lead on multi walled carbon nanotubes with different outer diameters and oxygen contents: kinetics, isotherms and thermodynamics. *Journal of Environmental Science*, 24, 195–203.

Zhou, Y., Lu, J., Zhou, Y., & Liu, Y. (2019) Recent advances for dyes removal using novel adsorbents: a review. *Environmental Pollution*, 252, 352–365.

Zhu, C., Luan, Z., Wang, Y., & Shan, X. (2007) Removal of cadmium from aqueous solutions by adsorption on granular red mud (GRM). *Separation and Purification Technology*, 57, 161–169.

Zhang, N., Gang, D., McDonald, L., Lin, L. (2018) Background electrolytes and pH effects on selenate adsorption using iron-impregnated granular activated carbon and surface binding mechanisms. *Chemosphere*, 195, 166–174.

5 Polymers as Coagulants for Wastewater Treatment

Sai Harsha Bhamidipati and
Dharani Prasad Vadlamudi
Indian Institute of Petroleum and Energy, Visakhapatnam, India

*Sudheshna Moka**
Shriram Institute for Industrial Research, New Delhi, India

CONTENTS

* Corresponding author.
Materials Science Division, Shriram Institute for Industrial Research, 19, University Road, near Hansraj College, Hansraj College, Delhi, 110007, India.
Email: mokas@shriraminstitute.org

DOI: 10.1201/9781003138303-5

5.1 INTRODUCTION

Pollution is one of the major causes of fresh water scarcity, apart from droughts and lack of rainfall. While on one hand the demand for fresh water is increasing due to increasing population and expansion of agriculture as well as industry, its availability is on a constant decline. Some of the major water pollutants include suspended organic matter, heavy metals, and organic compounds entering water bodies from various sources. Effective and safe methods of water treatment are thus indispensable to sustain life on Earth. In general, there are four steps in large-scale water treatment: coagulation and flocculation, sedimentation, filtration, and disinfection. Each of the steps employs various chemicals/materials such as coagulants, flocculants, sand, charcoal, etc. that bring about the desired effect. The coagulation and flocculation step is very crucial as it directly affects the efficiency of filtration process and hence the overall water purity achieved. This chapter describes different types of coagulants used for water treatment, with an emphasis on polymeric materials.

5.2 COAGULATION

Suspended particles and natural organic matter behave as colloidal particles and remain suspended in water due to their colloidal stability. The coagulation process destabilizes these particles, resulting in formation of small agglomerates. Flocculation aids in forming larger aggregates, called flocs that can easily settle down and can be decanted. Coagulation substantially reduces turbidity and natural organic matter in water. Over the ages various coagulants have been used for this step, the most famous being alum, or aluminum sulfate, and other aluminum salts and ferric salts. In ancient times crushed seeds were also known for their use in clarifying turbid river water.

Increasing awareness about the health problems caused by various pollutants as well as the residual toxicity of some of the conventionally used coagulants has driven the research for better coagulants. The efficacy of a coagulant is often determined by at least some of its properties such as ionic strength, charge density, pH, chemical structure, molecular composition, and molecular weight.

5.3 CLASSIFICATION OF COAGULANTS

Coagulants are broadly classified into organic and inorganic as mentioned in the following section. Inorganic coagulants are mostly metallic salts that have been in use in the industry for many years. Organic coagulants, which are further categorized into natural and synthetic, are mostly polymers. Fig. 5.1 gives a brief overview of the classification of coagulants in general. Polymers are increasingly preferred to metallic salts due to the many advantages they offer.

5.4 METALLIC SALTS

The most widely used metallic salts are aluminum sulfate/chloride, ferrous/ferric sulfate, etc. Inorganic polymers of these salts such as poly aluminum chloride (PAC), polyferric sulfate (PFS) etc. have also been in use and have been found to be better than the respective monomers. These salts form cationic complexes with water,

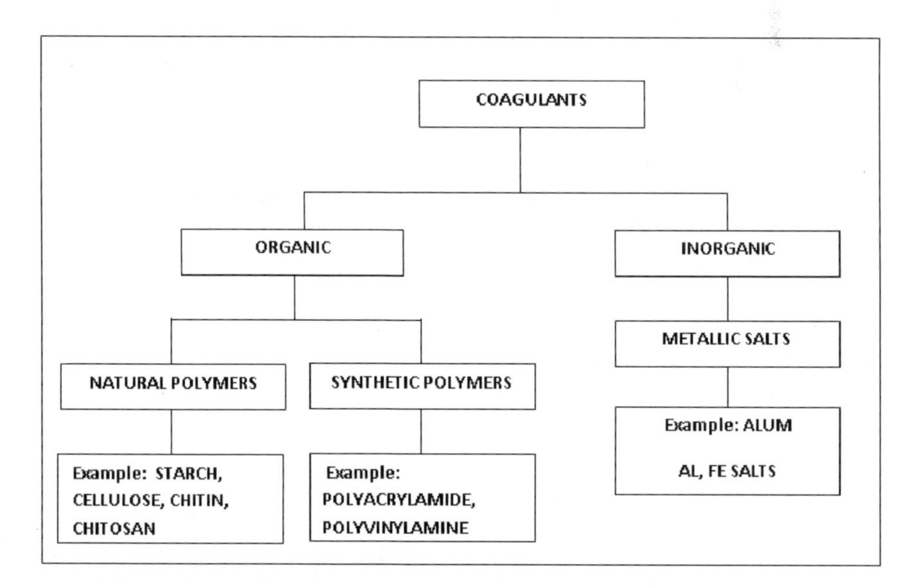

FIGURE 5.1 Classification of coagulants.

which get adsorbed by the negatively charged colloids of the natural organic matter (NOM) in the water, neutralizing and thus destabilizing them (Jiang & Graham, 1996; Sillanpää et al., 2018). Also, Fe salts form larger flocs than Al salts and are thus better. In fact, it is better to replace aluminum-based coagulants with iron-based salts because high residuals of aluminum in the water bodies damage the ecosystem and lead to human health issues (Packham, 1965; Snodgrass et al., 1984). Metallic salts also produce larger sludge volumes compared to polymeric materials.

5.5 POLYMERS IN COAGULATION — FLOCCULATION

A polymer is a substance composed of macromolecules (IUPAC, 1997a). A macromolecule is a molecule of high relative molecular mass, the structure of which essentially comprises the multiple repetition of units derived, actually or conceptually, from molecules of low relative molecular mass (IUPAC, 1997b). Polymeric coagulants require much lower dosages, produce less sludge volume, result in reduced ionic load and residual aluminum concentration in the treated water, and reduce operational costs by up to 30 percent (Nozaic et al., 2001; Rout et al., 1999). Formation of larger flocs also results in improved settling rates and hence increased capacity of the plant. Properties of polymers such as charge density, molecular weight, structure (homo polymers, random polymers, co-polymers, straight chain, branched, etc.) also affect the coagulation mechanism as well as efficiency.

5.6 MECHANISMS OF COAGULATION

Coagulation and flocculation consist of application of chemicals to destabilize the suspended particles and promote sedimentation in wastewater treatment. In various literature, the terms coagulation and flocculation are used interchangeably and equivalently. Here, coagulation is a process by which the particles are destabilized to produce small aggregates, and flocculation involves the conversion of smaller aggregates to larger flocs. The various mechanisms involved in the coagulation are mentioned below, where it is stated that no single mechanism governs the entire process. The mechanisms mentioned below may operate in combination or one might take predominance over others, depending on the type of coagulant in regard to its charge density, molecular weight, coagulant dosage, particle concentration, pH, mixing conditions, and solvent characteristics:

1. Double-layer compression
2. Adsorption and bridging
3. Charge neutralization and electrostatic patch mechanism
4. Sweep coagulation

5.6.1 Double-Layer Compression

In the solution, the suspended particles are charged mostly with a negative charge that attracts oppositely charged ions known as the counter-ions. These ions form the compact layer (Stern Layer) around the negatively charged colloidal particles held

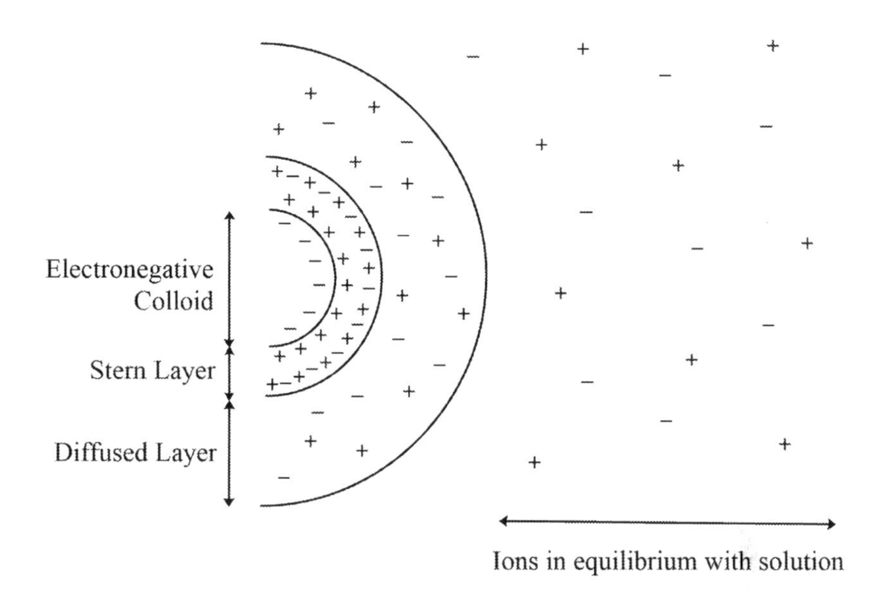

FIGURE 5.2 Double-layer negative colloid.

by electrostatic and Van der Waals forces. A diffusive layer is formed around the Stern Layer by the attraction of the negatively charged particles (co-ions of the primary colloid) onto the counter-ions. The schematic representation of the electrostatic double layer around the colloid is shown in Fig. 5.2 (Teh et al., 2016). Only part of the diffusive layer travels along with the colloid by shearing, and the potential at the shear surface is called zeta-potential, which signifies the charge of the colloidal particle. The stability of the colloidal particles that keeps them suspended in the solution is due to the difference in the repulsive (overlapping of the double layers) and attractive (Van der Waals forces) forces that constitute the energy barrier of the particles.

One of the mechanisms involved to destabilize the particles is double-layer compression. Here the electrostatic double layer is compressed by the addition of oppositely charged polyelectrolytes, mainly of cationic charge, to decrease the diffusive layer. As the thickness of the double layer is reduced, the addition of the counter-ions makes the Van der Waals forces overcome the repulsions between the double-layered colloids and results in the coagulation of the particles. Other mechanisms that will be discussed are found to dominate double-layer compression, particularly for polymeric coagulants.

5.6.2 Adsorption and Bridging

This mechanism is quite recognized in the coagulation process by polyelectrolytes. Mentioned below are the stages involved in this mechanism and schematically represented in Fig. 5.3 (Akers, 1972; Bratby, 2016):

- Polyelectrolyte dispersion in the suspension
- Adsorption at the particle surface

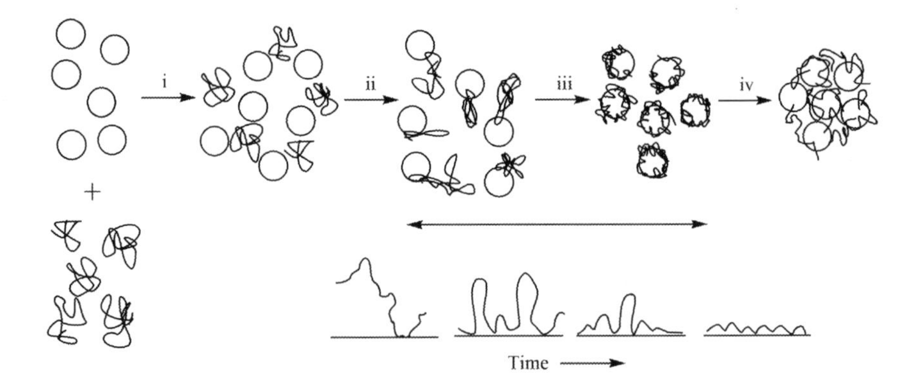

FIGURE 5.3 Different stages of bridging mechanism: (i) dispersion; (ii) adsorption; (iii) compression; and (iv) bridging.

- Adsorbed polyelectrolyte compression
- Collisions for Polymer Bridging

5.6.2.1 Polyelectrolyte Dispersion in the Suspension

Polyelectrolytes have high molecular weight, exhibiting low viscosity and leading to low diffusion rates in the suspension. Polymeric adsorption is irreversible because it is difficult for simultaneous desorption of long polymeric chains at various sites. Hence adsorption needs to occur evenly throughout the suspension for effective coagulation. The rate of adsorption is very high compared to the rate of diffusion; hence it is important in the initial stage to spread the polyelectrolytes in the suspension evenly and quickly. This dispersion is achieved by rapid mixing for a short period of time, making sure that the polyelectrolyte is evenly diluted in the suspension. The rate of adsorption is diffusion-rate limiting, therefore rapid mixing is an important initial stage (Gregory & Lee, 1990).

5.6.2.2 Adsorption at the Particle Surface

After diffusion of the polyelectrolytes into the suspension, a few functional groups get adsorbed onto the particle surface, with the rest polymer chain extending freely into the solution. As time progresses, the polymer chain gets attached at several adsorption sites on the particles, reducing the dangling ends (Fig. 5.3ii). In this configuration, multiple loops and dangling ends are extended into the suspension. The type of adsorption depends on the particle and polyelectrolyte characteristics. Different types of adsorption are:

- **Electrostatic interaction:** This type of adsorption is generally observed for oppositely charged polyelectrolytes to the suspended particles (generally cationic polyelectrolytes on the negatively charged particles in the suspension) due to electrostatic attraction.
- **Hydrogen bonding:** Certain polymers adsorb onto the surfaces of the particles at particular sites through hydrogen bonding. For example, hydrogen

bonding occurs between the oxygen double bonds or amide groups in the polyelectrolyte with aluminol, silanol, and metal hydroxides.

- **Ionic bonding:** This generally occurs when anionic polyelectrolytes interact with particles bearing a negative charge despite their electrostatic repulsion. Such adsorption occurs when there is sufficient concentration of divalent metal ions like Ca^{+2} ions (O'Gorman & Kitchener, 1974) acting as bridges for binding the polyelectrolyte onto the negatively charged adsorption sites on the particle.

5.6.2.3 Adsorbed Polyelectrolyte Compression

Adsorption progresses with time at various sites on the particle, and the polyelectrolytes reach an equilibrium stage where the polymeric chain is compressed onto the surface. It takes several seconds for high-MW polymers to attain equilibrium arrangement (Pelssers et al., 1990), and this time is crucial for bridging to happen between the adjacent surfaces in the suspension. Bridging is effective when the polymer chains are in the solution in an extended configuration before it reaches the equilibrium stage. If the polyelectrolyte attains a flat compressed configuration, then there won't be effective bridging due to the unavailability of polymeric loops or free chains extended into the solution.

5.6.2.4 Polymer Bridging

After adsorption, the polyelectrolyte loops extending into the solution are adsorbed onto adjacent particles during collisions leading to several polymeric bridges (Fig. 5.3iv). The essential criterion for bridging is the availability of adsorption sites on the particles for adjacent polyelectrolytes. This depends on the concentration of the polyelectrolyte added; if the dosage is high, then it leads to complete adsorption and non-availability of sites for inter-particle bridging, and if it is low, then that leads to poor adsorption and bridging. Hence there lies an optimum polymer dosage depending on the number of sites available for adsorption and therefore on the particle concentration. As discussed in the previous section, bridging is effective when it takes place before the equilibrium configuration; hence the relative difference between the collision time and reconfiguration time is very important. If the collision time is less compared to the reconfiguration, then the particles collide before the equilibrium stage when the polymer chains are extended into the solution. This gives effective polymeric bridges between the particles by adsorption of the polyelectrolytes onto adjacent particles.

Molecular weight and chain length characteristics of polyelectrolytes play a crucial role in the polymeric bridging mechanism. A longer chain length does not give a flat configuration on the surface, leading to extended loops into the suspension. Hence it is found that bridging is efficient for high molecular weight polymers (Gregory, 1993) and particularly for linear chained polymers.

Charge density (CD) is also an important factor in bridging. If charge density is very high, then that leads to compressed flat configuration and ineffective in terms of polymer bridging. However, CD is beneficial for polymer bridging up to a certain degree since repulsion between the polymeric segments of similar charge will extend the loops further, thereby improving the bridging mechanism. Hence, in terms of polymeric bridging as a predominant mechanism, the CD also has an upper limit, and optimum value is desired for enhancing coagulation.

Since the mechanism involves inter-particle bridging between the particles, the flocs formed are much stronger compared to other mechanisms. Such strong flocs under shear conditions grow even larger (Mühle, 1993) due to resistance from breakage. But the flocs breakage at high shear conditions is irreversible (Yoon & Deng, 2004), presenting a major disadvantage in this mechanism.

5.6.3 Charge Neutralization and Electrostatic Patch Mechanism

Cationic polyelectrolytes have a strong electrostatic interaction with the negatively charged suspended particles, resulting in charge neutralization. It leads to neutralization of the particle charge, hence making zeta potential zero. This type of mechanism is found to be effective for polyelectrolytes of a high CD showing a flat configuration and thereby giving little possibility for a bridging mechanism.

Another mechanism that arises within this is the electrostatic patch mechanism, which happens when there is a low density of negatively charged sites on the particle (Gregory, 1973).

When a polyelectrolyte of high CD adsorbs onto a particle of low CD, small patches of the polyelectrolyte are observed on the particle. Even though the particle achieves charge neutrality, each negatively charged site cannot be neutralized by the polymer segment as the charged surface sites are farther apart from each other than the length of the polymeric segments in its chain. These positive patches on the particle attract the negatively charged surface sites of the neighboring particle, leading to the formation of flocs as shown in the Fig. 5.4 (Bolto & Gregory, 2007).

Thus, flocs formed in this mechanism are strong compared to those formed by simple charge neutralization and by metallic salts but not as strong as those formed by bridging mechanism. One advantage is that the reformation of the flocs appears even after flocs breakage, in this case due to electrostatic interaction (Yoon & Deng, 2004).

5.6.4 Sweep Coagulation

When the particles dispersed in the suspension get enmeshed into the insoluble metal hydroxide matrix formed by the metallic coagulants, it is called sweep coagulation.

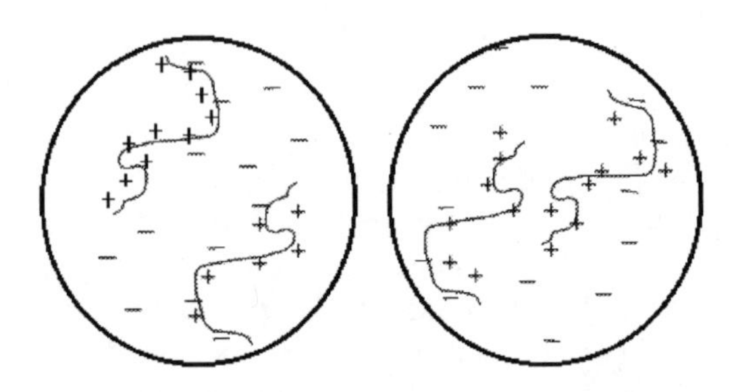

FIGURE 5.4 Electrostatic patch model.

This mechanism is better in comparison to charge neutralization, but the flocs formed are of huge volume and are weaker in comparison to those formed in the bridging mechanism.

It is generally found that metal coagulants operate through charge neutralization, double-layer compression, or sweep coagulation, not through bridging and adsorption. Polymeric coagulants, on the other hand, can operate in mechanisms other than sweep coagulation. Adsorption and bridging are most common for polymeric coagulants, but double-layer compression and charge neutralization can also be included based on the charge density of the polyelectrolyte.

5.7 CLASSIFICATION OF POLYMERIC COAGULANTS

As seen in Fig. 5.1, polymeric coagulants are classified into synthetic and natural. These are described briefly in the following sections.

5.7.1 Synthetic Polymers

Water treatment industry has witnessed the use of synthetic polymers as coagulant and flocculants aids since many years. Due to the presence of ionizable sites, they are also called polyelectrolytes and are classified mainly depending on the ionic charge — hence the names cationic polymers, anionic polymers, and non-ionic polymers. Synthetic polymers are advantageous compared to other coagulants, especially due to their better stability, purity, non-biodegradability, and colloidal bridging capability (Lapointe & Barbeau, 2020). They are also known for their low dosage requirement, stronger flocs formation, high removal efficiencies, high water solubility, and low sensitivity to pH, and less volume of sludge produced (Lee et al., 2014). Their molecular weight, charge density, and type of functional group can be modified and hence can be adapted to suit the different requirements, which may vary according on the type of pollutants present in the wastewater.

Natural organic matter, or NOM, is negatively charged; hence cationic polymers are used to treat NOM. The predominant mechanisms of colloidal particle destabilization are charge neutralization and bridging (Bolto & Gregory, 2007; Sillanpää et al., 2018). Cationic polymers have their charge density coming from functional groups such as the quaternary ammonium groups, and anionic polymers have carboxylic or sulfonic acid groups. Often the main polymer is modified by "grafting" the required functional groups for specific target molecules (Lapointe & Barbeau, 2020).

Some of the popularly used synthetic polymers are polyacrylamide (PAM), polyacrylic acid (PAA), polyethylene imine (PEI), polyvinyl alcohol (PVA), polyethylene oxide (PEO), polyvinyl sulfonic acid (PVSA), and so on. PAM is available as both cationic and anionic polymer with varying charge density and molecular weight (even up to 1.2×10^7) as required. It is very stable and can give very good results at low dosage (Lapointe & Barbeau, 2016, 2017, 2020; Leeman et al., 2007). PAA is highly anionic in nature due to the carboxylic acid groups, and the pH of the medium is known to affect its charge density. PAA also has very good water solubility and is very stable, available in huge molecular weights, and a range of charge densities (Bolto, 1995; Ravishankar & Pradip, 1995; Tripathy & De, 2006).

PEI is a cationic polymer, and it can also be produced at high molecular weights, though not as high as PAM, while PEO and PVA are non-ionic polymers. Detailed description, advantages, and disadvantages of each of these polymers along with a comprehensive list of earlier work on these polymers is given elsewhere (Lapointe & Barbeau, 2020).

5.7.2 Natural Polymers

Synthetic polymers are readily soluble in water, are not affected by pH of the medium, and produce much less sludge compared to inorganic coagulants. However, there are certain disadvantages with synthetic polymers. On the one hand, the non-biodegradability enhances the stability of these polymers, but their persistence in residual amounts in treated water as well as in the sludge poses a threat to human health; similarly the presence of monomers in treated waters is the cause for many severe health conditions (Bae et al., 2007; Bolto & Gregory, 2007; Renault et al., 2009; Yang et al., 2011). These problems have resulted in more stringent norms for the use of polymers in water treatment. Biopolymers, on the other hand, have gained wider attention because of the rising awareness of health problems due to use of synthetic polymers and the stringent regulations in many countries.

In addition, biopolymers are advantageous over synthetic coagulants as they are:

1. available in abundance and at low cost
2. safe, eco-friendly, non-toxic, and biodegradable nature
3. modifiable to enhance the properties and increase efficiency
4. functional groups that are naturally present, such as amino groups and hydroxyl groups (which can adsorb a wide variety of pollutants)

Chitosan, starch, cellulose, alginates, natural gums and mucilage, tannins, etc. are some of the popular biopolymers employed in wastewater treatment via coagulation. Among them chitosan, an amino-polysaccharide, has been the most researched for wastewater treatment containing dissolved or suspended particulate matter (organic as well as inorganic) (Renault et al., 2009). While both cellulose (Khiari et al., 2010) and starch (Xing et al., 2005; Zhang et al., 2004) have been used in a modified form for treating wastewater, starch has been used in the papermaking industry as well (Michele et al., 2005). Biopolymers sourced from plants have shown very good sensitivity toward metals and aromatic compounds and have been tested successfully on tannery effluents, landfill leachate, textile effluents, and sewage effluents. A brief summary of plant-based biopolymers preparation methods is given in Lee et al. (2014). A comparison of different biopolymers with respect to the flocculation efficiency, sedimentation time, optimum pH, optimum dosage of the coagulant and bio-flocculants is also given here (Lee et al., 2014).

Numerous seeds have been used for water purification for thousands of years. India's Nirmali seed is known to have been used for clarifying turbid water as early as 2000 BC (Yin, 2010). The –COOH and –OH groups in the Nirmali seed extract are quite effective for coagulation and destabilization through inter-particle bridging (Vijayaraghavan et al., 2011). Moringa oliefera seeds have been particularly effective

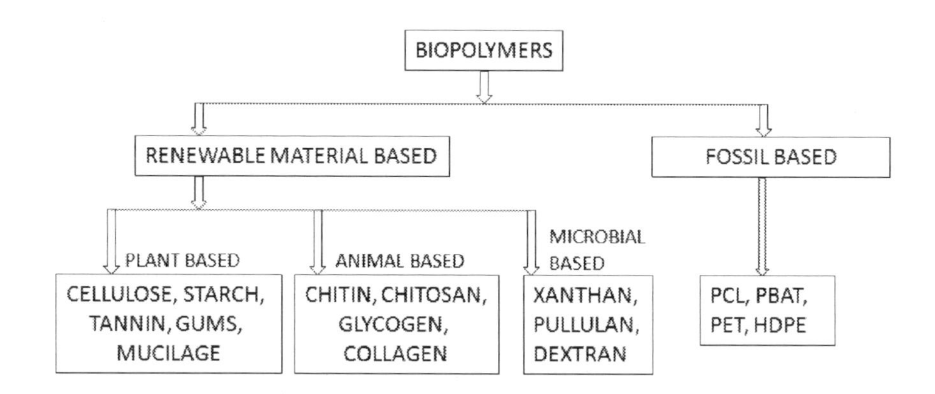

FIGURE 5.5 Classification of biopolymers.

in removing metals such as Fe, Cu & Cd, Pb in polluted waters (Shan et al., 2017). It has also shown very good removal efficiencies for coliform (97–98 percent), turbidity (83–90 percent), and others (Othmani et al., 2020; Ugwu et al., 2017). This seed has been successfully used to treat effluents from tanneries, laundries, dairies, palm oil mills, etc. (Othmani et al., 2020). Likewise, tamarind seed powder has also shown promise in turbidity removal (Ronke et al., 2016). Natural gums (guar, xantham) have been studied for applications in water treatment and have been proved efficacious for a good range of ionic strengths and for effluents from different types of industries (Gupta & Ako, 2005; Tripathy & De, 2006). Various accounts in literature have cited plant-based materials as effective coagulants for water treatment (Abiola, 2019; Agunbiade et al., 2016; Amagloh & Benang, 2009; Aminu et al., 2014; Butler, 1966; Ghimici & Nichifor, 2010; Gupta & Ako, 2005; Hemapriya et al., 2015; Hendrawati et al., 2016; Jayalakshmi et al., 2017; Khiari et al., 2010; Othmani et al., 2020; Ravikumar & Sheeja, 2013; Saranya et al., 2014).

5.7.3 Classification

Biopolymers may be classified based on the source of the polymer whether plant, animal, or fossil fuel based. A typical way of classifying biopolymers as suggested by Pandey (2020) is shown in Fig. 5.5.

5.8 ADVANCED MATERIALS

Coagulation is a crucial aspect of operation in the treatment of wastewater. The process of coagulation is primarily explained using the DLVO (Derjaguin–Landau–Verwey–Overbeek) theory. This theory states that coagulation is basically a process of overcoming the inter-particle repulsive energy barrier to separate impurities from water using external coagulants. Generally, until now, the most common coagulants have been conventional inorganic and organic coagulants such as aluminum and iron-based substances. These neutralize the surface charge of the particles in the wastewater, thus promoting the aggregation of these particles, so these particles can be removed easily using wide array of methods such as gravity separation.

Recently, various new types of materials known as hybrid coagulants are being produced. These have much more effectiveness and coagulation power and cost less than their conventional counterparts. As there is a lot of demand for more efficient coagulants in the market, hybrid materials have boomed in popularity in recent times.

Hybrid materials are synthesized by adding effective components into conventional coagulants to enhance the efficiency of the coagulant. The main reason for the improvement of the performance in the hybrid materials is due to the synergic effect of hybrid components in one material. One of the most important advantages in usage of hybrid materials is the reduction of operation time, mainly in industries that discharge large amounts of wastewater, as a result of various coagulant and flocculants components present in hybrid materials.

5.8.1 Classification of Hybrid Materials

According to Nanko (2009), hybrid materials have been divided majorly into three groups:

1. structurally hybridized, also known as composites
2. chemically bound
3. functionally hybridized

5.8.1.1 Structurally Modified Hybrid Materials or Composites

Composites are hybrid materials that are produced by combining various materials following the rule of mixtures at microscopic level (Nanko, 2009). This procedure is generally used for synthesizing organic–inorganic hybrid materials. Conventional coagulants are mixed through physical mixing/blending at either normal or higher temperatures. By changing the percentage of individual components in the mixture, the overall properties of the composite can be varied. An example of composite is PFC-PDMDAAC hybrid.

5.8.1.2 Chemically Bound Hybrid Materials

Chemically bound hybrid materials are different from composites in that the coagulant atoms are bonded at molecular levels using chemical bonds (Nanko, 2009). These materials are produced using chemical modification. A different chemical group is introduced and attached to the molecular chain to produce a compound that is chemically modified. Some common chemical modification methods used for the production are hydroxylation–pre polymerization, co-polymerization, and chemical grafting (also known as crosslinking). An example of a hybrid is PASiC (Gao et al., 2002), which is produced by chemically linking polysilicic acid to the molecular chain of PAC (poly aluminum chloride). Even though normal PAC has better performance than other inorganic coagulants, organic coagulants perform better due to their higher molecular weight. So, to increase the molecular weight of PAC, polysilicic acid is added to the molecular chain.

5.8.1.3 Functionally Modified Hybrid Materials

The concept of functionally modified hybrid materials is still relatively new and unknown compared to other hybridized materials. Nanko (2009) made an attempt to define functionally modified hybrid materials as materials that have harmonizing function and use of interface functions, which results in new functions or super functions. We can think of composites and chemically bound hybridized materials as a subset of functionally hybridized materials, as both of these hybrid materials integrate two or more functions such as flocculation and coagulation in the final synthesized aggregate (Lee, et al., 2011).

To elaborate the classification of hybrid materials further, a secondary classification can be done based on the type of materials that are combined to produce a hybrid material:

1. inorganic
2. organic
3. natural
4. biopolymer

Based on these materials, various combinations can be made, as seen in Fig. 5.6. Table 5.1 gives examples of some of these hybrid materials.

5.8.1.4 Inorganic–Inorganic Hybrid Materials

Hybrid materials of this type are known as inorganic polymeric coagulants (IPCs). Some of the examples of IPCs are modifications of polyaluminumchloride (PAC), polyferricsulfate (PFS), etc. (Moussas & Zouboulis, 2008, 2009; Tzoupanos & Zouboulis, 2011; Zouboulis & Moussas, 2008), which are produced by partial neutralization of inorganic salts such as $Fe_2(SO_4)_3$, $AlCl_3$, which control the basicity of the overall material.

IPCs are proven to have more effectiveness compared to conventional coagulating materials, mainly because of their resistance to pH and temperature of the wastewater. Regardless of this fact, the effectiveness of IPCs is much less when compared to organic polymers. This is due to the lower molecular weight and small size of IPCs, which lead to lower aggregating capacity. Some of the major IPCs that have been developed are aluminum and iron silicate complexes. A few researchers could also produce a single hybrid complex containing all aluminum, iron, and silicate groups to improve the aggregating capacity (Cheng et al., 2008; Gao et al., 2003; Zhang et al., 2004). Other well-known chemicals known for the production of IPCs are magnesium, zinc, phosphate, etc.

5.8.1.5 Organic–Inorganic Hybrid Materials

One of the major problems in the coagulant industry is the lower coagulation capacity of inorganic salts compared to organic counterparts. So, to overcome this limitation, organic polymers and inorganic salts were combined to prepare another type of hybrid material. Polyacrylamide (PAM) is a common polymer with high molecular weight and also costs less. PAM is versatile in that it is easy to chemically modify (Lee et al., 2008, 2009, 2011). So inorganic coagulants such as $Al(OH)_3$, $FeCl_3$, PAC,

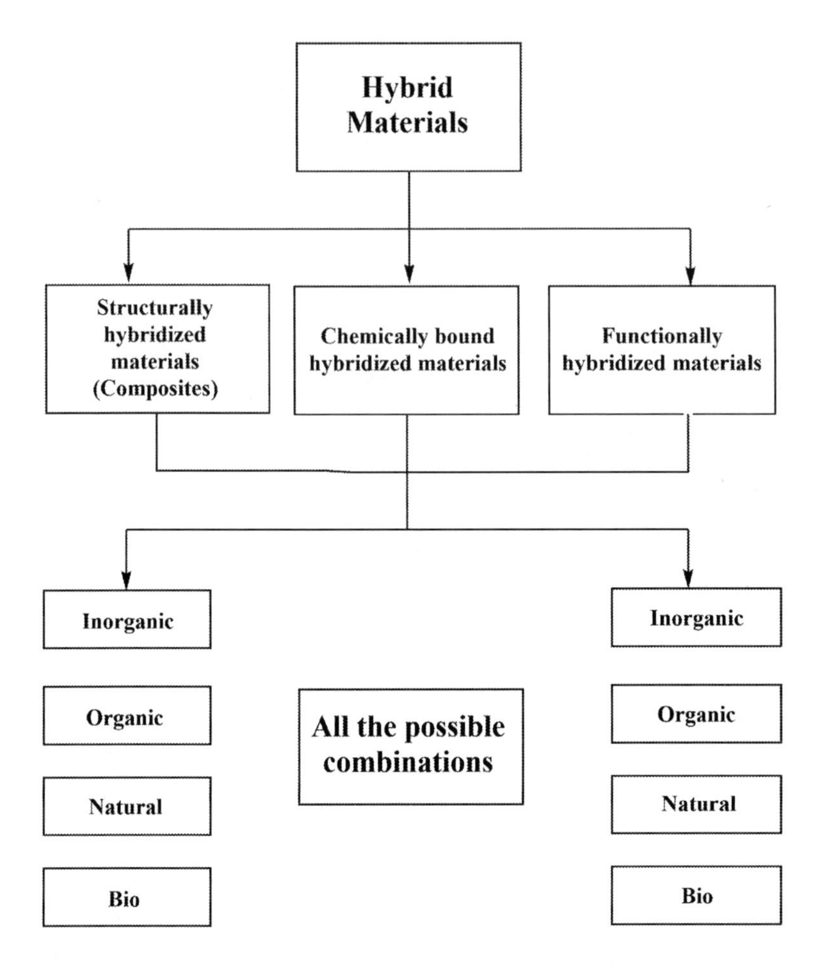

FIGURE 5.6 Classification of hybrid materials.

etc. were combined with PAM to form organic–inorganic hybrid materials. Due to this, the bridging mechanism is improved, which enhances the aggregating capacity.

Another organic polymer that is generally used is PDMDAAC (polydimethyldiallylammonium chloride), and various studies compose inorganic substances such as $Al_2(SO_4)_3$, $FeCl_3$, PAC, etc. to increase the coagulation effectiveness and reduce the dosage of inorganic substance. There is a lot of future prospect for water-soluble polymers such as polyamines, polyacrylic acid, polyamines, etc. to produce versatile hybrid compounds (Tripathy & De, 2006).

5.8.1.6 Inorganic–Natural Polymer Hybrid Materials

Natural polymer coagulants are becoming more popular day by day as they are environmentally friendly. One popular natural polymer that is used as coagulant is chitosan. Many modifications have been done in which several inorganic additives such as PAC, PSAF, or PAC-sodium silicate have been added to the chitosan molecule. Other

TABLE 5.1
Types of Hybrid Materials

Type of Hybrid Material	Examples
Inorganic–inorganic hybrid	PASiC (polyaluminum–silicate–chloride)
	PSiF (polysilicic–ferric)
	PAFSiC (polyaluminum–ferric–silicate–chloride)
	PFMS (poly–ferric–magnesium–sulfate)
	FeCl3–PAC (ferric chloride–polyaluminum chloride)
	Iron–aluminum polymer hybrid
Inorganic–organic hybrid	$Al(OH)_3$–PAM (aluminum hydroxide–polyacrylamide)
	$Mg(OH)_2$–PAM (magnesium hydroxide–polyacrylamide)
	PAC–PDMDAAC (polyaluminum chloride–polydimethyldiallylammonium chloride)
	Palygorskite–polyacrylamide (PGS–PAM)
Inorganic–natural polymer hybrid	Al2(SO4)3–CTS (aluminumsulfate–chitosan)
	CTS–PAC–Na2SiO3 (chitosan–polyaluminum chloride–sodium silicate)
	PSiAF–CTS (polysilicatealuminumferric–chitosan)
	Red mud–hydrochloric pickle liquor of bauxite
Inorganic–biopolymer hybrid	Pullulan–PAC (pullulan–polyaluminum chloride)
	MBF(Aspergillusniger)–zeolite
	MBFGA1–PAC (microbial flocculantGA1–polyaluminumchloride)
Organic–organic hybrid	Poly(acrylamide-co-acrylic acid)
Organic–natural polymer hybrid	SAG-g-PAM (sodium alginate grafted polyacrylamide)
	PAM-g-CMS (polyacrylamide grafted carboxymethylstarch)
	CMC–starch hybrid (carboxymethyl cellulose and starch)
	Chitosan-g-N,N-dimethylacrylamide
	Chitosan-g-N-vinyl formamide
Natural polymer–natural polymer hybrid	Cationic starch–chitosan crosslinking copolymer

than chitosan, many other polysaccharide-based (natural) coagulants such as guar gum, starch, sodium carboxymethyl cellulose (Tripathy & De, 2006), pectin (Yokoi et al., 2002) and amylopectin (Rath & Singh, 1998) can be composed by combining with inorganic substances to produce new type of hybrid materials.

5.8.1.7 Inorganic–Biopolymer Hybrid Materials

Apart from inorganic–natural polymer hybrid materials, another type of eco-friendly hybrid material occurs when a biopolymer is combined with an inorganic coagulant to produce inorganic–biopolymer hybrid materials. Some of the biopolymers are MBFGA1 and Pullulan. MBFGA1 represents microbial flocculant GA1, which is a flocculant-producing strain recognized as Paenibacillus polymyxa by 16SrDNA sequence. Pullulan is a water-soluble microbial polysaccharide with strains of Aureobasidium pullulans. Even though it has many advantages, such as innocuity, biodegradability, edibility, and doesn't cause pollution, it poses a limitation in the performance and also a higher production cost. Therefore, improving the performance and optimizing for a higher yield and lower cost is of utmost importance. One of the better methods is to make this polymer with an inorganic coagulant to decrease costs. There is much less research done on inorganic–biopolymer materials, and so there is a lot of scope and potential to produce more effective coagulants and flocculants.

5.8.1.8 Organic–Organic Hybrid Materials

Another type of hybrid material is synthesized when two different organic materials are merged to form a single hybrid material. Modified polymers such as copolymers, grafted-polymers, and chemically modified polymers are generally part of the organic–organic hybrid materials. An example is the introduction of poly (acrylamide-co-acrylic acid) into an organic polymeric chain.

5.8.1.9 Organic–Natural Polymer Hybrid Materials

By combining organic synthetic polymers with natural polymers, we get a desirable hybrid material that has the properties of both components. These hybrid materials can be produced by grafting organic groups onto the main chain of natural polymers. Some examples are SAG-g-PAM hybrid (Tripathy et al., 2001), chitosan-g-N,Ndimethylacrylamide (Tripathy et al., 2010), PAM-g-CMS (Sen et al., 2009), CMC-starch hybrid (Hebeish et al., 2010), starch-g-PAM (Mishra et al., 2011), chitosan-g-N-vinyl formamide (Mishra et al., 2008) and starch-g-PAM-co-sodium xanthate (Chang et al., 2008).

5.8.2 Preparation of Hybrid Materials

5.8.2.1 Hyroxylation Prepolymerization

This is one of the methods that is used for the preparation of inorganic–inorganic hybrid material, which is one type of chemically bound hybrid materials.

5.8.2.2 Physical Blending (at Atmospheric Temperatures)

Being one of the most popular methods of synthesizing organic–inorganic hybrid materials, the preparation process is simple compared to other methods. It usually involves blending of various components into a physical mixture at room temperature. This method is purely a physical process and doesn't involve any chemical reactions. Many hybrid materials have been developed based on this method, such as PAC–PDMDAAC, PFC–PDMDAAC, $FeCl_3$–PDMDAAC, PAC–chitosan, PAC–EPI–DMA, PAC–MBFGA1, MPFC, $FeCl_3$–PDMDAAC, PFC–EPI–DMA, $MgCl_2$–PAM, $Mg(OH)_2$–PAM, PFAC–PDMDAAC. Among these, PFC–PDMDAAC has received widespread attention and research as it can be used to treat various kinds of wastewater (Sun et al., 2012).

5.8.2.3 Elevated Temperature Blending

This method is similar to physical mixing but the blending is done at elevated temperatures (50–60°C) to prepare the organic–inorganic hybrid materials. Such mixing at elevated temperatures gives better stability and increases the homogeneity of the inorganic substance within the polymeric matrix by eliminating the presence any insoluble by-product (Lee et al., 2012a). Some hybrids produced by this method are PFSPDMDAAC and PFS–PAM, $CaCl_2$–PAM and $FeCl_3$–PAM.

5.8.2.4 Copolymerization

A copolymer is a polymer that is made up of two or more monomer species. Redox polymerization is a common method used for copolymerization. Redox polymerization is normally used for hybrid materials based on poly-acrylamide (Lee et al., 2012b; Qian et al., 2004; Yang et al., 2004). Many materials, such as $Al(OH)_3$-PAM, and PGS-PAM, are synthesized using this method in which acrylamide is polymerized in the colloidal solution of palygorskite (PGS) and $Al(OH)_3$ in the presence of various redox initiators.

5.8.2.5 Graft Polymerization

Graft copolymerization is mainly used to synthesize a novel, unique polymer with the best properties of both polymeric groups. Some of the methods that have been used successfully to produce grafting reactions are conventional redox, microwave-initiated, and microwave-assisted grafting methods. Conventional redox methods usually require a free radical initiator (Agarwal et al., 2002; Mishra et al., 2004, 2006) and also suffer from decreased yield due to competing parallel reactions. In this case, microwave-based techniques can mitigate the problems of conventional methods with higher yields, but that comes with its own problems. Microwave-based methods generally involve high production costs and thus can limit the production of many grafted flocculants. A simple representation of the grafting method is seen in Fig. 5.7. Usually, a graft copolymerization reaction consists of three steps:

1. the dissolution of the polymer backbone into a homogenous aqueous solution
2. reaction initiation
3. grafting the required amount of monomer with the polymer backbone

5.8.3 Chitosan-Based Advanced Materials for the Treatment of Wastewater

Chitosan, a very popular polymer, is used in many fields, including biotechnology, biomedicine, food processing, and wastewater treatment (Bolto, 1995; Crini & Badot, 2008; Muzzarelli & Muzzarelli, 2005; Reddy & Lee, 2013). It is a linear copolymer of N-acetyl-D-glucosamine (GlcNAc) and D-glucosamine (GlcN) that is produced by the alkaline deacetylation of chitin, the most abundant natural polymer in the world after cellulose. Its structure is shown in Fig. 5.8. Chitosan, being a naturally occurring polymer and having large molecular weight, acts as both as a coagulant

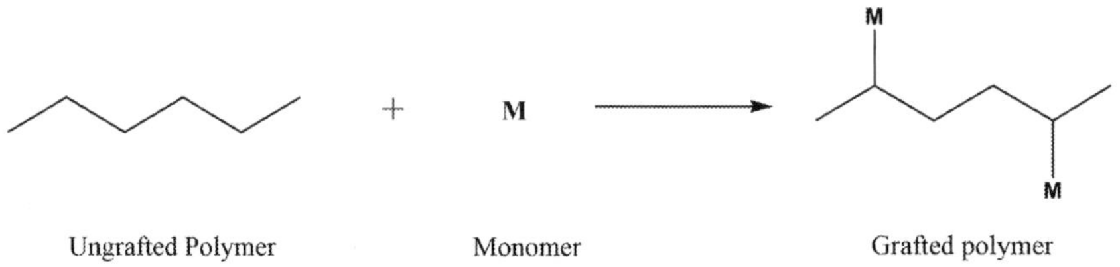

M
+ M →
M
Ungrafted Polymer Monomer Grafted polymer

FIGURE 5.7 Schematic representation of chemical grafting polymerization.

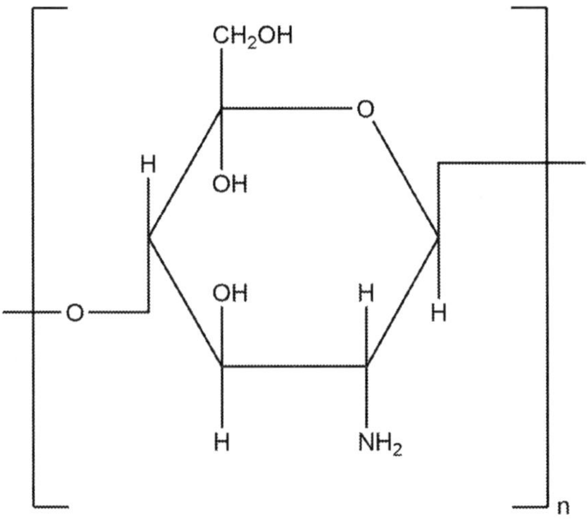

FIGURE 5.8 Structure of chitosan.

and flocculant. The coagulation and flocculation efficiency of chitosan is directly affected by its characteristics, such as the degree of deacetylation (which has an influence on the acidic and basic properties and charge neutralization) and the molecular weight (which has an influence on the flocculating properties). The main reason why chitosan-based coagulants are used is the reduced number of harmful effluents at the end compared to other coagulants and also the lower amount of sludge produced.

Chitosan can also be used to remove NOM using various modification methods such as quaternization of chitosan so that the increased positive charge in chitosan can electrostatically bind with the humic and fulvic acids of NOM. Another method is to hydrophobically modify chitosan, using carboxymethylation reactions and grafting reactions.

Even with such popularity, chitosan also has some defects that hinder its efficiency such as inactive chemical properties and poor solubility in alkaline or neutral aqueous solutions. So, for improving the coagulating properties, several modified forms have been formulated (Dao et al., 2016; Jiang et al., 2007; Muzzarelli & Muzzarelli, 2005; Rinaudo, 2006). These modifications have helped chitosan to act as a better flocculant and coagulant. It has also helped in diversifying the application of chitosan in removing many types of inorganic and organic impurities, biological contaminants, heavy metals, and bacteria, among many others.

Many people (Dao et al., 2016; Jiang et al., 2007; Muzzarelli & Muzzarelli, 2005; Rinaudo, 2006) have modified chitosan by using the free amines and hydroxyl groups on the chitosan backbone to combine with various functional groups. To remove inorganic suspended particles, their negative charge has been neutralized by introducing many cationic functional groups, such as quaternary ammonium salts, onto the chitosan backbone. Therefore, this enhanced charge neutralization increases the efficiency.

Some of the most common methods used to modify chitosan are etherification or amination (Bratskaya et al., 2009; Chang et al., 2009) and graft polymerization (Laue & Hunkeler, 2006; Wang et al., 2008; Yuan et al., 2010). In etherification, the

TABLE 5.2
Some Common Modifying Groups Used in Etherification/Amination Reactions for Synthesis of Modified Chitosan Flocculants

Type of the Group	Common Modifying Agents Used
Cationic groups	3 – Chloro-2-hydroxypropyltrimethyl ammonium chloride (CTA)
	2, 3-Epoxypropyltrimethyl ammonium chloride (ETA)
	N-methyl piperazine
	2,4 – bis(dimethylamino) – 6 – chloro - (1,3,5) – triazine
Carboxyalkyl groups	Monochloroacetic acid
	Acrylic acid
Others	Glycidyl methacrylate
	Chlorosulfonic acid

proton present in the -OH or -NH_2 present in chitosan is substituted via a nucleophilic reaction with the help of modifying agents such as CTA, ETA, acrylic acid etc. (see Table 5.2). As MW increases, chitosan-CTA becomes less efficient. However, the use of many agents other than CTA, ETA is still not popular due to complex synthesis techniques and higher cost. Chitosan can also be modified to enhance its ability to flocculate positively charged impurities by the addition of anionic groups to the chitosan backbone. Bratskaya et al. (2009) prepared various N-carboxyethylated chitosans (CECs) in which the degree of substitution ranges from 0.7 to 1.6 per glucosamine unit of chitosan. Similarly, Zhao et al. (2012) prepared two phosphorylated chitosan derivatives that have a similar degree of substitution (around 0.5 per glucosamine unit of chitosan) but with varied structures (i.e., O-phosphorylated chitosan and N-methylene phosphonic chitosan). Acidity plays an important role while selecting the phosphate groups for producing modified chitosan, and it determines the flocculation efficiency; so OPC shows much higher removal efficiency than NMPC.

Graft copolymerization is another important synthesizing method for production of modified coagulants. In grafting, synthetic functional polymers are introduced as side chains on the polymer backbone, thus enabling various molecular designs. Some of the monomers used for graft polymerization are listed in Table 5.3.

Chitosan-grafted copolymers have a wide variety of uses in wastewater treatment. The graft reaction is initiated with the help of radiation, which is either gamma ray or microwave (Singh et al., 2012), or ceric ammonium nitrate (CAN) and persulfates. Monomers that can be fed into the system may be either non-ionic, cationic, or anionic. The whole grafting process is done in an anon-reactive atmosphere. The dissolving medium is usually a dilute acidic medium such as acetic acid or HCl. One of the most common monomers that has been grafted onto chitosan is acrylamide, since the utility of PAM (Polyacrylamide) in water purification is well known. Wang et al. synthesized chitosan-g-PAM copolymer, which is initiated by gamma rays in an acid-water solution (Wang et al., 2008). Anionic chitosan-based flocculants like these have better water solubility and chelating effects in wastewater containing mostly positively charged pollutants.

Graft copolymerization and etherification reaction can also be carried out successfully to modify chitosan. Yang et al. used this to synthesize carboxymethyl chitosan (CMC) onto which they successfully grafted PAM, giving rise to composite

TABLE 5.3
Some Monomers Used in Graft Copolymerization Used for the Synthesis of Modified Chitosan Flocculants

Type of the Monomer	Common Modifying Agents Used
Non-ionic	Acrylamide
	N-vinyl formamide
	N, N-dimethylacrylamide
Cationic	(2-Methacryloyloxyethyl) trimethyl ammonium chloride
	3-(Acrylamide) propyl trimethylammonium bromide
	N-vinyl-2-pyrrolidone
	diallyl dimethyl ammonium chloride
Anionic	(Meth)acrylic acid
	2-Acrylamidoglycolic acid

flocculant CMC-g-PAM (Yang et al., 2012). Laue and Hunkeler prepared a series of chitosan copolymers that were grafted with DMC ((2-methacryloyloxyethyl) trimethylammoniumchloride) and AM (Acrylamide), using ceric ammonium nitrate as the initiator (Laue & Hunkeler, 2006).

5.8.4 Dextran-Based Advanced Materials for the Treatment of Wastewater

Dextran is a modified version of glucan (a polysaccharide produced from the condensation of glucose). Dextran is a branched poly-α-d-glucoside of microbial origin having glycosidic bonds predominantly $C_1 \rightarrow C_6$. Dextran chains are of varying lengths (3 to 2000 kilodaltons). Dextran derivatives have proven to be very efficient for separation of both organic and inorganic contaminants through the coagulation-flocculation process, which uses charge neutralization, charge patch, and bridging mechanism. Also, dextran has superior physicochemical properties that help in removing a wide array of toxic materials.

Tests have been conducted extensively on a huge number of hydrophilic and amphiphilic dextran-based polyelectrolytes to check their coagulation and flocculation efficiencies in many model wastewater treatments. Similar to chitosan-based materials, dextran derivatives can be either cationic or anionic, the type can be synthesized based on the type of wastewater being treated.

Cationic dextran derivatives are mostly dextran polysaccharides with quaternary ammonium groups produced by chemically modifying dextran (Nichifor et al., 2010, 2014). Anionic dextran derivatives generally used are mostly phosphorylated dextran derivatives and grafted dextran derivatives using acrylamide and sodium acrylate (Li et al., 2016; Suflet et al., 2010).

5.8.5 Plant-Based Grafted Bio Flocculants

Flocculants derived from plants are polysaccharides (natural), and their selectivity for metals and aromatic substances is superior to other flocculants; so they are very

TABLE 5.4
Properties of Some of the Grafted Bio flocculants derived from Plants

Name of Flocculent	Type of Treated Wastewater	Sedimentation Time	Flocculation Efficiency	Reference
polyacrylamide-grafted-Plantagopsyllium (psy-g-PAM)	Tannery and domestic wastewater	1 hour	>95%	93
polyacrylamide-grafted-Plantagoovata	kaolin suspension (0.25%)	15 min	59 to 22NTU	92
polyacrylamide-grafted-Tamarindusindica (Tam-g-PAM)	textile wastewater	10 min	43% using azo dye	41
polyacrylamide-grafted-tamarind kernel polysaccharide (TKP-g-PAM)	municipal sewage and textile industry wastewaters	15 min	125 to 6NTU	98, 99
polyacrylonitrile-grafted-Plantagopsyllium (Psy-g-PAN)	Textile effluent	1 hour	94% (SS)	95

efficient in the purification of wastewater. But conventional plant-based bio flocculants have been restricted in their feasibility as they cannot be stored for long and their flocculating properties are not as high as the synthetic polymers and hence require higher dosages. But recently grafted bio flocculants are claimed to resolve this issue as they have shown excellent flocculating power and biodegradability.

Chemical modification such as grafting of polysaccharide polymers can vastly improve the flocculating properties and overcome drawbacks such as short shelf life. Some of the natural bio flocculants derived from plants are Planta goovata, Planta gopsyllium, and Tamarindus indica. To these, polymers such as polyacrylamide have been grafted onto the backbone to produce grafted bio flocculants. Some of common acrylamide-grafted natural polymers are listed in Table 5.4.

Generally, conventional redox grafting methods are used to produce plant-based grafted copolymers, using radical initiators such as ceric ammonium nitrate and in the presence of inert gas such as nitrogen. As seen before, bio flocculants synthesized using microwave technology have better flocculating characteristics but suffer with higher production costs.

Similar to our expectations, experiments have shown that grafted bio flocculants are more efficient than non-grafted polymers. The higher hydrodynamic volume of grafted polymers is the main reason leading to this higher efficacy (Brostow et al., 2007). An example is the Plantagopsyllium mucilage, which has been grafted with polyacrylamide (Psy-g-PAM) copolymer; it has been found to be a better flocculant than the conventional psyllium, especially for treatment of wastewater for domestic purposes. Similarly, tamarind kernel polysaccharide (TKP-g-PAM) grafted with PAM by microwave-assisted grafting performed much better than TKP as well as polyacrylamide-based commercial flocculant (Rishfloc 226 LV) in flocculation tests (Ghosh et al., 2010).

Ongoing research on plant-based grafted bio flocculants is still being conducted at laboratory scale because of many factors. Monomers such as acrylamide, which is used for the production of bio flocculants, have been found to be toxic to human health (Shipp et al., 2006). Studies also show that acrylonitrile is carcinogenic in nature (Subramanian & Ahmed, 1995). These dangers prevent the usage of grafted biopolymers in the production of goods meant for human consumption. Another drawback is a time-consuming and energy-intensive process of synthesizing compared to those of conventional bio flocculants. Therefore there needs to be a lot of research done to synthesize much more environmentally friendly bio coagulants and to promote more industrial applications of the grafted coagulants.

5.8.6 Other Composite Bio Flocculants Used for the Removal of NOM

NOM is generally produced by algae and microbes as a product of various biological activities. Though not inherently toxic, when mixed with drinking water it poses a threat by changing the organoleptic properties. It could also act as a carrier of other toxic materials. And components such as fumic acids (FA), fulvic acids (FA), and humic acids (HA) form many complexes with heavy metals to increase toxicity. Furthermore, the efficiency of the coagulation step in the conventional water treatment process is reduced. So, removing NOM from drinking water is of utmost importance.

5.9 IMPROVEMENTS TO CONVENTIONAL WASTEWATER TREATMENT PLANTS

5.9.1 General Textile Effluent Treatment Processes

Textile industries generally discharge a lot of wastewater, and so there is a need to treat the wastewater in order to meet the pollution standards. With increased pollution standards and more consciousness of cost saving, textile manufacturers have integrated processes such as ultrafiltration and nano-filtration with conventional processes such as coagulation/flocculation to improve the efficiency. One such example where a conventional waste water treatment plant has been integrated with treatment processes for purifying textile effluent is given in Table 5.5.

5.10 RESOURCE RECOVERY

New techniques for advanced wastewater treatment using electrocoagulation, ultrafiltration, and UV disinfection seem encouraging for resource recovery (Rasi et al., 2007). Various nutrients can be recovered with the help of composting toilets, urine source separation, or fertigation and can be reused locally to fertilize plants. Resource-recovery technologies can be implemented at both small and large scale. In large-scale plants, it is cost effective to recover energy from the sludge by converting it to biogas and/or electricity, as the economics turn out to be better.

5.10.1 Carbon Redirection Technologies

Conventional wastewater treatment plants are being modified day by day as the importance of extracting resources such as nutrients and energy is increasing rapidly. The main intention for upgrading these treatment plants is to make an energy

TABLE 5.5
Improvements to Conventional Waste Water Treatment Plans

Treatment Processes	First Stage	Second Stage	Remarks
Physical treatment (2007)	Coagulation	Ultrafiltration	Using this process, substantial colloidal particle removal of more than 97% was achieved irrespective of the coagulant dosage and type that was used, but the type of coagulant affects the fouling of membrane (Choo et al., 2007).
Physical treatment (2005)	Coagulation/ flocculation	Nano-filtration	The first step of coagulation/flocculation can be a pretreatment to the second nano-filtration step as the permeate from the first step did not have sufficient quality. After NF, high-quality permeate was obtained (Suksaroj et al., 2005).
Physical/chemical treatment (1994)	Coagulation	Ozonation	This process is more effective than the conventional coagulation process: color removal (90%) and COD removal (20-25%) after the ozonation process (Tzitzi et al., 1994).
Physical/chemical treatment (1997)	Coagulation and electro-chemical oxidation	Ion exchange	The post-treatment water properties were found to be better (in terms of COD, turbidity, and color) but the reaction times of H_2O_2 and current of electrochemical treatment drastically changes the performance of the treatment (Lin & Chen, 1997).
Physical/chemical/ biological treatment (1996)	Coagulation and electro-chemical oxidation	Activated Sludge	This treatment process could achieve 24% cost savings over the conventional methods. But many operating variables can easily influence the performance of the treatment (Lin & Peng, 1996).

self-sufficient plant. Out of many efforts to improve, one of the novel technologies is the use of high-rate activated sludge to redirect organic compounds to produce energy (Sancho et al., 2019).

There are different carbon redirection technologies and we can classify these strategies based on their operating principle:

1. **Physical processes:**
    a. dynamic sand filtration
    b. dissolved air flotation
    c. membrane filtration
2. **Chemical processes:**
    a. chemically enhanced primary treatment
3. **Bio-process:**
    a. adsorption or bio-oxidation
    b. contact stabilization

More intricate carbon-redirection models have been discussed elsewhere (Sancho et al., 2019).

5.11 SUMMARY

Strict regulations have been set in various countries to restrict the use of metallic salts as well as polymeric coagulants due to the health hazards they pose. The rising demand for environmentally sustainable technology has driven the continual search for eco-friendly materials for use as coagulants and flocculants in the wastewater treatment industry. Various biopolymers have successfully demonstrated performance as eco-friendly coagulants in terms of high removal efficiencies and excellent selectivity for metals and toxic compounds. Biopolymers have the added advantage of being non-toxic, easily available from renewable resources, and biodegradable. A more recent area of research interest is the modification of biopolymers by "grafting" appropriate monomers to enhance or alter the properties of the biopolymers to overcome some of the disadvantages such as short shelf life, high dosage requirement, and variation in coagulating efficiency. Grafted biopolymers have also shown promising results at lab scale, but they are still at an early stage and there are a lot of research gaps as well as challenges in scaling up to industrial application. There is a need for more extensive research in this area to promote the use of biopolymers for safe and environment friendly ways of wastewater treatment.

ACKNOWLEDGEMENTS

The corresponding author would like to thank Shriram Institute for Industrial Research (SRI), New Delhi, India, for providing infrastructure facilities. The manuscript is SRI communication # SRI-MS#20210318-01.

REFERENCES

Abiola, O. N. (2019). Polymers for coagulation and flocculation in water treatment. 77–92. https://doi.org/10.1007/978-3-030-00743-0_4

Agarwal, M., Srinivasan, R., & Mishra, A. (2002). Synthesis of plantagopsyllium mucilage grafted polyacrylamide and its flocculation efficiency in tannery and domestic wastewater. *Journal of Polymer Research*, *9*(1), 69–73. https://doi.org/10.1023/A:1020658802755

Agunbiade, M. O., Pohl, C. H., & Ashafa, A. O. T. (2016). A review of the application of bioflocculants in wastewater treatment. *Polish Journal of Environmental Studies*, *25*(4), 1381–1389. https://doi.org/10.15244/pjoes/61063

Akers, R. J. (1972). Factors affecting the rate of the polyelectrolyte flocculation reaction. *Filtration and Separation*, 423–424.

Amagloh, F. K., & Benang, A. (2009). Effectiveness of *Moringaoleifera* seed as coagulant for water purification. *African Journal of Agricultural Research*, *4*(2), 119–123.

Aminu, I., Garba, M. D., & Abba, Z. Y. (2014). Biosorption potentials of Moringaoleifera seed in textile effluent treatment. *International Journal of Scientific & Engineering Research*, *5*(8), 1286–1292.

Bae, Y. H., Kim, H. J., Lee, E. J., Sung, N. C., Lee, S. -S., & Kim, Y. -H. (2007). Potable water treatment by polyacrylamide base flocculants, coupled with an inorganic coagulant. *Environment. Engineering Research*, *12*(1), 21–29.

Bolto, B., & Gregory, J. (2007). Organic polyelectrolytes in water treatment. *Water Research*, *41*(11), 2301–2324. https://doi.org/10.1016/j.watres.2007.03.012

Bolto, Brian A. (1995). Soluble polymers in water purification. *Progress in Polymer Science*, *20*(6), 987–1041. Pergamon. https://doi.org/10.1016/0079-6700(95)00010-D

Bratby, J. (2016). Treatment with polymers. In *Coagulation and Flocculation in Water and Wastewater Treatment*, Google Books, (3rd ed., pp. 247–286).

Bratskaya, S. Y., Pestov, A. V., Yatluk, Y. G., & Avramenko, V. A. (2009). Heavy metals removal by flocculation/precipitation using N-(2-carboxyethyl)chitosans. *Colloids and Surfaces A: Physicochemical and Engineering Aspects, 339*(1–3), 140–144. https://doi.org/10.1016/j.colsurfa.2009.02.013

Brostow, W., Pal, S., & Singh, R. P. (2007). A model of flocculation. *Materials Letters, 61*(22), 4381–4384. https://doi.org/10.1016/j.matlet.2007.02.007

Butler, G. B. (1966). Water soluble quaternary ammonium polymers (Patent No. 3,288,770).

Chang, Q., Hao, X., & Duan, L. (2008). Synthesis of cross-linked starch-graft-polyacrylamide-co-sodium xanthate and its performances in wastewater treatment. *Journal of Hazardous Materials, 159*(2–3), 548–553. https://doi.org/10.1016/j.jhazmat.2008.02.053

Chang, Q., Zhang, M., & Wang, J. (2009). Removal of Cu2+ and turbidity from wastewater by mercaptoacetyl chitosan. *Journal of Hazardous Materials, 169*(1–3), 621–625. https://doi.org/10.1016/j.jhazmat.2009.03.144

Cheng, W. P., Chi, F. H., Li, C. C., & Yu, R. F. (2008). A study on the removal of organic substances from low-turbidity and low-alkalinity water with metal-polysilicate coagulants. *Colloids and Surfaces A: Physicochemical and Engineering Aspects, 312*(2–3), 238–244. https://doi.org/10.1016/j.colsurfa.2007.06.060

Choo, K. H., Choi, S. J., & Hwang, E. D. (2007). Effect of coagulant types on textile wastewater reclamation in a combined coagulation/ultrafiltration system. *Desalination, 202*(1–3), 262–270. https://doi.org/10.1016/j.desal.2005.12.063

Crini, G., & Badot, P. M. (2008). Application of chitosan, a natural aminopolysaccharide, for dye removal from aqueous solutions by adsorption processes using batch studies: A review of recent literature. *Progress in Polymer Science (Oxford), 33*(4), 399–447. Pergamon. https://doi.org/10.1016/j.progpolymsci.2007.11.001

Dao, V. H., Cameron, N. R., & Saito, K. (2016). Synthesis, properties and performance of organic polymers employed in flocculation applications. *Polymer Chemistry, 7*(1), 11–25. Royal Society of Chemistry. https://doi.org/10.1039/c5py01572c

Gao, B., Yue, Q., & Miao, J. (2003). Evaluation of polyaluminium ferric chloride (PAFC) as a composite coagulant for water and wastewater treatment. *Water Science and Technology, 47*(1), 127–132. https://doi.org/10.2166/wst.2003.0033

Gao, B., Yue, Q., & Wang, B. (2002). The chemical species distribution and transformation of polyaluminum silicate chloride coagulant. *Chemosphere, 46*(6), 809–813. https://doi.org/10.1016/S0045-6535(01)00180-1

Ghimici, L., & Nichifor, M. (2010). Novel biodegradable flocculating agents based on cationic amphiphilic polysaccharides. *Bioresource Technology, 101*(22), 8549–8554. https://doi.org/10.1016/j.biortech.2010.06.049

Ghosh, S., Sen, G., Jha, U., & Pal, S. (2010). Novel biodegradable polymeric flocculant based on polyacrylamide-grafted tamarind kernel polysaccharide. *Bioresource Technology, 101*(24), 9638–9644. https://doi.org/10.1016/j.biortech.2010.07.058

Gregory, J. (1993). The role of colloid interactions in solid-liquid separation. *Water Science and Technology, 27*(10), 1–17. https://doi.org/10.2166/wst.1993.0195

Gregory, J., & Lee, S. (1990). The effect of charge density and molecular mass of cationic polymers on flocculation kinetics in aqueous solution. *Journal of Water SRT–Aqua, 39*(4), 265–274.

Gregory, John. (1973). Rates of flocculation of latex particles by cationic polymers. *Journal of Colloid and Interface Science, 42*(2), 448–456. https://doi.org/10.1016/0021-9797(73)90311-1

Gupta, B., & Ako, J. (2005). Application of guar gum as a flocculant aid in food processing and potable water treatment. *European Food Research & Technology, 221*(6), 746–751.

Hebeish, A., Higazy, A., El-Shafei, A., & Sharaf, S. (2010). Synthesis of carboxymethyl cellulose (CMC) and starch-based hybrids and their applications in flocculation and sizing. *Carbohydrate Polymers, 79*(1), 60–69. https://doi.org/10.1016/j.carbpol.2009.07.022

Hemapriya, G., District, P., Nadu, T., District, P., & Nadu, T. (2015). Textile effluent treatment using Moringaoleifera. *International Journal of Innovative Research & Development*, *4*(4), 385–390.

Hendrawati, Yuliastri, I. R., Nurhasni, Rohaeti, E., Effendi, H., & Darusman, L. K. (2016). The use of Moringaoleifera seed powder as coagulant to improve the quality of wastewater and ground water. *IOP Conference Series: Earth and Environmental Science*, *31*(1). https://doi.org/10.1088/1755-1315/31/1/012033

IUPAC. (1997a). Compendium of chemical terminology, (the "Gold Book"). In *Compendium of Chemical Terminology* (2nd ed.; Online). https://doi.org/doi:10.1351/goldbook.P04735

IUPAC. (1997b). Compendium of chemical terminology (the "Gold Book"). In *Compendium of Chemical Terminology* (2nd ed.; Online). https://doi.org/doi:10.1351/goldbook.M03667

Jayalakshmi, G., Saritha, V., & Dwarapureddi, B. K. (2017). A review on native plant based coagulants for water purification. *International Journal of Applied Environmental Sciences*, *12*(3), 469–487. http://www.ripublication.com

Jiang, H. L., Kim, Y. K., Arote, R., Nah, J. W., Cho, M. H., Choi, Y. J., Akaike, T., & Cho, C. S. (2007). Chitosan-graft-polyethylenimine as a gene carrier. *Journal of Controlled Release*, *117*(2), 273–280. https://doi.org/10.1016/j.jconrel.2006.10.025

Jiang, J. Q., & Graham, N. J. D. (1996). Enhanced coagulation using Al/Fe(III) coagulants: Effect of Coagulant chemistry on the removal of colour-causing NOM. *Environmental Technology*, *17*(9), 937–950. https://doi.org/10.1080/09593330.1996.9618422

Khiari, R., Dridi-Dhaouadi, S., Aguir, C., & Mhenni, M. F. (2010). Experimental evaluation of eco-friendly flocculants prepared from date palm rachis. *Journal of Environment Science*, *22*(10), 1539–1543.

Lapointe, M., & Barbeau, B. (2016). Characterization of ballasted flocs in water treatment using microscopy. *Water Resource*, *90*, 119–127. https://doi.org/10.1016/j. watres.2015.12.018.

Lapointe, M., & Barbeau, B. (2017). Dual starch–polyacrylamide polymer system for improved flocculation. *Water Resource*, *124*, 202–209. https://doi.org/10.1016/j. watres.2017.07.044.

Lapointe, Mathieu, & Barbeau, B. (2020). Understanding the roles and characterizing the intrinsic properties of synthetic vs. natural polymers to improve clarification through interparticle bridging: A review. *Separation and Purification Technology*, *231*(August 2019), 115893. https://doi.org/10.1016/j.seppur.2019.115893

Laue, C., & Hunkeler, D. (2006). Chitosan-graft-acrylamide polyelectrolytes: Synthesis, flocculation, and modeling. *Journal of Applied Polymer Science*, *102*(1), 885–896. https://doi.org/10.1002/app.24188

Lee, C., Chong, M., Robinson, J., & Binner, E. (2014). A review on development and application of plant-based bioflocculants and grafted bioflocculants. *Industrial and Engineering Chemistry Research*, *53*(48), 18357–18369. https://doi.org/10.1021/ie5034045

Lee, K. E., Morad, N., Poh, B. T., & Teng, T. T. (2011). Comparative study on the effectiveness of hydrophobically modified cationic polyacrylamide groups in the flocculation of kaolin. *Desalination*, *270*(1–3), 206–213. https://doi.org/10.1016/j.desal.2010.11.047

Lee, K. E., Morad, N., Teng, T. T., & Poh, B. T. (2012a). Development, characterization and the application of hybrid materials in coagulation/flocculation of wastewater: A review. *Chemical Engineering Journal*, *203*, 370–386. https://doi.org/10.1016/j.cej.2012.06.109

Lee, K. E., Morad, N., Teng, T. T., & Poh, B. T. (2012b). Kinetics and In situ rheological behavior of acrylamide redox polymerization. *Journal of Dispersion Science and Technology*, *33*(3), 387–395. https://doi.org/10.1080/01932691.2011.567176

Lee, K. E., Poh, B. T., Morad, N., & Teng, T. T. (2008). Synthesis and characterization of hydrophobically modified cationic acrylamide copolymer. *International Journal of Polymer Analysis and Characterization*, *13*(2), 95–107. https://doi.org/10.1080/10236660801905684

Lee, K. E., Poh, B. T., Morad, N., & Teng, T. T. (2009). Synthesis and characterization of hydrophobically modified cationic polyacrylamide with low concentration of cationic monomer. *Journal of Macromolecular Science, Part A: Pure and Applied Chemistry*, *46*(3), 240–249. https://doi.org/10.1080/10601320802637284

Lee, K. E., Teng, T. T., Morad, N., Poh, B. T., & Mahalingam, M. (2011). Flocculation activity of novel ferric chloride-polyacrylamide (FeCl3-PAM) hybrid polymer. *Desalination, 266*(1–3), 108–113. https://doi.org/10.1016/j.desal.2010.08.009

Leeman, M., Islam, M. T., & Haseltine, W. G. (2007). Asymmetrical flow field-flow fractionation coupled with multi-angle light scattering and refractive index detections for characterization of ultra-high molar mass poly(acrylamide) flocculants. *Journal of Chromatography A, 1172*(2), 194–203. https://doi.org/10.1016/j.chroma. 2007.10.006

Li, R. H., Zhang, H.bin, Hu, X. Q., Gan, W. W., & Li, Q. P. (2016). An efficiently sustainable dextran-based flocculant: Synthesis, characterization and flocculation. *Chemosphere, 159*, 342–350. https://doi.org/10.1016/j.chemosphere.2016.06.010

Lin, S. H., & Chen, M. L. (1997). Treatment of textile wastewater bychemical methods for reuse. *Water Research, 31*(4), 868–876. https://doi.org/10.1016/S0043-1354(96)00318-1

Lin, S. H., & Peng, C. F. (1996). Continuous treatment of textile wastewater by combined coagulation, electrochemical oxidation and activated sludge. *Water Research, 30*(3), 587–592. https://doi.org/10.1016/0043-1354(95)00210-3

Michele, M. M., John, J. T., & Paul, H. (2005). Starches for use in papermaking (Patent No. US6843888B2).

Mishra, A., Bajpai, M., Pal, S., Agrawal, M., & Pandey, S. (2006). Tamarindusindica mucilage and its acrylamide-grafted copolymer as flocculants for removal of dyes. *Colloid and Polymer Science, 285*(2), 161–168. https://doi.org/10.1007/s00396-006-1539-y

Mishra, A., Yadav, A., Agarwal, M., & Rajani, S. (2004). Polyacrylonitrile-grafted Plantagopsyllium mucilage for the removal of suspended and dissolved solids from tannery effluent. *Colloid and Polymer Science, 282*(3), 300–303. https://doi.org/10.1007/s00396-003-0895-0

Mishra, D. K., Tripathy, J., Srivastava, A., Mishra, M. M., & Behari, K. (2008). Graft copolymer (chitosan-g-N-vinyl formamide): Synthesis and study of its properties like swelling, metal ion uptake and flocculation. *Carbohydrate Polymers, 74*(3), 632–639. https://doi.org/10.1016/j.carbpol.2008.04.015

Mishra, S., Mukul, A., Sen, G., & Jha, U. (2011). Microwave assisted synthesis of polyacrylamide grafted starch (St-g-PAM) and its applicability as flocculant for water treatment. *International Journal of Biological Macromolecules, 48*(1), 106–111. https://doi.org/10.1016/j.ijbiomac.2010.10.004

Moussas, P. A., & Zouboulis, A. I. (2008). A study on the properties and coagulation behaviour of modified inorganic polymeric coagulant-polyferricsilicate sulphate (PFSiS). *Separation and Purification Technology, 63*(2), 475–483. https://doi.org/10.1016/j.seppur.2008.06.009

Moussas, P. A., & Zouboulis, A. I. (2009). A new inorganic-organic composite coagulant, consisting of polyferricsulphate (PFS) and polyacrylamide (PAA). *Water Research, 43*(14), 3511–3524. https://doi.org/10.1016/j.watres.2009.05.015

Mühle, K. (1993). Floc stability in laminar and turbulent flow. In M. Dekker (Ed.), *Coagulation and Flocculation*. New York: Marcel Dekker. (pp. 355–390).

Muzzarelli, R. A. A., & Muzzarelli, C. (2005). Chitosan chemistry: Relevance to the biomedical sciences. In: Thomas Heinze (Ed.), *Advances in Polymer Science* (Vol. 186, pp. 151–209). Springer, Berlin, Heidelberg. https://doi.org/10.1007/b136820

Nanko, M. (2009). Definitions and Categories of Hybrid Materials. *AZojomo, 6*(August), 1–8. https://doi.org/10.2240/azojomo0288

Nichifor, M., Mocanu, G., & Stanciu, M. C. (2014). Micelle-like association of polysaccharides with hydrophobic end groups. *Carbohydrate Polymers, 110*, 209–218. https://doi.org/10.1016/j.carbpol.2014.03.072

Nichifor, M., Stanciu, M. C., & Simionescu, B. C. (2010). New cationic hydrophilic and amphiphilic polysaccharides synthesized by one pot procedure. *Carbohydrate Polymers, 82*(3), 965–975. https://doi.org/10.1016/j.carbpol.2010.06.027

Nozaic, D. J., Freese, S. D., & Thompson, P. (2001). Long term experience in the use of polymeric coagulants at Umgeni water. *Water Science and Technology: Water Supply, 1*(1), 43–50. https://doi.org/10.2166/ws.2001.0006

O'Gorman, J. V., & Kitchener, J. A. (1974). The flocculation and de-watering of kimberlite clay slimes. *International Journal of Mineral Processing*, *1*, 33–49.

Othmani, B., Rasteiro, M. G., & Khadhraoui, M. (2020). Toward green technology: A review on some efficient model plant-based coagulants/flocculants for freshwater and wastewater remediation. *Clean Technologies and Environmental Policy*, *22*(5), 1025–1040. https://doi.org/10.1007/s10098-020-01858-3

Packham, R. F. (1965). Some studies of the coagulation of dispersed clays with hydrolyzing salts. *Journal of Colloid Science*, *20*(1), 81–92. https://doi.org/10.1016/0095-8522(65)90094-2

Pandey, J. (2020). Biopolymers and their application in wastewater treatment. 245–266. https://doi.org/10.1007/978-981-15-1390-9_11

Pelssers, E. G. M., Cohen Stuart, M. A., & Fleer, G. J. (1990). Kinetics of bridging flocculation. *Journal of the Chemical Society Faraday Transactions*, *86*(9), 1355–1361.

Qian, J. W., Xiang, X. J., Yang, W. Y., Wang, M., & Zheng, B. Q. (2004). Flocculation performance of different polyacrylamide and the relation between optimal dose and critical concentration. *European Polymer Journal*, *40*(8), 1699–1704. https://doi.org/10.1016/j.eurpolymj.2004.03.009

Rasi, S., Veijanen, A., & Rintala, J. (2007). Trace compounds of biogas from different biogas production plants. *Energy*, *32*(8), 1375–1380. https://doi.org/10.1016/j.energy.2006.10.018

Rath, S. K., & Singh, R. P. (1998). Crafted amylopectin: Applications in flocculation. *Colloids and Surfaces A: Physicochemical and Engineering Aspects*, *139*(2), 129–135. https://doi.org/10.1016/S0927-7757(98)00250-7

Ravikumar, K., & Sheeja, A. K. (2013). Heavy metal removal from water using Moringaoleifera seed coagulant and double filtration. *International Journal of Scientific & Engineering Research*, *4*(5), 10–13. http://www.ijser.org

Ravishankar, S. A., & Pradip, N. K. K. (1995). Selective flocculation of iron oxide from its synthetic mixtures with clays: A comparison of polyacrylic acid and starch polymers. *International Journal of Mineral Processing*, *43*(3–4), 235–247. https://doi.org/10.1016/0301-7516(95)00011-2

Reddy, D. H. K., & Lee, S. M. (2013). Application of magnetic chitosan composites for the removal of toxic metal and dyes from aqueous solutions. *Advances in Colloid and Interface Science*, 201–202, 68–93. https://doi.org/10.1016/j.cis.2013.10.002

Renault, F., Sancey, B., Badot, P. M., & Crini, G. (2009). Chitosan for coagulation/flocculation processes: An eco-friendly approach. *European Polymer Journal*, *45*(5), 1337–1348. https://doi.org/10.1016/j.eurpolymj.2008.12.027

Rinaudo, M. (2006). Chitin and chitosan: Properties and applications. In *Progress in Polymer Science (Oxford)*, *31*(7), 603–632. Pergamon. https://doi.org/10.1016/j.progpolymsci.2006.06.001

Ronke, R., Saidat, O., & Abdulwahab, G. (2016). Coagulation-flocculation treatment of industrial wastewater using Tamarind seed powder. *International Journal of ChemTech Research*, *9*(5), 771–780.

Rout, D., Verma, R., & Agarwal, S. K. (1999). Polyelectrolyte treatment: An approach for water quality improvement. *Water Science and Technology*, *40*(2), 137–141.

Sancho, I., Lopez-Palau, S., Arespacochaga, N., & Cortina, J. L. (2019). New concepts on carbon redirection in wastewater treatment plants: A review. *Science of the Total Environment*, *647*, 1373–1384. https://doi.org/10.1016/j.scitotenv.2018.08.070

Saranya, P., Ramesh, S. T., & Gandhimathi, R. (2014). Effectiveness of natural coagulants from non-plant-based sources for water and wastewater treatment: A review. *Desalination and Water Treatment*, *52*(31–33), 6030–6039. https://doi.org/10.1080/19443994.2013.812993

Sen, G., Kumar, R., Ghosh, S., & Pal, S. (2009). A novel polymeric flocculant based on polyacrylamide grafted carboxymethylstarch. *Carbohydrate Polymers*, *77*(4), 822–831. https://doi.org/10.1016/j.carbpol.2009.03.007

Shan, T. C., Al Matar, M. & Makky, E. A (2017). The use of Moringaoleifera seed as a natural coagulant for wastewater treatment and heavy metals removal. *Applied Water Science*, *7*(3), 1369–1376.

Shipp, A., Lawrence, G., Gentry, R., McDonald, T., Bartow, H., Bounds, J., Macdonald, N., Clewell, H., Allen, B., & Van Landingham, C. (2006). Acrylamide: Review of toxicity data and dose-response analyses for cancer and noncancer effects. In *Critical Reviews in Toxicology* (Vol. 36, pp. 481–608). Taylor & Francis. https://doi.org/10.1080/10408440600851377

Sillanpää, M., Ncibi, M. C., Matilainen, A., & Vepsäläinen, M. (2018). Removal of natural organic matter in drinking water treatment by coagulation: A comprehensive review. *Chemosphere*, *190*, 54–71. https://doi.org/10.1016/j.chemosphere.2017.09.113

Singh, V., Kumar, P., & Sanghi, R. (2012). Use of microwave irradiation in the grafting modification of the polysaccharides: A review. *Progress in Polymer Science (Oxford)*, *37*(2), pp. 340–364. Pergamon. https://doi.org/10.1016/j.progpolymsci.2011.07.005

Snodgrass, W. J., Clark, M. M., & O'Melia, C. R. (1984). Particle formation and growth in dilute aluminum (III) solutions. Characterization of particle size distributions at pH 5.5. *Water Research*, *18*(4), 479–488. https://doi.org/10.1016/0043-1354(84)90157-X

Subramanian, U., & Ahmed, A. E. (1995). Intestinal toxicity of acrylonitrile: In vitro metabolism by intestinal cytochrome P4502E1. *Toxicology and Applied Pharmacology*, *135*(1), 1–8. https://doi.org/10.1006/taap.1995.1202

Suflet, D. M., Chitanu, G. C., & Desbrires, J. (2010). Phosphorylated polysaccharides. 2. Synthesis and properties of phosphorylated dextran. *Carbohydrate Polymers*, *82*(4), 1271–1277. https://doi.org/10.1016/j.carbpol.2010.07.007

Suksaroj, C., Héran, M., Allègre, C., & Persin, F. (2005). Treatment of textile plant effluent by nanofiltration and/or reverse osmosis for water reuse. *Desalination*, *178* (1-3 SPEC. ISS.), 333–341. https://doi.org/10.1016/j.desal.2004.11.043

Sun, C., Yue, Q., Gao, B., Cao, B., Mu, R., & Zhang, Z. (2012). Synthesis and floc properties of polymeric ferric aluminum chloride-polydimethyldiallylammonium chloride coagulant in coagulating humic acid-kaolin synthetic water. *Chemical Engineering Journal*, *185–186*, 29–34. https://doi.org/10.1016/j.cej.2011.04.056

Teh, C. Y., Budiman, P. M., Shak, K. P. Y., & Wu, T. Y. (2016). Recent advancement of coagulation-flocculation and its application in wastewater treatment. *Industrial and Engineering Chemistry Research*, *55*(16), 4363–4389. https://doi.org/10.1021/acs.iecr.5b04703

Tripathy, J., Mishra, D. K., Yadav, M., & Behari, K. (2010). Synthesis, characterization and applications of graft copolymer (Chitosan-g-N,N-dimethylacrylamide). *Carbohydrate Polymers*, *79*(1), 40–46. https://doi.org/10.1016/j.carbpol.2009.07.026

Tripathy, T., Bhagat, R. P., & Singh, R. P. (2001). Flocculation performance of grafted sodium alginate and other polymeric flocculants in relation to iron ore slime suspension. *European Polymer Journal*, *37*(1), 125–130. https://doi.org/10.1016/S0014-3057(00)00089-6

Tripathy, T, & De, B. R. (2006). Flocculation: A new way to treat the waste water. *Journal of Physical Science*, *10*, 93–127.

Tzitzi, M., Vayenas, D. V., & Lyberatos, G. (1994). Pretreatment of textile industry wastewaters with ozone. *Water Science and Technology*, *29*(9), 151–160. https://doi.org/10.2166/wst.1994.0466

Tzoupanos, N. D., & Zouboulis, A. I. (2011). Preparation, characterization and application of novel composite coagulants for surface water treatment. *Water Research*, *45*(12), 3614–3626. https://doi.org/10.1016/j.watres.2011.04.009

Ugwu, S. N., Umuokoro, A. F., Echiegu, E. A., Ugwuishiwu, B. O., & Enweremadu, C. C. (2017). Comparative study of the use of natural and artificial coagulants for the treatment of sullage (domestic wastewater). *Cogent Engineering*, 4(1). https://doi.org/10.1080/23311916.2017.1365676

Vijayaraghavan, G., Sivakumar, T., & Kumar, A. (2011). Application of plant based coagulants for waste water treatment. *International Journal of Advanced Engineering Research and Studies*, *1*(1), 88–92.

Wang, J. P., Chen, Y. Z., Zhang, S. J., & Yu, H. Q. (2008). A chitosan-based flocculant prepared with gamma-irradiation-induced grafting. *Bioresource Technology*, *99*(9), 3397–3402. https://doi.org/10.1016/j.biortech.2007.08.014

Xing, G. -X., Zhang, S. -F., Ju, B. -Z., & Yang, J. Z. (2005). Recent advances in modified starch as flocculant. *Proc. 3rd Int. Conf. Funct. Mol.*

Yang, W. Y., Qian, J. W., & Shen, Z. Q. (2004). A novel flocculant of $Al(OH)_3$-polyacrylamide ionic hybrid. *Journal of Colloid and Interface Science*, *273*(2), 400–405. https://doi.org/10.1016/j.jcis.2004.02.002

Yang, Z., Shang, Y., Lu, Y., Chen, Y., Huang, X., Chen, A., Jiang, Y., Gu, W., Qian, X., Yang, H., & Cheng, R. (2011). Flocculation properties of biodegradable amphoteric chitosan-based flocculants. *Chemical Engineering Journal*, *172*(1), 287–295. https://doi.org/10.1016/j.cej.2011.05.106

Yang, Z., Yuan, B., Huang, X., Zhou, J., Cai, J., Yang, H., Li, A., & Cheng, R. (2012). Evaluation of the flocculation performance of carboxymethyl chitosan-graft-polyacrylamide, a novel amphoteric chemically bonded composite flocculant. *Water Research*, *46*(1), 107–114. https://doi.org/10.1016/j.watres.2011.10.024

Yin, C. Y. (2010). Emerging usage of plant-based coagulants for water and wastewater treatment. *Process Biochemistry*, *45*(9), 1437–1444.

Yokoi, H., Obita, T., Hirose, J., Hayashi, S., & Takasaki, Y. (2002). Flocculation properties of pectin in various suspensions. *Bioresource Technology*, *84*(3), 287–290. https://doi.org/10.1016/S0960-8524(02)00023-8

Yoon, S. Y., & Deng, Y. (2004). Flocculation and reflocculation of clay suspension by different polymer systems under turbulent conditions. *Journal of Colloid and Interface Science*, *278*(1), 139–145. https://doi.org/10.1016/j.jcis.2004.05.011

Yuan, B., Shang, Y., Lu, Y., Qin, Z., Jiang, Y., Chen, A., Qian, X., Wang, G., Yang, H., & Cheng, R. (2010). The flocculating properties of chitosan- *graft* -polyacrylamide flocculants (I)-effect of the grafting ratio. *Journal of Applied Polymer Science*, *117*(4), 1876–1882. https://doi.org/10.1002/app.32047

Zhang, K., Zhou, Q., & Wu, W. (2004). Wastewater treatment efficiency of combined aluminum-starch flocculant. *Journal of Applied Ecology*, *15*(8), 1443–1446.

Zhang, P., Hahn, H. H., Hoffmann, E., & Zeng, G. (2004). Influence of some additives to aluminium species distribution in aluminium coagulants. *Chemosphere*, *57*(10), 1489–1494. https://doi.org/10.1016/j.chemosphere.2004.08.066

Zhao, D., Xu, J., Wang, L., Du, J., Dong, K., Wang, C., & Liu, X. (2012). Study of two chitosan derivatives phosphorylated at hydroxyl or amino groups for application as flocculants. *Journal of Applied Polymer Science*, *125*(S2), E299–E305. https://doi.org/10.1002/app.36834

Zouboulis, A. I., & Moussas, P. A. (2008). Polyferric silicate sulphate (PFSiS): Preparation, characterization and coagulation behavior. *Desalination*, *224*(1–3), 307–316. https://doi.org/10.1016/j.desal.2007.06.012

6 Ultrasonic Wastewater Treatment

Gul Afreen and Sreedevi Upadhyayula
Department of Chemical Engineering, Indian Institute of Technology Delhi, New Delhi, India

CONTENTS

6.1 INTRODUCTION

Rapid urbanization and industrialization demands development of highly efficient wastewater treatment techniques. The existing primary techniques include physical, chemical, and biological treatment of wastewater [1]. Physical methods are unpredictable and ineffective, while chemical degradation often results in formation of

DOI: 10.1201/9781003138303-6

secondary pollutants. Biodegradation methods require stringent protocols for effective degradation. Hence, the ultrasound treatment technique has emerged as an effective and clean technology for the degradation of organic pollutants, which has the benefits of high degradation efficiency, low maintenance, short reaction time, etc. [2]. The ultrasound method has shown positive response in the treatment of reluctant, highly concentrated organic wastewater polluted by microbes, sludge, and industrial wastewater obtained after removing equipment scales [3]. The ultrasonic method has the advantage of mild reaction conditions and rapid treatment speed. This technology has great potential to be used as a hybrid method with other water treatment techniques [4]. This chapter describes the principle of ultrasound technique, important reaction parameters, and the degradation methods using ultrasound alone or in combination with other advanced techniques. Further, problems of organic pollutants removal by ultrasound method are also discussed and possible remedies are proposed.

6.2 ULTRASOUND DEGRADATION PRINCIPLE

Ultrasound is a pressurized sound wave in a frequency range of 20 kHz, as shown in Fig. 6.1. Ultrasound cavitation is the driving force for the chemical reactions involving pollutant degradation to occur during ultrasound treatment. When polluted water is exposed to ultrasonic waves of a particular frequency and intensity, several tiny bubbles are formed; these oscillate, grow, and collapse, resulting in many chemical and physical changes in the polluted water, which are exploited for pollutant removal [5]. This formation of bubbles in a liquid, typically, by the movement ultrasonic waves through it, is called cavitation [6].

At high ultrasonic velocities, there is a heavy drop in pressure, leading to the formation of cavitation bubbles. At this juncture, the average liquid molecular distance exceeds its critical value, which destroys the cohesive forces between molecules. The collapse of the cavitation bubbles creates instantaneous hot spots within and around them, releasing huge amounts of energy and raising the temperatures even up to 727°C and pressures to 1000 atm in a short span of time (Fig. 6.2). The degradation of organic pollutants is accelerated by a significant rate of temperature change of ~109°C/s [7, 8]. Cavitation bubble formation creates three reaction zones in the aqueous solution: within the cavitation bubble, the bubble/water interface, and the bulk solution. Each zone favors different kinds of pollutants decomposition. The cavitation bubbles and the interfaces favor the degradation of hydrophobic, nonpolar, or volatile organic compounds, whereas the bulk solution favors hydrophilic or non-volatile organic compounds.

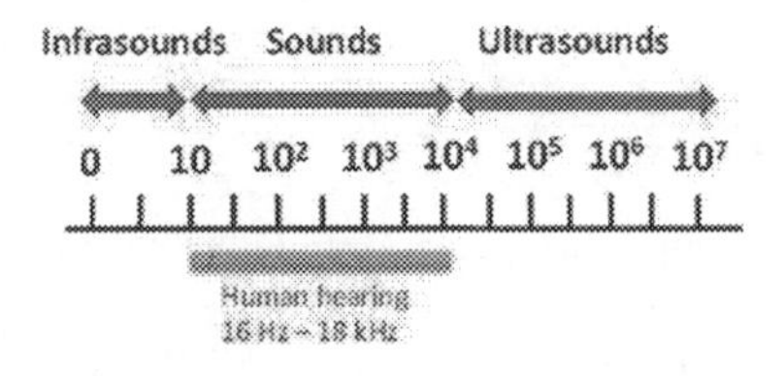

FIGURE 6.1 Frequency ranges.

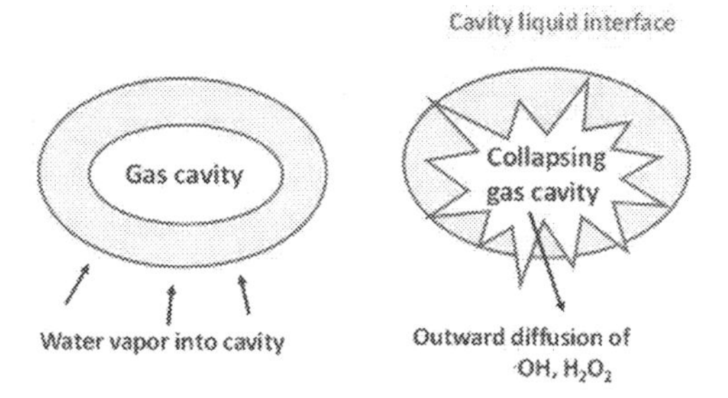

FIGURE 6.2 Schematic diagram of reaction system during ultrasonication.

Three reaction mechanisms following the ultrasonic cavitation method are listed in following sections.

6.2.1 High-Temperature Pyrolysis Reaction

Pyrolysis of volatile organic solutes inside the hot spot cavities leads to their sonochemical degradation into small eco-friendly harmless molecules [9]. For example, C_2HCl_3 can undergo following degradation reaction to form C_2Cl_2:

$$C_2HCl_3 \rightarrow C_2HCl_2^{\cdot} + Cl^{\cdot}$$

$$C_2HCl_3 + Cl^{\cdot} \rightarrow C_2Cl_3^{\cdot} + HCl$$

$$C_2HCl_3 \rightarrow C_2Cl_2 + HCl$$

6.2.2 Free Radical Oxidation

At the hot spot centers of cavitation bubbles, water molecules undergo oxidation to produce reactive species such as hydrogen peroxide and superoxide radicals:

$$H_2O+))) \rightarrow \cdot H + \cdot OH$$

$$2 \cdot OH \rightarrow H_2O_2$$

where))) symbolizes ultrasonic waves.

The collapse of cavitation bubbles generates shock waves, which release the reactive species into the solution. These reactive species are strong oxidants that efficiently degrade the organic pollutants in solution [10].

6.2.3 Supercritical Water Oxidation

The interface between the solution and the bubble is in supercritical state, i.e., above the critical point of water (374°C temperature and 22.1 MPa pressure). This region has gas-like properties such as lower viscosity and higher diffusion coefficients [11]. Also, it has the density of a liquid and high solubility toward organic compounds where they react with oxidants as follows:

$$RH + O_2 \rightarrow \cdot R + HO_2^{\cdot}$$

$$RH + HO_2^{\cdot} \rightarrow \cdot R + H_2O_2$$

$$\cdot R + O_2 \rightarrow \cdot RO_2$$

$$\cdot RO_2 + RH \rightarrow RHO_2 + \cdot R$$

This peroxide is unstable and breaks into smaller organic compounds such as aldehyde and acid which ultimately get oxidized to carbon dioxide and water.

6.3 FACTORS AFFECTING ULTRASONIC TREATMENT

6.3.1 Ultrasonic Frequency

The range of ultrasonic frequency is 20 kHz–10,000 kHz and is divided into low, high, and very high frequency ranges (Table 6.1).

In a free radical-based mechanism, maximum degradation of organic pollutants is observed at optimum frequency value. Increase of frequency does not necessarily increase the degradation efficiency. In case of a pyrolysis-based mechanism, increase of frequency beyond the cavitation threshold increases the degradation efficiency [12].

6.3.2 Ultrasonic Intensity

The degradation percent increases with the ultrasonic intensity in an aqueous solution. In a constant acoustic emission area, ultrasonic intensity is proportional to its amplitude. The hot spots, with high temperatures and pressures, generated by the bubble collapse are dependent on the amplitude of the waves [13]. The higher the amplitude, the more violent is the collapse, and hence higher temperatures and pressures are attained. This leads to improved free radical formation, rate of mass transfer, and dissolution of organic compounds.

TABLE 6.1
Frequency Ranges of Ultrasonication Process

Name	Ultrasound Range (kHz)
Very high	5000–10,000
High	200–1000
Low	20–100

Some studies show that above the optimal sound intensity, the degradation rate decreases with increase in intensity [14]. The reason for this behavior may be attributed to the higher number of bubbles formed at high ultrasound intensity, leading to wave scatter and energy loss or reduced availability of the sound field energy, due to the formation of large cavitation bubbles, leading to the inability to form an acoustic shield at high sound intensity.

6.3.3 Ultrasonic Power

The power supplied to the transducer affects the ultrasonic degradation. Increased power leads to high cavitation energy and high concentration of bubbles. Hence the volatile compounds that react inside the bubbles degrade more frequently. The rate constants of the degradation reaction increase simultaneously. Ultrasonic efficiencies are best observed in the range of 20–40 W.

6.3.4 Dissolved Gases

The introduction of dissolved gases in the solution enhances the formation of bubble nuclei during the ultrasonic degradation due to the following property effect [15]:

a. The specific heat ratio of monoatomic gases to polyatomic gases and air is higher; hence monoatomic gases in the aqueous solution improve the cavitation and pollutant degradation.
b. Increased thermal conductivity of the dissolved gas leads to higher transmission of the heat generated by bubble collapse to the surroundings, resulting in negative cavitation effect and organics degradation.
c. The higher the solubility of the gases, the more is their diffusion into the bubble, weakening the cavitation and the degradation. The solubility order of few of the frequently used gases is carbon dioxide > argon > oxygen > hydrogen > helium.
d. Contrary to point (c), the degradation of some organic pollutants increases with highly soluble oxygen, owing to the free radicals from oxidation, contributing to the ultrasonic degradation.

6.3.5 pH

Organic compounds exist as molecules in the solution at smaller pH value, which makes them convenient to evaporate and cross the gas–liquid interface of the cavitation and pyrolyze inside it. Further, free radicals are oxidized at the interface and in the bulk at lower pH, which enhances the degradation efficiency. On the other hand, at higher pH, organic compounds ionized in the solution are unable to evaporate and oxidize only at the interface and in the bulk solution with free radicals. This reduces the degradation performance. Therefore pH is an important parameter to adjust during the ultrasonic degradation of organic pollutants into neutral molecules, facilitating easy entrance into the cavitation bubbles [1, 11].

6.3.6 Temperature

Ultrasonic pollutant degradation is primarily driven by the cavitation effect even though the temperature has a large influence on the degradation reactions due to the accelerated reaction rates; yet, in this case, water boiling at high temperatures results in the reduction of pressure in the bubble with reference to the acoustic wave traversing it. Water vapor subsequently fills the cavitation bubble and decreases the hot spot temperature raised by the cavitation, resulting in the decrease of pollutant degradation rate. Hence ultrasonic-based treatments should be performed at room temperature [16].

6.4 ULTRASONIC HYBRID TECHNIQUES

The ultrasound method is often combined with other techniques to enhance degradation performances, called ultrasonic hybrid techniques. Such techniques include catalyst, ozone, biological, UV-TiO_2, and electrochemical methods in combination with ultrasonic techniques [11].

6.4.1 Ultrasound-Catalyst Hybrid Technique

The introduction of a catalyst in the aqueous solution improves the degradation of non-volatile pollutants such as phenols, dyes, benzene derivatives, etc. during ultrasound treatment. Most frequently used catalysts are metal oxides such as TiO_2, SnO_2, MnO_2, etc. Other examples of catalysts include sulfates of Ni and Cu, Fenton reagents, etc. Methyl orange degraded better with ultrasound treatment in the presence of silica doped TiO_2 compared to undoped TiO_2 and without catalysts [17, 18]. Further, the degradation efficiency against chroma was found to improve by 16 times in presence of Fenton reagent during ultrasound treatment.

6.4.2 Ultrasound-Ozone Hybrid Technique

The ultrasound-ozone hybrid method was used to degrade methyl tertiary butyl ether (MTBE) pollutant. The degradation increased twice in the presence ozone at 205 kHz frequency and 200 W/l power density [19]. The degradation of phenol was also increased during ultrasonication in the presence of ozone.

6.4.3 Ultrasonic-Biological Hybrid Technique

The toxicity of substituted phenols in sludge was found to decrease during 2 hour ultrasonic oscillation while keeping the bacteria safe. Further, the BOD/COD value of dye wastewater increased about 50 percent, and the aniline amount decreased significantly in the ultrasound-activated sludge reaction that favored biological treatment [15].

6.4.4 Ultrasonic-Ultraviolet-Titania Hybrid Technique

Aniline-containing wastewater was treated with ultrasonic/TiO_2 and UV/TiO_2 separately and showed degradation percentages of 30 percent and 35 percent, respectively. This degradation was significantly increased to 72 percent when the same wastewater was treated using a combined ultrasonic/UV/TiO_2 due to their synergistic effect [18].

6.4.5 Ultrasonic-Electrochemical Hybrid Technique

The combined ultrasound/electrolytic oxidation for violet dye degradation showed slight improvement in the degradation performance compared to ultrasound treatment. On the other hand, the degradation with combined ultrasonic/micro-electric field was noticeably high (~97 percent degradation in 2 hours). Additionally, the hybrid ultrasonic/electrochemical method showed good results in phenol degradation. A complete aniline degradation occurred in about 0.5 hour during this hybrid mechanism compared to 2 hours of electrochemical degradation [20]. The ultrasonic method acts either mechanically, where the cavitation constantly refreshes the electrodes surface, or by chemical degradation of pollutants in water by cavitation effect.

6.5 Ultrasonic Treatment Applications in Wastewater Treatment

6.5.1 Removal of Organic Dyes

An indigo dye in the water was removed by sonochemical oxidation where BOD/COD was found to increase by approximately two times in activated sludge batch reaction. Violet dye was degraded by ultrasound-electrolysis hybrid technique. The synergistic ultrasound and electrolytic effect improved the decolorization rate up to 99 percent in 1 hour. The ultrasonic removal rate is related to the structure of dyes such as reactive red, reactive black, azo dyes, etc. [20–23].

In case of aryl-azo-naphthol type mono-azo dyes, complex-structured active orange 16 with high molecular weight is more difficult to degrade than the small molecular weight and simple-structured acid orange 7. Ultrasonic and photocatalytic degradations were performed against acid orange 7 to observe the intermediate products by GC-MS analysis. Aliphatic compounds were formed as intermediates during ultrasonic treatment. Naphthalene ring compounds, aromatic compounds, and ring-opening compounds were formed as intermediates during photocatalytic treatment. The intermediate product analysis indicates that ultrasonic treatment is better than photocatalytic treatment [22, 23].

6.5.2 Removal of Pesticides

The degradation of metamorphos and methamidophos pesticides in wastewater by ultrasonication is shown to analyze the effects of reaction parameters such as frequency, intensity, pH, power, etc. on the degradation efficiency. The methamidophos degradation was noticeably affected by power, intensity, dissolved gases, and acidity, whereas it is slightly affected by ultrasound frequency [24–26]. The atrazine was removed by ultrasonication combined with photocatalytic degradation by TiO_2. The photocatalytic degradation by TiO_2 occurred in 4 hours, whereas the same degradation occurred in just 1 hour by using the hybrid method of TiO_2 plus ultrasonic degradation. Here, the rate of degradation was improved in the hybrid technique, but the mechanism remained the same. Rutile titania in combination with ultrasound treatment showed >90 percent degradation of parathion at 10 pH, 50 W power, 30~50 kHz frequency, and 2 h time [27].

6.5.3 Removal of Aromatic Hydrocarbons

Ultrasonic degradation of aromatic hydrocarbons was performed on monocyclic compounds such as benzene and its derivatives and polycyclics such as pyrene, biphenyl, etc. The ultrasonic frequency of 520 kHz was used for 100 minutes to degrade benzene, styrene, and chlorotoluene, and a first-order reaction kinetics was observed to be followed in this degradation reaction [28].

6.5.4 Removal of Alcohol

Ethanol and methanol were decomposed ultrasonically into different products. Ethanol is decomposed into acetic acid and formic acid. Methanol, in presence of argon forms hydrogen, carbon monoxide, formaldehyde, methane, ethane, and ethene when ultrasonicated. In the presence of oxygen, methanol ultrasonically forms carbon dioxide, carbon monoxide, formic acid, formaldehyde, and oxygen. The ratio of methanol to water determines the concentration of degraded products. The degraded products are higher than pure water at a lower methanol concentration (10 percent), whereas chemical reaction is negligible at a higher methanol concentration (80 percent) [29–31].

Table 6.2 lists the reported literature on ultrasonic treatment of various organic pollutants.

6.6 CONCLUSIONS

The degradation of organic pollutants in wastewater by ultrasonication treatment has been performed in recent years. It works on the combined mechanisms of advanced oxidation, pyrolysis, supercritical oxidation, and other treatment techniques. This technique is advantageous in terms of mild reaction conditions and short reaction time. Ultrasound treatment in combination with other techniques has also been reported as efficient wastewater treatment technology. It has the potential of development and application in the water pollution mitigation. However, currently this ultrasonic technology is immature and performed only at laboratory scale. To exploit this technique on a larger scale of water treatment technology, the following points need to be focused upon and improved:

1. Simultaneous degradation of multiple organic compounds and industrial wastewater by ultrasonic treatment and control of degraded organic products is needed to further broaden the ultrasonic scope.
2. Detailed reaction kinetics of the ultrasonic degradation of pollutants needs to be investigated thoroughly.

TABLE 6.2
Reported Literature on Removal of Pollutants by Ultrasonic Treatment [32]

Process	Pollutant	Operating Conditions	Reference
US	DCF/Synthetic water	Co DCF: 0.05 mM Frequency: 585 kHz Power intensity 160 W L^{-1} pH: 7 Situations: air saturation, argon, oxygen and nitrogen Temperature: 4°C Glass cylindrical reactor of 750 mL connected to transducer Working volume: 500 mL Treatment time: 60 min HO• scavenger agents: *Iso*propyl alcohol and terephthalic acid Co H_2O_2: 0.5 and 5 mM	[33]
US	Alachlor/Synthetic water	Co Alachlor: 100 µg L^{-1} Frequency: 575, 861, 1141 kHz Electric power: 45, 60, and 90 W Reactor: Glass cylindrical reactor of 500 mL Temperature: 25°C Treatment time: 90 min pH: 7	[34]
US	Rosaniline (PRA) and ethyl violet (EV)	[PRA] and [EV]: 10 ppm Frequency: 350 kHz Electrical Power: 60 W Treatment time: 30 min Presence of ions: Cl^-, NO^{3-}, SO_4^{2-}, CO_2^{3-}	[35]
US	Acetaminophen (ACP)/ Synthetic water and mineral water	Frequency: 600 kHz Electrical Power: 20–60 W Treatment volume: 300 mL [ACP]: 82.69 µM pH: 3–12 Temperature: 20 ± 1°C Addition: glucose, oxalic acid, propan-2-ol and hexan-1-ol	[36]

(*Continued*)

TABLE 6.2 (Continued)
Reported Literature on Removal of Pollutants by Ultrasonic Treatment [32]

Process	Pollutant	Operating Conditions	Reference
US US/UV	CBZ/Synthetic water	Co CBZ: 0.00625–0.1 mM Sonolytic Reactor: 500 mL Cylindrical glass beaker Frequency: 200 and 400 kHz Power: 20–100 W Temperature: 20°C pH: 2–11 Photolytic reactor: Camera with two low-pressure Hg lamps, 253.7 nm Combined reactor: Assembly of the sonolytic reactor inside the photolytic reactor	[37]
US/O_3	Benzophenone-3 (Bp3)/ Synthetic water	Frequency: 20 kHz Electrical power: 55.9 W Temperature: 25°C Working volume: 200 mL [Bp3]: 3.9 mg L-1 pH: 2, 6.5, and 10 O_3: 0.5 mL min^{-1} N_2 or O_2: 800 mL min^{-1} Presence of nitrate, chloride, and bicarbonate ions [5 mmol L^{-1}]	[38]
US O_3 O_3/US US/UV O_3/UV US/O_3/UV	Azo dyes (AD), Endocrine disrupting compounds (EDC) and pharmaceuticals (PHAC)/Synthetic water	Reactor 1: horn-type sonicator Capacity of 100 mL Frequency 20 kHz Power: 0.46 W mL^{-1} Reactor 2: plate-type sonicator Frequency: 577, 866, 1100 kHz Power intensity: 0.23 w mL^{-1} Reactor 3: Ultrasonic bath Frequency: 200 kHz Power: 0.07 W mL^{-1} Reactor 4: tailor-made hexagonal glass reactor coupled with 3 UV lamps (254 nm) Frequency: 520 kHz Power: 0.19 W mL^{-1}	[39]

3. Optimization of reaction parameters and the optimal design of reactors for enhanced degradation efficiency is necessary.
4. The hybrid ultrasonic process with other techniques is required at the industry level to develop an economic and technically feasible process to solve the actual degradation problems.
5. Analysis of intermediate and final degradation products and their environmental impact must be conducted rigorously, especially for complex organic compounds.

REFERENCES

[1] J. Wang, Z. Wang, C.L.Z. Vieira, J.M. Wolfson, G. Pingtian, S. Huang, Review on the treatment of organic pollutants in water by ultrasonic technology, Ultrason. Sonochem. 55 (2019) 273–278. https://doi.org/10.1016/j.ultsonch.2019.01.017.

[2] M.M. Ibrahim, S.A. El-Molla, S.A. Ismail, Influence of γ and ultrasonic irradiations on the physicochemical properties of CeO_2-Fe_2O_3-Al_2O_3 for textile dyes removal applications, J. Mol. Struct. 1158 (2018) 234–244. https://doi.org/10.1016/j.molstruc.2018.01.034.

[3] A.L. Camargo-Perea, A. Rubio-Clemente, G.A. Peñuela, Use of ultrasound as an advanced oxidation process for the degradation of emerging pollutants in water, Water (Switzerland). 12 (2020) 1–23. https://doi.org/10.3390/W12041068.

[4] A.A. Zewde, L. Zhang, Z. Li, E.A. Odey, A review of the application of sonophotocatalytic process based on advanced oxidation process for degrading organic dye, Rev. Environ. Health. 14 (2019) 1–10. https://doi.org/10.1515/reveh-2019-0024.

[5] M.R. Doosti, R. Kargar, M.H. Sayadi, Water treatment using ultrasonic assistance: A review, Ecology. 2 (2012) 96–110.

[6] C. Petrier, A. Francony, Incidence of wave-frequency on the reaction rates during ultrasonic wastewater treatment, Water Sci. Technol. 35 (1997) 175–180. https://doi.org/10.1016/S0273-1223(97)00023-1.

[7] S.A. Asli, M. Taghizadeh, Sonophotocatalytic degradation of pollutants by ZnO-based catalysts: A review, Chem. Select. 5 (2020) 13720–13731. https://doi.org/10.1002/slct.202003612.

[8] A.H. Mahvi, Application of ultrasonic technology for water and wastewater treatment, Iran. J. Public Health. 38 (2009) 1–17.

[9] N. Bounab, L. Duclaux, L. Reinert, A. Oumedjbeur, C. Boukhalfa, P. Penhoud, F. Muller, Improvement of zero valent iron nanoparticles by ultrasound-assisted synthesis, study of Cr(VI) removal and application for the treatment of metal surface processing wastewater, J. Environ. Chem. Eng. 9 (2021) 104773. https://doi.org/10.1016/j.jece.2020.104773.

[10] U.S. Bhirud, P.R. Gogate, A.M. Wilhelm, A.B. Pandit, Ultrasonic bath with longitudinal vibrations: A novel configuration for efficient wastewater treatment, Ultrason. Sonochem. 11 (2004) 143–147. https://doi.org/10.1016/j.ultsonch.2004.01.010.

[11] S. Anandan, V. Kumar Ponnusamy, M. Ashokkumar, A review on hybrid techniques for the degradation of organic pollutants in aqueous environment, Ultrason. Sonochem. 67 (2020) 105130. https://doi.org/10.1016/j.ultsonch.2020.105130.

[12] Z.H. Diao, F.X. Dong, L. Yan, Z.L. Chen, P.R. Guo, X.J. Xia, W. Chu, A new insight on enhanced Pb(II) removal by sludge biochar catalyst coupling with ultrasound irradiation and its synergism with phenol removal, Chemosphere. 263 (2021) 128287. https://doi.org/10.1016/j.chemosphere.2020.128287.

[13] N. Jaafarzadeh, A. Takdastan, S. Jorfi, F. Ghanbari, M. Ahmadi, G. Barzegar, The performance study on ultrasonic/Fe_3O_4/H_2O_2 for degradation of azo dye and real textile wastewater treatment, J. Mol. Liq. 256 (2018) 462–470. https://doi.org/10.1016/j.molliq.2018.02.047.

[14] L. Yang, J. Xue, L. He, L. Wu, Y. Ma, H. Chen, H. Li, P. Peng, Z. Zhang, Review on ultrasound assisted persulfate degradation of organic contaminants in wastewater: Influences, mechanisms and prospective, Chem. Eng. J. 378 (2019) 122146. https://doi.org/10.1016/j.cej.2019.122146.

[15] A.R. Mohammadi, N. Mehrdadi, G.N. Bidhendi, A. Torabian, Excess sludge reduction using ultrasonic waves in biological wastewater treatment, Desalination. 275 (2011) 67–73. https://doi.org/10.1016/j.desal.2011.02.030.

[16] T. Blume, U. Neis, Improved wastewater disinfection by ultrasonic pre-treatment, Ultrason. Sonochem. 11 (2004) 333–336. https://doi.org/10.1016/S1350-4177(03)00156-1.

[17] R. Zhang, Y. Ma, W. Lan, D.E. Sameen, S. Ahmed, J. Dai, W. Qin, S. Li, Y. Liu, Enhanced photocatalytic degradation of organic dyes by ultrasonic-assisted electrospray TiO_2/graphene oxide on polyacrylonitrile/β-cyclodextrin nanofibrous membranes, Ultrason. Sonochem. 70 (2021) 105343. https://doi.org/10.1016/j.ultsonch.2020.105343.

[18] C.Y. Teh, T.Y. Wu, J.C. Juan, An application of ultrasound technology in synthesis of titania-based photocatalyst for degrading pollutant, Chem. Eng. J. 317 (2017) 586–612. https://doi.org/10.1016/j.cej.2017.01.001.

[19] D.E. Kritikos, N.P. Xekoukoulotakis, E. Psillakis, D. Mantzavinos, Photocatalytic degradation of reactive black 5 in aqueous solutions: effect of operating conditions and coupling with ultrasound irradiation, Water Res. 41 (2007) 2236–2246.

[20] J.C. Yu, L. Zhang, J. Yu, Rapid synthesis of mesoporous TiO_2 with high photocatalytic activity by ultrasound-induced agglomeration, New J. Chem. 26 (2002) 416–420.

[21] S.G. Babu, R. Vinoth, B. Neppolian, D.D. Dionysiou, M. Ashok kumar, Diffused sunlight driven highly synergistic pathway for complete mineralization of organic contaminants using reduced graphene oxide supported photocatalyst, J. Hazard. Mater. 291 (2015) 83–92.

[22] P.S. Kumar, M. Selvakumar, S.G. Babu, S. Karuthapandian, S. Chattopadhyay, CdO nanospheres: Facile synthesis and bandgap modification for the superior photocatalytic activity, Mater. Lett. 151 (2015) 45–48.

[23] S.G. Babu, R. Vinoth, D.P. Kumar, M.V. Shankar, H.L. Chou, K. Vinodgopal, B. Neppolian, Influence of electron storing, transferring and shuttling assets of reduced graphene oxide at the interfacial copper doped TiO_2 p–n heterojunction for increased hydrogen production, Nanoscale 7 (2015) 7849–7857.

[24] R. Vinoth, S.G. Babu, R. Ramachandran, B. Neppolian, Bismuth oxy-iodide incorporated reduced graphene oxidenanocomposite material as an efficient photocatalyst for visible light-assisted degradation of organic pollutants, Appl. Surf. Sci. 418 (2017) 163–170.

[25] J. Zhang, Z. Xionga, X.S. Zhao, Graphene–metal–oxide composites for the degradation of dyes under visible light irradiation, J. Mater. Chem. 21 (2011) 3634–3640.

[26] G.D. Moon, J.B. Joo, I. Lee, Y. Yin, Decoration of size-tunable CuO nanodots on TiO_2 nanocrystals for noble metal-free photocatalytic H_2 production, Nanoscale 6 (2014) 12002–12008.

[27] D.P. Kumar, M.V. Shankar, M.M. Kumari, G. Sadanandam, B. Srinivas, V. Durgakumari, Nano-size effects on CuO/TiO_2 catalysts for highly efficient H_2 production under solar light irradiation, Chem. Commun. 49 (2013) 9443–9445.

[28] R. Vinoth, S.G. Babu, D. Bahnemann, B. Neppolian, Nitrogen doped reduced graphene oxide hybrid metal free catalysts for effective reduction of 4-nitrophenol, Sci. Adv. Mater. 7 (2015) 1443–1449.

[29] S.G. Babu, R. Karvembu, CuO nanoparticles: A simple, effective, ligand free, and reusable heterogeneous catalyst for N-arylation of benzimidazole, Ind. Eng. Chem. Res. 50 (2011) 9594–9600.

[30] Y. Son, M. Lim, J. Khim, M. Ashokkumar, Attenuation of UV Light in large-scale sonophotocatalytic reactors: The effects of ultrasound irradiation and TiO_2 concentration, Ind. Eng. Chem. Res. 51 (2012) 232–239.

[31] A. Zyoud, A. Zubi, M.H.S. Helal, D. Park, G. Campet, H.S. Hilal, Optimizing photomineralization of aqueous methyl orange by nano-ZnO catalyst under simulated natural conditions, J. Environ. Health Sci. Eng. 13 (2015) 46.
[32] A.L. Camargo-Perea, A. Rubio-Clemente, G.A. Peñuela, Use of ultrasound as an advanced oxidation process for the degradation of emerging pollutants in water, Water. 12 (2020) 1068.
[33] E. Nie, M. Yang, D. Wang, X. Yang, X. Luo, Z. Zheng, Degradation of diclofenac by ultrasonic irradiation: Kinetic studies and degradation pathways, Chemosphere. 113 (2014) 165–170.
[34] R. Kidak, S. Dogan, Degradation of trace concentrations of alachlor by medium frequency ultrasound. Chem. Eng. Process Process Intensif. 89 (2015) 19–27.
[35] M.P. Rayaroth, U.K. Aravind, C.T. Aravindakumar, Effect of inorganic ions on the ultrasound initiated degradation and product formation of triphenylmethane dyes. Ultrason. Sonochem. 48 (2018) 482–491.
[36] E. Villaroel, J. Silva-Agredo, C. Petrier, G. Taborda, R.A. Torres-Palma, Ultrasonic degradation of acetaminophen in water: Effect of sonochemical parameters and water matrix. Ultrason. Sonochem. 21 (2014) 1763–1769.
[37] Y. Rao, H. Yang, D. Xue, Y. Guo, F. Qi, J. Ma, Sonolytic and sonophotolytic degradation of carbamazepine: Kinetic and mechanisms. Ultrason. Sonochem. 32 (2016) 371–379.
[38] H. Zúñiga-Benítez, J. Soltan, G.A. Peñuela, Application of ultrasound for degradation of benzophenone-3 in aqueous solutions. Int. J. Environ. Sci. Technol. 13 (2016) 77–86.
[39] N.H. Ince, Ultrasound-assisted advanced oxidation processes for water decontaminaration. Ultrason. Sonochem. 40 (2018) 97–103.

7 Chemistry in Wastewater Treatment

Sonali Sengupta
Department of Chemical Engineering,
Indian Institute of Technology Kharagpur, India

Chandan Kumar Pal
Department of Chemistry, Scottish Church College,
University of Calcutta, Kolkata, India

CONTENTS

DOI: 10.1201/9781003138303-7

7.1 INTRODUCTION

Wastewater consists of used water by human beings, originating from household, agricultural, and industrial use. Other sources of wastewater are ground water infiltration, rain run-off, and animal wastes. Suspended or dissolved materials, which are roughly 1 percent by weight of wastewater, pollute water. These pollutants are mixtures of human and animal excreta, food particles, detergents, medicines, oil and grease, sand, animal and plant bodies, agricultural waste, pesticides, fertilizers, heavy metals, polymers, different salts, grits, etc.

The main aim of wastewater treatment is to transform the contaminants or pollutants present in wastewater into secure end products by feasible processes. The end products should be disposed of through safe and affordable techniques with minimum or no negative impact on the environment. Water treatment processes and disposal of wastes should comply with specific acts and legislation and must protect public health by continuous examination of the chemical and bacteriological composition of treated water. This treated water, after discharge to a receiving stream, is withdrawn for reuse by the downstream population.

7.1.1 General Description

The dissolved oxygen content in water is an important parameter to monitor the quality of water because low oxygen content is deleterious to desirable aquatic species. The oxygen content of water is reduced by the presence of microbiological contaminants and other pollutants. Undesirable water conditions such as eutrophication occur due to the presence of nitrogenous and phosphorous nutrients present in those contaminants and pollutants. Not only these nutrients but organic matter, composed principally of proteins, carbohydrates, fats and oils, present in wastewater, causes depletion of natural oxygen in water due to biological stabilization. Suspended solids in water may cause sludge formation and anaerobic conditions. In addition to these, priority pollutants, such as refractory organics (surfactants, pesticides, phenol), heavy metals (Pb, Hg, As, etc.), pathogenic organisms, and dissolved inorganic materials (Ca, sulfates) are known for their acute toxicity or carcinogenicity, and hence they should be removed if the water is to be reused.

Analysis of the content of oxygen-consuming matter present in wastewater is done by biochemical oxygen demand (BOD) and chemical oxygen demand (COD) tests. It is observed that the quality of aquatic habitat, such as that of fish, begins to decrease when the oxygen content of water drops below 4–5 mgL^{-1} (1). This is a direct measurement of oxygen-consuming capacity of the wastewater and is termed biochemical oxygen demand (BOD). The reaction can be written as:

$$Organic\ matter + nutrients + O_2 \xrightarrow{Biomass} CO_2 + H_2O + biomass$$

Here, organic matter is the growth substrate for generation of new biomass by the use of oxygen, and biomass is actually acting as catalyst for the reaction. Theoretically BOD can be expressed as:

$$BOD = \frac{Grams\ of\ Oxygen\ used}{Grams\ of\ substrate\ used}\ \mathrm{mgL^{-1}}$$

A BOD test typically runs for five days; hence the test is named BOD_5.

Another alternative test to BOD is chemical oxygen demand (COD), which measures the amount of oxygen required to degrade the organic matter present in a wastewater sample that can be digested by a digestion reagent. In the oxidation or digestion process, measured amounts of potassium dichromate and sulfuric acid are used at a high temperature of 150°C (2).

Measurement of solid content is another important point in wastewater analysis. Solid content represents the pollution load of wastewater as well as potential sediment formation in the receiving stream. Solids may be principally categorized as suspended solids and dissolved solids. Suspended solids are those that can be filtered out by 0.45–1.2 μ filter paper (1). Dissolved solids are the solids that are soluble in water, such as carbohydrates, proteins, fats and oils, surfactants, ammonia, urea, etc., with inorganic compounds that remain in soluble form in water. Again, solids can be classified as volatile solids and fixed solids. Volatile solids are those which combust at 550°C, and fixed solids do not. Volatile solids are often considered the organic part of the pollutants that are associated with biomass and can be used as a measure of the microbial population.

7.1.2 Chemical Treatments Based on Chemical Reactions

One of the common practices of wastewater treatment and waste stabilization is neutralization, or acid-base reaction, and it is done before proceeding to other treatment processes such as, physical, chemical, and biological. In many chemical treatment processes, the pH of water plays a crucial role. Hence, adjustment of the water pH is necessary for processes such as precipitation of metals, coagulation, phosphorous precipitation, water softening, etc. Here comes the concept of solubility product (3). Another important factor for precipitation is the pH of the solvent. In chemical treatment of industrial wastewater, neutralization of excess acidity or alkalinity of treated water is often necessary to meet the required specifications for adding it to the receiving stream. It is necessary to identify the substances responsible for causing acidity or alkalinity of the water or wastewater before neutralization (4). Water hardness is soluble cations other than Na^+, K^+ and NH_4^+, but chiefly due to calcium and magnesium carbonate and bicarbonate in water, with sulfates, chlorides, and silicates of these metals to a lesser extent. The removal of these soluble compounds from water is done by precipitating the metal hydroxides by changing the pH of water. This process is called water softening and is done by using lime (calcium oxide), a combination of Lime (CaO) and ferrous sulfate or alum ($KAl(SO_4)_2.12\ H_2O$), slaked lime ($Ca(OH)_2$), and soda ash (Na_2CO_3). When hard water is treated with a combination of slaked lime and soda-ash (lime-soda process), calcium is precipitated as calcium carbonate ($CaCO_3$) and magnesium as magnesium hydroxide ($Mg(OH)_2$). When alum is used to soften water with calcium and magnesium bicarbonate, it forms insoluble aluminum hydroxide precipitate. Insoluble metal sulfides are formed by the addition of ferrous sulfate and lime. All these metal salt precipitates are pH dependent (5).

Fluoride contamination in water is one of the major concerns to the environment, as fluoride is toxic not only to human beings but also to the diverse species of animals, causing osteo-dental fluorosis. Aluminum melting, glass, phosphate fertilizer, brick manufacturing, and coal-based thermal processes are the sources of fluoride pollution to water (6). Fluorinated wastewater can be treated by precipitation, coagulation and precipitation, adsorption, electrocoagulation, reverse osmosis, and ion exchange methods. Generally $CaCl_2$ and CaO are added as precipitating agents for fluoride in wastewater. The coagulation and flocculation method precipitates fluoride by adding coagulant such as iron salt, polyferric sulfate and aluminum salt, or polyaluminum sulfate to form a gelatinous substance; single use of iron salt is not that effective. Recently a polyaluminum chloride containing coagulant Actifluo has been used successfully (7).

Phosphorous, which comes to the surface water from fertilizer, must be removed from wastewater. Phosphorous in the form of phosphate PO_4^{-3} can be easily removed by using alum or sodium aluminate. Aluminum combines with phosphate to form aluminum phosphate $AlPO_4$. Optimum pH for removal of phosphate by aluminum is 5.5–6.5. Except aluminum, other cations, such as iron and calcium can also be used to remove phosphate from water. Iron, in the form of either ferric phosphate (pH 7–8) or ferrous phosphate (pH 4.5–5), and calcium in the form of hydroxyapatite ($pH < 9.5$) remove phosphate (1).

A versatile range of metals and heavy metals can be removed by precipitating their hydroxides or sulfides. Hydroxide precipitation is effective for arsenic, cadmium, trivalent chromium, copper, iron, manganese, nickel, lead, and zinc, whereas sulfide precipitation acts better for the removal of cadmium, cobalt, copper, iron, mercury, manganese, nickel, silver, tin, and zinc. Metal sulfide precipitation has several advantages over hydroxide precipitation. The chemicals used for sulfide precipitation are sodium sulfate, sodium sulfide, and ferrous sulfate. Precipitation by chemical oxidation-reduction reaction often becomes effective in removal of metal ions. Reduction of hexavalent Cr^{6+} with Cr(VI) and Cr^{3+} with Cr(III) and the oxidation of manganous (Mn^{2+}) to manganese (Mg^{3+}) and ferrous (Fe^{2+}) to ferric (Fe^{3+}) produce insoluble precipitates. In some processes, sodium borohydride is used to reduce heavy metals to their elemental state, which yields a compact precipitate with metals such as cadmium, silver, lead, and mercury (1).

7.2 CHEMICAL PRECIPITATION

Chemical precipitation is a conventional physicochemical wastewater treatment technique that is mainly used to treat wastewater, either acid or neutral, to remove heavy metals, such as chromium, copper, nickel, lead, cadmium, etc., as well as phosphate, cyanide, fluoride, softening and stabilizating hard water (1).

Precipitation is different from coagulation. Coagulation is a rapid mixing process where non-settleable particles that are dispersed in water in the form of colloids or suspended solids produce larger particles, either by reaction with or by adsorption on a coagulant, and form microflocs. In the next stage, flocculation, these microfloc particles become bigger on slow stirring and separated by filtration or sedimentation (8). Usually, coagulation and flocculation steps come after precipitation, but sometimes the processes are associated together for suitable wastewater treatment.

The precipitation process involves the addition of any of the reagents—alkali, sulfide, carbonate, coagulant—or other chemicals to induce a reaction with the metal cation or inorganic anion contaminants and form insoluble precipitate by altering the ionic equilibrium of water. Alkalis used are sodium hydroxide (NaOH), calcium hydroxide or hydrated lime ($Ca(OH)_2$), quick lime (CaO), and magnesium hydroxide ($Mg(OH)_2$). Carbonate reagents used are sodium carbonate (Na_2CO_3), calcium carbonate ($CaCO_3$), or CO_2 under pressure. Sulfide compounds used in precipitation of pollutants are mainly iron sulfide (FeS), sodium hydrogen sulfide (NaHS), sodium sulfide (Na_2S), calcium sulfide (CaS), etc. The coagulants include alum ($KAl(SO_4)_2$), ferric hydroxide ($Fe(OH)_3$), or ferric chloride ($FeCl_3$).

The precipitation process can be classified into two broad types: chemical precipitation and coprecipitation or adsorption. In chemical precipitation, pH, ionic equilibrium, and temperature are important factors. Exceeding the solubility product of cation and anion at favorable pH and temperature causes precipitation. Chemical precipitation proceeds through three stages: a) nucleation, b) crystal growth, and c) flocculation (9). Coprecipitation, or adsorption, is a process in which the normally soluble ions or contaminants in water, at the condition of precipitation, are adsorbed on the nuclei or crystals of solid precipitate and removed as a single phase with the precipitate.

7.2.1 Reactions in Chemical Precipitation

The chemical precipitation process is one of the major water treatment processes, in which unwanted heavy metal cations and some anions are precipitated by the reaction with precipitating reagents. One or combination of the following reactions are involved in precipitation method:

i. **Hydroxide precipitation**

Addition of an alkali to wastewater causes an increase in pH, and this condition causes the dissolved heavy metal salts, such as Cd, Cu, Pb, Fe, Mn, Ni, Zn, etc., to form hydroxide precipitates. The precipitation reaction can be written as:

$$M^{2+} + 2OH^- \rightarrow M(OH)_2$$

If the pH is above the optimum pH for precipitation, $M(OH)_2$ will form a precipitate, but if the pH is below the optimum, it will result in a soluble metal hydroxide (9).

$$M^{2+} + 2OH^- \rightarrow M(OH)^+$$

The degree or extent of precipitation depends on the solubility product (K_{sp}) of the metal hydroxide, type and concentration of the ions present in water, pH of water, etc.

For a precipitation reaction, where there is one product, which is the solid precipitate, the reaction can be written as

$$mA^{n+} + nB^{m-} = A_mB_n$$

The solubility product is $K_{sp} = [A^{n+}]^m\,[B^{m-}]^n$ here the activity of the solid precipitate is considered as 1.0 by convention.

Usually hydroxide precipitation is effective in the pH range of 9–12. After precipitation, the solids are separated from the cleaned effluent in a clarifier. The hydroxide sludge generally contains lots of bound water, which is difficult to extract; hence reuse of that water has limited potential (10).

The hydroxide precipitation method has the advantages of automatic pH control, ease of operation, and low cost reagents, but it involves several drawbacks as well: the precipitate may tend to resolubilize if the pH of water changes; it is disadvantageous for the presence of several metals in wastewater because of the difference in solubility of different metal hydroxides at a definite pH; it forms a large amount of hydroxide sludge, which is difficult to dewater to a substantial level; the presence of cyanide in water hinders the metals from forming hydroxide precipitate, etc. It has been observed that Cr cannot be removed by this process (11).

ii. **Sulfide precipitation**

Addition of a sulfide reagent produces sulfide precipitate of metal ions. Metal sulfides are less soluble than metal hydroxides, so sulfide precipitation is more effective than hydroxide precipitation for removal of many metal ions.

Sulfide precipitation can be done by two ways: soluble sulfide precipitation (SSP) and insoluble sulfide precipitation (ISP). In the first method, i.e. SSP, a sulfide reagent is added to the wastewater in the form of a water solution, such as sodium sulfide (Na_2S) or sodium bisulfide (NaHS) solution. This process can be operated in batch or continuous mode. In the ISP process, a water suspension or slurry of a sparingly soluble sulfide, such as ferrous sulfide (FeS), is added to form metal sulfide precipitate. Most of the heavy metal salts are less soluble than FeS; hence metals are easily precipitated as metal sulfides after the reaction with FeS. The sulfide precipitates do not usually resolubilize, and the sulfide sludge is easier to dewater than hydroxide sludge (12). Sulfide sludge can be easily processed to recover metals. But one of the important concerns of sulfide precipitation is the generation of hydrogen sulfide gas, which restricts the application in some of the metals removal.

iii. **Calcium precipitation**
Metals from wastewater can be removed by carbonate precipitation, and the major reagent for that is sodium carbonate, or soda ash (Na_2CO_3). Sodium bicarbonate ($NaHCO_3$) is also used for precipitation, but owing to its low alkalinity, it is not as effective as other alkalis. Sodium bicarbonate can neutralize excess acidity of water and also can precipitate some metals in a narrow range of pH. Metals, such as zinc, are difficult to precipitate, but the addition of a mixture of sodium carbonate, sodium bicarbonate, and lime can precipitate zinc hydroxide. It is easier to dewater carbonate sludge than the hydroxide one (12).

iv. **Coprecipitation or adsorption**
In this process, some soluble impurities in wastewater are removed by adsorption on the precipitate. This removal may occur by different ways, such as surface adsorption, in which the impurities are adsorbed on the surface of the precipitate without being incorporated into the crystal lattice, or by the inclusion into the crystal lattice structure of the precipitate. Another type of coprecipitation is the combined precipitation of different metals by forming hydroxide and sulfide precipitate together and then removing.

7.3 COAGULATION AND FLOCCULATION

The chemical purification process includes coagulation and flocculation, which are among the primary steps to remove recalcitrant suspended solids from wastewater. The suspended solids remain in suspension until proper coagulation and flocculation techniques are applied (8). Coagulation and flocculation are done successively, followed by sedimentation. If coagulation is not complete because of either inadequate addition of coagulant or lack of other factors, flocculation step becomes unsuccessful and so does the next step, sedimentation. Both coagulation and flocculation should be carried out to get sufficiently clean water.

7.3.1 Coagulation

A coagulant, when added to the wastewater, dissociates to form positively charged ions, which neutralize the suspended particles and form invisible small particles called microflocs. This stage is also called destabilization. Formation of microflocs should be completed with proper adjustment of the coagulant amount (8). The mixing of coagulant is done with rapid stirring to ensure even dispersion of the coagulant in water and appropriate collision with the particles. The rapid mixing contact time is typically kept to 1–3 minutes. This stage takes care of coagulation and the early stage of flocculation. In addition to normal coagulation, enhanced coagulation is the step that includes the addition of an extra coagulant dose and the lowering of pH, and/or development of new coagulants with improved property and performance.

The conventional coagulants used in water treatment are mainly aluminum or iron salts. These are inorganic coagulants, and major examples are aluminum chloride, aluminum sulfate, sodium aluminate, etc., as aluminum-based coagulants and ferric chloride, ferric sulfate, ferrous sulfate, etc., as iron-based coagulants. When these coagulants are directly added to raw water, they dissociate to release Al^{3+} and Fe^{3+} ions. These ions, in conjunction with water molecules, form complex ions,

$Al(H_2O)_6^{3+}$ and $Fe(H_2O)_6^{3+}$, which afterward are transformed into soluble $Al(OH)^{2+}$ and $Fe(OH)^{2+}$. These positively charged ions strongly attract the negatively charged non-settleable particles and dissolved organic species to form flocs. Charge neutralization occurs at low pH (< 6) and low ionic strength water. This generally requires low coagulant doses and produces less sludge (13). Sometimes the dosage of Al or Fe coagulant is more than the amount needed for complete neutralization in the usual treatment process. In that case, more non-settleable contaminants are enmeshed to the $Al(OH)_3$ or $Fe(OH)_3$ precipitations when they are formed and settled. This process is termed "sweep coagulation." Dissolved organic species, such as humic acid and fulvic acid originated from natural organic matter (NOM) and heavy metals, may adsorb on the hydroxide precipitate. Phosphate and fluoride ions also adsorb on the hydroxide and coprecipitate. Sweep coagulation is particularly useful for water with low turbidity, because here the number of collisions and chance of contact are less, and hence the floc volume is not appreciable compared to that for other water types. Use of a higher dose of coagulant enhances the flocculation rate by increasing the number of collisions and ensures the formation of settleable flocs. The disadvantages of sweep coagulation are the need for more coagulant and large sludge formation. Large sludge formation has a merit in that it can trap bacteria as they settle (14).

Some examples of reaction of undesired materials with Al and Fe-based coagulants are shown below (8):

$l_2(SO_4)_3$ +	$3Ca(HCO_3)_2$ →	$2\ Al(OH)_3$ +	$3CaSO_4$ +	$6CO_2$
Aluminum Sulfate (Alum)	Calcium Bicarbonate (in water)	Aluminum Hydroxide	Calcium Sulfate	Carbon Dioxide
$2FeCl_3$ +	$3Ca(HCO_3)_2$ →	$2\ Fe(OH)_3$ +	$3CaSO_4$ +	$6CO_2$
Ferric Chloride	Calcium Bicarbonate (in water)	Ferric Hydroxide	Calcium sulfate	Carbon Dioxide

Table 7.1 compiled different common coagulants, their advantages and disadvantages.

Although treatment with conventional coagulants offers cheap and easy operation, there are several disadvantages to their long-term usage. These chemicals may alter the physicochemical characteristics of water, consume alkalinity of water, and generate huge volumes of sludge, and it is important to recover valuable metals or toxic metals from sludge before disposal. All these factors present serious concerns in terms of ecosystems. These limitations of using inorganic metal-based coagulants lead to orienting the focus of water treatment toward using pre-polymerized coagulants or organic polymers, either alone or in combination (coagulant aid) with inorganic metal coagulants (8).

Pre-polymerized coagulants are mainly polyaluminum chloride (PACL) and polyferric chloride (PFCL). Both of these coagulants are prepared by partial hydrolysis of acid aluminum chloride or ferric chloride in a specific reactor. The purpose behind using pre-polymerized coagulants is to enhance the charge neutralization or destabilization of the suspended matter. Because the coagulants are already polymerized before they are added to water, $Al(OH)_3$ or $Fe(OH)_3$ solids precipitation are slowed down, which favors charge neutralization and adsorption. As pre-polymerized

TABLE 7.1
Advantages and Disadvantages of Inorganic Coagulants (8)

Name	Advantages	Disadvantages
Aluminum sulfate (Alum) $Al_2(SO_4)_3 \cdot 18H_2O$	Easy to handle and apply; most commonly used; produces less sludge than lime; most effective between pH 6.5 and 7.5	Adds dissolved solids (salts) to water; effective over a limited pH range
Sodium aluminate $Na_2Al_2O_4$	Effective in hard waters; small dosage usually needed	Often used with alum; high cost; ineffective in soft waters
Polyaluminum chloride (PAC) $Al_{13}(OH)_{20}(SO)_4C_{115}$	In some applications, the floc formed is denser and faster-settling than alum	Not commonly used; little full-scale data compared to other aluminum derivatives
Ferric sulfate $Fe_2(SO_4)_3$	Effective between pH 4–6 and 8.8–9.2	Adds dissolved solids (salts) to water; usually need to add alkalinity
Ferric chloride $FeCl_3.6H_2O$	Effective between pH 4 and 11	Adds dissolved solids (salts) to water; consumes twice as much alkalinity as alum
Ferrous sulfate $FeSO_4 \cdot 7H_2O$	Not as pH sensitive as lime	Adds dissolved solids (salts) to water; usually need to add alkalinity
Lime $Ca(OH)_2$	Commonly used; very effective; may not add salts to effluent	pH-dependent; produces large quantities of sludge; overdose can result in poor effluent quality

species, such as $Al_{13}O_4(OH)^{7+}$, have a higher charge than monomeric species, these coagulants are expected to provide better charge neutralization and adsorption than the traditional ones. Several other examples of pre-polymerized coagulants are polyferric sulfate $[Fe_2(OH)_{0.6}(SO_4)_{2.7}]$, polyaluminum chlorohydrate $[Al_2(OH)_5Cl]$, and polyaluminum silicosulfate $[Al(OH)_{3.24}Si_{0.1}(SO_4)_{1.58}]$ (15).

Polymers are superior to inorganic metal coagulants for several reasons. Lower dose requirement, formation of smaller volumes of more concentrated, rapidly settling floc, and cleaner effluent water make the polymers more attractive coagulants; however, they are more expensive.

7.3.1.1 Jar Test

The jar test is done in a laboratory by taking a specific amount of wastewater in a series of beakers, usually five or six, and adding an appropriate volume of coagulant solutions. The mixtures are stirred, followed by settling for sufficient time. Then the top layer of water from each beaker is collected and the pH and turbidity by absorbance can be determined. To study the test parameters, researchers can vary the stirring speed, time of stirring, pH of solution and coagulant dosages. The pH of the solution of each jar can be adjusted at different pH levels at the beginning of the experiment by using 1N sulfuric acid. The turbidity values at each pH can be measured, and optimum pH can be noted. Stirring speed in each beaker is varied from rapid to slow to mimic the actual plant mixing operation. Residual turbidity versus coagulant dose can be plotted and optimum condition is determined (16).

7.3.2 Flocculation

Flocculation is a gentle mixing stage in which the contact time ranges from 15 minutes to one hour or more, depending up on the quality of water and other parameters. Flocculation increases the particle size from invisible submicroscopic microfloc by collision between them to produce larger visible particles. These visible flocs are called pinflocs. Mixing velocity is carefully adjusted to avoid the tearing out or shearing of the floc particles because once torn out, the particles cannot regain their original size and strength (15). Continuous growth of Pinfloc size occurs by collision and interaction with additional inorganic coagulants or organic polymers. High-molecular-weight polymers, called coagulant aids, are added to form a bridge, bind, and impart strength to the floc particles as well as increase their size and weight. Once optimum size and weight are reached, the particles undergo sedimentation.

Conventional wastewater treatment plants have separate coagulation and flocculation stages. They have different feed points for coagulants, flocculants, polymers, and other chemicals. There is conservative retention time and rise rate in each stage (17). Retention (detention) time is defined as the amount of time that water spends in a particular stage. The liquid volume in the basin of the equipment is expressed in gallons, and plant flow rate is in gallons per minute (gpm).

$$Retention\ time = \frac{Basin\ volume\ (gallons)}{gpm\ flow}$$

$$Rise\ rate = \frac{gpm\ flow}{net\ water\ surface\ area\ of\ the\ basin,\ ft^2}$$

7.3.3 Removal of Some Important Contaminants from Water by Coagulation-Flocculation

7.3.3.1 Removal of Natural Organic Matter

Natural organic material (NOM) is usually organic substances created by natural aqueous extraction of the vegetation, either living or decaying, degradation products of non-living organisms, and natural soil organic matters that mix with the natural source of water. These are mainly humic substances and are objectionable for a variety of reasons. These substances are responsible for color in water; they are the carriers of a substantial number of toxic substances and micro-pollutants, including heavy metals and organic pollutants. These pollutants react with chlorine to form carcinogenic by-products. Hence, NOM has detrimental effects on ecosystems (15). The degree of removal of NOM is related to the amount of dissolved organic carbon (DOC) in water. The specific ultraviolet absorption (SUVA) is used to get the value of NOM removal in terms of DOC removal. The relation between SUVA and DOC is as follows (15):

$$\text{SUVA} = \text{UV254/DOC}\ (\text{l/mg.m})$$

where, UV 254 is the ultraviolet absorbance measure at a wavelength of 253.7 nm, after filtration through 0.45-μm filters; DOC in water is measured after filtration through 0.45-μm filters (mg/l).

An enhanced coagulation process has long been used to remove NOM efficiently. This process removes turbidity too, along with NOM to an appreciable extent. These dual objectives can be met to the maximum level by choosing the appropriate coagulant, its dose, and the best value of pH of water for coagulation. In some cases, for water containing lower-molecular-weight organics, activated carbon is used for adsorption along with metal coagulants for improving overall removal. In general, lower-molecular-weight species, such as fulvic acids, are more difficult to remove by coagulation than higher-molecular-weight humic acids (14).

7.3.3.2 Removal of Inorganics

Inorganic substances are removed to a considerable extent by coagulation. Heavy metals, such as arsenic, chromium, copper, mercury, and other inorganic compounds, including fluoride and phosphate, are successfully removed from wastewater. Iron-based coagulants often perform slightly better than aluminum-based coagulants for heavy-metal removal due to the activity at wider pH range and the larger surface area of resulting flocs (18). As arsenic and fluoride are among the most toxic contaminants in ground or surface water and detrimental to human health, the removal of these compounds are discussed here.

7.3.3.3 Arsenic

Arsenic is a commonly occurring toxic element in groundwater, and long-term exposure to arsenic is injurious to health. The WHO's provisional guideline value for arsenic in drinking water is 0.01 mg/l (10 μg/l) (19). The permissible limit of arsenic in India, in the absence of an alternative source, is 0.05 mg/l (50 μg/l) (20).

Arsenic occurs mostly in the forms of arsenite, As(III) AsO_3^{3-} and arsenate, As(V) AsO_4^{3-} in groundwater. As(III) is usually found in oxygen-free groundwater, but As(V) is more common in aerobic waters. In general, As(V) is more readily removed than As(III) (15).

Iron-based coagulants are generally more effective than aluminum coagulants for arsenic removal. Iron coagulants hydrolyze to form ferric hydroxide, which produces Fe^{3+} ions. This positive-charged ion formation is a function of pH; with a decrease in pH, the number of positively charged sites on the ferric hydroxide increases. Anionic arsenate, As(V) AsO_4^{3-} adsorbs to the positively charged ferric hydroxide particles by complex ion formation. Arsenic removal is generally optimized at pH values of less than 7 (21).

7.3.3.4 Fluoride

The major source of fluoride pollution in a developing country such as India is the use of phosphate fertilizers. In addition, aluminum melting industries, glass industries, coal-based thermal processes, and mineral-processing industries are responsible for fluoride pollution. Fluoride pollution in water poses health dangers such as dental and skeletal fluorosis (22). The permissible limit of fluoride in drinking water set by WHO as well as the Bureau of Indian Standards is 1.5 mg/l maximum (23).

Fluoride can be removed to a great extent by combining activated alumina adsorption and reverse osmosis. But in some cases, coagulation using alumina coagulant is effective and economical. The removal is highly dependent on the pH of water and the coagulant dose. The optimum pH for this operation is in the range of 6–7.5. Fluoride ions reacting with aluminum hydroxide produce complex aluminum-fluoride ions, which generate flocs. Some fluoride ions adsorb on the aluminum flocs rather than producing complex ions. In a recent study, it was observed that fluoride removal efficiency was increased with increasing alum coagulant concentration and contact time. The maximum removal rate (93.3 percent) was achieved with 300 mg/l alum concentration, 45 minute contact time, and pH value of 6 when fluoride concentration in water was 3 mg/l (24). Also, removal efficiency is decreased by increasing fluoride concentration. Experimental results reveal that aluminum sulfate coagulant shows an acceptable efficiency with easy operation.

7.4 COAGULATION SLUDGE MANAGEMENT

Water treatment by coagulation produces a large amount of sludge, although the amount depends on the quantity of the wastewater treated, quality of its content, color intensity, coagulant dosage, etc. India also produces huge amount of sludge from different treatment steps in water treatment plants, and because of the lack of sludge management strategies, most treatment plants in India discharge their sludge into nearby drains, which ultimately meet a river or other water source. Sometimes the sludge is used for landfills (25). These simple methods of sludge disposal are not a proper and sustainable solution because of the large quantity of contaminants present in them, and this affects environment adversely. The common coagulants used are aluminum salts or iron salts, which hydrolyze in water to form their respective hydroxide precipitates. By the coagulation process, the colloidal and suspended impurities such as sand, silt, clay, and humic particles present in the crude water are removed by charge neutralization, sweep floc process, and adsorption onto hydroxide precipitates, and all these together constitute sludge. The moisture content of the wet sludge is generally above 80 percent. The development of sustainable sludge management strategies under stringent rules is highly desirable (26).

After dewatering the sludge, it may be reused for agricultural and ecological purposes, for landfilling, or for brick and cement production processes. It has been reported that first-class brick and roof tiles can be made by adding 10–20 percent of sludge into the brick kiln. The heavy metals present in the sludge are sintered at the high temperature of the kiln and form a bridge among themselves. This gives strength to the bricks and tiles and ensures an extremely low chance of heavy metal leaching (25).

The coagulants can be recovered after water treatment and can be reused again at water and wastewater treatment plants. For instance, sludge may be used as a coagulant due to its own composition. In one study it was observed that dewatered alum sludge can be used as a low-cost phosphorous sorption medium in water treatment plants. In addition, the sludge has effective use as adsorbent of hydrogen sulfide, boron, fluorides, perchlorate, mercury, etc.

Sludge stabilization by aerobic digestion or a self-oxidation process allows the reduction of organic matter content by conversion into CO_2, H_2O, or NH_3 and limits the odor emissions (26).

7.5 CHEMICAL OXIDATION

Chemical oxidation of polluted water treatment involves an oxidative chemical process to chemically convert the Contaminants (plural) via oxidation and lower their concentrations in the polluted water. Oxidizing agents oxidize the pollutants, whereby these yield less-hazardous by-products or end products that can be separated or removed and thus produce water with improved quality in terms of contaminants. It has been under review for over two decades (27, 28). In this method, chemical oxidants are used for the oxidation of reduced inorganic species such as ferrous iron, Fe(II); manganous manganese, Mn(II); or the like and hazardous synthetic organic compounds. Apart from the removal of the contaminants, these are also effective in removing odor and color and in restoring tastelessness, which is a primary property of pure water. Another benefit is the removal of bio-contaminants, thereby lowering BOD. The most common oxidant in use is chlorine, in its elementary and various oxidized forms such as hypochlorite or chlorine dioxide; others are permanganate, ozone, etc. Use of several of these in combination with various forms of energy in terms of irradiation of UV or ultrasound produces strong oxidants, which are discussed in Section 7.6, "Advanced Oxidation."

7.5.1 Oxidants Used in Water Treatment

7.5.1.1 Chlorine

Chlorine is one of the oldest and the most commonly used oxidants in water treatment. Gaseous chlorine, upon reaction with water, undergoes dismutation to generate hypochlorous acid (HOCl) and the chloride ion (Cl^-):

$$Cl_2 + H_2O \rightarrow HOCl + H^+ + Cl^-$$

If chlorination is performed by adding liquid sodium hypochlorite, it undergoes the following change:

$$NaOCl \rightarrow Na^+ + OCl^-$$

$$OCl^- + H_2O = HOCl + OH^-$$

If it is initiated by dissolving calcium hypochlorite granules in water, hypochlorite ions are generated, which react with water to form hypochlorous acid in a manner similar to the one stated above:

$$Ca(OCl)_2 \rightarrow Ca^{2+} + 2OCl^-$$

In whatever form chlorine is added to water, it generates hypochlorite, hypochlorous acid, and molecular chlorine. Chlorine gas generates protons and thus increases the acidity and lowers the pH of the medium. On the other hand, sodium hypochlorite and calcium hypochlorite, being bases, will raise alkalinity and the pH of the treated water.

Hypochlorite and its conjugate hypochlorous acid both incorporate chlorine in the same oxidation state (+1); however, hypochlorous acid acts as the stronger one, which is again corroborated by the fact that chlorine is a better oxidant at lower pH values.

Like the better oxidizing ability of chlorine under acidic conditions, the disinfecting properties of chlorine are more effective under lower pH. A liquid sodium hypochlorite solution is usually introduced from a concentrated NaOCl bulk storage directly to the main plant flow using a pump. Alternatively, calcium hypochlorite granules may be added directly or as a concentrated solution using the same means as in the case of liquid sodium hypochlorite. The handling of gaseous chlorine poses a potential hazard.

7.5.1.2 Chlorine Dioxide

Chlorine dioxide is one of the chlorine derivatives incorporating an intermediately quasi stable chlorine oxidation state and hence is a strong oxidant and has several advantages over the use of elementary chlorine. Unlike chlorine, it does not react with ammoniated contaminants or ammonia; nor does it get involved in substitution reaction with dissolved organic matter; hence a much lower amount of it is necessary compared to raw chlorine. It also reacts with a number of metal ions, including reduced iron and manganese. However, the primary use of chlorine dioxide is for removal of odor and taste in drinkable water. There are two points to be noted for chlorine dioxide usage: most of the chlorine dioxide gets converted to chlorite, and residual chlorite, if any, reacts with chlorine in the system to yield low concentrations of chlorite and chlorate.

$$HOCl + 2ClO_2^- \rightarrow 2ClO_2 + Cl^- + OH^-$$

$$HOCl + ClO_2^- \rightarrow ClO_3^- + Cl^- + H^+$$

Generation of chlorate seems hazardous because there is no feasible method is in practice to get rid of this unwelcome reaction product.

7.5.1.3 Potassium Permanganate

Potassium permanganate ($KMnO_4$) is a very strong oxidant with considerable solubility in water; it stands as a potential reagent of choice in the chemical oxidation processes in the treatment of wastewater. Below are the various chemical reactions that depend on the pH of the medium:

$$MnO_4^- + 4H^+ + 3e^- \rightarrow MnO_2(s) + 2H_2O$$

When added as a reagent of in situ chemical oxidation (ISCO) processes the reactions may be recast as

$$MnO_4^- + 2H_2O + 3e^- = MnO_2(s) + 4OH^- (pH\,3.5-12)$$

$$MnO_4^- + 8H^+ + 5e^- = Mn^{2+} + 4H_2O(pH < 3.5)$$

$$MnO_4^- + e^- = MnO_4^{2-} (pH > 12)$$

The most commonly produced by-product, manganese dioxide appears as a black precipitate. If it is allowed to remain in the system it grows as dark particulate matter, clogging the supply line. However, simple techniques such as clarification or filtration effectively clear the feed. Most often, removal of MnO_2(s) is achieved by conventional clarification or filtration processes. Addition or intake of permanganate is designed as close to the plant as possible to accomplish its complete reduction to solid MnO_2, thus making it easily separable during subsequent filtration. The manganese dioxide obtained as a product has significant adsorbing ability thus clears a number of inorganic ions in course of reaction. Some of the remarkable ions are Fe(II), Mn(II), and Ra(II). It also binds some of the organic contaminants, which appears to be an additional benefit.

7.5.1.4 Ozone

7.5.1.4.1 General Ozone Oxidation and Related Chemical Reactions

Ozone decomposes readily in an aqueous medium. It is very reactive and is well known for its spontaneous self-decomposition. Initiated by hydroxide ion, the auto decomposition of ozone in various forms is illustrated below:

$$OH^- + O_3 \rightarrow HO_2 + O_2^-$$

$$HO_2 = H^+ + O_2^-$$

$$O_2^- + O_3 \rightarrow O_2 + O_3^-$$

$$O_3^- + H^+ \rightarrow HO_3$$

$$HO_3 \rightarrow O_2 + OH$$

$$OH + O_3 \rightarrow HO_2 + O_2$$

Direct oxidation

$$O_3 + C_2HCl_3 + H_2O = 2CO_2 + 3H^+ + 3Cl^-$$

OH Formation

$$O_3 + H_2O = O_2 + 2 \cdot OH(Slow)$$

$$2O_3 + 3H_2O_2 = 4O_2 + 2 \cdot OH + 2H_2O(Fast)$$

It is to be noted that the hydroxyl radical is one of the strongest oxidants; this opens up a scope to use a combination of these and is separately explored in AOPs.

7.5.2 In Situ Chemical Oxidation

In situ chemical oxidation (ISCO) is remedial measure that is practiced with an aim to transform groundwater or soil contaminants into less harmful chemical species. It involves the introduction of a chemical oxidant into the subsurface (29, 30). The most common oxidants used in chemical oxidation are permanganate (MnO_4^-), hydrogen peroxide (H_2O_2), iron (Fe: Fenton-driven, or H_2O_2-derived oxidation), persulfate ($S_2O_8^{2-}$), and ozone (O_3). The success of the technique depends on the persistence of the reagent in the injected subsurface and its diffusive transport toward the targeted affected zone. Unlike usual chemical oxidation there is only limited scope of removal of the transformed oxidized product here, which is also to be taken into consideration. Like all other technological innovations, this too has its advantages and limitations. The major indicators in these arena are listed in the following sections.

7.5.2.1 Scope and Limitations: Future of ISCO

ISCO works on a wide range of contaminants and transforms them in situ. It involves lower remedial cost than other measures that employ huge mechanical arrangements. Another advantage is that it affects all phases of the contaminant—aqueous, sorbed, and non-aqueous—at a faster pace. It uses enhanced mass transfer and better energy management and helps restore natural control through post-oxidation microbial activity.

At the same time, its limitations include oxidant delivery problems due to reactive transport and aquifer heterogeneities. Its effectiveness is affected by variations in natural oxidant demand, especially at its higher end in the ground ambience. The short life of some oxidants over the fast reaction rates in the sub-surface also lowers its efficacy and might even pose a threat to public health in case of strong oxidants. The mobility of the contaminants and reduction in the permeability are other major issues. It is also seriously restricted in its application at heavily contaminated sites.

7.6 ADVANCED OXIDATION

The age-old wastewater treatment methods such as precipitation, coagulation, sedimentation, and chemical oxidation are quite useful, but some contaminants could still remain and pose a serious threat to water quality (31, 32), and it is a challenge

TABLE 7.2
Relative Oxidizing Ability of Various Conventional Oxidants (41)

Oxidant	Oxidation Potential (v)	EOP Relative to Chlorine
Fluorine	3.03	2.25
Hydroxyl radical (OH*)	2.80	2.05
Oxygen (Atomic)	2.42	1.78
Ozone	2.08	1.52
Hydrogen peroxide	1.78	1.30
Hypochlorite	1.49	1.10
Chlorine	1.36	1.00
Chlorine dioxide	1.27	0.93
Molecular oxygen	1.23	0.90

to get rid of them. Newer methods put a question mark on the utility and usage of conventional methods (33, 34). The advanced oxidation processes (AOPs) are a timely and well-deserved improvement over chemical oxidation methods (35–37). This emerging water treatment technology provides the scope to get rid of stubborn contaminants with remarkably poor biodegradability or unusual stability to withstand chemical oxidations (38, 39).

In AOPs a hydroxyl free radical (HO*) is generated. It is one of the most powerful oxidants, second only to the elementary fluorine in terms of potential (Table 7.2). Its powerful oxidizing ability oxidizes organic compounds that are otherwise non-degradable by conventional oxidants such as chlorine, ozone, and even atomic oxygen (40).

The hydroxyl radical is non-selective and destroys almost all organic chemicals because as a radical electrophile (42), it attacks electron-rich sites in organic compounds through fast kinetics. A very high value of E^0 provides a thermodynamic edge toward yield of the reaction (43). The in situ-generated hydroxyl radicals seek organic chemicals and destroy them through rapid electron transfer, hydrogen abstraction, and radical combination (44). In the ultimate degradation or mineralization, carbon dioxide is produced from organic carbon compounds and an unacceptable degradation yields toxic quotient compounds compared to the actual contaminants; hence this is not advised to be practiced. (45–48).

The broad classification of AOPs may be done on the basis of miscibility, as homogeneous and heterogeneous. The homogeneous process may further be divided on the basis of energy use as those that require energy and those that go without energy. This includes ozone in an alkaline medium; a combination of ozone and hydrogen peroxide; hydrogen peroxide and catalyst. The processes that are driven by energy are divided into three major categories: a) UV radiative processes: ozone/UV, hydrogen peroxide/UV, ozone-hydrogen peroxide/UV, photo-Fenton/UV; b) ultrasound (US) energy processes: ozone/US, hydrogen peroxide/US; c) electrical-energy–based processes: electrochemical oxidation, anodic oxidation, and electro-Fenton.

7.6.1 Heterogeneous Advanced Oxidation Processes

The heterogeneous advanced oxidation process (HTAOP) involves catalytic intervention that includes ozonation, photocatalytic ozonation, and photocatalysis. In heterogeneous AOPs, degradation of the contaminants is affected by solid catalysts, which break down water to generate short-lived radical electrophiles (OH*) in situ. The kinetics is fast, for it involves rapid electron transfer. One of the advantages is that it occurs in ambient conditions wherein the oxygen gets partially reduced to generate superoxide in course of reaction.

7.6.1.1 Photocatalysts

The photocatalyst that is abundantly used is titanium dioxide (TiO_2); its very high efficiency, relative abundance, and considerably low cost makes it a catalyst of choice (49). A good photocatalyst should be not only photoactive in the Vis-near UV range and photo stable; it should also be biologically inert and chemically non-interfering, with no toxicity and, above all, be commercially viable. Some other examples are ZnO, ZnS, CdS, $SrTiO_3$, SnO_2, Fe_2O_3, WO_3, WSe_2, and Si (50).

7.6.1.2 Titanium Dioxide

TiO_2 (Anatase) absorbs UV or normal daylight and releases a pair of electrons, leaving holes. The photoexcited electrons move to the conduction band of the titanium dioxide, leaving positive holes in place of the electrons. The pH, amount of catalyst, and concentration of the substrates are the deciding factors for the course and the kinetics of the reaction (51).

7.6.2 Homogenous Advanced Oxidation Processes

7.6.2.1 Photolytic Ozonation (O_3/UV)

In O_3/UV AOPs, the system is made saturated with ozone followed by UV irradiation at 254 nm. This method is particularly effective for the degradation of chlorophenols (CPs). It is evidently more effective than UV photolysis and ozonation (52–55).

7.6.2.2 Hydrogen Peroxide and Ultraviolet Radiation (H_2O_2/UV)

Hydrogen peroxide generates the radical OH* as a result of primary photo absorption in the UV range through a number of chain reactions. The elementary reaction is cited here under

$$H_2O_2 + h\nu \rightarrow 2HO*$$

The cleavage of hydrogen peroxide primarily depends on the exposure time of UV and the concentration of H_2O_2, and the increase of pH increases the rate of dissociation of H_2O_2. (56). The absorption of UV is competitive between the polluted water and the catalyst H_2O_2, with the latter having moderately low absorptivity, which poses a major limitation to the process (57). One of the key disturbing features is that it involves higher cost in terms of the operating devices and energy demand. Further, the hydrogen peroxide that remains in excess in the system might further initiate biological regeneration in the treated substrate (58).

7.6.2.3 UV Photolysis

7.6.2.3.1 Fenton and Photo-Fenton Oxidation

The Fenton process is one of the most efficient and most used AOPs for effluents that are hugely contaminated. However, the iron sludge that is generated in the course of the process poses a significant setback. The photo-Fenton process is considered an improvement as it uses photo emissions from the sun and UV that reduce Fe(III) to Fe(II), which effectively controls sludge production. The conversion to Fe(II) in the process is a welcome departure from the Fenton process because it has further scope to react with hydrogen peroxide. The control and optimization of pH appears tricky. The optimum range falls between 2.6 and 3; levels of $pH > 3$ lead to the increase in the quantity of sludge, lowering efficacy of the process. On the other hand, at $pH < 2.6$, lower light absorption leads to the decrease in the reaction rate and the yield. The hydrogen peroxide concentration on the Fenton and photo-Fenton processes affects the extent and rate of degeneration of the contaminants. However, excess hydrogen peroxide beyond the optimum limit has no influence on the system. The catalytic ferrous ion concentration $[Fe^{2+}]$ is important for the Fenton and photo-Fenton processes.

7.6.2.3.2 Electro-Fenton

Electro-Fenton (EF) technology generates H_2O_2 in a continuous manner at a suitably designed cathode, which requires continuous supply of air/O_2. Fenton's reaction leads to the formation of OH* in the system as an iron catalyst is added to the treated solution (59). The notable advantages are the in situ production of H_2O_2, the continuous regeneration of Fe(II) at the regeneration of Fe(II) at the cathode which effects in the greater degradation in the greater degradation rate of contaminants, and reduced amount of sludge (60).

7.7 ELECTRICAL ENERGY-BASED PROCESSES

7.7.1 Anodic Oxidation

The anodic oxidation type of AOP involves an accelerated electrochemical change that is promoted by the natural oxide coat of the aluminum anode. As the reaction advances, the oxide layer grows thicker over the natural oxide layer. This appears advantageous for proven resistance against corrosion leading to long-term protection (59).

7.7.2 Scope of AOPs

AOPs using a combination of oxidants and/or catalysts and energy are evidently more effective than the processes with any of the single reactants used (for example, hydrogen peroxide, UV or Ozone) in a process. The cost involvement is relatively less for it utilizes less expensive hydrogen peroxide or ozone in most of the cases. The radical electrophile (OH*) generated, have large scope to transform the organics through hydrogen abstraction, radical addition, radical combination and electron transfer (61).

7.8 CONCLUDING REMARKS

AOPs open up an emerging horizon in the treatment of wastewater. They surpass the efficacy of almost all the conventional methods, which fail to get rid of almost non-removable toxic contaminants. The working principle of the AOPs is to generate the radical electrophile (HO*) in situ. Being one of the strongest oxidants, it attacks the contaminants leading to its degradation in a rapid process producing easily eliminable product. The scope of AOPs is yet to be explored, and it has a promising possibility to meet the challenges of future threats in water purification.

REFERENCES

1. Chemistry of Waste Water, Timothy G. Ellis, Environmental and Ecological Chemistry, Vol-II, Edited by A. Sabljic, Encyclopedia of Life Support System, UNESCO, Google book, 2009. ISBN: 978-1-84826-206-5 (eBook) ISBN: 978-1-84826-693-3 (Print Volume)
2. Study of Chemical Oxygen Demand (COD) in Relation to Biochemical Oxygen Demand (BOD) in Tigris River at Missan Governorate, lecture by Dr. Nezar Hassan Mohammed, Tech. Inst. of Amara and Haleem Khalaf Alewi, Tech. Inst. of Shatra, 2019.
3. Principles of Water Treatment, K. J. Howe, D. W. Hand, J. C. Crittenden, R. R. Trussell, G. Tchobanoglous, John Wiley, NJ. ISBN 978-0-470-40538-3.
4. Handbook of Environmental Engineering, Volume 3: Physicochemical Treatment Processes: Chemical Precipitation, Chapter 5, Edited by L. K. Wang, Y.-T. Hung, and N. K. Shammas, 2005, The Humana Press Inc., Totowa, NJ.
5. Waste Water Technology Fact Sheet Chemical Precipitation, USEPA, EPA 832-F-00-018, September 2000.
6. Fluoride Contamination of Water in India and Its Impact On Public Health, Mahipal Singh Sankhla and Rajeev Kumar, ARC J. Forensic Sci., 3(2), 10–15, 2018.
7. Precipitation Methods Using Calcium-Containing Ores for Fluoride Removal in Wastewater, Li Wang, Ye Zhang, Ning Sun, Wei Sun, Yuehua Hu and Honghu Tang, Minerals, 9, 511–523, 2019; doi:10.3390/min9090511.
8. Application of Organic Coagulants in Water and Wastewater Treatment, E. K. Tetteh and S. Rathilal, 2019, DOI: http://dx.doi.org/10.5772/intechopen.84556, InTechOpen Publication.
9. Fundamental Chemistry of Precipitation and Mineral Scale Formation, A. W. Rudie and P. W. Hart, Proceedings of the 2005 TAPPI engineering, pulping & environmental conference. 2005 August 28–31; Philadelphia, PA. Norcross GA: TAPPI Press: 10 p. ISBN 1-59510-095-4.
10. Technology Overview: Chemical Precipitation, ITRC (Interstate Technology & Regulatory Council). 2010, Washington, D.C.: Interstate Technology & Regulatory Council, Mining Waste Team. www.itrcweb.org.
11. Advantages and Disadvantages of Techniques Used for Wastewater Treatment, G. Crini and E. Lichtfouse, Environmental Chemistry Letters, Springer Verlag, 17 (1), 145–155, 2019. 10.1007/s10311-018-0785-9. hal-02082890.
12. Separation of Heavy Metals: Removal from Industrial Wastewaters and Contaminated Soil. In Emerging Separation Technologies for Metals and Fuels, R. W. Peters and L. Shem (Edited by Laksmanan V. I., Bautista R. G. and Somasundaran P.), pp. 3–64. 1993. The Min., Metals, Mater. Soc., Warrendale, PA.

13. Chemical Water and Waste Water Treatment, Ed. by Hermann H. Hahn Rudolf Klute, Proceedings of the 4th Gothenburg Symposium 1990 October 1–3, 1990 Madrid, Spain, Springer-Verlag.
14. Development of Coagulation Theory and Pre-Polymerised coagulants for Water Treatment, Jia-Qian Jiang, Separation and Purification Methods, 30(1), 127–141, 2001. DOI: 10.1081/SPM-100102986
15. Coagulation and Flocculation in Water and Wastewater Treatment, John Bratby, Third Edition, **ISBN13:** 9781780407494, IWA Publishing, 2006
16. Waste Water Treatment by Coagulation and Flocculation, N.B. Prakash, Vimala Sockan, P. Jayakaran, International Journal of Engineering Science and Innovative Technology (IJESIT) 3(2), 479–484, 2014.
17. Coagulation and Flocculation Process Fundamentals: Why They Are Used, Abhilash Chetty, https://www.mrwa.com/WaterWorksMnl/Chapter%2012%20Coagulation.pdf
18. Chemical Coagulation Process for the Removal of Heavy Metals from Water: A Review, Xiaomin Tang, Huaili Zheng, Houkai Teng, Yongjun Sun, Jinsong Guo, Wanying Xie, Qingqing Yang and Wei Chen, Desalination and Water Treatment, 57(4), 1733–1748, 2016. DOI: 10.1080/19443994.2014.977959
19. Guideline for Drinking Water Quality, 4th edition, WHO, 2011
20. Indian Standards for Drinking Water, second revision of IS 10500, 2004.
21. Enhanced Removal of Heavy Metals in Primary Treatment Using Coagulation and Flocculation, Pauline D. Johnson, Padmanabhan Girinathannair, Kurt N. Ohlinger, Stephen Ritchie, Leah Teuber, Jason Kirby, Water Environment Research, 80(5), 472–479, 2008.
22. Fluoride Contamination of Water in India and its Impact on Public Health, Mahipal Singh Sankhla, Rajeev Kumar, ARC Journal of Forensic Science, 3(2), 10–15, 2018. ISSN 2456-0049
23. WHO, 1993.
24. Study the Efficiency of Alum Coagulant in Fluoride Removal from Drinking Water, Abdollah Dargahi, Zahra Atafar, Mitra Mohammadi, Ali Azizi, Ali Almasi, Mir Mohammad Hoseini Ahagh, International Jo of Pharmacy & Technology, 8(3), 16772–16778, 2016.
25. Characterization of Water Treatment Plant's Sludge and its Safe Disposal Options, T. Ahmad, K. Ahmad, M. Alam, Procedia Environmental Sciences, 35, 950–955, 2016.
26. Post-Coagulation Sludge Management for Water and Wastewater Treatment with Focus on Limiting Its Impact on the Environment, Economic and Environmental Studies, Łukasiewicz, Ewelina, 16(4), 831–841, 2016. ISSN 2081-8319.
27. Water Quality and Treatment: A Handbook of Community Water Supplies, By American Water Works Association, Raymond D. Letterman, Technical Editor, Chapter 12: Chemical Oxidation, Phillip C. Singer, David A. Reckchow. McGraw-Hill, Inc., Fifth Edition, 1999, ISBN 0070016593.
28. Chemical Oxidation Applications in Industrial Waste Waters Olcay Tünay, Işik Kabdaşli, Idil Arsalan-Alaton and Tuğba Ölmez-Hanci, IWA Publishing, First Published 2010.
29. Sustainable Remediation of Contaminated Soil and Groundwater, Materials, Processes, and Assessment, Chapter 7 — Application of slow-release materials for in situ and passive remediation of contaminated groundwater, Chih-Ming Kao, Yih-Terng Sheu, Jiun-Hau Ou, Wei-Han Lin, 169–199, 2020.
30. In-situ Chemical Oxidation, Engineering Issue, United States Environmental Protection Agency USEPA), Scott G. Hauling, Bruce E. Pivetz, Report date 01 August 2006.
31. Decolorization of Dye Waste Waters by Biosorbents: A Review, A. Srinivasan, T. Viraraghavan, Journal of Environmental Management, 91, 1915–1929. 2010.

32. Adsorption of Brilliant Red 2BE Dye From Water Solutions by a Chemically Modified Sugarcane Bagasse Lignin, L. Silva, R. Ruggiero, P. Gontijo, R. Pinto, B. Royer Chemical Engineering Journal, 168, 620–628, 2011.
33. Heterogeneous Water Phase Catalysis as an Environmental Application: A Review, K. Pirkanniemi, M. Sillanpää, Chemosphere, 48, 1047–1060, 2002.
34. Adsorption of Phenolic Compounds by Activated Carbon: A Critical Review, A. Dąbrowski, P. Podkościelny, Z. Hubicki, M. Barczak, Chemosphere, 58, 1049–1070, 2005.
35. Advanced Oxidation Processes for Wastewater Treatment: Formation of Hydroxyl Radical and Application, Jian Long Wang & Le Jin Xu Critical Reviews in Environmental Science and Technology, 42(3):251–325, February 2012 DOI: 10.1080/10643389.2010.507698
36. Recent Developments in Homogeneous Advanced Oxidation Processes for Water and Wastewater Treatment, M. Muruganandham, R. P. S. Suri, Sh. Jafari, M. Sillanpää, Gang-Juan Lee, J. J. Wu and M. Swaminathan, International Journal of Photoenergy, Volume 2014, Article ID821674, https://doi.org/10.1155/2014/821674
37. Advanced Oxidation Processes for the Removal of Antibiotics from Water. An Overview, Eduardo Manuel Cuerda-Correa, María F. Alexandre-Franco and Carmen Fernández-González, Water, 12, 102–153, 2020, doi:10.3390/w12010102
38. Comparison of the Advanced Oxidation Processes (UV, UV/H_2O_2 and O_3) for the Removal of Antibiotic Substances During Wastewater Treatment, A.K. Bin, S. Majed, Ozone: Science and Engineering, 34, 136–139, 2012.
39. Degradation of Endocrine Disrupting Chemicals by Ozone/AOPs, B. Ning, N. Graham, Y. Zhang, M. Nakonechny, M. El-Din, Ozone: Science and Engineering, 29, 153–176, 2007.
40. Advanced Oxidation Processes: Current Status and Prospective, R. Munter, Proceedings of the Estonion Academy of Sciences. Chemistry, 50, 59–80, 2001.
41. Advanced Oxidation Processes (AOPs) for Wastewater Treatment and Reuse: A Brief Review, Ahmed Al Mayyahi and Dr. Hamid Ali AbedAl-Asadi, Asian Journal of Applied Science and Technology (AJAST) 2(3), 18–30, 2018.
42. The Chemistry of Water Treatment Processes Involving Ozone, Hydrogen Peroxide and UV-Radiation, W. H. Glaze, J. Kang, D. H. Chapin, Journal of Ozone: Science and Engineering, 9, 335–352, 1987.
43. Wastewater Engineering: Treatment and Resource Recovery, G. Tchobanoglous, Metcalf & Eddy, 4th Ed. 2013.
44. The UV/oxidation Handbook, Solarchem Environmental Systems, Markham Ontario Canada 1994.
45. Advanced Oxidation Processes for Wastewater Treatment in the Pulp and Paper Industry: A Review, L. G. Covinich, D. I. Bengoechea, R. J. Fenoglio, M. C. Area, American Journal of Environmental Engineering, 4, 56–70, 2014.
46. Post-Treatment of Pulp and Paper Industry Wastewater by Advanced Oxidation Processes, M. F. Sevimli, Ozone: Science and Engineering, 27, 37–43, 2005.
47. Treatment Technologies for Petroleum Refinery Effluents: A Review, B. H. Diya'uddeen, W. M. A. W. Daud, A.R. Aziz, Process Safety and Environmental Protection, 89, 95–105, 2011.
48. Decontamination of Wastewaters Containing Synthetic Organic Dyes by Electrochemical Methods: A General Review, C. A. Martínez-Huitle, E. Brillas, Applied Catalysis B: Environmental, 86, 105–145, 2009.
49. Titanium Dioxide as a Catalyst Support in Heterogeneous Catalysis, S. Bagheri, N. M. Julkapli, S. B. Abd Hamid, The Scientific World Journal, 2014.
50. Recent Developments in Photocatalytic Water Treatment Technology: A Review, M. N. Chong, B. Jin, C. W. Chow, C. Saint, Water Research, 44, 2997–3027, 2010.

51. Combination of Advanced Oxidation Processes and Biological Treatments for Wastewater Decontamination: A Review, I. Oller, S. Malato, and J. A. Sanchez-Perez, Science of the Total Environment, 409, 4141–4166, 2011.
52. Fe^{3+} and UV-enhanced Ozonation of Chlorophenolic Compounds in Aqueous Medium, K. Abe, K. Tanaka, Chemosphere, 35, 2837–2847, 1997.
53. Ozone/H_2O_2 Performance on the Degradation of Sulfamethoxazole, R. C. Martins, R. F. Dantas, C. Sans, S. Esplugas, R. M. Quinta-Ferreira, Ozone: Science and Engineering, 37, 509–517, 2015.
54. UV-O_3, UV-H_2O_2, UV-TiO_2 and the Photo-Fenton reaction—Comparison of AOP's for Wastewater Treatment, G. Ruppert, R. Bauer, G. Heilser, Chemosphere, 28, 1447–1454, 1994.
55. Synergistic Effects of Combination of Photolysis and Ozonation on Destruction of Chlorophenols in Water, W. S. Kuo, Chemosphere, 39, 1853–1860, 1999.
56. Applicability of Fluidized Bed Reactor in Recalcitrant Compound Degradation Through Advanced Oxidation Processes: A Review, F. Tisa, A. A. Abdul Raman, and W. M. A. Wan Daud, Journal of Environmental Management 146, 260–275, 2014.
57. Advanced Oxidation Processes (AOPs) Involving Ultrasound for Waste Water Treatment: A Review with Emphasis on Cost Estimation, N. N. Mahamuni and Y. G. Adewuyi, Ultrasonics Sonochemistry, 17, 990–1003, 2010.
58. Wastewater Engineering Treatment and Reuse, G. Tchobanoglous, F. L. Burton, and D. H. Stensel, 4th ed. New York: McGraw-Hill Companies, 2003.
59. Trends in Electro-Fenton Process for Water and Wastewater Treatment: An Overview, P. V. Nidheesh, R. Gandhimathi, Desalination, 299, 1–15, 2012.
60. The Role of Ferrous Ion in Fenton and Photo-Fenton Processes for the Degradation of Phenol, V. Kavitha, K. Palanivelu, Chemosphere, 55, 1235–1243, 2004.
61. Advanced Oxidation Technologies: Sustainable Solutions for Environmental Treatment, M. I. Litter, R. J. Candal, J. M. Meichtry, Taylor and Francis Group, 348, 2017.

8 Advanced Oxidation Processes for Wastewater Treatment

Gunjan Deshmukh and Haresh Manyar
School of Chemistry and Chemical Engineering,
Queen's University Belfast, Belfast, Northern Ireland, UK

CONTENTS

8.1 INTRODUCTION

Rapidly increasing global population, coupled with increasing demand for energy and materials to support a better lifestyle, has resulted in the increased need for clean water and has led to the subsequent scarcity of potable water. The increase in demand for clean water is burdening our natural resources, and it has become a major global challenge to ensure the supply of clean water [1]. The available natural

DOI: 10.1201/9781003138303-8

water resources are finite and must be used sustainably. The water bodies have been severely polluted with industrial effluents, containing a variety of pollutants categorized into metal, organic and inorganic chemicals, and microplastics, making water unsuitable for direct consumption and causing a harmful effect on the environment and public health [2]. Oxidation process technologies are the most promising approach for water purification and are constantly practiced and investigated for development of better catalysts and reactor designs suitable for easy scale-up. Water oxidation is widely classified into physical, chemical, and biological processes. Biological processes are economically feasible, but have limited potential in treatment of water containing hazardous pollutants. The thermal decomposition processes are effective for removal of chemical pollutants, but energy intensive, thus limiting the economic feasibility. The chemical oxidation processes are versatile, well developed and overcome most of the above-mentioned limitations. The catalytic oxidation technologies have emerged as preferred tools to efficiently tackle water purification by removal of organic pollutants [3]. Advanced oxidation processes (AOPs) have become an integral study for waste water treatment. They were begun in 1987 by Glaze et al. [4] and since then have been a topic of developing chemistry and technologies over three decades. In a general photocatalytic water purification process, an oxidant (H_2O_2 or O_3) is utilized under ultraviolet light as an energy source to generate reactive oxygen species, which further react to decompose the pollutants in the water. In heterogeneous catalysis the catalytic oxidation process follows different mechanisms to generate reactive oxygen species. Metal oxides such as TiO_2 or ZnO act as semiconductors wherein photons with equal or more energy than the material's band gap are adsorbed on the surface of the catalyst, which results in the formation of pair of conduction band electron (e^-_{cb}) and valence band hole (e^+_{vb}). TiO_2 is extensively studied as a heterogeneous catalyst in the photo oxidation process. A typical TiO_2 catalyzed mechanism for complete photo oxidation process is represented in Fig. 8.1. The AOP for wastewater treatment generally proceeds through use of different oxidizing agents. The oxidizing agents in order of their decreasing oxidation potential is given as $OH^- > O_3 > H_2O_2 > KMnO_4 > HOCl > HClO_2 > O_2$. The preferred and most powerful oxidizing agent is OH^- ranging down to free O_2. The degradation process of organic pollutants involves redox reaction decomposition, resulting in CO_2 and H_2O as by products in ambient conditions.

Constant evolution and changing lifestyles have led to an increase in water contaminants from the pharmaceutical industry, consumer products, textile industry, dyes, etc. There is constant need to develop new technologies and innovative remedies to remove these hazardous substances from water. Various technological developments in AOPs have been achieved over the past three decades. The more recent developments build on rational design criteria such as the structure of catalyst surface, active sites, band gap, electron-hole pair separation and lifetime, and plasmonics to develop more effective processes for water purification and to develop robust, efficient, and stable catalysts that are easy to recover, recycle, and reuse. For instance, new magnetic materials are fabricated for easy separation of the catalyst incorporating the magnetic properties. This chapter briefly reviews the catalytic oxidation processes and summarizes the developments in the field of photocatalysis in water purification based on the type of pollutants, technologies, and the operating conditions.

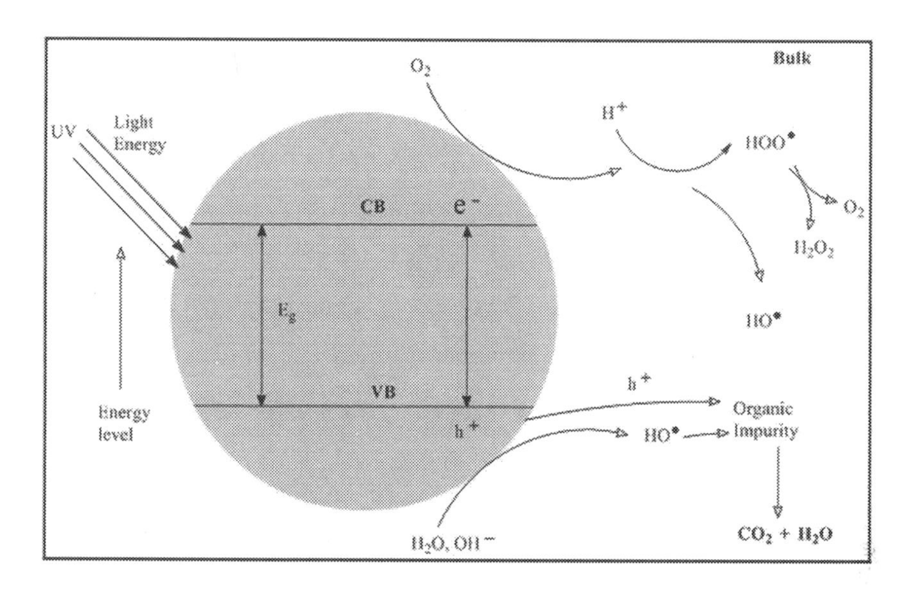

FIGURE 8.1 Schematic representation of a photocatalytic process.

8.2 DIFFERENT AOPs FOR WATER PURIFICATION

8.2.1 Photocatalytic AOPs

In photocatalytic AOPs, the main aim is to generate a hydroxyl radical having high oxidation potential (2.8 eV) just below the fluorine atom in the series. This hydroxyl radical can be generated via different pathways, providing them with an oxygen source and a suitable catalyst to facilitate the reaction in either presence or absence of light. The overall process can be broadly classified into different combination of these parameters as given in Fig. 8.2. Further, the photocatalytic process is subdivided into homogeneous catalytic systems and heterogeneous catalytic systems. In a homogeneous catalytic system the catalyst remains in the same phase, with uniform distribution, while in a heterogeneous catalytic system the catalyst and the reaction medium are in different phases. This fundamental difference in the catalytic system affects the mode of reaction mechanism and the activity of the catalyst.

8.2.1.1 Photocatalytic AOPs Based on Homogeneous Catalysis

Transition metals have been widely studied as homogeneous catalytic systems in AOPs. The activity of homogeneous catalysts is well known, and these catalysts are mainly involved in Fenton's process, photo-Fenton's process and electrochemical oxidations integrated with O_3 and UV. The most commonly studied transition metals involved in the above mentioned AOPs are Fe, Zn, Mn, Ti, Cr, Cu, Co, Ni, Cd, and Pb salts in their possible multivalent forms [5]. Among these elements, Fe is the most extensively studied element in Fenton and photo-Fenton processes to generate hydroxyl radicals. Further, Fe salts are widely used in catalytic ozonolysis processes, which help to enhance the activity of overall process [6]. An ideal homogeneous

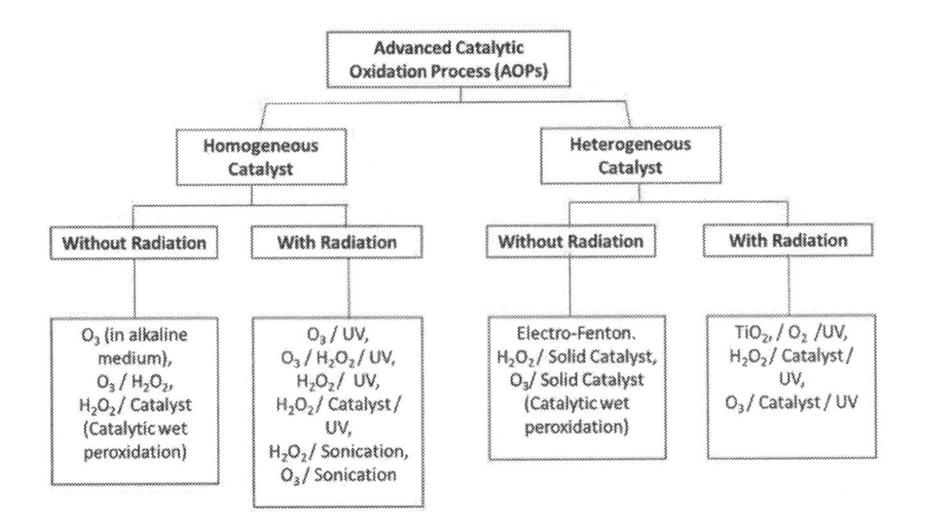

FIGURE 8.2 Classification of advanced catalytic oxidations processes.

photocatalyst should be effective in a wide range of pH, and it was observed that the degradation efficiency of pollutants is higher at acidic pH with Fe as a catalyst compared to a catalyst-free process [7]. Further, it can also be noted that higher Fe catalyst concentration leads to Fe^{3+} ion formation, which increases the scavenging effect and decreases the efficiency of the catalyst. There are disadvantages of homogeneous catalysts, which can be listed as narrow pH range, short life span of the catalyst, difficult isolation and recovery cost, scavenging effect, no reusability, formation of by-products, and use of large excess of oxidants. To overcome these barriers, heterogeneous catalysts are advantageous in current science for water purification.

8.2.1.2 Photocatalytic AOPs Based on Heterogeneous Catalysis

An ideal heterogeneous catalyst in water purification should be hydrothermally stable, highly effective in wide range of pH, easily recoverable and reusable, cost effective, photosensitive, and highly reactive and should not leach or deactivate in reaction conditions. Further, a heterogeneous catalyst prohibits scavenging effect and sludge formation. Several metal oxides, supported metal oxides, mixed metal oxides, zeolites, and noble metals supported on metal oxides have been reported. Among them, TiO_2 is the most-studied metal oxide. The anatase phase of TiO_2 has become the most active for generation of reactive oxygen species, and it is well established as a catalyst in water treatment processes [6]. Other than TiO_2, Fe and its metal oxide, other metal oxides such as Nd, Cr, Ce, Ag, Zn, Mn, Cu, Co, Ni, and Cd were supported on different available supports such as activated carbon, TiO_2, graphene, etc. The recognized catalysts for AOPs are TiO_2, nano-size Fe and Fe minerals (ferrihydrite, α-Fe_2O_3 lepidocrocite (γ-FeOOH), and α-FeOOH), activated carbon, magnetite, and pyrite [8, 9]. The activity of these metal oxides can be enhanced by supporting them on different frameworks having stable structure and high surface area. Supports such as Nafion, graphene, carbon aerogel, carbon nanotube (CNT), clays, polymers, alumina, fly ashes, and zeolite are studied widely. These supports enable the activity of the catalyst over wide range of pH, and the porous nature of catalyst exposes the more active sites to increase the efficiency of the catalyst. These supports are preferably porous with high surface area, which helps in catalyst stability and enhances activity [6, 10].

Commercially available catalysts such as zeolites and pillared clay-based catalysts have proven to have the best catalytic activity. Zeolites have been efficiently used in degradation of organic pollutants in the presence of UV light. They were used as a photocatalyst in a continuous reactor with 70 percent color degradation and 57 percent COD reduction [11]. Similarly, pillared clay having larger pore size was efficiently used as an effective catalyst in photo-Fenton processes [12]. It can be concluded that the type of metal oxide and the support can be chosen based on the pH of the solution, operating temperature, type of the pollutant, source of energy, concentration of the pollutant, etc.

8.2.2 Fenton and Fenton-Like AOPs

Fenton process is second most studied method for elimination of water pollutants; in this, H_2O_2 is used as an oxidation reagent, using Fe as a catalyst which is referred as Fenton's reagent. Though the reagent was discovered at the beginning of 19th century, it was used in 1980s. Fenton wastewater purification is an effective method for removing organic impurities with CO_2 and H_2O as final products and inorganic salts. This process also proceeds through generation of hydroxyl radicals, which further attack and destroy organic pollutants. This process may involve two oxidizing agent, such as H_2O_2 and O_2, along with a metal oxide catalyst. Further, incorporation of light (wavelength up to 600 nm) in the process is known as the photo-Fenton process, which helps in degradation of pollutants and boosts the rate of decomposition. The reaction proceeds through a combination of Fe (II) and H_2O_2 with the generation of hydroxyl radicals ($OH^{\cdot}$, $HO_2^{\cdot}$), as shown in Scheme 8.1.

$$Fe^{2+} + H_2O_2 \rightarrow Fe^{3+} + OH^- + OH^{\cdot}$$
$$Fe^{3+} + H_2O_2 \rightarrow Fe^{2+} + HO_2^{\cdot} + H^+$$

SCHEME 8.1 Fenton process.

The hydroxyl radicals further undergo chain reaction and react with organic pollutants, resulting in their degradation. The coagulated residual salt is settled and filtered. The Fenton process is used for purification of industrial waste containing organic impurities such as phenols, aldehydes, plastic additives, rubber additives, and pharmaceutical compounds. Although the Fenton process is widely studied, there are some limitation for this process, such as optimization of H_2O_2 quantity with higher efficacy, limited pH range, and separation and recovery of the catalyst.

8.2.3 Sonocatalytic AOPs

Sonocatalysis is an emerging technology that has attracted significant interest over the past two decades. It is modification of AOP in which energy generated from ultrasound waves is used to produce hydroxyl radicals using H_2O_2 (Scheme 8.2).

$H_2O(g)$ + ultrasound $\rightarrow \bullet OH(aq) + \bullet H(aq)$
$\bullet OH(aq) + \bullet OH(aq) \rightarrow H_2O_2\ (aq)$
$Fe^{2+}(aq) + H_2O_2\ (aq) \rightarrow Fe^{3+}(aq) + \bullet OH(aq) + OH^-(aq)$
$Fe^{2+}(aq) + \bullet OH(aq) \rightarrow Fe^{3+}(aq) + OH^-(aq)$
$Fe^{3+}(aq) + H_2O_2\ (aq) \rightarrow Fe^{2+}(aq) + \bullet OOH(aq) + H^+(aq)$
Fe^0 (s) + ultrasound $\rightarrow Fe^{2+}(aq) + 2e^-$

SCHEME 8.2 Generation of reactive oxygen species in ultrasound conditions.

Fe-based metal oxide catalysts are used with ultrasound energy or waves. Extensive use of H_2O_2 in the Fenton process limits application in safety, storage, and hazards. Ultrasound energy is applied as an efficient tool for degradation of organic pollutants at higher rate. In sonocatalysis, liquid under ultrasound application undergoes rapid pressure change, resulting in the formation of cavities that eventually undergo implosive collapse. The overall process happens in a fraction of a second, releasing a high magnitude of energy helping to generate hydroxyl radicals.

In sonocatalysis, frequencies in the range of 20 kHz to 1000 kHz can be used effectively. The higher ultrasound wave frequency helps in effective destruction of pollutants. The efficacy of the process also depends on the concentration of pollutants, and it can be diluted to optimum concentration for best results. The low frequency of the ultrasound waves mechanically disturbs the surface of the catalyst to clean it off with adsorbed impurities. The chemical effects are observed at higher frequencies.

Fe-based catalysts are mostly reported with higher efficiency in this process [13]. Catalyst particles suspended in water can act as a nucleus for cavitation bubbles. Once the catalyst particle sizes are in the same order of magnitude as the size of the cavitation bubbles, catalyst particles can form an extra nucleus of cavitation bubbles. The extra nucleus generates more cavitation bubbles. This cascade effect causes stronger cavitation effects, leading to higher degradation efficiency. Sonocatalytic degradation (20 kHz, 125 W, 35°C) of phenol derivatives was studied using $FeSO_4$ as a homogeneous catalyst efficiently at 35°C [14]. The use of a heterogeneous catalyst along with ultrasound treatment is more effective. The nanocrystalline TiO_2 was efficiently used as a catalyst in degradation of chlorophenol at 20–25°C [15]. This process still requires further technological development and design modifications to facilitate an easier scale-up. This process has some disadvantages, such as high energy consumption, rise in turbidity of water, high maintenance cost for replacement of ultrasound probes, and lack of water disinfection ability.

8.2.4 Ozonation AOPs

Ozone is a highly reactive molecule that breaks down to dioxygen and oxygen radicals. It is a powerful oxidizing agent with $E^0 = +2.07$ eV. Ozone readily reacts with a double bond capable of degrading several organic molecules containing double bonds. Ozone may react with water to produce a hydroxyl radical as a source of reactive oxygen molecule. The decomposition of ozone leads to the formation of H_2O_2, as elaborated in Scheme 8.3. The literature reports different mechanisms based on the catalyst system used. The efficiency of ozonolysis is pH-, temperature-, and pressure-dependent.

Initiation

$O_3 + H_2O = 2\ HO\cdot + O_2$

$O_3 + OH^- = O_2\cdot - + HO_2\cdot$

Propagation

$O_3 + HO\cdot = O_2 + HO_2\cdot$

$O_2 + HO_2\cdot = O_2\cdot^- + H^+$

$O_3 + HO_2\cdot = 2O_2 + HO\cdot$

Termination

$2\ HO_2\cdot = O_2 + H_2O_2$

SCHEME 8.3 Generation of reactive oxygen species in ozonolysis.

The activity of ozone is controlled at low pH to prevent discriminative reaction of ozone with organic and inorganic constituents of the mixture. Further, the process demands optimum pressure and temperature to keep a check on the dissolution of ozone in the water. This directly affects the rate of the desired reaction.

8.2.5 Wet Air Oxidation

Catalytic wet air oxidation (CWAO) is one of the oldest and most commonly practiced methods when the pollutants are too dilute to incinerate. This process operates at high temperature and pressure conditions in the presence of a suitable catalyst. This process also follows the free radical mechanism to eliminate the organic pollutants, resulting in H_2O and CO_2 as final products along with the salts. Temperature is the key parameter in this process, as temperature is related to the concentration of dissolved oxygen and hence the reaction rate of the AOP process. The design of the reactor to maintain the pressure is also important, considering the economy and safety requirements of the process. Several homogeneous as well as heterogeneous catalysts are reported for CWAO process. Recent studies focus more on the use of heterogeneous catalysts because of higher efficiency and reusability of the catalysts. Several catalytic systems have recently been reported for the CWAO; they include metal oxides, supported metal oxides with noble metals, and carbon-based catalysts [16]. The CWAO process is often expressed in a two-step oxidation process in which organic pollutants are converted into intermediate compounds that subsequently undergo further oxidative degradation (Fig. 8.3).

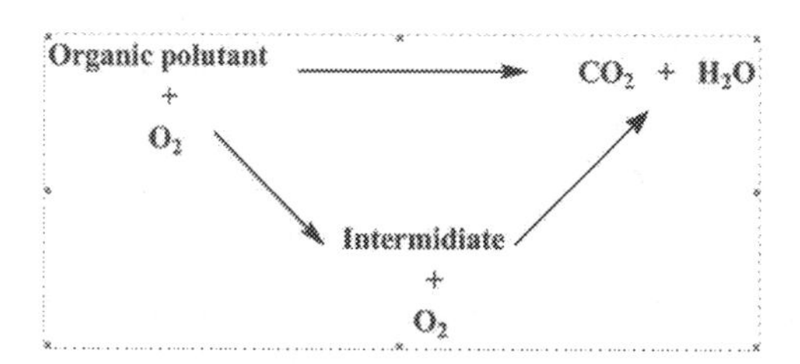

FIGURE 8.3 Wet air oxidation of organic pollutants in water.

8.3 FACTORS AFFECTING PHOTOCATALYSIS

The efficiency of the photocatalytic process in generation of hydroxyl radicals is governed by different operating conditions, which also decide the economic feasibility and scalability of the process. These parameters are optimized according to the AOPs and selected catalysts. There are several parameters, such as pH of the system, concentration of the catalysts, concentration of the oxidants, system temperature, type of the catalysts, residence time (RT) in case of continuous flow processes, agitation speed in batch and CSTR processes, pressure of air or O_2, source of the light, intensity and wavelength, concentration and the type of the pollutant, type of buffer used, etc. Some of these parameters are extensively critical, and these parameters are briefly discussed below.

8.3.1 pH Range

The pH of the solution is one of the most important factors; it decides the efficiency of the process and the photocatalytic activity of the catalyst. The pH range of the system may vary according to the adapted catalyst, oxidant, and type of pollutant. In the case of ozonation, higher pH is considered to improve the selectivity of the recalcitrant pollutants [12]. In contrast, generation of OH^- in photo-Fenton process is favorable at lower pH in decomposition of H_2O_2. It should be noted that pH also plays a critical part in photodegrading activity. The pH of the solution directly affects the surface charge of the semiconductor catalyst, flat band potential, and the rate of dissociation [17]. A change in pH during the process also affects the degradation and decolorization of the dyes (methyl orange) due to change in adsorption of the dye on the catalyst surface. An increase in the pH reduces the adsorption of the dye on the catalyst surface [18]. The point of zero charge (PZC) is a point at which the surface of the catalyst is unchanged. When the pH value is smaller than PZC, the catalyst surface is positively charged and vice-versa. Thus, activity of the catalyst is largely dependent on charge of the catalyst surface and can be described based on PZC.

8.3.2 Temperature

Several studies have been performed to relate the activity of photocatalytic oxidation to the process temperature when TiO_2 was used as catalyst. It was observed that an increase in the photocatalytic reaction temperature (above 80°C) promotes the recombination of charge carriers and does not favor the adsorption of organic pollutants onto the surface of the catalyst. The rate of the photocatalytic reaction above 80°C was interpreted by the Langmuir-Hinshelwood (LHHW) mechanism. It was observed that nanostructured TiO_2 showed increase in the rate constant which was promoted by 8 minutes when temperature was increased from 38°C to 100°C [19].

8.3.3 Initial Concentration of Pollutants

Initial concentration of pollutants plays a significant role in their degradation in AOPs. In photocatalytic degradation, increased concentration of pollutants raises the collision between hydroxyl radicals and the organic pollutants increase the degradation up to a certain point. After the optimum concentration, the degradation efficiency decreases [20]. In case of the dyes, the increase in the concentration of dyes results in the deactivation of catalyst surface, due to excess adsorption. This results in adsorption of oxidants, whether O_2 or OH^-. This consequently reduces the concentration of reactive oxygen species. Furthermore, the increased concentration of pollutants blocks the light (UV or sunlight), which subsequently reduces the number of photons, leading to a reduction in the rate of photocatalytic reaction. Thus the overall process is dependent on the optimized concentration of each species present in the system.

8.3.4 Photocatalyst Concentration

The concentration of catalysts affects the rate of the degradation process. It is widely accepted that an increase in the catalyst concentration is directly proportional to the rate of the reaction. However, the ideal concentration is required for an efficient process, as excess catalyst leads to unfavorable scattering of light and restricts the penetration of light in the solution.

8.3.5 Effect of Light Intensity

The hydroxyl radical is responsible for degradation of pollutants. UV radiation has a direct effect on the formation of reactive oxygen species generated from the oxidants (H_2O_2/O_3). High photocatalytic activity for water purification is achieved by high intensity of UV radiation to adequately provide the light on the catalytic surface. The nature or form of the light does not affect the reaction pathway [21]. Researchers concluded that the band-gap sensitization mechanism does not matter in photocatalytic degradation. Unfortunately, only 5 percent of total irradiated natural sunlight is available to cause effective photosensitization. Also, there is loss in energy due to the light reflection and transmission [22]. The quanta absorbed by the photocatalyst or reactant in the overall process can be given by the equation below [23].

$$\Phi_{overall} = \frac{rate\ of\ reaction}{rate\ of\ absorption\ of\ radication}$$

8.4 ADVANTAGES AND DISADVANTAGES OF COMMERCIAL AOPs

The advantages and disadvantages of the conventional AOPs are summarized in Table 8.1.

8.5 RECENT ADVANCES AND FUTURE PERSPECTIVES IN PHOTOCATALYTIC AOPs

A wide range of studies have reported on photocatalytic AOPs, but industrial application of these advanced process is challenging due to the economic feasibility, scalability and robustness, stability, and longevity of the catalysts. UV radiation is a commercially accepted process in purification of groundwater and drinking water,

TABLE 8.1
Advantages and Limitations of Conventional AOPs

	Advantages	Disadvantages
O_3	Strong oxidation power; easily performed; short reaction time; no remaining sludge; all residuals of O_3 easily decomposed	Higher operation cost; energy intensive process; needs pretreatment
O_3/UV	Higher efficiency; more efficiency at generating •OH; more effective than O_3 alone or UV alone	Not cost-effective; energy intensive; mass transfer limitation; sludge production; turbidity
H_2O_2/UV	Disinfectant; simple process	Can interfere with the penetration of light H_2O_2/UV; turbidity can interfere with the penetration of light; less efficient in generating •OH
O_3/H_2O_2	More efficient than O_3 or H_2O_2 alone; effective for H_2O treatment	bromate formation; excess usage of H_2O_2; not cost-effective; energy intensive
O_3/H_2O_2/UV	Nonselective with all species in solution; degradation of aromatics and polyphenols significantly faster	Expensive; COD removal not complete; sludge production
Fenton-based processes	Rapid reaction rates; small footprint; cost and energy effective; generate strong •OH; degrade a wide range of recalcitrant components; no mass transfer; recycling of ferrous catalysts by reduction of Fe^{3+}	Not full-scale application exists; small of sludge production; acidic environment

but the use of visible light and direct sun light is much more desirable. There is need for development of energy-efficient processes with use of visible light. TiO_2 has shown very promising catalytic activity in several photocatalytic AOPs but is still associated with numerous limitations and needs further development. Anderson and co-workers have shown improvements by using advanced mixed-metal oxide composites, such as Cu_2O/TiO_2 and Cu_2O/TiO_2-AC composite photocatalysts [24]. Mixed Cu_2O/TiO_2 and Cu_2O/TiO_2-AC composite catalysts showed superior activity in simultaneous removal of nitrate and oxalic acid from aqueous environments with high selectivity to N_2, which otherwise was not possible using the TiO_2 alone. The authors have extensively studied the potential of Cu_2O/TiO_2-AC composite photocatalysts for adsorptive behavior of nitrate and organics from wastewater, as well as removal using visible light [25]. Manyar and co-workers have developed a highly efficient continuous flow photocatalytic AOP for photodegradation of phenols using Taylor-flow regime and visible light in presence of very stable and robust Au-Pd/TiO_2 photocatalysts [26]. In future, photocatalytic AOPs need further development both from reaction engineering as well as materials science perspectives. On one hand, it is important to develop more efficient, robust, economic, stable, and recyclable photocatalysts that are active in visible light, while it is also essential to develop proper reactor design for efficient light penetration, enhanced mass transfer, and increased reaction rates to achieve higher efficiency for photocatalytic AOPs using visible spectrum and direct solar light. The key for easy scale-up and potential commercialization is in the development of new photocatalysts with innovative structural modifications that can be combined with niche reactor designs because of complicated multiphase reactions.

8.6 CONCLUSIONS

The growing number of publications in the area of wastewater treatment and AOPs in the 21th century is evidence of the value of wastewater treatment because of the water scarcity. Current research thrives on the development of novel, cheaper, and highly effective methods applicable on wide range of wastewater problems. However, it is very difficult to find a universal solution for domestic as well as industrial wastewater treatment, the latter being more difficult. This chapter highlights recent developments in AOP technologies for water treatment and their classification. AOPs have further emerged as wide-range solutions with constant modifications and evolution as a result of the new challenges. Even though there are number of publications on the topic, very few AOPs are used industrially on a large scale considering economic feasibility, catalyst stability, robustness, and longevity. The literature data further suggest that oxidative degradation using photocatalytic processes is being continuously developed. The structural development of semiconductor catalysts and their comparative efficiency in industrial wastewater treatment is summarized. AOPs are widely practiced methods due to their higher treatment efficiency and multiplicity.

REFERENCES

[1] M.N. Chong, B. Jin, C.W.K. Chow, C. Saint, Recent developments in photocatalytic water treatment technology: A review, Water Res. 44 (2010) 2997–3027. https://doi.org/10.1016/j.watres.2010.02.039.

[2] C.S.D. Rodrigues, R.M. Silva, S.A.C. Carabineiro, F.J. Maldonado-Hódar, L.M. Madeira, Wastewater treatment by catalytic wet peroxidation using nano gold-based catalysts: A review, Catalysts. 9 (2019). https://doi.org/10.3390/catal9050478.

[3] D. LI, J. QU, The progress of catalytic technologies in water purification: A review, J. Environ. Sci. 21 (2009) 713–719. https://doi.org/10.1016/S1001-0742(08)62329-3.

[4] W.H. Glaze, J.W. Kang, D.H. Chapin, The chemistry of water treatment processes involving ozone, hydrogen peroxide and ultraviolet radiation, Ozone Sci. Eng. 9 (1987) 335–352. https://doi.org/10.1080/01919518708552148.

[5] A. Buthiyappan, A.R. Abdul Aziz, W.M.A. Wan Daud, Recent advances and prospects of catalytic advanced oxidation process in treating textile effluents, Rev. Chem. Eng. 32 (2016) 1–47. https://doi.org/10.1515/revce-2015-0034.

[6] P.R. Gogate, A.B. Pandit, A review of imperative technologies for wastewater treatment I: Oxidation technologies at ambient conditions, Adv. Environ. Res. 8 (2004) 501–551. https://doi.org/10.1016/S1093-0191(03)00032-7.

[7] M. Hammad Khan, J.Y. Jung, Ozonation catalyzed by homogeneous and heterogeneous catalysts for degradation of DEHP in aqueous phase, Chemosphere. 72 (2008) 690–696. https://doi.org/10.1016/j.chemosphere.2008.02.037.

[8] K. Rusevova, F.D. Kopinke, A. Georgi, Nano-sized magnetic iron oxides as catalysts for heterogeneous Fenton-like reactions-Influence of Fe(II)/Fe(III) ratio on catalytic performance, J. Hazard. Mater. 241–242 (2012) 433–440. https://doi.org/10.1016/j.jhazmat.2012.09.068.

[9] R. Matta, K. Hanna, S. Chiron, Fenton-like oxidation of 2,4,6-trinitrotoluene using different iron minerals, Sci. Total Environ. 385 (2007) 242–251. https://doi.org/10.1016/j.scitotenv.2007.06.030.

[10] R. Gonzalez-Olmos, M.J. Martin, A. Georgi, F.D. Kopinke, I. Oller, S. Malato, Fe-zeolites as heterogeneous catalysts in solar Fenton-like reactions at neutral pH, Appl. Catal. B Environ. 125 (2012) 51–58. https://doi.org/10.1016/j.apcatb.2012.05.022.
[11] S. Apollo, M.S. Onyango, A. Ochieng, Integrated UV photodegradation and anaerobic digestion of textile dye for efficient biogas production using zeolite, Chem. Eng. J. 245 (2014) 241–247. https://doi.org/10.1016/j.cej.2014.02.027.
[12] E.G. Garrido-Ramírez, B.K.G. Theng, M.L. Mora, Clays and oxide minerals as catalysts and nanocatalysts in Fenton-like reactions: A review, Appl. Clay Sci. 47 (2010) 182–192. https://doi.org/10.1016/j.clay.2009.11.044.
[13] N. Zhang, G. Xian, X. Li, P. Zhang, G. Zhang, J. Zhu, Iron Based Catalysts Used in Water Treatment Assisted by Ultrasound: A Mini Review, Front. Chem. 6 (2018) 1–6. https://doi.org/10.3389/fchem.2018.00012.
[14] M. Papadaki, R.J. Emery, M.A. Abu-Hassan, A. Díaz-Bustos, I.S. Metcalfe, D. Mantzavinos, Sonocatalytic oxidation processes for the removal of contaminants containing aromatic rings from aqueous effluents, Sep. Purif. Technol. 34 (2004) 35–42. https://doi.org/10.1016/S1383-5866(03)00172-2.
[15] Z. Dai, A. Chen, H. Kisch, Efficient sonochemical degradation of 4-chlorophenol catalyzed by titanium dioxide hydrate, Chem. Lett. 34 (2005) 1706–1707. https://doi.org/10.1246/cl.2005.1706.
[16] G. Jing, M. Luan, T. Chen, Progress of catalytic wet air oxidation technology, Arab. J. Chem. 9 (2016) S1208–S1213. https://doi.org/10.1016/j.arabjc.2012.01.001.
[17] A. Nezamzadeh-Ejhieh, M. Amiri, CuO supported Clinoptilolite towards solar photocatalytic degradation of p-aminophenol, Powder Technol. 235 (2013) 279–288. https://doi.org/10.1016/j.powtec.2012.10.017.
[18] M. Muruganandham, M. Swaminathan, Decolourisation of Reactive Orange 4 by Fenton and photo-Fenton oxidation technology, Dye. Pigment. 63 (2004) 315–321. https://doi.org/10.1016/j.dyepig.2004.03.004.
[19] Q. Hu, B. Liu, Z. Zhang, M. Song, X. Zhao, Temperature effect on the photocatalytic degradation of methyl orange under UV-vis light irradiation, J. Wuhan Univ. Technol. Sci. Ed. 25 (2010) 210–213. https://doi.org/10.1007/s11595-010-2210-5.
[20] A. Nezamzadeh-Ejhieh, Z. Salimi, Solar photocatalytic degradation of o-phenylenediamine by heterogeneous CuO/X zeolite catalyst, Desalination. 280 (2011) 281–287. https://doi.org/10.1016/j.desal.2011.07.021.
[21] C. Karunakaran, S. Senthilvelan, Photooxidation of aniline on alumina with sunlight and artificial UV light, Catal. Commun. 6 (2005) 159–165. https://doi.org/10.1016/j.catcom.2004.11.014.
[22] L. Yang, Z. Liu, Study on light intensity in the process of photocatalytic degradation of indoor gaseous formaldehyde for saving energy, Energy Convers. Manag. 48 (2007) 882–889. https://doi.org/10.1016/j.enconman.2006.08.023.
[23] H. Kim, S. Lee, Y. Han, J. Park, Preparation of dip-coated TiO2 photocatalyst on ceramic foam pellets, J. Mater. Sci. 41 (2006) 6150–6153. https://doi.org/10.1007/s10853-006-0574-x.
[24] H. Adamu, A. J. McCue, S. R. F. Taylor, H. G. Manyar, J. A. Anderson, Simultaneous photocatalytic removal of nitrate and oxalic acid over Cu_2O/TiO_2 and Cu_2O/TiO_2-AC composites, Appl. Catal. B. Environ. 217 (2017) 181–191.
[25] H. Adamu, M. Shand, S.R.F. Taylor, H. G. Manyar, J. A. Anderson, Use of carbon-based composites to enhance performance of TiO_2 for the simultaneous removal of nitrates and organics from aqueous environments, Environ. Sci. and Poll. Res., 25 (2018) 32001–32014.
[26] M. T. Yilleng, E. C. Gimba, G. I. Ndukwe, I. M. Bugaje, D. W. Rooney, H. G. Manyar, Batch to Continuous Photocatalytic Degradation of Phenol using TiO_2 and Au-Pd nanoparticles supported on TiO_2, J. of Environ. Chem. Eng. 6, 5 (2018) 6382–6389.

9 Catalytic Ozonation Processes for Wastewater Treatment

Jose A. Lara-Ramos and
Fiderman Machuca-Martinez
GAOX, CENM, Escuela de Ingeniería Química,
Universidad del Valle, Ciudad Universitaria Meléndez,
760032, Cali, Colombia

Miguel A. Mueses
Department of Chemical Engineering, Universidad
de Cartagena, Cartagena A.A., Colombia

Jennyfer Diaz-Angulo
Research Group in Development of Materials and Products
GIDEMP, CDT ASTIN SENA, Tecnoparque, Cali, Colombia;
Escuela de Ingeniería Agroindustrial, Instituto Universitario
de la Paz, Barrancabermeja, Colombia

CONTENTS

DOI: 10.1201/9781003138303-9

9.1 INTRODUCTION

Ozonation, as an advanced oxidation process, has received increasing interest in water treatment. Ozone molecules were discovered in 1840 by Schönbein, and 46 years later its germicide potential was harnessed for the first time to sterilize water (Wang & Chen 2020). Later, ozonation pilot tests were implemented in Niza, and in 1906 was applied to drinking water disinfection and has continued until our days (Beltrán 2003; Gottschalk et al. 2010; Wang and Chen 2020).

Ozonation has been applied to the disinfection of wastewater and degradation of different compounds (Oller et al. 2011; Wang and Bai 2017a; Wang and Chen 2020). The main advantages of ozone are its easy implementation and automation at the industrial level and its short reaction time; it decays rapidly in water, removes color, taste, and odor, and it does not require chemicals for its generation (Beltrán 2003; Wang and Bai 2017a; Li et al. 2020a). However, the relatively low solubility of ozone can increase the cost of application. Besides, ozone is selective to react with some pollutants, which leads to incomplete mineralization (Shammas & Wang 2005). To overcome these limitations, catalytic ozonation emerges as a promising alternative because allowing the ozone decomposition in hydroxyl radicals that have higher oxidant power reaches good mineralization percentages and increases the ozone utilization efficiency (Nawrocki 2013; Lara-Ramos et al. 2021).

This chapter aims to describe the outlook of the ozonation and catalytic ozonation processes regarding the mechanism, the main factors that influence the processes, economic aspects, types of reactors used, and some real applications.

9.1.1 Fundamentals

9.1.1.1 Ozone

9.1.1.1.1 Characteristics and Properties

Ozone (O_3) is an inorganic molecule; it is composed of three oxygen atoms (triatomic form of oxygen) less stable than molecular oxygen, O_2. The oxygen atoms in ozone have the following electronic configuration (Beltrán 2003; Sanchez-Polo & Rivera-Utrilla 2004): $1s^2\ 2s^2\ 2p_x^2\ 2p_y^1\ 2p_z^1$. Under normal conditions, ozone is a pale blue gas, has a pungent odor, and is unstable. It is a powerful oxidant with a weak polarity (0.53 Debyes). Other properties are shown in Table 9.1 (Beltrán 2003).

9.1.1.1.2 Reaction Mechanism

Ozone has an oxidizing potential of 2.07 V, which is why it can destroy or degrade organic pollutants in two ways: first by a direct electrophilic attack by molecular ozone (direct reaction mechanism), second by indirect attraction with the radical hydroxyl (OH•) as the primary oxidant (indirect reaction mechanism), which were generated through the ozone decomposition process (Beltrán 2003; Gottschalk et al. 2010). Fig. 9.1 shows the indirect and direct pathways and their interaction.

The electronic configuration of the oxygen atoms attributes a high reactivity to the ozone molecule, since an electrophilic character of the O_3 molecule is generated by the electrons' absence in one of these oxygen atoms within the resonance structure.

TABLE 9.1
Physicochemical Properties of Ozone (Morsch 2019; Wang et al. 2020)

Property	Value	Views, Bonding in Ozone
Molecular weight (g/mol)	48	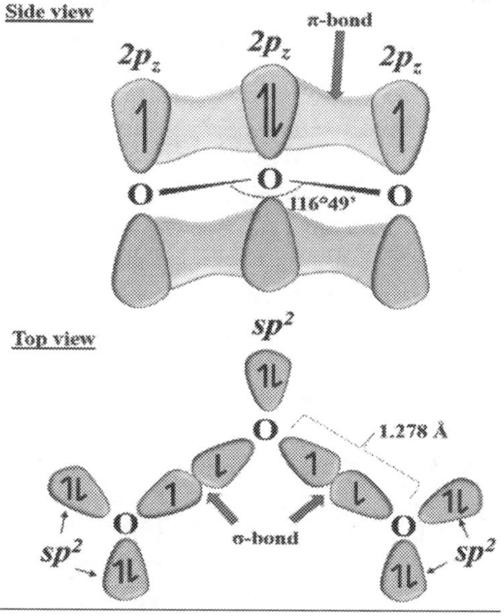
Melting point (°C)	−251	
Boiling point (°C)	−112	
Critical pressure (atm)	54.62	
Critical temperature (°C)	−12.1	
Water solubility (g/100 mL)	0.105 (0°C)	
Specific gravity	1.658 higher than air, 1.71 gcm^{-3} at −183°C	
Critical density (kg m^{-3})	436	
Heat of vaporization (cal mol^{-1})	2.980	
Heat of formation (cal mol^{-1})	33.880	
Free energy of formation (cal mol^{-1})	38.860	
Oxidation potential (V)	2.07	

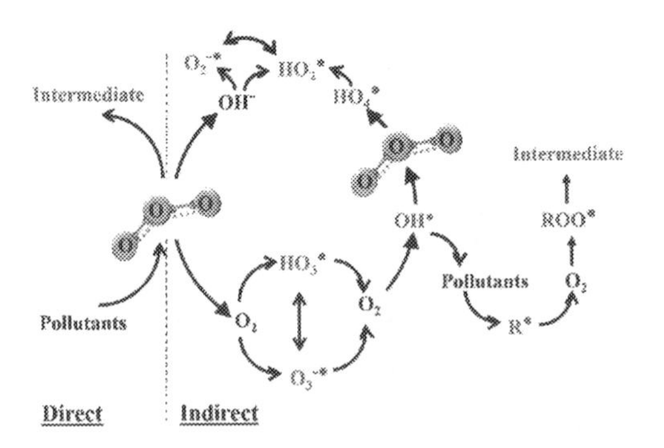

FIGURE 9.1 Chain reaction mechanism for indirect and direct ozonation. The species in red are reaction products, and R• is an intermediate that can function as a radical.

Since the other two oxygen atoms have a nucleophilic behavior due to the excess negative charge (see the top view, Table 9.1), all of the above makes ozone reactive.

9.1.1.2 Types of Ozone Attack

The ozone molecule's direct reactions with pollutants can be divided into oxidation-reduction reaction, cycloaddition reaction, electrophilic substitution reaction, and nucleophilic reaction (Beltrán 2003; Gottschalk et al. 2010).

9.1.1.2.1 Oxidation-Reduction Reaction

The O_3 molecule can react with numerous contaminants via oxidation-reduction reactions, due to its high standard redox potential (2.7 V). The redox reactions of O_3 and HO_2^- (or $O_2^{-\bullet}$) proceed primarily by electron transfer (Wang & Chen 2020).

$$O_3 + HO_2^- \rightarrow O_2^{-\bullet} + HO_2^{\bullet} \quad (1)$$

$$O_3 + O_2^{-\bullet} \rightarrow O_3^{-\bullet} + O_2 \quad (2)$$

9.1.1.2.2 Cycloaddition Reaction

Cycloaddition reactions in ozonation to form intermediates occur between an electrophilic compound (such as O_3) and a pollutant that has double bonds or π electrons (unsaturated) in its structure (Beltrán 2003). An example of cycloaddition reactions is the reaction of ozone with olefinic substances, which occurs based on a three-step mechanism; (I) formation of primary ozonide; (II) generation of the zwitterion; (III) Reactions in an aqueous solution of the zwitterion with the olefinic ones to form acids, ketones or aldehydes (Wang & Chen 2020).

9.1.1.2.3 Electrophilic Substitution Reaction

Ozone is an electrophilic agent that can attack the nucleophilic positions of organic compounds and replace a part of the organic compound. Some organic pollutant molecules have in their structures activating and deactivating groups that favor or not electrophilic substitution reaction (Beltrán 2003; Gottschalk et al. 2010). For example, the following groups present in contaminating molecules are classified as deactivating groups based on their importance: weak (-F, -Cl, -Br, -I), intermediate (-C≡N, -CHO, and -COOH), and strong (-NO_2 and -NR_3^+). Furthermore, the activating groups for the electrophilic substitution reactions, depending on their importance, are weak (-C_6H_5 and -Alkyl), intermediate (-OR and -NHCOR), and strong (-OH-, -O-, -NH_2, -NHR, and -NR_2) (Beltrán 2003; Wang & Chen 2020).

9.1.1.2.4 Nucleophilic Reaction

The O_3 molecule has a negative charge on one of its terminal oxygen atoms (see Table 9.1). Therefore ozone has nucleophilic properties and can react with carbonyl-containing compounds or C=N or C≡N bonds (electrophilic compounds) (Wang & Chen 2020).

9.1.1.2.5 Ozone reactions with Different Types of Compounds

Ozone demand is usually defined as ozone reactions with organic and inorganic compounds in water or wastewater. The ozone demand is vital in designing the ozone treatment system because it allows defining the ozone available and necessary during a disinfection application. Numerous research papers have been reported describing reactions with ozone and various organic and inorganic compounds (Shammas & Wang 2005; Wang & Bai 2017a). Some of these reactions are shown in Table 9.2.

9.1.1.3 Kinetic Regimes and Mass Transfer in Ozonation

The ozone concentration profile in water depends on the affinity to react to the contaminants present in solution with the O_3 molecule (Lara-Ramos et al. 2021). Another significant factor in ozone concentration profiles in water is the mass transfer of the ozone that is being transferred from the bulk of the gas to the bulk of the liquid (Gottschalk et al. 2010). This is illustrated in Fig. 9.2. Based on the ozone concentration profiles' trends, five kinetic reaction regimes can be presented, which are very slow, diffusional, moderate, fast, and instantaneous (Beltrán 2003; Gottschalk et al. 2010).

Therefore, going from a very slow to an instantaneous regime would mean that the ozone concentration at the liquid interface (δ film, water volume in contact with the gas) would suffer a rapid decrease (Beltrán 2003). The above behavior results in film transfer coefficients and general mass (k_L, see Eq. 6) different from those without reaction. Instead of changing the k_L, an improvement factor E is introduced, which allows describing the increase in mass transfer due to chemical reactions (Gottschalk et al. 2010).

9.1.1.3.1 Very Slow

In this regime, the mass transfer is independent of the chemical reaction, whereby the individual mass transfer coefficient and reaction rate constant can be calculated independently (Gottschalk et al. 2010). The pollutant reactions with ozone are very slow (and occur in the bulk of the liquid) compared to ozone transfer, which is very high (Lan et al. 2008). This regimen is common in treating drinking water with ozone, where the concentration of pollutants is usually low (Beltrán 2003; Lan et al. 2008).

TABLE 9.2

Reactions with Organic and Inorganic Compounds (Gottschalk et al. 2010; Shammas and Wang 2005; Von Gunten 2003; Wang and Bai 2017a)

Reaction with/ Compound	Remark
Organic Compounds	
Aromatic	Ozone can easily oxidize phenol in aqueous solution. In addition, xylenols and cresols oxidize faster than phenol. O_3 molecule oxidizes phenanthrene, naphthalene, and pyrene by rupture their rings. On the other hand, oxalic and acetic acids are relatively resistant to ozone attack.
Aliphatic	Unsaturated aliphatic or alicyclic compounds are susceptible to attack by the ozone molecule.
Pesticides	2,4,5-trichlorophenoxyacetic acid (2,4,5-T) and aldrin are easily oxidized by the attack of the O_3 molecule, while chlordane, lindane, diedrin, and endosulfan are slightly affected by ozone attack. Parathion and malathion have been treated with ozone, but their intermediates (paraoxon and malaoxon) are more toxic than the initial compounds.
Humic acids	Ozonated organic matter is usually more biodegradable than non-oxidized. The ozonation of humic acids, followed by chlorination, can reduce trihalomethanes formation in some cases. Humic acids could be resistant to ozone attack and therefore require long treatment times.
Inorganic Compounds	
Sulfide	The ozone molecule can attack the sulfide ion and oxidize it to sulfur, sulfite, and sulfate. For organic sulfides, the treatment is slower than sulfide ions because it depends on the amount of ozone and reaction time. The oxidation of organic sulfides produces sulfones, sulfoxides, and sulfonic acids.
Nitrogen	The ozone molecule can easily oxidize nitrite ion to nitrate. The O_3 molecule attack on organic nitrites, nitroso compounds, and hydroxylamines depends on the contact conditions and reaction time.
Iron and manganese	The ozone molecule attack on ferrous and manganous ions produces insoluble precipitates, ferric oxide, and manganic oxide, respectively.
Cyanide	Ozone can easily oxidize toxic cyanide ions, but toxic cyanate ions to a lesser extent

9.1.1.3.2 *Diffusional*

The diffusional regime is a specific case of the very slow kinetic regime, in which the absorption rate is equal to the maximum physical absorption rate (Gottschalk et al. 2010). Meanwhile, to describe the ozone absorption rate, Henry's law is used, which leads to $P_{O3g} = H_e C_{O3}$, where P_{O3g} is the ozone partial pressure in the gas, H_e is the Henry's Law constant, and C_{O3} is the ozone concentration in the liquid (Beltrán 2003; Shammas & Wang 2005).

9.1.1.3.3 *Moderate and Fast*

When the ozone concentration decreases considerably in the δ film, the pollutant degradation occurs both in the δ film and the liquid bulk, then the kinetic regime should be considered moderate. In this regime, the pollutant's degradation rate is

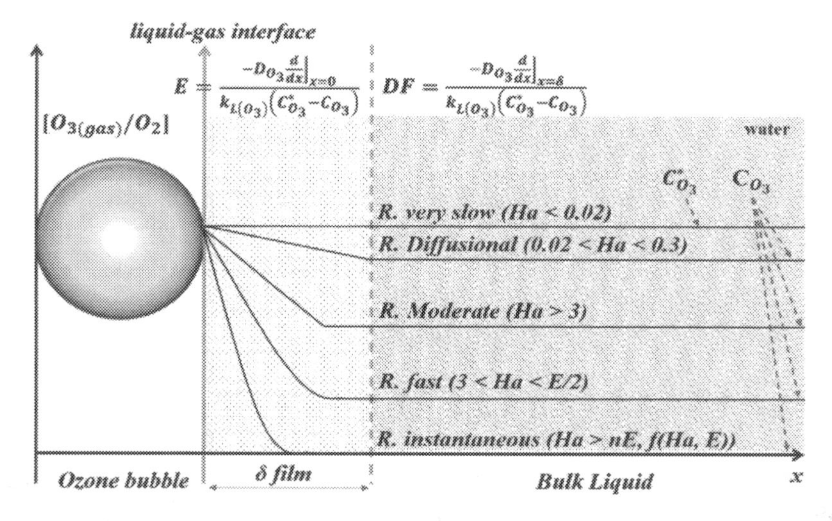

FIGURE 9.2 The different kinetic mass transfer regimes (and their relationship with Hatta) with simultaneous reaction of ozone.

controlled by the reaction rate with ozone and by the ozone mass transfer (Beltrán 2003; Lara-Ramos et al. 2021). On the other hand, pollutant degradation by direct ozone attack occurs in the δ film where all the ozone is consumed (O_3 concentration in bulk is zero). In this regime, the improvement factor is defined (see Eq. 4) (Gottschalk et al. 2010; Lucas et al. 2009).

9.1.1.3.4 Instantaneous

In this kinetic reaction regime, the pollutant reaction rate with ozone is much higher than the ozone mass transfer to the bulk. Since the reaction rate does not depend on the ozone transfer through the δ film, the ozone reactions with the pollutant occur very close to the bubble surface. On the other hand, if a high pollutant's concentration is found in the wastewater concerning the applied ozone (by a factor of 10 or more), a fast and instantaneous kinetic regime is presented (Beltrán 2003; Gottschalk et al. 2010).

The mass transfer flow of ozone (*N*) transferred out of the gas phase is described by Eq. 3, in which E (term for enhancement, see Eq. 4) is included. E is defined as the ratio between the actual flux with chemical reaction and the flux due to mass transfer from physical absorption alone at the gas-liquid interface ($x = 0$ or $\delta = 0$) (Beltrán 2003; Gottschalk et al. 2010).

$$N = k_{L(O_3)} E \left(C_{O_3}^* - C_{O_3} \right) \tag{3}$$

$$E = \frac{-D_{O_3} \left(dC_{O_3} / dx \right)_{x=0}}{k_{L(O_3)} \left(C_{O_3}^* - C_{O_3} \right)} = \left(\frac{\textit{rate with reaction}}{\textit{rate for mass transfer alone}} \right) \tag{4}$$

where D_{O3} is the ozone diffusivity in liquid, $k_{L(O3)}$ is the individual mass transfer coefficient of ozone for the liquid phase, C_{O3}* is the ozone concentration in equilibrium with the bulk gas phase, and C_{O3} is the bulk ozone concentration in the liquid.

Another important parameter is the depletion factor (*DF*, see Eq. 5), which is analogous to E (Beltrán 2003). *DF* is defined as the ratio between the flux out of the laminar film into the bulk liquid and the flux due to mass transfer from physical absorption alone at the film/bulk liquid surface ($x = \delta$) (Beltrán 2003; Gottschalk et al. 2010).

$$DF = \frac{-D_{O_3}\left(dC_{O_3}/dx\right)_{x=\delta}}{k_{L(O_3)}\left(C_{O_3}^{*} - C_{O_3}\right)} \tag{5}$$

The Hatta (Ha) dimensionless number establishes the relative importance of the chemical reaction and the mass transfer rates (Beltrán 2003; Lara-Ramos et al. 2021).

$$Ha_n = \frac{\sqrt{kD_{O3}\left(C_{O3}^{*}\right)^{1-n}}}{k_{L(O_3)}} \tag{6}$$

where k is the nth order reaction rate constant, and n is nth order kinetics.

9.2 OZONE GENERATION SYSTEM

Ozone generation from oxygen (or air) commonly occurs using electrical energy in advanced oxidation processes (AOPs). The electrical energy process can occur in two ways: applying a high voltage current between two electrodes placed in a dielectric or employing an electric discharge field as in corona discharge type ozone generators (Wang et al. 2017) (see Fig. 9.3a). Other types of ozone generators are the UVs, which use ultraviolet radiation of high-energy irradiation (see Fig. 9.3c). Besides these commercial methods, ozone can also be produced by electrolytic and chemical reactions in an electrochemical reactor (Lara-Ramos et al. 2020b) (see Fig. 9.3b). It is important to note that in the case of electrochemical reactors, ozone generation occurs in situ or within the reaction area or volume in which the pollutants are degraded (Lara-Ramos et al. 2020). Finally, a commercial ozonation system requires oxygen or clean, dry air to pass through generations of ozone generators or reactors.

9.3 THE SCALE OF REACTORS AND APPLICATIONS

Ozonation is an AOP due to the production of $OH^{\bullet}$ by homogeneous decomposition of ozone. Ozone treatments' efficiency depends on the type of pollutant (chemical and/or biological properties), physicochemical properties of contaminated water, contact time, and ozone dose applied (Beltrán 2003; Gottschalk et al. 2010; Wang & Chen 2020). The components and types of systems used in ozonation are shown in Fig. 9.4. Ozonation systems, whether on a laboratory or industrial scale, include four components for safe and proper operation (see Fig. 9.4a). These components

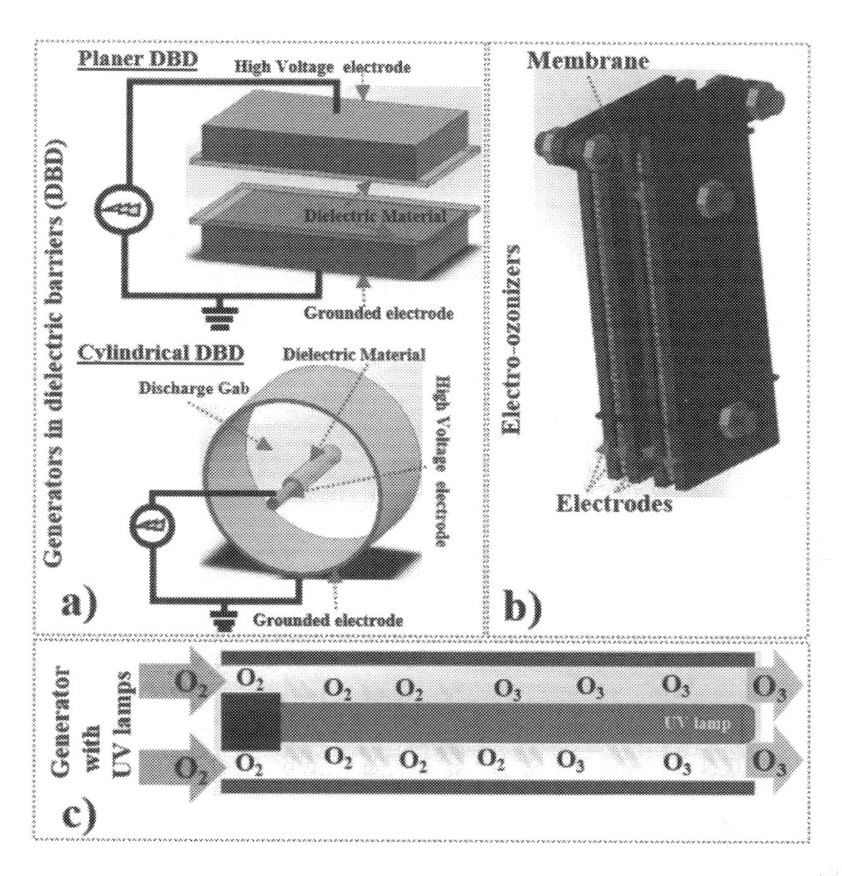

FIGURE 9.3 Schemes of ozone generation systems: (a) use of electrical energy (modified from Ansari et al. 2021); (b) electrochemical methods (Lara-Ramos et al. 2020); and (c) with the use of a UV lamp.

work together during water treatment: the ozone generator (principal component), the gas supply (oxygen or air), the ozone contactor, and the ozone destruction system (Gottschalk et al. 2010; López-Vinent et al. 2020; Wang & Chen 2020).

Water disinfection has been achieved by ozonation because ozone can destroy many bacteria, viruses, and protozoa (Shammas & Wang 2005; Werschkun et al. 2012). Furthermore, ozonation is an effective treatment to degrade organic pollutants (such as drugs, pesticides, some hydrocarbons, micropollutants, and landfill leachates) and oxidize metal ions (Barbosa et al. 2016; Shammas & Wang 2005; Wang & Bai 2017a). Before implementing the ozonization process on an industrial scale, tests to laboratory scale are necessary to determine the ozone demand, reaction kinetics of ozone, volume, and reaction time as a critical activity for system design (Beltrán 2003; Gottschalk et al. 2010; Shammas & Wang 2005) (see Fig. 9.4b).

Ozone has multiple uses in drinking water treatment, water tower cooling, wastewater renewal (industrial waste), and groundwater remediation (Shammas & Wang 2005). However, ozonation as a process has advantages and disadvantages that

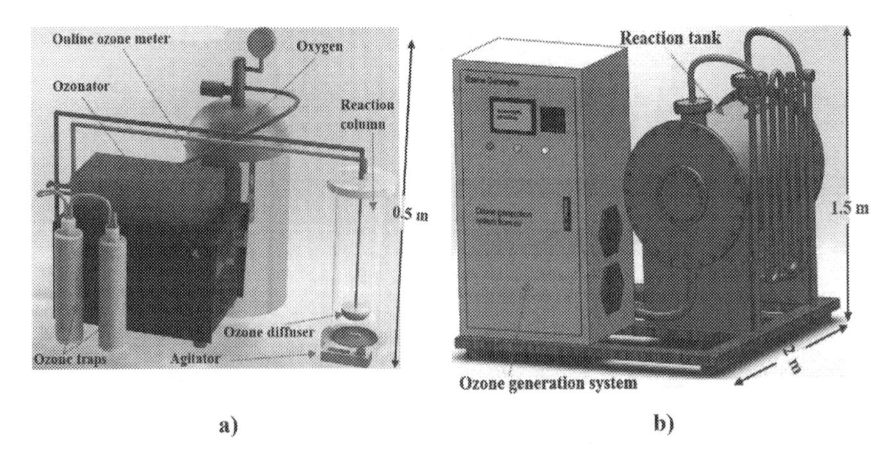

FIGURE 9.4 Diagrams of the conventional ozonation systems: (a) laboratory scale and (b) industrial scale.

should be considered for its application or use. Among the advantages that should be considered are the following (Nawrocki and Kasprzyk-Hordern 2010; Valdés & Zaror 2006):

- Ozone has strong oxidizing power and sometimes requires a short reaction time, which allows pathogens to be eliminated in a matter of seconds. Additionally, it helps in coagulation by destabilizing certain types of turbidity.
- Ozone removes color, taste, and odor.
- Ozone oxidizes (iron and manganese), reacts with, and removes organic matter and algae
- Ozone does not require chemicals for its generation and decays rapidly in water.

Ozone treatment is an effective oxidation process; however, several disadvantages limit its application (Shammas & Wang 2005; Wang & Bai 2017b), such as

- The solubility and stability of ozone in water are relatively low. Also, the cost of applying ozonation can be high compared to chlorination.
- Reactions between ozone and pollutants can be selective, and because of this, the process can be slow and with incomplete oxidation of pollutants (low TOC removal).
- Ozone toxicity is proportional to concentration and exposure time.
- Ozone can produce undesirable aldehydes and ketones by reacting with certain organics.
- It is necessary to install an ozone depletion device in the ozone reactor exhaust to prevent toxicity, and the installation of ozonation systems can be complicated.

9.4 CATALYSTS USED WITH OZONE

The use of catalysts improves the ozone decomposition and hydroxyl radicals formation during the ozonation process. Therefore, with the application of catalytic ozonation in water treatment, some disadvantages of ozonation can be overcome. Metal oxides (especially iron-based) are commonly used as catalysts for catalytic ozonation due to their abundance in nature, good catalytic performance, and easy preparation. For comparative purposes, Fig. 9.5 shows the number of publications on catalytic ozonation from 2010 to 2020, in which the catalyst was prepared (or synthesized) by sol-gel, precipitation, hydrothermal method, thermal-decomposition, electrochemical methods, biomineralization, and wet impregnation. The information was collected from the ScienceDirect and Scopus databases using keywords such as "catalytic ozonation," "catalyst synthesis methods," etc. It can be concluded from Fig. 9.5 that the sol-gel and precipitation method are the most used in the past ten years, while the electrochemical and wet impregnation synthesis methods are the least common in the reports of catalyst preparation for catalytic ozonation.

The catalytic ozonation process can be classified in two ways depending on the catalyst phase. In the homogeneous catalytic ozonation process the catalyst is

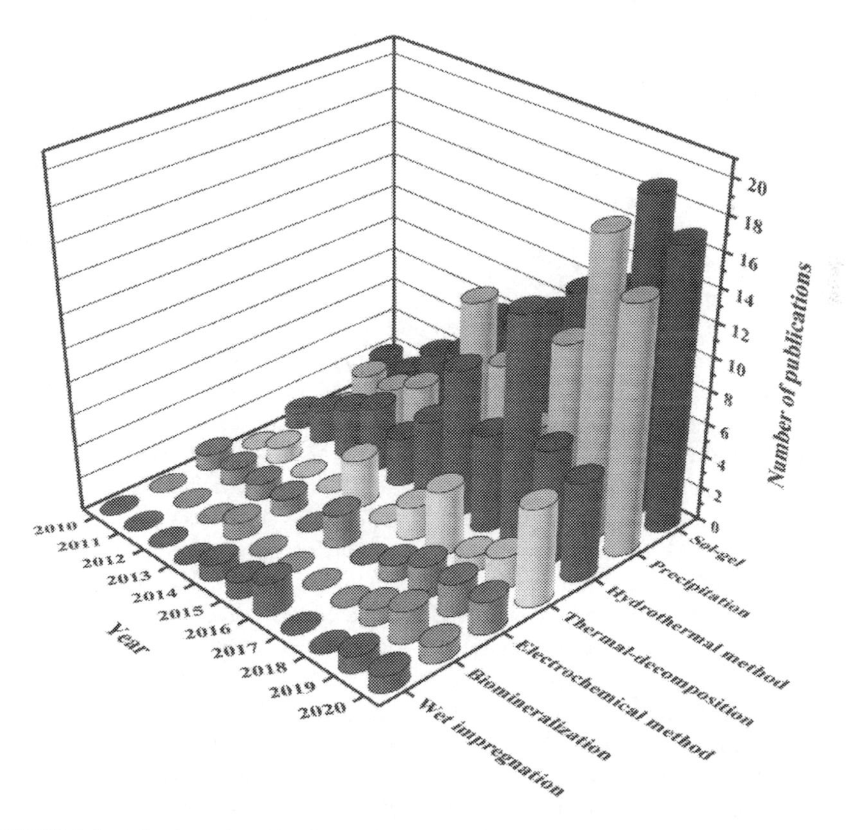

FIGURE 9.5 Number of publications on catalytic ozonation where catalysts are synthesized by six different preparation methods from 2010 to 2020.

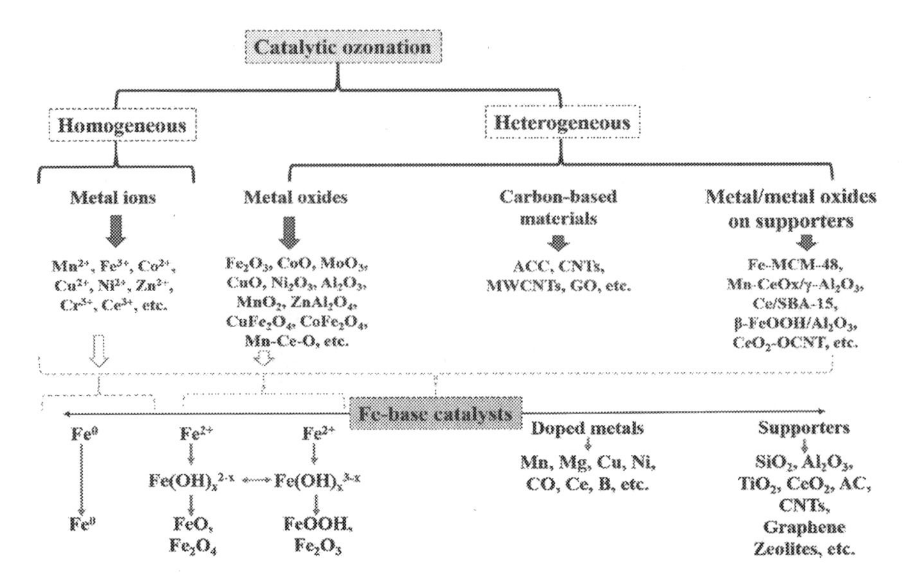

FIGURE 9.6 A variety of homogenous, heterogeneous, and Fe-based catalysts are applied in the catalytic ozonation process. Taken and modified from Wang and Bai (2017b) and Wang and Chen (2020).

dissolved in an aqueous solution; metal ions are usually used. On the other hand, heterogeneous catalytic ozonation is based on solid-phase catalysts for the ozone activation. Some heterogeneous catalysts that are commonly used are metal oxides, doped, or on supports (Beltrán 2003; Wang & Bai 2017b; Wang & Chen 2020), as shown in Fig. 9.6.

In a particular case, iron (Fe) is a material that is highly promoted in nature and is very important in ecological systems and organisms of living beings (Wang & Bai 2017b; Yu et al. 2019). Furthermore, it is a material with numerous applications and uses in catalytic ozonation, due to its characteristics and properties, such as its valence states and contribution to the crystalline structure of materials (Beltrán 2003; Wang & Bai 2017b; Yu et al. 2019). Various Fe-based catalysts that have been evaluated in catalytic ozonation are shown in Fig. 9.6. For example, FeO (it is unstable and can be easily oxidized to other iron oxides), Fe_2O_3 (it has many acid sites on the surface), Fe_3O_4 (it shows a super paramagnetic property), and FeOOH (it has a high density of hydroxyl radicals on its surface) (Wang & Bai 2017b). Recently, the doping of iron-based materials has been investigated in order to improve the electron transfer of the catalysts and increase the surface area of the catalysts. In Fig. 9.6 some metals used in doping are shown.

In the literature, works on catalytic ozonation carried out using supported catalysts are widely described. Catalysts used with ozone include metals (Cu, Fe, Ce, Cr, Mn, Ni, and Zn) or metal oxides (Fe_2O_3, CoO, MnO_3, MgO, TiO_2, Al_2O_3, CeO_2, etc.) impregnated on supports catalytic (activated carbon, zeolites, alumina, titania, silica, ceria, graphene, etc.) (Ghuge & Saroha 2018). Minerals such as magnetite and goethite have also been used as catalysts in catalytic ozonation (Wang & Bai 2017a).

TABLE 9.3
Catalysts, Synthesis Method, Pollutant, and Results Reported on Catalytic Ozonation

Catalyst	Synthesis Method	Pollutant	Results	Ref.
MgO	Sol-gel	Acetaminophen	100% degradation; 94% mineralization	(Mashayekh-Salehi et al. 2017)
WO_3	Thermo-decomposition	Ibuprofen	100% degradation; 87% TOC removal	(Rey et al. 2015)
$MgFe_2O_4$	Sol-gel	Acid orange II	90% degradation	(Lu et al. 2015)
Fe-SBA-15	Hydrothermal	Oxalic acid	86.6% degradation	(Yan et al. 2016)
Mg-Ce-MCM41	Hydrothermal	Oxalic acid	90.3% TOC removal	(Jeirani and Soltan 2016)
MnO_2/CeO_2	Impregnation	Sulfosalicylic acid	97% TOC removal	(Xing et al. 2016)
IS-FeOOH	Precipitation	p-Chloronitro benzene	99.8% degradation	(Liu et al. 2017)
Zn-Fe_2O_4	Hydrothermal	Phenol	92.6% degradation	(Zhang et al. 2015)
MgO/AC	Impregnation	Phenol	88.5% degradation; 83.5% COD removal	(Zhou et al. 2020)
Biogenic-Fe_3O_4	Biomineralization	para-Chlorobenzoic acid	10.5% degradation	(Jung et al. 2008)
Ce/Al_2O_3	Wet impregnation	Phenol	46.3% degradation	(Li et al. 2020b)
Fe^0-CNTs	Electrophoresis deposition	Methylene blue	100% degradation	(Zhang et al. 2013)

The catalyst synthesis methods presented in Fig. 9.5 lead to diverse products (doped, supported materials, etc.) that may show different catalytic performance during the application of catalytic ozonation. Thus Table 9.3 presents a review of various catalysts, their synthesis methods, pollutants treated, and the main results obtained for the catalytic ozonation of various degraded organic pollutants, COD removal, or TOC removal.

The transition metals used as catalysts can be especially harmful to the environment and human health (Savvina et al. 2019). Therefore norms have been adopted to establish a maximum permissible concentration limit for drinking water (Table 9.4). Therefore catalysts, which can be used as catalysts and potentially leach metal ions, must be post-treated to remove metal ions from the treated effluent effectively. This is undoubtedly a drawback to consider when applying homogeneous catalytic ozonation (especially for drinking water treatment) (Savvina et al. 2019). Therefore some countries establish limits on the concentrations of transition metal ions. For example, the United States through the Environmental Protection Agency (EPA) 440/5-86-001 establishes limits of quality criteria for water (EPA 1986); Iran establishes a limit according to (Zahrani & Ayati 2020); and for the countries of the European Union legislation is established, according to Savvina et al. (2019). Table 9.4 shows the limit values for transition metal ions in water for the European Union and the countries mentioned.

TABLE 9.4
Maximum Permissible Concentration Limits for Some Catalysts Used in Water

Transition (Heavy) Metal	Concentration Limit (mg/L)		
	European Union	Iran	United States
Copper	2	1	0.83
Chromium	0.05	0.5	0.993
Iron	0.2	3	–
Lead	0.01	–	0.951
Manganese	0.05	–	–
Cobalt	–	1	–
Nickel	–	–	0.998

9.5 GENERALITIES OF THE KINETICS OF THE GAS-LIQUID-SOLID REACTION

The reactions in a gas-liquid-solid catalytic ozonation system are heterogeneous systems in which a series of parallel stages of mass transfer and chemical reactions can occur on the catalyst surface. Fig. 9.7 shows the steps of this process. The catalyst's surface properties are essential to improving pollutant mineralization and reaction rate (Beltrán 2003; Rekhate & Srivastava 2020). The pollutant reacts on the surface catalyst according to the following steps: It begins with the external diffusion of O_3 molecules from the gas bulk to the gas–liquid interface and then reaches the bulk of the liquid (Beltrán 2003). After that, the O_3 and P molecules diffuse from the liquid bulk to the surface catalyst. Also, ozone and pollutant molecules can react on the

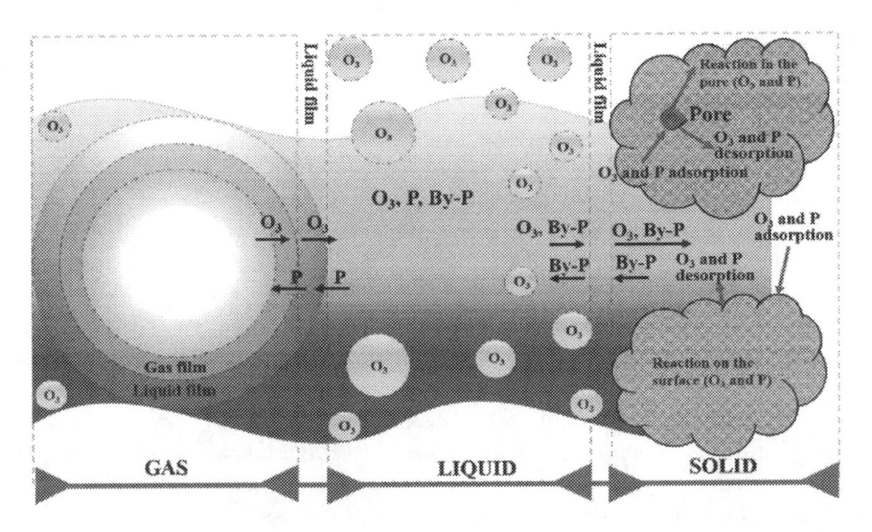

FIGURE 9.7 Physical processes and mechanism of a gas-liquid-solid catalytic reaction.

catalyst surface and diffuse internally toward catalyst pores. The reaction on the catalyst involves three consecutive steps (Gottschalk et al. 2010; Wang & Bai 2017b):

a. Adsorption of reactive molecules to the active centers on the catalyst surface
b. Surface reaction of adsorbed molecules, to produce the adsorbed products
c. Desorption of adsorbed P

Now, for reversible reactions, there are also two steps. First, the P molecules' internal diffusion from the catalyst pores toward the external catalyst surface and second, P's external diffusion from the surface catalyst towards the bulk of the liquid (Beltrán 2003). When the external and internal diffusions are very fast steps, the mass transfer is negligible, compared to the surface reaction step. The reaction rate equation depends exclusively on the slowest step. Therefore, adsorption, chemical surface reaction, or desorption can control the reaction rate in the catalytic ozonation process (Beltrán 2003; Gottschalk et al. 2010; Wang & Bai 2017b).

9.5.1 Homogeneous Catalytic Ozonation

Although the ozonation is a process successfully applied since 1906 to disinfect wastewater (Wang & Chen 2020), it presents some disadvantages to industrial application: high cost, low ozone utilization, and in most cases, incomplete mineralization because the undesirable formation of by-products even more dangerous than the initial contaminant or the inability of ozone molecule by attacking acids or aldehydes produced by direct oxidation of organic compounds (Lara-Ramos et al. 2019; Shammas & Wang 2005; Wang & Chen 2020). Among the technologies investigated to overcome the ozonation obstacles is the catalytic ozonation, classified into homogeneous and heterogeneous depending on the catalyst used (Ghuge & Saroha 2018).

The homogeneous catalytic ozonation was used for the first time in 1972 by Hawis and Davinson (Wang & Chen 2020). They investigated municipal wastewater treatment by ozonation using transition metal ions and found that some transition metals could improve the degradation and mineralization of organic compounds and ozone utilization efficiency.

Different mechanisms have been proposed in order to explain the effect of transition metal in ozonation. The most accepted is the ozone decomposition to generate hydroxyl radical that oxidizes organic contaminants described by Eqs. 7–10 (Wang & Chen 2020). According to Chung-Hsin Wu et al. (W2008) and Wang and Hai Chen (2020), ozone decomposition strongly depends on the catalyst concentration and the pH of the solution. That is, at pH < 4 the direct reaction of ozone molecule with organic matter dominates in the range of pH 4 to 9 both indirect and direct reactions take place, while at pH > 9 the indirect reactions are predominant. Although independent of pH value, the indirect reaction occurs in the catalytic ozonation process.

$$M^{n+} + O_3 + H^+ \rightarrow M^{n+1} + HO^{\bullet} + O_2 \quad (7)$$

$$O_3 + HO^{\bullet} \rightarrow HO_2^{-\bullet} + O_2 \quad (8)$$

$$M^{n+1} + HO_2^{-\bullet} + OH^- \rightarrow M^n + H_2O + O_2 \quad (9)$$

$$M^{n+} + HO^{\bullet} \rightarrow M^{n+1} + OH^- \quad (10)$$

According to Sánchez-Polo et al., not all transition metals can promote ozone decomposition, only those susceptible to oxidation by ozone. Thus, the good catalytic activity of the transition metals ions is related to their reduction potential. The 1,3,6,-naphthalenetrisulfonic remotion efficiency in the wastewater is improved in the presence of Fe(II), Ni(II), Zn(II) Mn(II), and Cr(III), while ions as Sr(II), Cu(II), Cd(II), and Hg(II) showed no catalytic activity (Sanchez-Polo & Rivera-Utrilla 2004).

Another mechanism proposed by Pines and Reckhow (2002) and corroborated by Beltran et al. (2005) suggested that the transition metals form a complex with the organic compound and then the ozone molecule attacks it. This behavior was demonstrated for Co(II) and Fe(III) through scavenger tests using t-butanol. The authors report that oxalic acid degradation is not affected by scavenger presence. Even Pines and Reckhow confirm the formation of hydroxyl radical, but it does not intervene. Eqs. 11 to 16 describe the mechanism proposed by Beltran et al., for oxalic acid removal at acid pH (Nawrocki 2013).

$$Fe^{3+} + C_2O_4^{2-} \rightarrow FeC_2O_4^{+} \quad (11)$$

$$FeC_2O_4^{+} + C_2O_4^{2-} \rightarrow Fe(C_2O_4)_2^{-} \quad (12)$$

$$Fe(C_2O_4)_2^{-} + C_2O_4^{2-} \rightarrow Fe(C_2O_4)_3^{3-} \quad (13)$$

$$FeC_2O_4^{+} + O_3 \rightarrow 2CO_2 + Fe^{3+} + 2O_3^{-\bullet} \quad (14)$$

$$Fe(C_2O_4)_2^{-} + O_3 \rightarrow 2CO_2 + FeC_2O_4^{+} + 2O_3^{-\bullet} \quad (15)$$

$$Fe(C_2O_4)_3^{3-} + O_3 \rightarrow 2CO_2 + Fe(C_2O_4)_2^{-} + 2O_3^{-\bullet} \quad (16)$$

The transition metal ions mainly used were collected in Table 9.5. Chung-Hsin Wu et al. (2008) evaluated the discoloration of Reactive Red 2 by homogeneous catalytic ozonation using different metal ions (Mn(II), Fe(II), Fe(III), Zn(II), Co(II), and Ni(II). Complete decolorization was achieved in all cases in less time than the ozonation process alone. However, this is not necessarily linked to mineralization (Wu et al. 2008).

Psaltou et al. (Savvina et al. 2019) evaluated the role of Co(II), Fe(II), Mn(II), Ni(II) and Zn(II) in the p-chlorobenzoic acid (p-CBA) degradation and found that Co(II) and Fe(II) present better behavior as catalysts. Degradations of 95.5 percent and 92.5 percent were obtained for Co(II) and Fe(II), respectively, which is attributed to the formation of oxy-hydroxides with low solubility and a positive charge at pH 7 that not only favor the p-CBA negative ionic species adsorption, also improve the ozone decomposition in hydroxyl radicals. Fu et al. (2020) investigated the

TABLE 9.5
Use of Homogeneous Catalytic Ozonation for the Treatment of Organic Pollutants

Catalyst	Catalyst (mM)	Pollutant	$Pollutant_{initial}$/ $TOC_{initial}$ (mg/L)	Efficiency	Ref.
Mn(II)	1	Reactive Red	100	100% decoloration, 28% TOC removal	(Wu et al. 2008)
	6×10^{-2}	Humic acid	10.6	62.3% mineralization	(Gracia et al. 1996)
	5×10^{-2}	NTS	0.045	100% degradation, 66% TOC removal	(Sanchez-Polo and Rivera-Utrilla 2004)
	1.8×10^{-2}	p-CBA	6×10^{-4}	92% degradation	(Savvina et al. 2019)
	1.1×10^{-3}	Oxalic acid	20	10% degradation	(Sun et al. 2014)
	3.6×10^{-3}	Simazine	5	90% degradation	(Rivas et al. 2001)
Co(II)	1	Reactive Red	100	100% degradation	(Wu et al. 2008)
	6×10^{-2}	Humic acid	11.1	30.6% TOC removal	(Gracia et al. 1996)
	1.6×10^{-2}	p-CBA	6×10^{-4}	97% degradation	(Savvina et al. 2019)
	1×10^{-2}	Oxalic acid	720	70% degradation	(Savvina et al. 2019)
Ni(II)	1	Reactive Red	100	100% degradation	(Wu et al. 2008)
	5×10^{-2}	NTS	0.045	100% degradation, 66% TOC removal	(Sanchez-Polo and Rivera-Utrilla 2004)
	8×10^{-3}	p-CBA	6×10^{-4}	100% degradation	(Savvina et al. 2019)
Fe(II)	1	Reactive Red	100	100% degradation, 19% TOC removal	(Wu et al. 2008)
	1.8×10^{-1}	Aerofloat	75	44.3% TOC removal	(Fu et al. 2020)
	6×10^{-2}	Humic acid	10.4	57.3% TOC removal	(Gracia et al. 1996)
	5×10^{-2}	NTS	0.045	100% degradation, 76% TOC removal	(Sanchez-Polo and Rivera-Utrilla 2004)
	1.8×10^{-2}	p-CBA	6×10^{-4}	93% degradation	(Savvina et al. 2019)
	1.8×10^{-2}	2-chlorophenol	100	80% degradation	(Savvina et al. 2019)
Fe(III)	1	Reactive Red	100	100% degradation, 18% TOC removal	(Wu et al. 2008)
	1.8×10^{-1}	Aerofloat	75	40.3% TOC removal	(Fu et al. 2020)
	6×10^{-2}	Humic acid	10.3	51% TOC removal	(Gracia et al. 1996)
Cu(II)	1.6×10^{-1}	Aerofloat	75	39% TOC removal	(Fu et al. 2020)
	6×10^{-2}	Humic acid	11	55.1% TOC removal	(Gracia et al. 1996)
Zn(II)	1	Reactive Red	100	100% degradation, 35% TOC removal	(Wu et al. 2008)
	1.5×10^{-1}	Aerofloat	75	65% degradation	(Fu et al. 2020)
	6×10^{-2}	Humic acid	11	52.5% degradation	(Gracia et al. 1996)
	7.6×10^{-3}	NTS	0.045	68% TOC removal	(Sanchez-Polo and Rivera-Utrilla 2004)
	3.8×10^{-3}	p-CBA	6×10^{-4}	87% degradation	(Savvina et al. 2019)

Abbreviations: NTS: 1,3,6-naphthalenetrisulfonic; p-CBA: p-chlorobenzoic acid.

mineralization of Aniline Aeroflot using Fe(II), Fe(III), Cu(II), Pb(II), and Zn(II). The transition metal with better results was Fe(II), achieving 44.3 percent of mineralization and 78 percent of COD removal.

Most transition metals are harmful to the environment, so their use as a catalyst is limited. Table 9.4 shows the maximum permissible concentration limits of the metal transitions most used. Some alternatives to overcome the limitations presented in the homogeneous catalytic ozonation process is coupled with other processes such as membrane distillation (Zhang et al. 2016), heterogeneous catalytic with activated carbon (Valdés & Zaror 2006), or support the catalyst in alumina (Ghuge & Saroha 2018).

9.5.2 Heterogeneous Catalytic Ozonation

The main quality of heterogeneous catalytic ozonation is that the catalyst is in the solid phase. Therefore the pollutant reactions with ozone could occur in the water or on the catalyst surface (Malik et al. 2020; Wang & Bai 2017b). It is important to mention that catalytic ozonation is considered when the influence of O_3 and the catalyst is much greater on the process than the sum of ozonation alone and adsorption of pollutant on the catalyst surface at the same operating conditions (temperature, pH, volume solution, etc.) (Lara-Ramos et al. 2019, 2020a). Additionally, it is expected that the hydroxyl radicals formation is due to ozone decomposition on the catalyst surface (Malik et al. 2020). A catalytic effect on the ozonation process is possible when at least one of the three conditions in Fig. 9.8 is met. First condition: the O_3

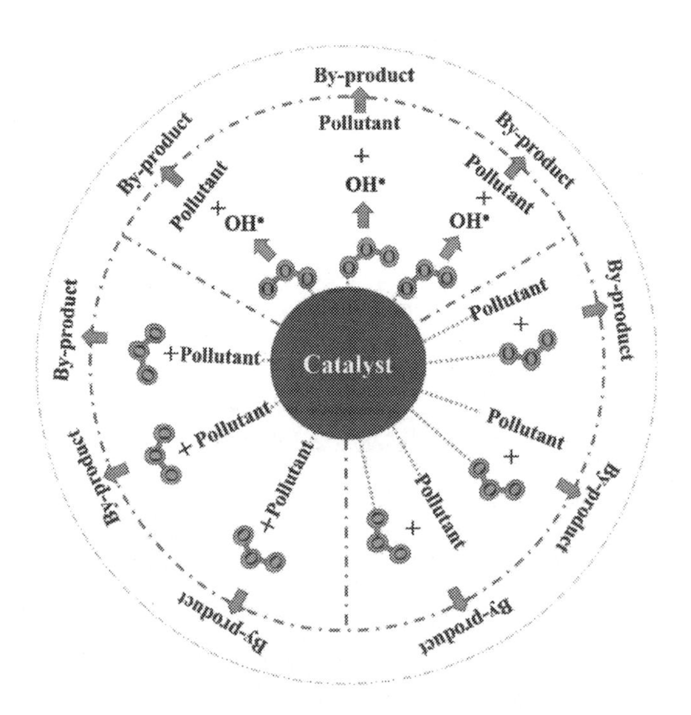

FIGURE 9.8 Three possible cases of heterogeneous catalysis ozonation.

molecule is adsorbed on the catalyst surface; second condition: the contaminant is adsorbed on the catalyst surface; third and final condition: the O_3 molecule and/or contaminant is adsorbed on the catalyst surface (Nawrocki 2013).

9.5.2.1 Mechanisms of Heterogeneous Catalytic Ozonation

Under certain conditions, the transition metals (M) have various valence numbers. This characteristic of the (M) allows the decomposition of ozone in several radical species of oxygen. In the specific case of iron-based catalysts, the possible mechanisms are shown in the reactions given by Eqs. 17 to 22, reported by Nawrocki (2013), Wang & Bai (2017b, and Wang & Chen (2020). The metal ions on the surface of the oxide are assigned as triple bond M. The difference of the previous mechanism described with the homogeneous is that the metal ions are dissolved, while in the oxides, the metal ions are undissolved (Ghuge & Saroha 2018; Malik et al. 2020; Wang & Bai 2017b).

$$\equiv M^{(n-1)+} + O_3 \rightarrow \equiv M^{n+} + O_3^{-\bullet} \tag{17}$$

$$O_3^{-\bullet} + H^+ \rightarrow O_2 + OH^\bullet \tag{18}$$

$$\equiv M^{(n-1)+} + O_3 \rightarrow \equiv MO^{(n-1)+} + O_2 \tag{19}$$

$$\equiv M^{(n-1)+} + H_2O \rightarrow \equiv M^{n+} + OH^- \tag{20}$$

$$\equiv M^{(n-1)+} + OH^\bullet \rightarrow \equiv M^{n+} + OH^- \tag{21}$$

$$\equiv M^{n+} + O_3 + H_2O \rightarrow \equiv MO^{(n-1)+} + OH^\bullet + H^+ + O_2 \tag{22}$$

Pollutant degradation can take place in different ways: firstly, the pollutant molecule is adsorbed on the catalyst surface, and then it can react with the O_3 or $OH^\bullet$ molecule in bulk (Lara-Ramos et al. 2021). Secondly, the O_3 molecule onto the catalyst surface can adsorb or decompose to form hydroxyl radical, and both species can react with the contaminant in bulk (Rekhate & Srivastava 2020). Thirdly, the contaminant and the O_3 molecule are adsorbed onto the catalyst surface and then react. Finally, oxidant species can be produced by interacting with the ozone molecule with hydroxyl ions (chemisorbed water) on the catalyst surface (Wang & Bai 2017b).

Activated carbon has been used as catalyst support in the catalytic ozonation reactions of cerium and iron. Furthermore, alumina (Al_2O_3) has been used to support metallic ruthenium (Ghuge & Saroha 2018). Meanwhile, several research works have reported using various materials as heterogeneous catalysts in ozonation, such as metal oxides, compatible metals, oxyhydroxide, and some porous materials to treat organic pollutants. (Beltrán 2003; Ghuge and Saroha 2018; Sui et al. 2010).

In the past 15 years (2005–2019), numerous catalytic ozonation studies have been published with different types of materials used as catalysts, some doped or supported, as shown in Fig. 9.9. Fig. 9.9 was edited and updated from the work of Ghuge

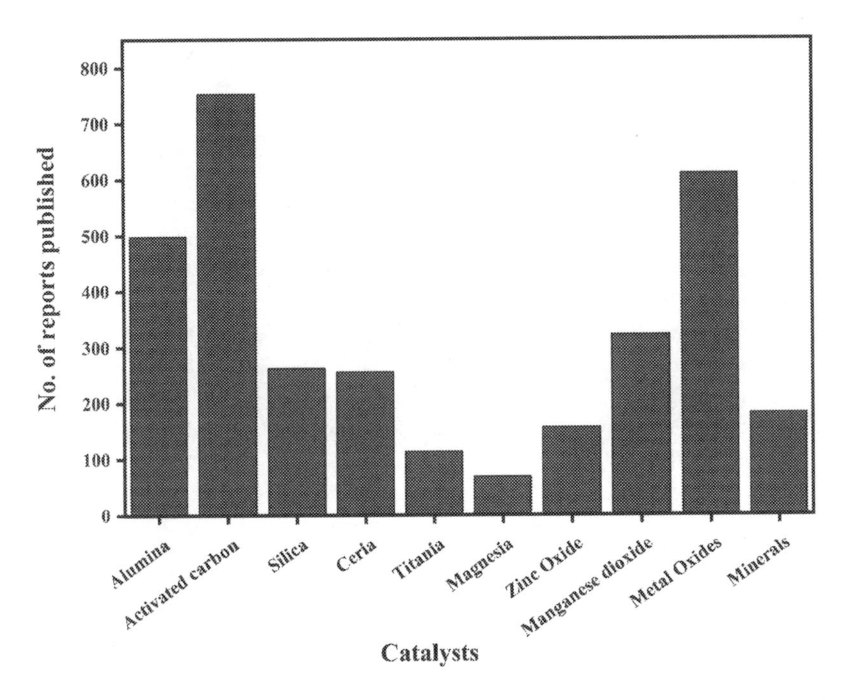

FIGURE 9.9 Distribution of catalysts used in homogeneous and heterogeneous catalytic ozonation.

and Saroha (2018). The information was updated with publications collected from the ScienceDirect and Scopus database using keywords such as "catalytic ozonation" and "catalysts," etc. It can be concluded from Fig. 9.9 that activated carbon (AC), metal oxides, and alumina are the most used catalysts in catalytic ozonation. While manganese oxide, silica, and ceria were reported and used about 300 times each. Meanwhile, minerals zinc oxide, titania, and magnesia were less reported than other catalysts.

9.5.3 Photocatalytic Ozonation

Ozonation and photocatalysis are effective oxidation processes, and their coupling can have a synergistic effect on both oxidation and mineralization efficiency. Even with the increased consumption of additional electrical energy due to ozone generation, it has been shown that photocatalytic ozonation can be profitable for use in water treatment, due to much shorter treatment times than in separate processes (Lara-Ramos et al. 2019, 2020a; Malik et al. 2020; Mecha et al. 2016; Valério et al. 2020). Furthermore, a synergistic effect on the rate of degradation and mineralization of various substances has been observed (Valério et al. 2020; Xiao et al. 2015). The higher efficiency of photocatalytic ozonation can be explained by the considerably higher electron affinity of ozone compared to oxygen (Lara-Ramos et al. 2019; Valério et al. 2020).

Meanwhile, the most widely used photocatalyst is titanium dioxide (TiO_2), as it has the following advantages: it is cheap, inert, and has prominent photocatalytic activity. Photocatalyst particles can be supported or immobilized on some surfaces by various techniques, such as dip coating, embedding the particles, electrophoretic deposition, among others (Cardoso et al. 2016; Lara-Ramos et al. 2019, 2020a; Simon et al. 2018; Xiao et al. 2015). The mentioned methods use prefabricated photocatalysts, but there are processes, such as the sol-gel process or chemical vapor deposition, in which the photocatalyst is formed in situ; this because its properties are difficult to control (Wang & Bai 2017b; Xiao et al. 2015). There is a diversity of photocatalysts with varied morphologies and compound components, but their application in photocatalytic ozonation is limited (Malik et al. 2020; Xiao et al. 2015). The most used are shown in Table 9.6 (updated and modified from the article by Xiao et al. 2015). Photocatalysts can be divided into UV-driven and visible-light—driven. Also, photocatalysts can be suspended or immobilized materials (Xiao et al. 2015).

TABLE 9.6
Various Catalysts are Applied in Heterogeneous Photocatalytic Ozonation

Catalyst	Light Source	Synthesis Method	Ref.
Suspended TiO_2			
Degussa P25	UV	Commercial	[a]
TiO_2	UV	Hydrothermal	(Xiao et al. 2015)
TiO_2 fiber	UV	Commercial	(Giri et al. 2008)
TiO_2/activated carbon	UV	Sol-gel	(Xiao et al. 2015)
TiO_2 nanotubes	UV- B	Electrochemical anodization	(Cardoso et al. 2016)
TiO_2/montmorillonite-(MMT) nanocomposite	UV	Hydrothermal	(Hassani et al. 2016)
Suspended non-TiO_2			
Copper ferrite nanoparticle ($CuFe_2O_4$)	UV	Co-precipitation	(Xiao et al. 2015)
Nickel-zinc ferrite magnetic nanoparticle	UV	Co-precipitation	(Mahmoodi et al. 2012)
Multiwalled carbon nanotube	UV	Commercial	(Xiao et al. 2015)
ZnO nanorods	UV	Pyrolysis	(Wu et al. 2017)
WO_3 powder	Visible	Commercial	(Nishimoto et al. 2010)
Bi_2O_3 and Au/Bi_2O_3 nanorods	Visible	Microwave assisted	(Xiao et al. 2015)
Immobilized			
TP-2 TiO_2-coated tube	UV	Commercial	(Xiao et al. 2015)
TiO_2 on Pilkington Active™ glass sheet	UVA	Commercial	(Mehrjouei et al. 2011)
Carbon-black-modified nano-TiO_2 thin films on Al sheet	UV	Sol-gel	(Xiao et al. 2015)
TiO_2 film on Al sheet	UV	Commercial	(Xiao et al. 2015)
Aeroxide P251® (TiO_2)-ceramic papers	UV	Commercial	(Simon et al. 2018)

Source: Edited and updated from Xiao et al. (2015).
Note:
[a] Lara-Ramos et al. (2019, 2020a), Simon et al. (2018), and Xiao et al. (2015).

9.5.3.1 Mechanisms of Photocatalytic Ozonation

The use of photocatalysts in ozonation changes the oxidation mechanisms completely. In this sense, for photocatalytic ozonation, the reactions begin with the photocatalyst's photoexcitation by UV radiation, which can provide the energy to generate the electron-hole pair (Eq. 23) (Malik et al. 2020). In parallel, ozone molecules may be adsorbed on the catalyst surface, and this interaction produces active oxygen radicals (•O) (see Eq. 24). The •O reacts with water molecules to produce hydroxyl radicals (Eq. 32), a critical species in the photocatalytic ozonation process (Malik et al. 2020; Xiao et al. 2015) (see Fig. 9.10).

Ozone and hydrogen peroxide molecules can produce oxidizing species (Eqs. 25 and 26) when they are irradiated with wavelengths less than 300 nm. The production of H_2O_2 is caused as an intermediate species of the ozone decomposition chain reactions (Eqs. 36, 37, and 38). The photogenerated electrons (Eq. 23) react with oxygen and ozone (it is an electron acceptor) (see Eqs. 25 and 28); these reactions decrease recombination. The recombination of the electron-hole pair affects the efficiency and performance of the process. The 2.1 eV of ozone is 4.7 times greater than that of oxygen (approx. 0.44 eV), so the ozone presence could improve the process's photocatalytic activity. Besides, the H_2O_2 that is generated can also with the electrons photogenerated on the surface of the catalyst and produce hydroxyl radicals (Eq. 29) (Xiao et al. 2015).

The holes generated (Eq. 23) can react directly with the pollutant adsorbed on the photocatalyst (Eq. 30), or they can produce $OH^{\bullet}$ (Eq. 31) from the reaction with water ($pH < 7$) or with hydroxide anions ($pH > 7$). Additionally, it is presumed that other oxidant species, such as oxygen atom radicals and superoxide and ozonide radical anions, are generated as intermediates of the photocatalytic ozonation reactions and could degrade or reinforce the production of hydroxyl radicals that oxidize pollutants, as shown in the reactions in chain given by Eqs. 32–44. Hydroxyl radicals destroy

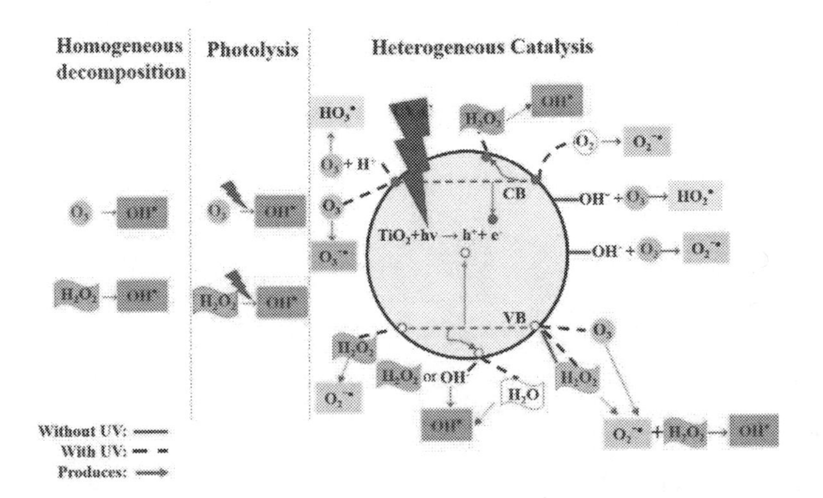

FIGURE 9.10 Proposed mechanism for photocatalytic ozonation and hydrogen peroxide, ozone, and TiO2 couplings. Taken from Lara-Ramos et al. (2020a).

pollutant molecules non-selectively (R22). The three possible attack mechanisms of $OH^{\bullet}$ are electron transfer, radical addition, and hydrogen abstraction. The reaction mechanism proposed by Xiao et al. is described by Eqs. 32–44 (Xiao et al. 2015).

$$Photocatalyst + hv \rightarrow Photocatalyst + e^{-} + h^{+} \quad (23)$$

$$O_3 + Photocatalyst \rightarrow \cdot O + O_2 \quad (24)$$

$$O_3 + hv \rightarrow \cdot O + O_2 \quad (25)$$

$$H_2O_2 + hv \rightarrow 2OH^{\bullet} \quad (26)$$

Reactions of photogenerated electrons (e^-) on the surface of photocatalyst:

$$O_{2(ads)} + e^{-} + \cdot O_2^{-} \quad (27)$$

$$O_{3(ads)} + e^{-} + \cdot O_3^{-} \quad (28)$$

$$H_2O_{2(ads)} + e^{-} \rightarrow OH^{\bullet} + OH^{-} \quad (29)$$

Reactions of holes (h^+) on the surface of photocatalyst:

$$Pollutant_{(ads)} + h^{+} \rightarrow Byproducts \quad (30)$$

$$H_2O_{(ads)} + h^{+} \rightarrow OH^{\bullet} + H^{+} \left(pH < 7\right) \; or \quad (31)$$

$$OH^{-}{}_{(ads)} + h^{+} \rightarrow OH^{\bullet} \left(pH > 7\right)$$

Further chain reactions:

$$\cdot O + H_2O \rightarrow 2OH^{\bullet} \quad (32)$$

$$\cdot O_2^{-} + H^{+} \leftrightarrow HO_2^{\bullet} \quad (33)$$

$$\cdot O_3^{-} + H^{+} \leftrightarrow HO_3^{\bullet} \quad (34)$$

$$HO_3^{\bullet} \rightarrow O_2 + OH^{\bullet} \quad (35)$$

$$O_3 + OH^{-} \rightarrow \cdot O_2^{-} + HO_2^{\bullet} \quad (36)$$

$$O_3 + OH^{\bullet} \leftrightarrow HO_4^{\bullet} \rightarrow O_2 + HO_2^{\bullet} \quad (37)$$

$$2HO_2^{\bullet} \rightarrow H_2O_2 + O_2 \tag{38}$$

$$H_2O_2 + \cdot O_2^- \rightarrow OH^{\bullet} + OH^- + O_2 \tag{39}$$

$$O_3 + HO_2^{\bullet} \rightarrow 2O_2 + OH^{\bullet} \tag{40}$$

$$O_3 + \cdot O_2^- \rightarrow \cdot O_3^- + O_2 \tag{41}$$

$$O_3 + H_2O_2 \rightarrow OH^{\bullet} + HO_2^{\bullet} + O_2 \tag{42}$$

$$Pollutant + O_3 \rightarrow Byproducts + \cdot O_3^- \tag{43}$$

$$Pollutant_{(ads)} + OH^{\bullet} \rightarrow Byproducts + H_2O \tag{44}$$

9.5.4 Influence Factors in the Catalytic Ozonation Process

The factors that can affect the catalytic ozonation process are discussed below.

9.5.4.1 pH

The pH of the solution determines the effectiveness of the catalytic ozonation process. The influence of the pH is manifested in two ways (Wang & Bai 2017a). The pH solution affects ozone decomposition; basic pH is the most favorable for ozone decomposition, and acidic pH is the least beneficial for ozone decomposition. Poznyak et al., reported the O_3 decomposition when varying pH (Poznyak et al. 2018) (see Fig. 9.11) pH is of vital importance to promoting the interactions between catalytic sites and organic molecules. It influences the charge of active catalyst sites, the charge of the molecules (ionic or ionizable, and possible poisons of the catalyst). Finally, pH is a parameter that must be investigated to establish the mechanism of interaction of ozone and the catalyst in the catalytic ozonation process (Wang & Bai 2017b; Ziylan & Ince 2015).

Fig. 9.11 shows the effect of pH on the degradation, removal of COD, TOC, and dissolved O_3 for ozonation and catalytic ozonation process. Yan et al., reported the degradation of iohexol, which presented a higher percentage of degradation at basic conditions with ozone/α-$Fe_{0.9}Mn_{0.1}OOH$ (Yan et al. 2021). In addition, An et al. reported a similar behavior in the degradation of phenol. However, for pH > 8 the degradation decreased by 16 percent for the O_3/g-C_3N_4/MgO/Vis system (An et al. 2020). Similarly, in the catalytic ozonation with activated carbon, it showed better performance in removing COD (chemical oxygen demand) at basic pH conditions based on the data reported by Zhou et al. (2020). Xiong et al. reported an insignificant contribution in removing TOC in the degradation of phenol when varying the pH on the process with O_3/Carbon-Activado (Xiong et al. 2020). Meanwhile, Shen et al. demonstrated that with increasing pH, dissolved O_3 decreases due to its possible decomposition into ROS (such as $OH^{\bullet}$), which could explain the favorable performance in degradation, removal of COD and TOC (Shen et al. 2020).

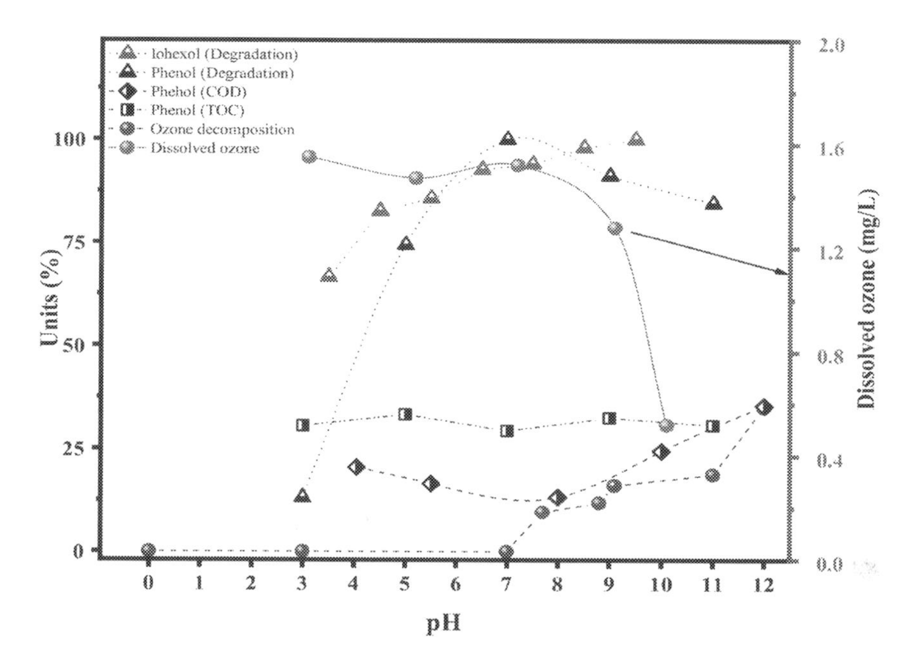

FIGURE 9.11 Effect of pH on degradation, removal of TOC and COD on contaminants, and dissolved and decomposition O_3.

9.5.4.2 Ozone Dosage

In the catalytic ozonation process, multiphase reactions (gaseous, liquid, and solid) occur; this means that the concentration and flow rate of the ozone that enters the system must be mentioned. Therefore, the excess ozone is adverse to the degradation of organic and inorganic contaminants, as explained by Eqs. 45 and 46. Therefore, ozone production and use must be enhanced to improve the catalytic ozonation process efficiency (Sui et al. 2010; Wang & Bai 2017b).

$$2HO^{\bullet} + O_3 \rightarrow 2O_2 + H_2O \tag{45}$$

$$HO^{\bullet} + HO_2^{\bullet} \rightarrow O_2 + H_2O \tag{46}$$

Lara-ramos et al. evaluated the influence of the effect of ozone dose and ozone concentration on caffeine degradation. Also, they reported the effect on the degradation of the interaction with the variation in the applied ozone concentration ($[O_3]_T$) and catalyst load (Lara-Ramos et al. 2021) (see Fig. 9.12).

9.5.4.3 Catalyst Dosage

The use of catalysts in ozonation can provide active sites for ozone catalytic reactions, in addition to the adsorption of water, O_3, and organic compounds. Consequently, the surface area and the number of active sites increase with increasing catalyst loading (Oputu et al. 2015). The catalytic ozonation process could obtain a higher speed

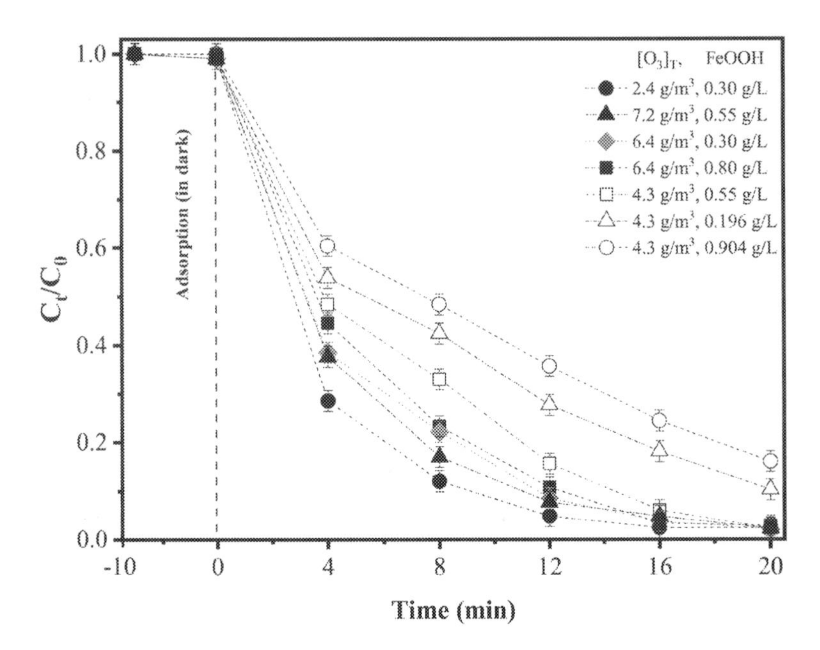

FIGURE 9.12 Effect of catalyst loading and applied ozone concentration on caffeine degradation. Taken from Lara-Ramos et al. (2021).

in the degradation of organic compounds than ozonation. However, an excessive increase in the catalyst load can decrease the concentration of pollutants and ozone per unit area.

Consequently, the catalytic efficiency in catalytic ozonation could decrease (Valdés et al. 2012). Therefore, it is necessary to evaluate the catalyst dosage in the future.

Meanwhile, metal leachates in heterogeneous catalytic ozonation systems generally range from 1 μg/L to 10 mg/L in the solutions subjected to the process. For all these reasons, it is necessary to optimize the catalyst load in the catalytic ozonation process. To this, Lara-Ramos et al. studied the influence of the goethite load (FeOOH) and applied ozone concentration on caffeine degradation (Lara-Ramos et al. 2021), and found that 0.3 g/L of FeOOH and 2.4 g/m^3 of ozone showed the best performance in degradation.

9.5.4.4 Initial Concentration of Contaminants

A high initial concentration of pollutants can affect the catalyst's active sites since the pollutant molecules would be occupying the active sites, which is not favorable to the formation of $OH^•$ by adsorption and decomposition of ozone. Additionally, it is important to mention that as the reaction progresses, small molecule organic acids accumulate on the catalyst surface during catalytic ozonation (Valdés et al. 2012). Ye et al. evaluated four atrazine concentrations and reported the degradation percentages. The initial atrazine concentrations were 5 mg/L, 10 mg/L, 20 mg/L, and 30 mg/L, the degradation were 98.2 percent, 93.9 percent, 87.3 percent, and

68.4 percent, respectively (Ye et al. 2020). The results indicated that the degradation percentages gradually declined with atrazine concentration (Ye et al. 2020).

9.5.4.5 Temperature

Ozone solubility is inversely proportional to the temperature inside the reactor. Furthermore, an increment in the temperature positively affects the catalytic ozonation process (Jung et al. 2017; Wang & Bai 2017b) (see Fig. 9.13). In most of the work reported in the literature (Huang et al. 2011; Naddeo et al. 2009; Rey et al. 2012; Rezaei et al. 2016; Wang & Bai 2017a; Wu et al. 2017), it is reported that the temperature condition of 25°C is the appropriate reaction temperature.

Many authors reported the effect of temperature on the processes of catalytic ozonation and ozonation. For example, Beltrán et al. reported the degradation rate constants (pseudo first-order, k') of oxalic acid (8×10^{-3} M, pH = 2.5) at different temperatures by catalytic ozonation with 1.25 g L^{-1} Fe_2O_3/Al_2O_3 (Beltrán et al. 2005). Additionally, data of the decrease in dissolved ozone concentration increased the temperature up to 40°C (see Fig. 9.13). On the other hand, Yarahmadi et al. reported for four temperatures (5, 15, 21, and 35 °C) the rate constants of second-order reactions (k'') of the ozonation of five hormones (testosterone, progesterone, medroxyprogesterone, levonorgestrel, and norethindrone, worked concentration of 10 µg/L and pH = 6) for the degradation with 2 mg O_3/L (Yarahmadi et al. 2019); also, ozone decomposition with and without scavenger are reported (see Fig. 9.13).

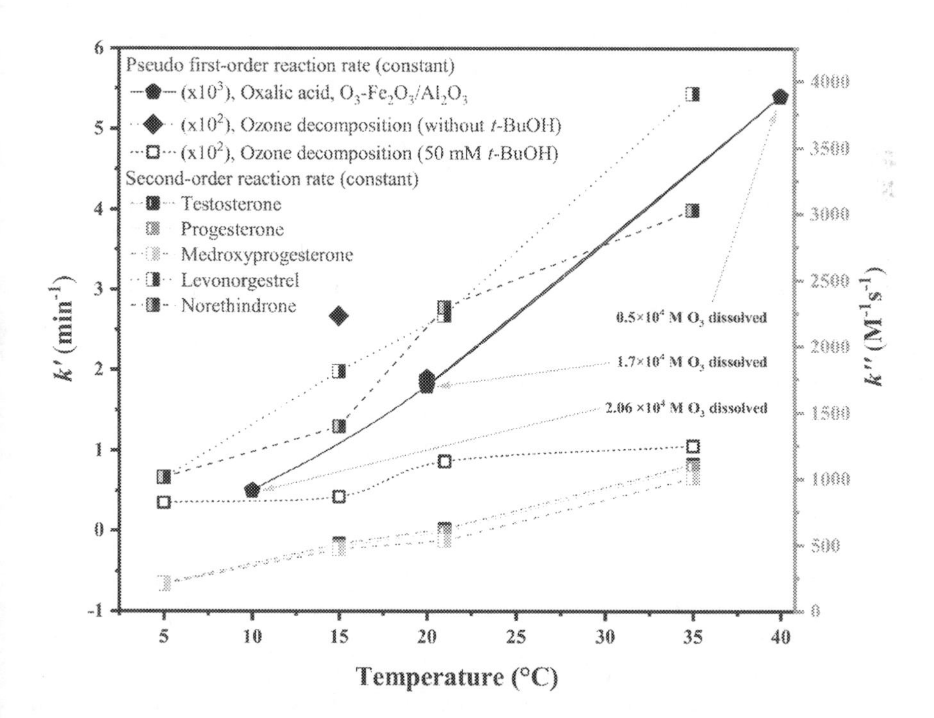

FIGURE 9.13 Influence of solution temperature on the reaction rate of organic compounds, ozone decomposition, and dissolved ozone concentration for catalytic ozonation processes.

Testosterone, progesterone, and medroxyprogesterone had similar trends in rate constants k'' with increasing temperature, which could be due to the formation of similar reactive intermediates due to the similarity in both the substituents in the single bonds C-C as is its chemical structure (Yarahmadi et al. 2019). Levonorgestrel and norethindrone had higher reaction rates than in the case of the three previous hormones with increasing temperature (see Fig. 9.13). Therefore, higher values in k'' for the ozonation of norethindrone and levonorgestrel could be explained by the affinity of the ethynyl group for the O_3 molecule and less steric hindrance at the C = C bond (due to the presence of the -H substituent instead of $-CH_3$) towards ozonation (Yarahmadi et al. 2019). Meanwhile, the ozone decomposition constants k'' with and without collector increased with increasing temperature, see Fig. 9.13.

Other parameters influence catalytic ozonation efficiency. For example, the composition of the waters to be treated are complex systems because, in addition to organic compounds, there may be inorganic ions (for example, chloride ions, sulfates, carbonates, phosphates, bromates, etc.). These inorganic ions could affect the catalyst's active sites, which produces an inhibition or decrease of the catalyst's catalytic activity (Wang & Bai 2017a). Pressure is also a factor present during catalytic ozonation. However, the effect of pressure has not been widely reported in the literature. A possible cause of this is that catalytic ozonation can be applied at atmospheric pressure conditions and obtain a complete degradation and mineralization of pollutants in aqueous solution. However, Ghuge and Saroha reported a positive influence by increasing pressure (from 0 to 0.02 MPa) on the degradation of reactive network X-3B (from 93 percent to 99.3 percent) d and e by catalytic ozonation with a ceramic honeycomb of Mn-Fe as a catalyst (Ghuge & Saroha 2018). However, the authors consider it necessary to study the effect of pressure on catalytic ozonation.

9.5.5 Reactor Configuration, Catalysts, and their Application in Water Treatment

In Fig. 9.14 each reactor's schematic diagrams that can be used for ozonation and catalytic ozonation are presented. Based on the objective of applying the ozone and catalyst, the type of reactor must be chosen (wastewater, multi-chamber reactors are usual). One aspect to bear in mind is the type of reaction regime. For example, for slow reactions, the contact time must be improved (packed bed reactors are preferred), while for moderate and fast regimes, the hydrodynamics of the reactor and ozone mass transfer must be improved (fast reactions, high rotation reactors are used). In almost all ozone applications to treat water, ozone is supplied in the form of bubbles, as shown in Fig. 9.14. The chemical and physical composition of water (including contaminants) can lead to drawbacks such as mass transfer limitations, higher operating, and energy costs. Furthermore, in some applications, it is necessary to ensure the control of the ozone dose and chemical analysis of the treated water in order to verify that by-products more dangerous than the original ones have not been formed.

Heterogeneous catalytic ozonation has shown to be a promising AOP for removing organic contaminants, including antibacterial agents, in drinking water and wastewater (Sui et al. 2012; Ranjbar Vakilabadi et al. 2017). Wang and Bai (2017b) carried out a review of different articles on treatment with catalytic ozonation and

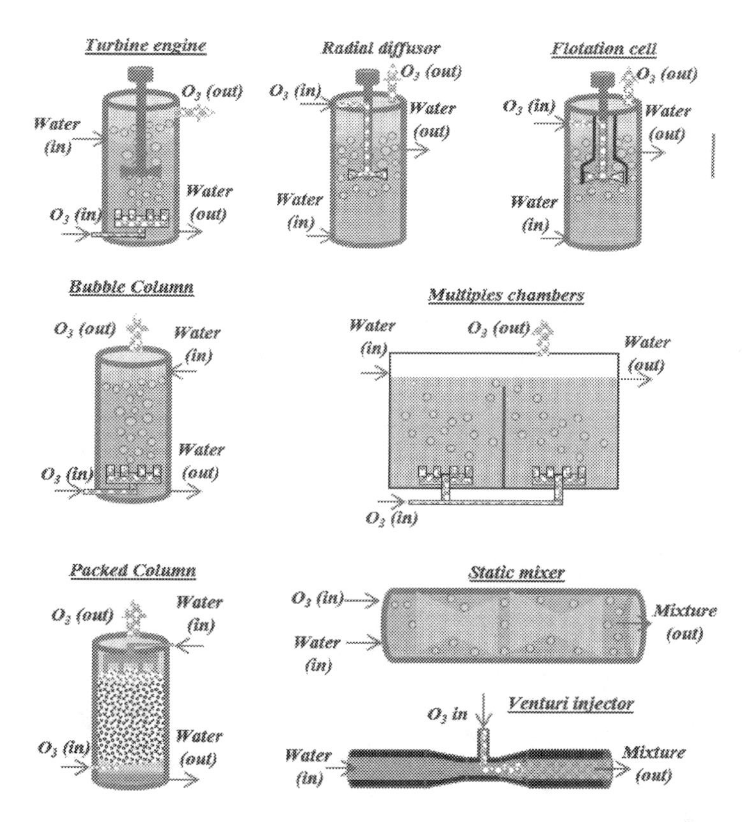

FIGURE 9.14 Conventional reactor schemes for ozonation and catalytic ozonation. Adapted from Schmitt et al. (2020).

reported different organic pollutants, catalysts, treatment efficiency, and operating conditions of the articles studied. Based on the information collected by Wang and Bai (2017b), Table 9.7 was prepared and edited. In the list shown in Table 9.7, some works are presented in heterogeneous catalytic ozonation with different catalysts and types of reactors applied in the treatment of pesticides, herbicides, pharmaceutical products, phthalic acid, colorants, and phenolics at different operating conditions.

9.5.6 Applications of Catalytic Ozonation in Combination Processes

9.5.6.1 Ozone/UV Radiation

The combination of ozone and ultraviolet radiation as a process is initiated by the photolysis of ozone (UV radiation less than 310 nm). Photodecomposition of ozone leads to the formation of $OH^{\bullet}$ and hydrogen peroxide (Rivas et al. 2012; Wang & Chen 2020). In this sense, the reaction rate between the ozone molecule and hydrogen peroxide is reported to be a constant of 6.5×10^{-2} $M^{-1}s^{-1}$. On the other hand, the application of ozone and UV radiation for wastewater treatment in recent years has been used to treat 17-ß-estradiol (E2) and bisphenol A in aqueous medium

TABLE 9.7
Degradation of Pesticides, Herbicides, Pharmaceuticals, Phthalic Acid, Colorants, Phenolics, and Wastewater by Catalytic Ozonation

Pollutants	Catalyst/Reactor	Efficiency/Reaction Time	Operating Conditions
2,4-Dichlorophenoxyacetic acid (2,4-D)	Co/Mn/γ-Fe_2O_3/Column with fluidized catalyst	93% TOC removal, 40 min	pH: 6; Cat.: 1 g/L; [O_3]: 20 mg/L; flow rate: 12 L/h; [2,4-D]$_0$: 20 mg/L
2,4-Dichlorophenoxyacetic acid (2,4-D)	Fe_3O_4/Al_2O_3/Column with fluidized catalyst	95% TOC removal, 40 min	pH: 6; Cat.: 1 g/L; [O_3]: 30 mg/L; flow rate: 12 L/h; [2,4-D]$_0$: 20 mg/L
2,4-Dichlorophenoxyacetic acid (2,4-D)	Fe-Ni/AC/Tubular glass with fluidized catalyst	72% TOC removal, 60 min	pH: 4.18; Cat.: 0.5 g/L; [O_3]: 50 mg/h; [2,4-D]$_0$: 10 mg/L
Omethoate (OMT)	Fe(III)/AC/Glass cylinder with fluidized catalyst	50% TOC removal, 120 min	pH: 8; Cat.: 0.2 g/L; [O_3]$_0$: 15 mg/min; [OMT]$_0$: 10 mg/L
Ibuprofen (IBU)	β-FeOOH/Al_2O_3/Column with fluidized catalyst	100% degradation, 9 min	pH: 7; Cat.: 1.5 g/L; [O_3]: 30 mg/L; flow rate: 12 L/h; [IBU]$_0$: 10 mg/L
Ciprofloxacin (CPFX)	β-FeOOH/Al_2O_3/Column with fluidized catalyst	80% degradation, 88% TOC removal, Nd	pH: 7; Cat.: 1.5 g/L; [O_3]: 30 mg/L; flow rate: 12 L/h; [CPFX]$_0$: 10 mg/L
Acetylsalicylic acid (ASA)	Fe_3O_4/SiO_2/CeO_2/Cylindrical glass with porous spheres	81.0% degradation, 60 min	Not defined, (Nd)
Ibuprofen (IBU)	Fe_2O_3/Al_2O_3@SBA-15/Beaker stirrer (magnetic)	90% TOC removal, 60 min	pH: 7.0; Cat.: 1.5 g/L; [O_3]: 30 mg/L; flow rate: 0.2 L/min; [IBU]$_0$: 10 mg/L
Phenacetin (PNT)	$CuFe_2O_4$/Glass column flask	90% degradation, 180 min	pH: 7.72; Cat.: 2.0 g/L; [O_3]: 0.36 mg/min; [PNT]$_0$: 0.2 mM
Sulfamethoxazole (SMX)	Fe_3O_4/Shaken conical flask	100% degradation, 5 min	pH: 7; Cat.: 0.3 g/L; [O_3]: 2g/h; [SMX]$_0$: 50 mg/L
Dimethyl phthalate (DMP)	Fe_2O_3/SBA-15/Column with fluidized catalyst	100% degradation, 30 min	pH: 5.7; Cat.: 0.28 g/L; [O_3]: 50 mg/h; flow rate: 0.81 L/min; [DMP]$_0$: 10 mg/L
Dibutyl phthalate (DBP)	FeOOH/AC/Column with fluidized catalyst	63% degradation, 60 min	pH: 6; Cat.: 10 mg/L; [O_3]: 0.15 mg/L; [DBP]$_0$: 2 mg/L

(*Continued*)

TABLE 9.7 (Continued)

Degradation of Pesticides, Herbicides, Pharmaceuticals, Phthalic Acid, Colorants, Phenolics, and Wastewater by Catalytic Ozonation

Pollutants	Catalyst/Reactor	Efficiency/Reaction Time	Operating Conditions
Diethyl phthalate (DEP)	$Pr/SiO_2@Fe_3O_4$/Column with micromembranes	98.1% degradation, 62.3% TOC removal, 90 min	pH: 6.4; Cat.: 0.5 g/L; $[O_3]$: 60 mg/L; flow rate: 60 mL/min; $[DBP]_0$: 20 mg/L
Methylene blue (MB)	Fe^0-CNTs/Column with fluidized catalyst	89% TOC removal, 125 min	pH: 3; Cat.: 13.6 mg/L; $[O_3]$: 5 g/h; $[MB]_0$: 0.025 mM
Acid Red B (ARB)	Fe-Cu-O/Column with fluidized catalyst	90% degradation, 20 min	pH: 6.36; Cat.: 1 g/L; $[O_3]$: 6 mg/min; $[ARB]_0$: 200 mg/L
Acid Orange II (AOII)	$MgFe_2O_4$/Column with fluidized catalyst	94.1% degradation, 40 min, 48.1% TOC removal, 160 min	pH: 6.36; Cat.: 0.1 g/L; $[O_3]$: 0.5 mg/min; $[AOII]_0$: 50 mg/L
Phenol	Fe_3O_4/AC/Column with fluidized catalyst	98.5% degradation, 60 min	pH: 8; Cat.: 2 g/L; $[O_3]$: 33 mg/min; $[Phenol]_0$: 500 mg/L
p-Nitrophenol (PNP)	$MnFe_2O_4$/Column with fluidized catalyst	95.7% degradation, 30 min	pH: 6.5; Cat.: 0.2 g/L; $[O_3]$: 2 mg/min; $[PNP]_0$: 10 mg/L
Coal gasification wastewater	FeO_x/Activated Carbon Based Sewage Sludge/Fluidized Column	78.1% DQO removal, 60 min 64.5% TOC removal, 60 min	pH: 7.5; Cat.: 1 g/L; $[O_3]$: 15 mg/L; flow rate: 0.5 L/min; $[COD]_0$: 180 mg/L; $[TOC]_0$: 70 mg/L; $[TN]_0$: 40 mg/L
Oil refining wastewater	Fe/AC/Column with porous catalyst plates	DBO_5/DQO: 0.337, Nd	pH: 7,8; Cat.: 5 g/L; $[O_3]$: 10 g/L; flow rate: 45 L/h; $[COD]_0$: 80.8 mg/L; $[DBO_5/DQO]_0$: 0.171
Oil refinery wastewater	$Mn\text{-}Fe\text{-}Cu/Al_2O_3$/Pressurized with supported catalyst	DBO_5/DQO: 0.330, 60 min	pH: 8.2; Cat.: 7 g/L; $[O_3]$: 2.19 g/h; $[COD]_0$: 2825 mg/L; $[DBO5/DQO]_0$: 0.098
Wastewater dyeing and finishing	Iron shavings/Porous plates	48% COD removal, Nd	pH: 7.37; Cat.: 20 g/L; $[O_3]$: 10.8 g/L; flow rate: 500 mL/min; $[COD]_0$: 142 mg/L

Source: Taken and edited from Wang and Bai (2017a).

Note: TOC is total organic carbon removed; $[O_3]$ is the ozone dosage during the process; Cat. is the initial catalyst concentration; COD is the chemical demand; COD is the chemical oxygen demand; BOD is biological oxygen demand; subscripts 0 represents initial condition.

(100 percent degradation); synthetic samples of citrus wastewater (68.4 percent color removal); winery wastewater (21 percent COD removal); terephthalic acid wastewater (44 percent COD removal); disinfection by-product precursors from raw surface water (50 percent TOC removed) etc. as reported in the work of Malik et al. (2020).

9.5.6.2 Ozone/Hydrogen Peroxide

In recent years, hydrogen peroxide (H_2O_2) has been used frequently to treat industrial wastewater (Wang & Chen 2020). The combination of ozone and H_2O_2 (known as Peroxone) allows acceleration of the decomposition of ozone into $OH^•$ through the electron transfer mechanism that allows H_2O_2 as an electron acceptor (Lara-Ramos et al. 2020a; Malik et al. 2020). Therefore, the degradation of organic pollutants is improved. For this reason, Peroxone has been used to treat wastewater such as textile wastewaters (60.12–63.41 percent decolorization), 1,4-dioxane (100 percent degradation), and fresh surface water samples fortified with wild strains of Escherichia coli. (inactivation of E. coli, around 6.80 log) as reported in the work of Malik et al. (2020).

9.5.6.3 Ozone/Biological Treatment

Biological treatments sometimes cannot eliminate or degrade organic pollutants present in wastewater due to their high chemical stability and low biodegradability (Oller et al. 2011). On the other hand, ozone treatment is usually more efficient in removing organic compounds that are stable to conventional treatment. Therefore, investigations of the combination of biological treatment with ozone have been reported in the literature. In these reports, the combination of the aforementioned treatments is divided into two types: pre-treatment (ozonation/aerobic biological batch system, among others) and post-treatment (biological treatment of activated sludge/ozonation, among others), as reported by Oller et al. (2011) and Wang and Chen (2020).

9.5.6.4 Ozone/Electrochemical Processes

Electrochemical processes use a reactor consisting of an electrolytic cell with an anode and a cathode (electrodes). During the reactor operation, a current is applied to the electrodes to produce hydroxyl radicals or the same ozone in the case of electro-ozonizers, as reported by Lara-Ramos et al. (2020b). On the other hand, iron and aluminum electrodes are used for electrocoagulation when an electric current is applied to increase the concentration of metal ions in the solution. These metal ions allow the coagulation or precipitation of pollutants present in wastewater since the ions act as destabilizing (or neutralizing) agents of the particle's charge. At an industrial level, this combination of ozone with electrochemical processes has been used to treat wastewater such as distillery industrial effluent (100 percent of color and COD removal), gray wastewater (91.31 percent TOC removal), ferrous solutions deriving from the color fixation stage of ripe olive processing (reduction of phenol concentration from 80 mg/L to 10 mg/L), and industrial wastewater (90 percent removal of color; 60 percent COD removal) as reported in the work of Malik et al. (2020).

9.5.7 Economic Aspects

The approximate costs for an ozonation system are described below through four sections: Energy Consumption, Equipment Costs, Operating and Maintenance Costs, and Cost of Drinking Water and Wastewater Treatment.

9.5.7.1 Energy Consumption

The consumption of electrical energy is a very important economic factor to consider in the application of AOPs, and this factor may be the main operating cost to consider during the application of the catalytic ozonation (Abdallah et al. 2020; Cardoso et al. 2016; Lara-Ramos et al. 2019, 2020b; Salsabil et al. 2010). The expression shown in Equation 47 is used to calculate the electrical energy in order (EE/O, kW h m^{-3} $order^{-1}$), where the concentration of pollutants or even COD units, color, and turbidity can be used. This should not be a problem since the $\log(C_0/C_t)$ expression has no dimensions (Cardoso et al. 2016).

$$EE/O = \frac{P \times t \times 1000}{V \times 60 \times \log(C_0/C_t)} \tag{47}$$

$$\ln(C_0/C_t) = k't \tag{48}$$

where P is the electrical power (kW) for ozonation; t is the reaction time (min); V is the volume (L) of the reactor; C_0 and C_t are the initial and final concentration of pollutants or even COD units, color, turbidity, etc., and k' is the pseudo-first-order velocity constant (min^{-1}) for the decrease in pollutant concentration. Combining Eqs. (47) and (48), EE/O becomes

$$EE/O = \frac{38.4 \times P}{V \times k'} \tag{49}$$

9.5.7.2 Equipment Costs

Among the equipment necessary for the assembly and start-up of an ozonation system, the ozone generator, air preparation equipment, ozone destruction unit, ozone contactor, turbulence, or catalyst mixing equipment, instrumentation, and control could be mentioned (see Table 9.8). However, it is essential to note that there are many ways to pretreat the air, cause ozone contact with the solution, waste exhaust gas handling, and mix the catalyst (Gottschalk et al. 2010; Shammas & Wang 2005). Therefore, the equipment costs shown in this section should be considered only as a general price criterion. Table 9.8 presents a cost estimation list of equipment obtained from two provisioners (A and B) for an ozonation system applied to small water-supply systems. All costs shown in the table correspond to costs in 2003 US dollars (Shammas & Wang 2005).

Based on the information reported in Table 9.8, Suppliers A and B estimate that the average installation costs of an ozone production unit would be $14,560 to $24,260 (in 2003 dollars, for Supplier A) and $19,130 to $31,880 (in 2003 dollars, for Supplier B) this information was reported by Shammas and Wang 2005.

TABLE 9.8
Equipment Cost of Ozonation Systems for Small Water Plants: Supplier A and B

	Supplier A			Supplier B		
	Size of the treatment plant					
	500000 GPD[a]	350000 GPD	180000 GPD	0.1 MGD[b]	0.3 MGD	0.5 MGD
Maximum dosage of ozone (mg/L) at peak flow	5	5	5	3	3	3
Daily ozone requirement	9.53 kg	6.35 kg	3.18 kg	1.36 kg/day	3.18 kg/day	6.35 kg/day
Contact chamber diameter (m)	1.83	1.52	1.22	–	–	–
Equipment Costs						
Air preparation and ozone generation unit	$47,250	$37,500	$33,000.0	$26,250.0	$52,500.0	$70,200
Contact chamber with diffusers	$17,250	$15,300	$14,850.0	$12,750.0	$24,000.0	$43,500
Monitoring instrumentation or ozone generation	$22,500	$22,500	$22,500.0	$12,750.0	$24,000.0	$43,500
Ozone destruction unit or monitor exhaust	$10,050	$7,500	$6,300	$6,000	$6,000	$6,000
Total Equipment Costs	$97,050	$82,800	$76,650	$79,350	$88,800	$125,250
Power Requirement (kWh)	13.3	10.1	5.0	20	13.5	13.5

Source: This information was reported by Shammas and Wang (2005).
Notes:
[a] 1 MGD = 0.044 m^3/s.
[b] 1 GPD = 0.003785 m^3/day.

9.5.7.3 Operating and Maintenance Costs

Some of the operating costs that could be involved in the implementation of a catalytic ozonation system based on those considered by Shammas and Wang (2005) are

- Ozone dose and catalyst load required.
- Use of oxygen or from the air, the latter requires dehumidification devices and air preparation methods.
- Ozone generation system cooling.
- Catalyst and ozone contact methods: static mixers and diffuser contactors do not require electrical power like turbine mixers, fluidized bed mixers, cyclone, impeller, etc.

- Catalyst residue disposal and exhaust gas handling. In the case of gases, it is common to use thermal destruction, catalytic destruction, potassium iodide solutions, or adsorption with activated carbon.

The maintenance costs of a catalytic ozonation process necessarily include cleaning, repairs, and replacement of parts and material damaged by use (Shammas & Wang 2005). Maintenance costs due to the ozone generation system would involve cleaning and replacing filters in the air or oxygen adaptation system, maintaining the ozone generator and pipes (of the generation system and contact with the reactor), and performing periodic maintenance required because of the use and replacement of the catalyst. The half-life of the catalyst, deactivation by poisoning, wear by mechanical effect, and loss on the catalyst's dissolution must be taken into account. Lastly, there is general maintenance (including equipment that allows ozone and catalyst contact) and labor.

9.5.7.4 Cost of Drinking Water and Wastewater Treatment

Fig. 9.15 shows some operating costs for the treatment of drinking and wastewater. The cost in dollars to produce 1 kg of ozone ranged from \$1.89 –\$2.52 (price in dollars for 2006). It is important to mention that operating costs were reported by Gottschalk et al. (2010) in Euros but are presented in dollars in Fig. 9.15, with an exchange rate of 1.26 (for the year 2006) reported by World Economic Outlook (World Economic Outlook 2006). The operating cost of drinking water depends on the treatment (disinfection, color, COD, Particle removal, etc.) and ranges between <1 and 7 cents per m^3. COD removal (biodegradable or not) is presented as the most expensive (Σ0.07 $\$m^{-3}$), while particle removal is the most economical (0.005 $\$m^{-3}$), with an ozone dose of 0.02 kg of O_3 per m^3, similar in all drinking water treatments (see Fig. 9.15).

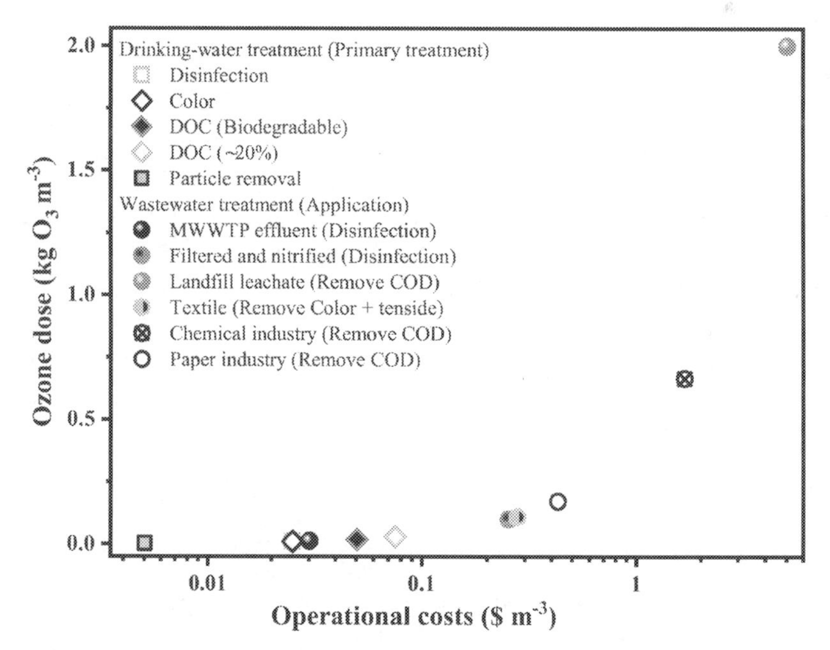

FIGURE 9.15 Operating costs versus ozone dose in drinking and wastewater treatment. Information reported by Gottschalk et al. (2010).

On the other hand, the operating costs of the ozonation of WWTP effluent depend on the type of application (WWTP effluent, filtered and nitrified, landfill leachate, etc.) and, to a greater extent, on the type of treatment, whether disinfection or elimination of contaminants that can be oxidized (Lucas et al. 2010). Additionally, the cost of treating industrial wastewater depends mostly on its composition. The price can range from 1 to 66 cents per m^3 to treat industrial wastewater filtered and nitrified (disinfection) and chemical industry (remove COD), respectively.

9.6 CONCLUSIONS

Some key aspects were analyzed and showed some catalytic ozonation applications as treatment of drinking and wastewater. Possible reaction mechanisms were also presented, using homogeneous, heterogeneous, and photocatalytic catalysts in the ozonation process. Numerous investigations have been carried out at the literature level to describe each element that intervenes in catalytic ozonation. However, it is still necessary to continue investigating and evaluating the application of the process on an industrial scale to treat pollutants present in wastewater. Additionally, the following scientific parameters and aspects must be taken into account in order to justify the relevance from the scientific to the economic of catalytic ozonation: the scientific, technical, and economic criteria used to evaluate whether to choose a catalytic ozonation system in the treatment of drinking and wastewater are; achieve the proposed objective with the treatment (reduce contamination to legal limits); reduce the production of by-products (and not toxic produce by-products); develop ozonation systems that are safe to operate; minimize the use of ozone; and lastly, keep investment and energy costs low.

ACKNOWLEDGMENTS

The author thanks Tecnoparque Nodo Cali and Universidad del Valle. Diaz-Angulo and Lara-Ramos thank the Ministerio de Ciencia Tecnología e Innovación for the Ph.D. Scholarship 647, 727 and for the financial support to produce this work through the project CI. 2987. Machuca-Martinez thanks to Sistema General de Regalías Colombia, Grant BPIN 2018000100096.

REFERENCES

Abdallah M, Shanableh A, Elshazly D, Feroz S (2020) Techno-economic and environmental assessment of wastewater management systems: Life cycle approach. Environ Impact Assess Rev 82:106378. https://doi.org/10.1016/j.eiar.2020.106378

An W, Tian L, Hu J, et al. (2020) Efficient degradation of organic pollutants by catalytic ozonation and photocatalysis synergy system using double-functional $MgO/g-C_3N_4$ catalyst. Appl Surf Sci 534:147518. https://doi.org/10.1016/j.apsusc.2020.147518

Ansari M, Sharifian M, Ehrampoush MH, et al. (2021) Dielectric barrier discharge plasma with photocatalysts as a hybrid emerging technology for degradation of synthetic organic compounds in aqueous environments: A critical review. Chemosphere 263. https://doi.org/10.1016/j.chemosphere.2020.128065

Barbosa MO, Moreira NFF, Ribeiro AR, et al. (2016) Occurrence and removal of organic micropollutants: An overview of the watch list of EU Decision 2015/495. Water Res 94:257–279. https://doi.org/10.1016/j.watres.2016.02.047

Beltrán FJ (2003) Ozone Reaction Kinetics for Water and Wastewater Systems, 1st Edition. CRC Press; 1 edition (December 29, 2003), London

Beltrán FJ, Rivas FJ, Montero-De-Espinosa R (2005) Iron type catalysts for the ozonation of oxalic acid in water. Water Res 39:3553–3564. https://doi.org/10.1016/j.watres.2005.06.018

Cardoso JC, Bessegato GG, Boldrin Zanoni MV (2016) Efficiency comparison of ozonation, photolysis, photocatalysis and photoelectrocatalysis methods in real textile wastewater decolorization. Water Res 98:39–46. https://doi.org/10.1016/j.watres.2016.04.004

EPA (1986) Quality Criteria for Water 1986

Fu P, Lianghua Wang, Gen Li, et al. (2020) Homogenous catalytic ozonation of aniline aerofloat collector by coexisted transition metallic ions in flotation wastewaters. J Environ Chem Eng 8:103714

Ghuge SP, Saroha AK (2018) Catalytic ozonation for the treatment of synthetic and industrial effluents — application of mesoporous materials: A review. J Environ Manage 211:83–102. https://doi.org/10.1016/j.jenvman.2018.01.052

Giri RR, Ozaki H, Takanami R, Taniguchi S (2008) A novel use of TiO_2 fiber for photocatalytic ozonation of 2,4-dichlorophenoxyacetic acid in aqueous solution. J Environ Sci 20:1138–1145. https://doi.org/10.1016/S1001-0742(08)62161-0

Gottschalk C, Libra JA, Saupe A (2010) Ozonation of Water and Waste Water: A Practical Guide to Understanding Ozone and Its Applications. Wiley-VCH, Weinheim Germany. ISBN: 978-3-527-62893-3.

Gracia R, Aragües JL, Ovelleiro JL (1996) Study of the catalytic ozonation of humic substances in water and their ozonation byproducts. Ozone Sci Eng 18:195–208. https://doi.org/10.1080/01919519608547326

Hassani A, Khataee A, Karaca S, Fathinia M (2016) Heterogeneous photocatalytic ozonation of ciprofloxacin using synthesized titanium dioxide nanoparticles on a montmorillonite support: Parametric studies, mechanistic analysis and intermediates identification. RSC Adv 6:87569–87583. https://doi.org/10.1039/c6ra19191f

Huang R, Yan H, Li L, et al. (2011) Environmental catalytic activity of Fe/SBA-15 for ozonation of dimethyl phthalate in aqueous solution. Applied Catal B, Environ 106:264–271. https://doi.org/10.1016/j.apcatb.2011.05.041

Jeirani Z, Soltan J (2016) Ozonation of oxalic acid with an effective catalyst based on mesoporous MCM-41 supported manganese and cerium oxides. J Water Process Eng 12:127–134. https://doi.org/10.1016/j.jwpe.2016.07.004

Jung H, Kim J, Choi H, et al. (2008) Synthesis of nanosized biogenic magnetite and comparison of its catalytic activity in ozonation. Appl Catal B Environ J 83:208–213. https://doi.org/10.1016/j.apcatb.2008.02.016

Jung Y, Hong E, Kwon M, Kang JW (2017) A kinetic study of ozone decay and bromine formation in saltwater ozonation: Effect of O_3 dose, salinity, pH, and temperature. Chem Eng J 312:30–38. https://doi.org/10.1016/j.cej.2016.11.113

Lan BY, Nigmatullin R, Li Puma G (2008) Ozonation kinetics of cork-processing water in a bubble column reactor. Water Res 42:2473–2482. https://doi.org/10.1016/j.watres.2008.01.030

Lara-Ramos JA, Diaz-angulo J, Machuca-Martínez F (2021) Use of modified flotation cell as ozonation reactor to minimize mass transfer limitations. Chem Eng J 405:126978. https://doi.org/10.1016/j.cej.2020.126978

Lara-Ramos JA, Llanos-Diaz GD, Diaz-Angulo J, Machuca-Martínez F (2020a) Evaluation of caffeine degradation by sequential coupling of $TiO_2/O_3/H_2O_2/UV$ processes. Top Catal. https://doi.org/10.1007/s11244-020-01316-w

Lara-Ramos JA, Saez C, Machuca-Martínez F, Rodrigo MA (2020b) Electro-ozonizers: A new approach for an old problem. Sep Purif Technol 241:. https://doi.org/10.1016/j.seppur.2020.116701

Lara-Ramos JA, Sánchez-Gómez K, Valencia-Rincón D, et al. (2019) Intensification of the O_3/TiO_2/UV advanced oxidation process using a modified flotation cell. Photochem Photobiol Sci 18:920–928. https://doi.org/10.1039/c8pp00308d

Li X, Ma J, He H (2020a) Recent advances in catalytic decomposition of ozone. J Environ Sci (China) 94:14–31. https://doi.org/10.1016/j.jes.2020.03.058

Li Y, Wu L, Wang Y, et al. (2020b) γ-Al_2O_3 doped with cerium to enhance electron transfer in catalytic ozonation of phenol. J Water Process Eng 36:101313. https://doi.org/10.1016/j.jwpe.2020.101313

Liu Y, Wang S, Gong W, et al. (2017) Heterogeneous catalytic ozonation of p-chloronitrobenzene (pCNB) in water with iron silicate doped hydroxylation iron as catalyst. Catal Commun 89:81–85. https://doi.org/10.1016/j.catcom.2016.10.022

López-Vinent N, Cruz-Alcalde A, Gutiérrez C, et al. (2020) Micropollutant removal in real WW by photo-Fenton (circumneutral and acid pH) with BLB and LED lamps. Chem Eng J 379:122416. https://doi.org/10.1016/j.cej.2019.122416

Lu J, Wei X, Chang Y, et al. (2015) Role of Mg in mesoporous $MgFe_2O_4$ for efficient catalytic ozonation of Acid Orange II. J Chem Technol Biotechnol 91:985–993. https://doi.org/10.1002/jctb.4667

Lucas MS, Peres JA, Lan BY, Li Puma G (2009) Ozonation kinetics of winery wastewater in a pilot-scale bubble column reactor. Water Res 43:1523–1532. https://doi.org/10.1016/j.watres.2008.12.036

Lucas MS, Peres JA, Li Puma G (2010) Treatment of winery wastewater by ozone-based advanced oxidation processes (O_3, O_3/UV and $O_3/UV/H_2O_2$) in a pilot-scale bubble column reactor and process economics. Sep Purif Technol 72:235–241. https://doi.org/10.1016/j.seppur.2010.01.016

Malik SN, Ghosh PC, Vaidya AN, Mudliar SN (2020) Hybrid ozonation process for industrial wastewater treatment: Principles and applications: A review. J Water Process Eng 35. https://doi.org/10.1016/j.jwpe.2020.101193

Mahmoodi NM, Bashiri M, Moeen SJ (2012) Synthesis of nickel-zinc ferrite magnetic nanoparticle and dye degradation using photocatalytic ozonation. Mater Res Bull 47:4403–4408. https://doi.org/10.1016/j.materresbull.2012.09.036

Mashayekh-Salehi A, Moussavi G, Yaghmaeian K (2017) Preparation, characterization and catalytic activity of a novel mesoporous nanocrystalline MgO nanoparticle for ozonation of acetaminophen as an emerging water contaminant. Chem Eng J 310:157–169. https://doi.org/10.1016/j.cej.2016.10.096

Mecha AC, Onyango MS, Ochieng A, et al. (2016) Synergistic effect of UV–vis and solar photocatalytic ozonation on the degradation of phenol in municipal wastewater: A comparative study. J Catal 341:116–125. https://doi.org/10.1016/j.jcat.2016.06.015

Mehrjouei M, Müller S, Möller D (2011) Degradation of oxalic acid in a photocatalytic ozonation system by means of Pilkington Active™ glass. J Photochem Photobiol A Chem 217:417–424. https://doi.org/10.1016/j.jphotochem.2010.11.016

Morsch L (2019) UIS: CHE 267 – Organic Chemistry I (Morsch). In Libr. https://chem.libretexts.org/@go/page/40268. Accessed 13 Oct 2020.

Naddeo V, Belgiorno V, Ricco D, Kassinos D (2009) Degradation of diclofenac during sonolysis, ozonation and their simultaneous application. Ultrason Sonochem 16:790–794. https://doi.org/10.1016/j.ultsonch.2009.03.003

Nawrocki J (2013) Catalytic ozonation in water: Controversies and questions. Discussion paper. Appl Catal B Environ 142–143:465–471. https://doi.org/10.1016/j.apcatb.2013.05.061

Nawrocki J, Kasprzyk-Hordern B (2010) The efficiency and mechanisms of catalytic ozonation. Appl Catal B Environ 99:27–42. https://doi.org/10.1016/j.apcatb.2010.06.033

Nishimoto S, Mano T, Kameshima Y, Miyake M (2010) Photocatalytic water treatment over WO3 under visible light irradiation combined with ozonation. Chem Phys Lett 500: 86–89. https://doi.org/10.1016/j.cplett.2010.09.086

Oller I, Malato S, Sánchez-Pérez JA (2011) Combination of advanced oxidation processes and biological treatments for wastewater decontamination: A review. Sci Total Environ 409:4141–4166. https://doi.org/10.1016/j.scitotenv.2010.08.061

Oputu O, Chowdhury M, Nyamayaro K, et al. (2015) Catalytic activities of ultra-small B-FeOOH nanorods in ozonation of 4-chlorophenol. J Environ Sci (China) 35:83–90. https://doi.org/10.1016/j.jes.2015.02.013

Pines DS, Reckhow DA (2002) Effect of dissolved cobalt(II) on the ozonation of oxalic acid. Environ Sci Technol 36:4046–4051. https://doi.org/10.1021/es011230w

Poznyak TI, Oria IC, Poznyak AS (2018) Ozonation and biodegradation in environmental engineering: Dynamic neural network approach. Elsevier

Ranjbar Vakilabadi D, Hassani AH, Omrani G, Ramavandi B (2017) Catalytic potential of Cu/Mg/Al-chitosan for ozonation of real landfill leachate. Process Saf Environ Prot 107:227–237. https://doi.org/10.1016/j.psep.2017.02.013

Rekhate CV, Srivastava JK (2020) Recent advances in ozone-based advanced oxidation processes for treatment of wastewater: A review. Chem Eng J Adv In Press 3:100031. https://doi.org/10.1016/j.ceja.2020.100031

Rey A, Mena E, Chávez AM, et al. (2015) Influence of structural properties on the activity of WO_3 catalysts for visible light photocatalytic ozonation. Chem Eng Sci 126:80–90. https://doi.org/10.1016/j.ces.2014.12.016

Rey A, Qui DH, Álvarez PM, et al. (2012) Environmental Simulated solar-light assisted photocatalytic ozonation of metoprolol over titania-coated magnetic activated carbon. 112:246–253. https://doi.org/10.1016/j.apcatb.2011.10.005

Rezaei F, Moussavi G, Bakhtiari AR, Yamini Y (2016) Toluene removal from waste air stream by the catalytic ozonation process with MgO/GAC composite as catalyst. J Hazard Mater 306:348–358. https://doi.org/10.1016/j.jhazmat.2015.11.026

Rivas FJ, Beltrán FJ, Encinas A (2012) Removal of emergent contaminants: Integration of ozone and photocatalysis. J Environ Manage 100:10–15. https://doi.org/10.1016/j.jenvman.2012.01.025

Rivas J, Rodríguez E, Beltrán FJ, et al. (2001) Homogeneous catalyzed ozonation of simazine. Effect of Mn(II) and Fe(II). J Environ Sci Heal - Part B Pestic Food Contam Agric Wastes 36:317–330. https://doi.org/10.1081/PFC-100103572

Salsabil MR, Laurent J, Casellas M, Dagot C (2010) Techno-economic evaluation of thermal treatment, ozonation and sonication for the reduction of wastewater biomass volume before aerobic or anaerobic digestion. J Hazard Mater 174:323–333. https://doi.org/10.1016/j.jhazmat.2009.09.054

Sanchez-Polo M, Rivera-Utrilla J (2004) Ozonation of 1,3,6-naphthalenetrisulfonic acid in presence of heavy metals. J Chem Technol Biotechnol 79:902–909. https://doi.org/10.1002/jctb.1077

Savvina P, Apostolis K, Manassis M, Anastasios Z (2019) The role of metal ions on p-CBA degradation by catalytic ozonation. J Environ Chem Eng 7:103324. https://doi.org/10.1016/j.jece.2019.103324

Schmitt A, Mendret J, Roustan M, Brosillon S (2020) Ozonation using hollow fiber contactor technology and its perspectives for micropollutants removal in water: A review. Sci Total Environ 729:138664. https://doi.org/10.1016/j.scitotenv.2020.138664

Shammas NK, Wang LK (2005) In Physicochemical Treatment Processes. Humana Press, Totowa, NJ

Shen T, Zhang X, Lin KYA, Tong S (2020) Solid base Mg-doped ZnO for heterogeneous catalytic ozonation of isoniazid: Performance and mechanism. Sci Total Environ 703:134983. https://doi.org/10.1016/j.scitotenv.2019.134983

Simon G, Gyulavári T, Hernádi K, et al. (2018) Photocatalytic ozonation of monuron over suspended and immobilized TiO_2–study of transformation, mineralization and economic feasibility. J Photochem Photobiol A Chem 356:512–520. https://doi.org/10.1016/j.jphotochem.2018.01.025

Sui M, Sheng L, Lu K, Tian F (2010) FeOOH catalytic ozonation of oxalic acid and the effect of phosphate binding on its catalytic activity. Appl Catal B Environ 96:94–100. https://doi.org/10.1016/j.apcatb.2010.02.005

Sui M, Xing S, Sheng L, et al. (2012) Heterogeneous catalytic ozonation of ciprofloxacin in water with carbon nanotube supported manganese oxides as catalyst. J Hazard Mater 227–228:227–236. https://doi.org/10.1016/j.jhazmat.2012.05.039

Sun Q, Li L, Yan H, et al. (2014) Influence of the surface hydroxyl groups of MnOx/SBA-15 on heterogeneous catalytic ozonation of oxalic acid. Chem Eng J 242:348–356.

Valdés H, Tardón RF, Zaror CA (2012) Role of surface hydroxyl groups of acid-treated natural zeolite on the heterogeneous catalytic ozonation of methylene blue contaminated waters. Chem Eng J 211–212:388–395. https://doi.org/10.1016/j.cej.2012.09.069

Valdés H, Zaror CA (2006) Heterogeneous and homogeneous catalytic ozonation of benzothiazole promoted by activated carbon: Kinetic approach. Chemosphere 65:1131–1136. https://doi.org/10.1016/j.chemosphere.2006.04.027

Valério A, Wang J, Tong S, et al. (2020) Synergetic effect of photocatalysis and ozonation for enhanced tetracycline degradation using highly macroporous photocatalytic supports. Chem Eng Process - Process Intensif 149:107838. https://doi.org/10.1016/j.cep.2020.107838

Von Gunten U (2003) Ozonation of drinking water: Part I. Oxidation kinetics and product formation. Water Res 37:1443–1467. https://doi.org/10.1016/S0043-1354(02)00457-8

Wang J, Bai Z (2017a) Fe-based catalysts for heterogeneous catalytic ozonation of emerging contaminants in water and wastewater. Chem Eng J 312:79–98. https://doi.org/10.1016/j.cej.2016.11.118

Wang J, Bai Z (2017b) Fe-based catalysts for heterogeneous catalytic ozonation of emerging contaminants in water and wastewater. Chem Eng J 312:79–98. https://doi.org/10.1016/j.cej.2016.11.118

Wang J, Chen H (2020) Catalytic ozonation for water and wastewater treatment: Recent advances and perspective. Sci Total Environ 704:135249. https://doi.org/10.1016/j.scitotenv.2019.135249

Wang J, Sun Y, Jiang H, Feng J (2017) Removal of caffeine from water by combining dielectric barrier discharge (DBD) plasma with goethite. J Saudi Chem Soc 21:545–557. https://doi.org/10.1016/j.jscs.2016.08.002

Wang T, Song Y, Ding H, et al. (2020) Insight into synergies between ozone and in-situ regenerated granular activated carbon particle electrodes in a three-dimensional electrochemical reactor for highly efficient nitrobenzene degradation. Chem Eng J 394:124852. https://doi.org/10.1016/j.cej.2020.124852

Werschkun B, Sommer Y, Banerji S (2012) Disinfection by-products in ballast water treatment: An evaluation of regulatory data. Water Res 46:4884–4901. https://doi.org/10.1016/j.watres.2012.05.034

World economic Outlook (2006) World economic and financial surveys. Washington

Wu CH, Kuo CY, Chang CL (2008) Homogeneous catalytic ozonation of C.I. Reactive Red 2 by metallic ions in a bubble column reactor. J Hazard Mater 154:748–755. https://doi.org/10.1016/j.jhazmat.2007.10.087

Wu D, Li X, Tang Y, et al. (2017) Mechanism insight of PFOA degradation by ZnO assisted-photocatalytic ozonation: Efficiency and intermediates. Chemosphere 180:247–252. https://doi.org/10.1016/j.chemosphere.2017.03.127

Xiao J, Xie Y, Cao H (2015) Organic pollutants removal in wastewater by heterogeneous photocatalytic ozonation. Chemosphere 121:1–17. https://doi.org/10.1016/j.chemosphere.2014.10.072

Xing S, Lu X, Liu J, et al. (2016) Catalytic ozonation of sulfosalicylic acid over manganese oxide supported on mesoporous ceria. Chemosphere 144:7–12. https://doi.org/10.1016/j.chemosphere.2015.08.044

Xiong W, Cui W, Li R, et al. (2020) Mineralization of phenol by ozone combined with activated carbon: Performance and mechanism under different pH levels. Environ Sci Ecotechnology 1:100005. https://doi.org/10.1016/j.ese.2019.100005

Yan H, Chen W, Liao G, et al. (2016) Activity assessment of direct synthesized Fe-SBA-15 for catalytic ozonation of oxalic acid. Sep Purif Technol 159:1–6. https://doi.org/10.1016/j.seppur.2015.12.055

Yan P, Chen Z, Wang S, et al. (2021) Catalytic ozonation of iohexol with α-$Fe_{0.9}Mn_{0.1}OOH$ in water: Efficiency, degradation mechanism and toxicity evaluation. J Hazard Mater 402. https://doi.org/10.1016/j.jhazmat.2020.123574

Yarahmadi H, Vo Duy S, Barbeau B, et al. (2019) Effect of temperature on oxidation kinetics of testosterone and progestogens by ozone. J Water Process Eng 31:100879. https://doi.org/10.1016/j.jwpe.2019.100879

Ye G, Luo P, Zhao Y, et al. (2020) Three-dimensional Co/Ni bimetallic organic frameworks for high-efficient catalytic ozonation of atrazine: Mechanism, effect parameters, and degradation pathways analysis. Chemosphere 253:126767. https://doi.org/10.1016/j.chemosphere.2020.126767

Yu D, Wu M, Hu Q, et al. (2019) Iron-based metal-organic frameworks as novel platforms for catalytic ozonation of organic pollutant: Efficiency and mechanism. J Hazard Mater 367:456–464. https://doi.org/10.1016/j.jhazmat.2018.12.108

Zahrani AA, Ayati B (2020) Improving Fe-based heterogeneous Electro-Fenton nano catalyst using transition metals in a novel orbiting electrodes reactor. Chemosphere 256:127049. https://doi.org/10.1016/j.chemosphere.2020.127049

Zhang F, Wei C, Hu Y, Wu H (2015) Zinc ferrite catalysts for ozonation of aqueous organic contaminants : Phenol and bio-treated coking wastewater. Sep Purif Technol 156:625–635. https://doi.org/10.1016/j.seppur.2015.10.058

Zhang S, Wang D, Quan X, et al. (2013) Multi-walled carbon nanotubes immobilized on zero-valent iron plates (Fe^0-CNTs) for catalytic ozonation of methylene blue as model compound in a bubbling reactor. Sep Purif Technol 116:351–359. https://doi.org/10.1016/j.seppur.2013.05.053

Zhang Y, Zhao P, Li J, et al. (2016) A hybrid process combining homogeneous catalytic ozonation and membrane distillation for wastewater treatment. Chemosphere 160:134–140. https://doi.org/10.1016/j.chemosphere.2016.06.070

Zhou L, Zhang S, Li Z, et al. (2020) Efficient degradation of phenol in aqueous solution by catalytic ozonation over MgO/AC. J Water Process Eng 36:101168. https://doi.org/10.1016/j.jwpe.2020.101168

Ziylan A, Ince NH (2015) Catalytic ozonation of ibuprofen with ultrasound and Fe-based catalysts. Catal Today 240:5–8. https://doi.org/10.1016/j.cattod.2014.03.002

10 Biological/Biochemical Processes in Wastewater Treatment

Mushtaq Ahmad Rather
Chemical Engineering Department, National Institute of Technology Hazratbal, Srinagar, Kashmir, India

Parveena Bano
Division of Entomology, Sher-e-Kashmir University of Agricultural Sciences & Technology of Shalimar, Kashmir, India

CONTENTS

DOI: 10.1201/9781003138303-10

10.1 BIOLOGICAL CHARACTERISTICS AND BIOCHEMISTRY

Biological wastewater treatment involves the harnessing of microbes. Due to involvement of microbes, wastewater is the medium for microbial growth both in aerobic or anaerobic processes. The main groups of pathogenic organisms (direct) and indicator organisms (indirect, indicating the presence of pathogens) found in wastewater are:

1. Bacteria
2. Protozoa
3. Fungi
4. Viruses
5. Algae
6. Rotifers
7. Nematodes

10.1.1 BACTERIA

Bacteria are prokaryotic organisms found most abundantly on Earth. The existence of life itself cannot be imagined without the bacteria, as these are involved in most of the essential activities that sustain life and balance in ecosystems. Some of such activities are capture of nitrogen from the atmosphere, decomposition of organic matter, and photosynthesis. Some other important examples of the role of various bacteria are as follows:

a. *Pseudomonas* reduces NO_3 to N_2.
b. Zoogloea form large flocs via slime production in the aeration tanks.
c. Bdellovibrio kill pathogens in biological treatment.
d. *Acinetobacter* removes phosphate by storing it under aerobic conditions and then release same under anaerobic conditions.
e. *Nitrosomonas* changes NH_4 into NO_2.
f. *Nitrobacter* changes NO_2 to NO_3.
g. Coliform bacteria, e.g. *Escherichia coli*, indicate the presence of pathogens.

10.1.2 PROTOZOA

These are unicellular organisms that exist in varying sizes and shapes. One common protozoan, amoeba, is capable of changing to paramecium, altering its fixed shape and complex structure. Protozoa thrive in fresh water, sea, and soil. Some parasitic protozoa cause many diseases by living in hosts. Cryptosporidium is a common protozoan, while an endemic, cyclospora, is limited to communities found where adequate water treatment (such as filtration) and sanitation (sewage treatment) is lacking. Some protozoa feed on bacteria, thus helping in the purification of waste water (Arrowood et al. 2006).

10.1.3 Fungi

About 144,000 species of fungus are found in the kingdom fungi; these include yeasts, mildews, molds, rusts, smuts, and mushrooms (https://www.britannica.com/science/fungus). Fungi are responsible for decomposing the organic matter to simpler forms. Sometimes fungi can prove more beneficial than bacteria in wastewater treatment processes, thus having the merit of being a distinct form of biomass. This significantly changes the wastewater treatment economics (Sankaran et al. 2010).

10.1.4 Viruses

The term virus is derived from a Latin word that means "poison" or "slimy liquid. Viruses are infectious agents that reproduce only in host cells (of animals, plants, or bacteria). Viruses are neither truly plants nor animals; nor are they prokaryotic bacteria. For this reason viruses are placed in their own distinct kingdom. In a strict sense, viruses even cannot be considered as true organisms as they are not living freely. They not even are able to carry out their own metabolic activities and reproduction without thriving in a host cell. In viruses we find usually the nucleic acids, viz. deoxyribonucleic acid (DNA) and ribonucleic acid (RNA) and a protein sheath. For each virus, the genetic information is contained in nucleic acid that is distinct. The infective part of a virus is called the "virion," which is the extracellular form of virus (www.britannica.com/science/virus). Viruses lead to many dreadful infectious diseases, thus risking public health. Some viruses thrive for about a month in water and wastewater at about 20°C.

10.1.5 Algae

Algae are the microorganisms living in water that are capable of assimilating carbon heterotrophically and mixotrophically by the phenomena of photosynthesis. Due to the lack of a vascular system, algae live in solitary cells, colonies, filaments, or primitive vegetative bodies. Algae are cryptogams that propagate by so-called concealed or hidden reproductive methods, which are in contrast to the phanerogams (i.e., plants producing seeds; Likens et al. 2009). Algae are responsible for causing eutrophication phenomena but are useful in oxidation ponds. Upon decaying, algae also cause taste and other problems in the medium.

10.1.6 Rotifers

Although their bodies do not resemble typically to worms, rotifera are freshwater invertebrates or worms that belong to the subphylum Trochhelminthes. Rotifers possess a corona, a cilia at the anterior end, and a pharynx, or jaw. The body size varies from 45 μm to about 2.5 mm. Respiration takes place via their whole body surface, so survival in anaerobic milieu is difficult. About 2000 species of rotifers are found, of which 1700 species (i.e., less than 5 percent) thrive in brackish and marine conditions (Sladecek 1983).

10.1.7 Nematodes

Nematodes are a group of non-segmented worm-like invertebrates found in habitats like fresh and salt water, soil, plants, and animals. Nematodes are found in most types of aerobic wastewater treatment and play a significant part in predation on bacteria and on other animals in such processes. The most common forms are the micrphagus rhabditids and diplogasterids (whose burrowing and feeding keep biological beds porous and accessible to oxygen and also help to prevent clogging by encouraging the necessary sloughing of the biological films (Calaway 1963).

Among the pathogenic organisms which exist in water and as discussed above, bacteria and protozoa constitute the key microbes. In the presence of dissolved oxygen, bacteria transform soluble organic matter into fresh cells and inorganics. These substrates in turn lead to a reduction in the organic matter for higher orders of life. Wastewater of domestic origin contains bacteria as dominant microbial species, with concentrations ranging from 105 to 108 per ml (Muttamara 1996).The quality of water of a receiving body is regulated by the various biological interactions taking place there.

10.2 BIOCHEMICAL OXYGEN DEMAND, BIOKINETICS, YIELD, AND ENERGETICS

10.2.1 Biochemical Oxygen Demand

Biochemical oxygen demand (BOD) is one of the versatile criteria for the assessment of water quality. It is an index of the ready fraction of organic matter in water that is biodegradable. BOD is the amount of oxygen taken up by the growing microorganisms due to the organics in the sample of water (or effluent) divided by the system volume, which is incubated at a specific temperature (usually 20°C) for a definite period of time (usually five days, BOD5). Thus BOD5 is a time-consuming analytical method with an incubation period of 5 days. The results may vary from laboratory to laboratory (typically with 20 percent variation) due to the microbial/inoculum diversity used (Jouanneau et al. 2014).

It is the measure of organic pollutant having tendency of biological degradation. BOD is expressed as milligrams O_2 per liter. BOD5 is an index of the severity of treatment procedure to be adopted, and it regulates the prevailing norms of wastewater discharge. The fraction of an effluent that is biodegradable is indicated by the ratio of BOD5 and COD in the wastewater treatment plants. The size of a wastewater treatment plant required for a specific process and location is set by the ratio of COD/BOD5, and this ratio carries a lot of significance.

10.2.2 Biokinetics, Yield and Energetics in Wastewater Treatment

Wastewater treatment reactions are biochemical processes/reactions that are characterized by various factors such as retention time, concentration of substrate, concentration of biomass, concentration of cell mass, and growth rate. The success and kinetics of biological reactions are determined by the substrate utilization and cell growth rate. Kinetics of substrate utilization is usually explained by the Monod growth kinetics model. Many studies are available on biochemical reactions and their kinetics. Use of mathematical modeling has benefited us in better understanding of the role of various factors that affect biokinetics. Researchers have used the Monod kinetic model for the microbial growth and Michaelis–Menten kinetics for the substrate utilization in order to evaluate kinetic parameters. Proper maintenance of appropriate pH in the BOD reduction is important. In some cases, investigators have reported that two times more substrate removal capacity is achieved by the moving bed biofilm reactor in comparison to the activated sludge process. The bacterial population specific growth rates are also functions of population density. The effectiveness and kinetics of biological reactions define the substrate utilization and cell growth (Kulkarni 2016).

Many kinetic models post optimization have been used such as that of Nadeem et al. (2019);

1. First order model
2. Second order Grau model
3. Stover–Kincannon modified model

10.3 OVERVIEW OF TRADITIONAL BIOLOGICAL PROCESSES

10.3.1 General Scheme of Wastewater Treatment

As described in Fig. 10.1, a modern purification process generally contains five successive stages. The various steps involved are:

1. Pre-treatment or preliminary treatment, which is physical and a mechanical process.
2. Primary treatment (a physicochemical or pure in chemical nature).
3. Secondary treatment (chemical and biological in nature).
4. Tertiary treatment (a physical and chemical process).
5. Sludge treatment carried out by supervised tipping, recycling, or incineration.

In third world countries only steps 1–3 are usually practiced in traditional processes, while sludge is usually disposed of without any further processing.

However, modern treatment processes in some advanced countries involve tertiary treatment processes also, such as adsorption and advanced oxidation processes (step 4) and sludge treatment (step 5).

Fig. 10.1 summarizes conventional, established, and some emerging water treatment methods.

Fig. 10.2 shows the details of types of processes involved in preliminary (or primary), secondary, and tertiary treatment steps of treatment of water.

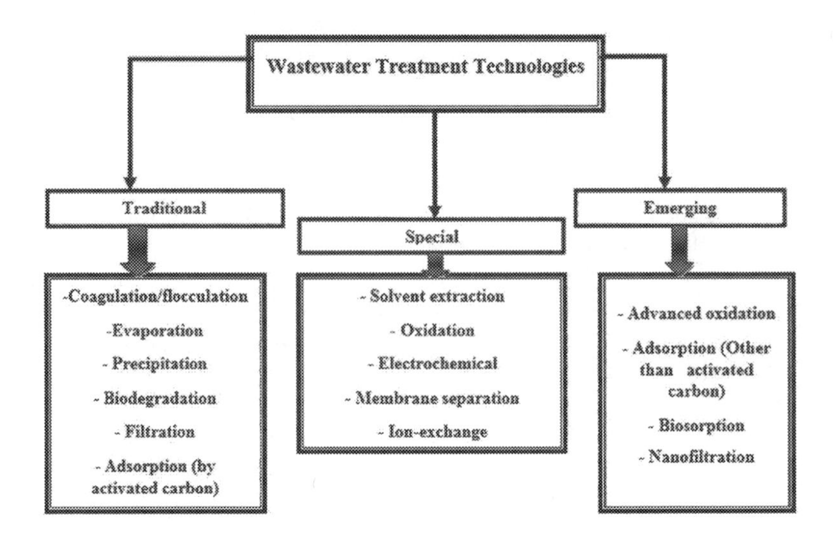

FIGURE 10.1 Technologies available for wastewater treatment.

10.3.2 Traditional Biological Processes

Domestic, agricultural, and industrial processes have led to release of inorganic and organic substances into the environment, causing organic and inorganic pollution. Conventionally, carbon adsorption, chemical precipitation, evaporation, and ion exchange and membrane processes are successful in treating the waste and wastewater.

Wastewater treatment methods were either purely biological, or non-biological, or mixed in nature (e.g., reverse osmosis membrane bioreactor, and the photo-Fenton

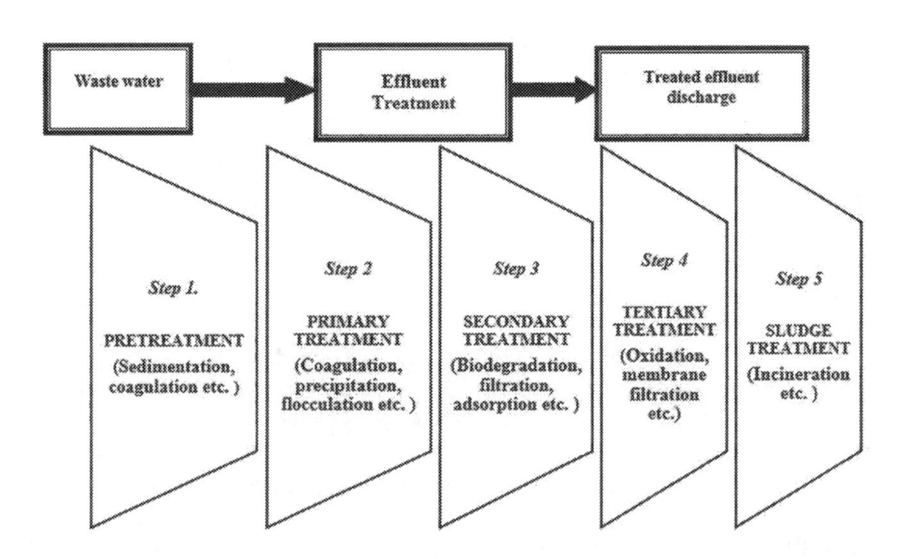

FIGURE 10.2 Preliminary (or primary), secondary, and tertiary steps of treatment of water.

reaction). Some processes have been observed to be more efficient for addressing COD issues while others were observed to be less effective. However, recently biological treatment processes have been found to be promising in removing toxic and other harmful substances. Algae also have been used in the treatment processes of wastewater because of various reasons such as

i. reduction of BOD
ii. removing N and P (individually or together)
iii. coliform inhibition
iv. removal of heavy metals

Algal biomass can further be employed for the production of methane, liquid fuels, and fine chemicals. They are also useful in composting and animal feed (Rajasulochana & Preethy 2016).

Biological processes in form of biosand filters (BSFs) are columns of finely crushed rock or sand upon which many microorganisms sustain. BSFs are designed for an optimal environment to sustain different types of aquatic microorganisms for hunting pathogens. Sand filter has advantages in that it separates not only the suspended solids and particles from the water but also other chemical constituents. These include nitrogen compounds, micro-pollutants, pesticides, heavy metals, and biological contaminants (pathogens such as bacteria and viruses).

In case of biological water treatment, two types of BSF exist, viz. slow and rapid sand filter. A few of the modern biological treatment processes for wastewater treatment are summarized below (Hasan et al. 2020);

10.3.2.1 Biological Activated Carbon

Activated carbon (AC) is covered with a biofilm or a microbial film. AC's large surface area allows it to adsorb pollutants of organic nature on its surface texture. This also helps to serve as the major sites/factors for bacterial attachment.

10.3.2.2 Trickling Filter

For attaching biofilm, trickling filter (TF) contains a packing medium. Thus this serves as a fixed-film biofilter. On the packing medium development of biofilm occurs, which contains microorganisms. The microrganisms are useful for biological degradation of the pollutants or contaminants.

10.3.2.3 Biological Aerated Filter

The BAC, SSF, and RSF operate differently in comparison to the biological aerated filter (BAF) system. It must be clear that the BAC, SSF, and RSF operate without any aeration. For microorganisms to thrive under aerobic conditions in the BAF system and assist in the degradation and remediation of pollutants, they need proper aeration. The BAF is also known for its performance in wastewater treatment.

10.3.2.4 Membrane Bioreactor

This technology synergizes the use of a membrane and a bioreactor for the purpose of treating and separating the organic contaminants. It supports the sustenance and growth of useful microorganisms in the bioreactor. These microorganisms not only degrade the organic pollutants but also keep pathogens under check. This way treated water is free of biodegradable organic contaminants.

10.3.2.5 Moving-Bed Biofilm Reactor

This type of reactor operates in a well-mixed and continuous operation with capability of withstanding a high density of biomass. The process is advantageous as it prevents biomass sticking in the reactor. It also aids the sludge recycling process.

10.3.2.6 Fluidized-Bed Biofilm Reactor

In fluidized-bed biofilm reactor (FBBR) technology, the contaminated water is pumped up through the biological granular media bed at a sufficient velocity corresponding to fluidized state. However, FBBR technology is not found to be very useful in drinking water treatment.

10.3.2.7 Biological Activated Carbon-Membrane Bioreactor

The combination of BAC and MBR has been observed not only to improve the quality of the treated water but also BAC has been seen to addresses the issue of membrane fouling.

10.3.2.8 Integrated/Combined Technologies

Integrating two different technologies is always an advisable option for enhancing drinking water treatment technology. This ensures that the treated water meets the prescribed discharge limits as regulated by the national or regional norms. The concept is a modern trend that allows benefits from the desired features of both bio and non-bio processes.

10.4 BIODEGRADATION OF WASTEWATER ORGANICS

Because of improper water treatment processes that have been adopted, various types of organic micro pollutants have been found in surface water and drinking water supplies. Water analysis is focusing on macropollutants that are related to extensive agricultural and industrial activities and also on micro pollutants such as pharmaceutical residues, pesticides, industrial chemicals, and personal care products, which are detected in very small amounts. In the micro pollutant category, biologically active substances (e.g., endocrine disrupting compounds and pharmaceutical residues) have been found to have harmful effects on living organisms. Some types of such micro pollutants are steroid and xeno-estrogens, etc. Processes have been devised to treat these pollutants. For efficiency of conventional treatment plants (CTP) and membrane bioreactors (MBR), treatment factors such as sludge retention time (SRT), biomass concentration, pH value, temperature, etc., need to be taken care of and well understood.

The treatment of wastewater also concerns the removal of organic matter such as proteins, carbohydrates, and nutrients in addition to the already-mentioned usual pollutants. Sorption, biological degradation of organics, and assimilation of inorganics by means of activated sludge process in the case of CTP and MBR processes takes place. In mixed aerated tanks of both of the above systems, activated sludge mainly contains flocculation microorganisms, which are in suspension and in contact with wastewater. In the case of CTP, larger particles are removed from the wastewater, which is first fed into a mechanical treatment step. Water is flowed at low velocity through large tanks after the primary sedimentation process, after which the wastewater is routed to a bio-activated sludge tank. Finally, separation is achieved additionally by gravity sedimentation by means of an external clarifier.

In the MBR process, primary sedimentation and mechanical treatment are not carried out. The sludge-liquid separation step is thus the main difference. Here the activated sludge tank incorporates a filtration step by a micro- or nano-size filtration membrane by which solid particles are retained in an aeration tank. The quality of wastewater effluent is highly influenced by the biomass separation technique. CTPs are operated commonly in the range of 1–5 g/l of mixed liquor suspended solids (MLSS), while as in the case of MBR this concentration is much higher, varying from 8–25 g/l or even higher. MBR technology allows sewage treatment at high MLSS concentration, which is attributed to the membrane separation step. This is not limited by the secondary clarifier sedimentation capacity. To maintain a constant concentration of microorganisms in the tank, biomass and excess sludge have to be removed from the system with its continuous growth. SRT is one of the main parameters of activated sludge processes, which are controlled via the removal of excess sludge. High performance of the wastewater treatment process is indicated by high SRTs in relation to the COD removal. Usually in MBR, SRT up to 25 or 80 days is applied, while for CTP typical values exist in the range from 8 to 25 days.

10.4.1 Factors Responsible for the Removal of Micro-pollutants in Wastewater Treatment

10.4.1.1 Chemical Properties of Micro-pollutants

Hydrophobicity has a huge role, leading fat and particulate matter to sorption to the sludge, in the wastewater treatment. During CTP or MBR treatment, the elementary composition and the chemical structure of a compound can affect its removal rates from wastewater.

10.4.1.2 CTP and MBR Process Parameters

SRT in the case of CTP and MBR systems, is the mean residence time of the microorganisms. Many studies have shown that a sufficiently high SRT is important for degradation and removal of the micro-pollutants from wastewater. High SRT allows for the enrichment of slow growing bacteria. It also leads to the establishment of a more diverse biocoenosis, being able to degrade a variety of micro-pollutants.

10.4.1.3 Influence of Biomass Characteristics

Biomass characteristics are important factors for biodegradation in the case of CTP and MBR treatment, and they differ in both systems (Brindle and Stephenson, 1996). At the higher STP, the possibility for microorganisms to genetically mutate and to assimilate persistent organic compounds increases.

10.4.1.4 pH Value and Temperature

As we know, the pH value of an aqueous solution is the measure of the acidity or alkalinity, with a capacity to influence the removal of organic micro-pollutants from the wastewater. It influences both the physiology of microorganisms (optimum pH of microbial enzyme activities) and the solubility of micro pollutants existing in wastewater.

By virtue of applied temperature conditions, because the microbial growth rate strongly varies it influences the microbial activity in case of both of CTP and MBR (Price and Sowers, 2004). Adsorption equilibriums and degradation rate have been observed to reach earlier and the bacterial growth faster with the increase in temperatures.

REFERENCES

Alexopoulos, Constantine John, David Moore, and Vernon Ahmadjian. "Fungus". Encyclopedia Britannica, 27 Feb. 2020, https://www.britannica.com/science/fungus. Accessed 11 May 2021.

Arrowood, Michael J., Ynes R. Ortega, Lihua X. Xiao, and Ronald Fayer. "Emerging Food- and Waterborne Protozoan Diseases." *Emerging Infections 7* (2006): 283–308. https://doi.org/10.1128/9781555815585.ch15

Calaway, Wilson T. "Nematodes in Wastewater Treatment." *Journal (Water Pollution Control Federation)* (1963): 1006–1016.

Hasan, H. A., M. H. Muhammad, and NurIzzati Ismail. "A Review of Biological Drinking Water Treatment Technologies for Contaminants Removal from Polluted Water Resources." *Journal of Water Process Engineering.* 33 (2020). doi:10.1016/j.jwpe.2019.101035.

Jouanneau, S., Loic Recoules, M. J. Durand, Ali Boukabache, Vincent Picot, Y. Primault, A. Lakel, M. Sengelin, B. Barillon, and Gerald Thouand. "Methods for Assessing Biochemical Oxygen Demand (BOD): A review." *Water research* 49 (2014): 62–82.

Kulkarni, Sunil J. "A Review on Research and Studies on Dissolved Oxygen and Its Affecting Parameters." *International Journal of Research and Review* 3, no. 8 (2016): 18–22.

Likens, Gene, M. E. Benbow, T. M. Burton, E. Van Donk, J. A. Downing, and R. D. Gulati. "Encyclopedia of inland waters." (2009). https://pure.knaw.nl/portal/en/publications/e40443e4-ba44-4eac-941f-ac0730473d12.

Muttamara, Samorn. "Wastewater Characteristics Resources, Conservation and Recycling" 16, no.1-4 (1996): 145–159.

Price, P. B. and T. Sowers. "Temperature Dependence of Metabolic Rates for Microbial Growth, Maintenance, and Survival." *Proceedings of the National Academy of the Sciences of United States of America* (2004):4631–4636. https://doi.org/10.1073/pnas.0400522101.

Rajasulochana, P., and V. Preethy. 2016. "Comparison on Efficiency of Various Techniques in Treatment of Waste and Sewage Water: A Comprehensive Review" *Resource-Efficient Technologies 2*, 175–184. doi:10.1016/j.reffit.2016.09.004.

Sankaran, Sindhuja, Samir Kumar Khanal, Nagapadma Jasti, Bo Jin, Anthony L. Pometto III, and J. Hans Van Leeuwen. "Use of Filamentous Fungi for Wastewater Treatment and Production of High Value Fungal Byproducts: A Review." *Critical reviews in environmental science and technology* 40, no. 5 (2010): 400–449. https://doi.org/10.1080/10643380802278943

Sladecek, Vladimir. "Rotifers as Indicators of Water Quality." *Hydrobiologia* 100, no. 1 (1983): 169–201. https://doi.org/10.1007/BF00027429.

Ullah, Nadeem, S. Tafazul, S., Farooqib, Isac, Hussain A., Fazlollah Changanie, and Astha Dhingra. "Kinetics Performance Evaluation of Column-SBR in Paper and Pulp Wastewater Treatment: Optimization and Bio-Kinetics." *Desalination and Water Treatment* 156 (2019): 204–219. doi:10.5004/dwt.2019.23775.

Wagner, Robert R. and Robert M. Krug. "Virus". Encyclopedia Britannica, 12 Nov. 2020, https://www.britannica.com/science/virus. Accessed 11 May 2021.

11 Role of Biochar in Water Treatment

Biswajit Saha and Sonali Sengupta
Chemical Engineering Department, IIT Kharagpur, Kharagpur, India

CONTENTS

11.1 INTRODUCTION

In recent decades the world has been facing the crisis of fresh water due to significantly growing population and modernization of industries and agriculture. According to a survey of the United Nations Department of Economic and Social Affairs, about 700 million people in 43 countries are facing water scarcity, and in 2025 1.8 billion people will live with absolute water crisis [Water for Life 2005–2015]. This crisis is of great importance to the treatment of various wastewaters originating from multiple industrial sectors such as pharmaceutical, textile, refinery, lather, battery, cement, power plant, rail manufacturing, etc., as well as from municipalities, agriculture, and hospitals. Reuse and recycling of the treated wastewater from these

DOI: 10.1201/9781003138303-11

sources are very much sought after for various appropriate uses, such as formation of wetlands for wildlife, firefighting, flushing of toilets, agriculture, industrial cooling, and park or playground watering etc. [Inyang et al. 2015; Wang et al. 2019].

Industrial, agricultural, or municipal wastewater often contains a wide range of heavy metals such as cadmium, copper, uranium, nickel, mercury, arsenic, chromium, cobalt, zinc, etc., beyond the permissible limits, as well as alarming concentrations of other contaminants and pollutants such as various antibiotics such as chloramphenicol, erythromycin, sulfadiazine, tetracycline, norfloxacin, penicillin, carbadox, etc.; different harmful dye chemicals such as malachite green, methylene blue, eriochrome black-T, congo red etc.; disease-causing microbes such as *Escherichia coli*, *Staphylococcus aureus*, and *Pseudomonas aeruginosa*, etc. [Inyang et al. 2015; Wang et al. 2019]. All these substances directly pollute water and soil to a dangerous level and thus enter the food chain. Those contaminants pose fearful effects on human health and aquatic life and destroy the balance of our ecosystem cycle. One of example of these wide-ranging effects is that when wastewater containing a threatening range of antibiotics is provided without any treatment to use as drinking water, it promotes antibiotic resistance in the human body, and medicine will not effectively work on bacterial infection [Ahmed et al. 2015; Inyang et al. 2015; Wang et al. 2019]. Table 11.1 shows the effects of different possible contaminants of wastewater on human health [Lellis et al. 2019; Tatarchuk et al. 2019].

According to a recent survey, appreciable concentrations of heavy metal contamination in wastewater is another major concern in some countries, particularly Turkey, China, India, and Bangladesh [Hassan et al. 2020]. Therefore removal of those contaminants is highly essential to save the environment and human life. Wastewater treatment is mainly divided into several well-known processes, such as chemical precipitation, ion exchange, ultrafiltration, reverse osmosis, electro-dialysis, coagulation, flotation, and adsorption. Among them, adsorption is an easy-to-operate,

TABLE 11.1
Effect of Contaminants of Wastewater on Human Health [Lellis et al. 2019; Tatarchuk et al. 2019]

Pollutant	Health Issue
Cu	Infection in nose, mouth, and eye
U	Mental retardation and estrogenic effect
Zn	Cause of metal fume fever
Ni	Chronic bronchitis, disorder in bones and lung function
Cr	Lung tumor and allergic dermatitis
Pb	High blood pressure, joint and muscle pain, and kidney problems
Cd	Lung fibrosis, weight loss, and kidney problem
Hg	Corrosion in skin, eye, kidney problem, respiratory disorders
Ar	Cancer, liver tumor, and skin gastrointestinal problems
Pesticides	Cancer
Nitrate	Methemoglobinemia
Textile dye	Bladder cancer, disorder of central nervous system, skin and eye irritation

cost-efficient, and eco-friendly processes, whereas other processes are expensive and produce harmful waste [Hassan et al. 2020; Wang et al. 2019]. Several popular adsorbents have been reported that work effectively for wastewater treatment [Ahmed et al. 2015]. Biochar (BC) is one widely used adsorbent obtained from vegetable or animal waste after some processing; it is a carbon-rich porous material with high aromatic content and is used for water treatment purposes. Biochar is produced by thermal pyrolysis (300 to 700°C) of vegetable waste under oxygen-free or limited oxygen conditions [Liu et al. 2020]. The feedstock so far reported for biochar production are pine sawdust, waste tea residue, food waste, peanut shells, rapeseed straw, seaweed biomass, rice straw, bamboo sawdust, pomelo peels, rabbit feces, date palm fronds, litchi peels, wheat straw residue, chicken manure, fruit peels, municipal solid waste, etc. [Tan et al. 2016]. All these raw materials are easily available at very low cost or no cost at all; as a result, the cost of the biochar ($246 per ton) is remarkably low compared to other widely used adsorbents, such as activated carbon ($1500 per ton) [Inyang et al. 2015; Wang et al. 2019]. It was found that biochar worked efficiently for removing contaminants such as heavy metals, microbes, antibiotics, pesticides, dyes, organic materials, etc., from wastewater [Gwenzi et al. 2017]. Surface functional groups of biochar, such as phenolic, hydroxyl, carboxyl and anhydride, perform as active sites of the adsorbent [Li et al. 2019a]. The wide production and use of biochar are mainly conducted in United States, Australia, South America, China, and Europe [Gwenzi et al. 2015]. In this chapter, we discuss biochar production procedures, applications, wastewater treatment mechanisms, kinetics, effects of different process parameters on removal efficiency of contaminants by biochar, its regeneration efficiency, challenges, and future perspectives of biochar for wastewater treatment.

11.2 SYNTHESIS PROCEDURE OF BIOCHAR

The synthesis of biochar is mainly carried out by one of three processes: pyrolysis, hydro-carbonization, and gasification. The pyrolysis method is robust, simple, and cost effective. This process can be done in different ways, which are classified as slow (<700°C), microwave (400 to 500 W), and fast pyrolysis (<1000°C). Hydro-carbonization is operated in the temperature range of 120 to 260°C and is more suitable for high-moisture–content feedstock. Gasification is a high-temperature process, operated at greater than 800°C, and the biochar obtained from this method contains high levels of alkali salts [Deng et al. 2017; Xiang et al. 2020]. A literature review indicates that researchers used mostly the pyrolysis and hydro-carbonization methods for biochar production as the yield of biochar by gasification is very low [Xiang et al. 2020]. Among all pyrolysis and hydro-carbonization methods, the slow pyrolysis process is widely used because biochar produced by this method has high aromaticity as well as more polar surface oxygen functional groups, which act as active sites of biochar to remove contaminants from wastewater [Gwenzi et al. 2017]. But at the same time, some reports show that the fast pyrolysis, or carbonization, method produces biochar with higher surface area and pore volume than slow pyrolysis. Bandara et al. [2020] synthesized biochar from sugar gum wood using the slow pyrolysis (550°C) method. The surface area and pore volume of this biochar were

429 m^2/g and 63 cm^3/kg respectively with composition: carbon 72.35 wt%, hydrogen 1.10 wt% and oxygen content 23.57 wt%. Choudhury et al. [2020] reported that biochar prepared from cladodes of *Opuntia ficus-indica* (OFI) cactus by slow pyrolysis (400°C) has low surface area (1.15 m^2/g), with pore volume 0.018 cm^3/g and average pore size 27.46 nm. Wang et al. [2020a] prepared biochar from grape fruit peels by the slow pyrolysis (400°C) method; it contains carbon and oxygen (wt. %) 41.57% and 38.11% respectively. It was found that the surface area, pore volume, and average pore size of this biochar were 20.73 m^2/g, 0.111 cm^3/g and 20.89 nm respectively. Zhang et al. [2020a] prepared biochar from bamboo sawdust using slow pyrolysis (450°C); the surface area, carbon, and oxygen contents were reported as 140.51 m^2/g, 80.08 wt% and 17.67 wt% respectively, whereas Keerthanan et al. [2020] synthesized biochar from waste tea residue by fast pyrolysis (700°C); it showed higher surface area (576 m^2/g) and pore volume (0.109 cm^3/g). Tomul et al. [2020] reported that the surface area and pore volume of biochar prepared from peanut shells by fast pyrolysis (800°C) were 571 m^2/g and 0.328 cm^3/g respectively. Another researcher, Zhang et al. [2019a] observed that biochar prepared from Medulla Tetrapanac is by fast pyrolysis has high surface area (246.85 m^2/g) but low carbon (44.61 wt%) and oxygen (8.76 wt%) content compared to biochar prepared from slow pyrolysis [10.02 m^2/g, C(wt%) 58.13, O(wt%) 15.92]. Wu et al. [2020a] prepared biochar from litchi peels by using hydrothermal carbonization (180°C), which produces high surface area (1006 m^2/g) and pore volume (0.588 cm^3/g). Li et al. [2020a] synthesized biochar from corncob using an ultrasound-assisted carbonization process (450°C); it has high surface area (2368 m^2/g) with carbon and oxygen content 75.68% and 23.44 wt% respectively. Huang et al. [2018] developed biochar from rabbit feces using a fast pyrolysis (700°C) process. The surface area and pore volume of this biochar were 91.52 m^2/g, 0.237 cm^3/g, respectively, with 67.22 wt% of carbon and 4.15 wt% of oxygen. All these data are presented for the sake of readers' convenience to understand the processes and differences in properties of the biochar obtained.

11.3 APPLICATION OF BIOCHAR AS AN ADSORBENT FOR WASTEWATER TREATMENT

The role of biochar for wastewater treatment is highly significant. It effectively works to remove different contaminants (heavy metals, pesticides, pharmaceuticals, plasticizers, dyes, organic materials, inorganic materials, etc.) present in wastewater. Biochar is also used as a filter for microbe removal from wastewater. In a work reported by Mercado et al. [2019] biochar was used to filter bacteriophages, *E. coli*, *Enterococcus*, and *Saccharomyces cerevisiae* microbes from municipal wastewater. Chen et al. [2020] used a biochar-assisted membrane bioreactor for removal of absorbable organic halogen from pharmaceutical wastewater, with 73.2% removal efficiency. Table 11.2 presents the performances of biochar from different sources and surface modified biochar by impregnation with different metals or treatment by acids, oxidants, or bases for removal of contaminates from wastewater. Biochar is not only used for wastewater treatment but has different other applications also, namely, carbon sequestration, reduction of greenhouse gases, soil property improvement, etc. [Qambrani et al. 2017].

TABLE 11.2
Applications of Biochar for Wastewater Treatment

Reference	Adsorbent/Catalyst/Filter and Source	Contaminants	Performance (Adsorption Capacity in mg/gor Efficiency in %)	Wastewater Type
		Heavy Metals		
Bandara et al. 2020	Biochar (Sugar gum wood)	a. Cd (II) b. Cu(II)	a. 6.28 mg/g b. 18 mg/g	Acidic mine water
Ali et al. 2020	Biochar (Almond shells)	a. As(V) b. As(III)	a. 90% b. 94%	Simulated wastewater
Huang et al. 2020a	Biochar (Chicken manure)	Cd^{2+}	60.69 mg/g	Simulated wastewater
Younis et al. 2020	Biochar (Waste rice straw)	a. Ba(II) b. Sr(II)	a. 73.9 mg/g b. 59.8 mg/g	Simulated saline water such as oilfield wastewater
Hanandeh et al. 2016	Biochar (Olive husks)	Hg^{2+}	104.59 mg/g	Simulated wastewater with yttrium salt
Mishra et al. 2017	Biochar (Eucalyptus wood)	U (VI)	27.2 mg/g	Simulated wastewater
Ho et al. 2017	Biochar (Anaerobic digestion sludge)	a. Pb^{2+} b. Cu^{2+} c. Cd^{2+} d. Ni^{2+} e. Zn^{2+} f. Cr^{6+}	a. 51.2 mg/g b. 21.2 mg/g c. 16.2 mg/g d. 22.5 mg/g e. 18.1 mg/g f. 8.3 mg/g	Simulated wastewater
Park et al. 2019	Biochar (Pine tree residue)	Cd	85.8 mg/g	Simulated wastewater
Zhang et al. 2019b	Biochar (Crayfish shells)	Pb^{2+}	135 mg/g	Simulated wastewater
Pap et al. 2018	Biochar (Plum kernels)	a. Pb(II) b. Cr(III)	a. 28.8 mg/g b. 14 mg/g	Simulated wastewater

(Continued)

TABLE 11.2 (Continued)
Applications of Biochar for Wastewater Treatment

Reference	Adsorbent/Catalyst/Filter and Source	Contaminants	Performance (Adsorption Capacity in mg/gor Efficiency in %)	Wastewater Type
Zhang et al. 2018a	Biochar (Tobacco petiole)	a. Cr(VI) b. Ni c. Zn d. Pb e. Co	a. 66.7% b. 27.3% c. 48.1% d. 21.1% e. 16.8%	Electroplating effluent from battery industries
Shakya et al. 2019	Biochar (pineapple peel)	Cr(VI)	41.67 mg/g	Simulated wastewater
Zhang et al. 2019a	Biochar (*Medulla tetrapanacis*)	a. Cu^{2+} b. Pb^{2+}	a. 458.72 mg/g b. 1031.23 mg/g	Simulated wastewater
Jalayeri et al. 2019	**Biochar (pistachio green hull biomass)**	Cu(II)	19.84 mg/g	Simulated wastewater
Jang et al. 2018	Biochar (Rice straw)	Sr (Radioactive)	175.95 mg/g	Simulated wastewater
Hong et al. 2019	Biochar (Sewage sludge and corncob)	Pb^{2+}	31.25 mg/g	Simulated wastewater
Fan et al. 2017	Biochar (Sewage sludge and tea waste)	Cd	94%	Simulated wastewater
Hong et al. 2020	Fe/Biochar (Food waste)	Se(VI)	11.72 mg/g	Simulated wastewater
Johansson et al. 2016	Fe/Biochar (Fresh water macroalgae Oedogonium)	a. As b. Mo c. Se	a. 80.7 mg/g b. 67.4 mg/g c. 36.8 mg/g	Simulated wastewater
Son et al. 2018	Fe/Biochar (Kelp: marine macroalgae biomass)	a. Cd^{2+} b. Cu^{2+} c. Zn^{2+}	a. 23.16 mg/g b. 55.86 mg/g c. 22.22 mg/g	Simulated wastewater
He et al. 2018	Fe/Biochar (Corn Straw)	As(V)	6.8 mg/g	Simulated wastewater
Alqadami et al. 2018	Fe/Biochar (Camel bone)	a. Pb(II) b. Cd(II) c. Co(II)	a. 344.8 mg/g b. 322.6 mg/g c. 294.1 mg/g	Simulated wastewater

(Continued)

TABLE 11.2 (Continued)
Applications of Biochar for Wastewater Treatment

Reference	Adsorbent/Catalyst/Filter and Source	Contaminants	Performance (Adsorption Capacity in mg/gor Efficiency in %)	Wastewater Type
Niu et al. 2020	Mn-Zn ferrite/Biochar (Pine sawdust)	a. Pb^{2+} b. Cd^{2+}	a. 98.5% b. 97.8%	Industrial wastewater (electronic company)
Jia et al. 2020	Mn/Biochar (Rapeseed straw)	a. Sb (III) b. Sb (V)	a. 0.94 mg/g b. 0.73 mg/g	Simulated wastewater
Imran et al. 2020	a. Magnetite nanoparticle/Biochar b. HNO_3/Biochar [Quinoa crop waste/biomass]	Cr(VI)	a. 77.35 mg/g b. 55.85 mg/g	Simulated wastewater
Zhou et al. 2017	H_3PO_4/Biochar (Banana peel)	Pb(II)	359 mg/g	Simulated wastewater
Biswas et al. 2019	H_2SO_4/Biochar (Pine cone)	Pb^{2+}	321 mg/g	Simulated wastewater
Jin et al. 2018	a. HNO_3/Biochar (Wheat straw) b. HNO_3/Biochar (Cow manure)	U (VI)	a. 355.6 mg/g b. 73.3 mg/g	Simulated wastewater
Gao et al. 2020	$KMnO_4$/Biochar (Rape straw)	Pb	1343 mmol/kg	Simulated wastewater
Li et al. 2019b	$KMnO_4$/Biochar (*Ficus microcarpa* aerial root)	U (VI)	27.29 mg/g	Simulated wastewater
An et al. 2019	$KMnO_4$-KOH/Biochar (Peanut shell)	Ni(II)	87.15 mg/g	Simulated wastewater
Wang et al. 2020b	Mg-Al double hydroxide layered biochar from pinewood sawdust	a. Pb^{2+} b. CrO_4^{2-}	a. 421.5 mg/g b. 362.5 mg/g	Electroplating wastewater
Jeon et al. 2020	Elemental S/Biochar (Pine tree needles)	Hg(II)	48.2 mg/g	Simulated wastewater
Chaukura et al. 2016	a. Hydroxylated biochar (Paper pulp sludge) b. Sulphonated biochar (Paper pulp sludge)	Zn^{2+}	a. 60.30 mg/g b. 76.92 mg/g	Simulated wastewater
Wu et al. 2017	Red Mud/Biochar (Rice straw)	As(V)	5923 µg/g	Simulated wastewater
Banerjee et al. 2017	Steam/Biochar (*Colocasia esculenta* root)	Fe^{2+}	72.96%	Simulated coal mine wastewater
Paranavithana et al. 2016	Soil/Biochar (Coconut shells)	a. Cd^{2+} b. Pb^{2+}	a. 99.9% b. 92.5%	Simulated wastewater
Lin et al. 2019	Fe-Mn-La/Biochar (Corn stem powder)	As(III)	14.9 mg/g	Simulated wastewater

(*Continued*)

TABLE 11.2 (Continued)
Applications of Biochar for Wastewater Treatment

Reference	Adsorbent/Catalyst/Filter and Source	Contaminants	Performance (Adsorption Capacity in mg/gor Efficiency in %)	Wastewater Type
Wang et al. 2018a	Fe_2O_3-SiO_2 Layered double hydroxide/ Biochar (Leaf)	Pb(II)	146.84 mg/g	Simulated wastewater
Wang et al. 2018b	Mn-Al-layered double hydroxide/ Biochar (Oil tea camellia shells)	Cu(II)	74.07 mg/g	Simulated wastewater
Luo et al. 2019b	TiO_2/Biochar (Corncob)	a. Cd(II) b. As(V)	a. 72.62 mg/g b. 118.06 mg/g	Simulated wastewater
Tang et al. 2019	Amino (NH_2)/Biochar (Sewage Sludge)	Cu(II)	74.51 mg/g	Simulated wastewater
Mian et al. 2018	N-Fe/Biochar (Agar biomass)	Cr(VI)	142.86 mg/g	Simulated wastewater
Gao et al. 2018	S-Fe/Biochar [Herbal residue] (*Astragalus membranaceus*)	Cr(VI)	126.12 mg/g	Simulated wastewater
Ding et al. 2018	Al/Biochar (Abandoned Tetra Pak aluminum foil)	a. As(III) b. As(V)	a. 24.2 mg/g b. 33.2 mg/g	Simulated wastewater
Shi et al. 2018	Fe_3O_4-SiO_2-NH_2/Biochar (Phoenix tree leaf)	Cr(VI)	27.2 mg/g	Simulated wastewater
Yu et al. 2018	ZnO/Biochar (Water hyacinth biomass)	Cr(VI)	43.48 mg/g (>95%)	Simulated wastewater
		Pharmaceuticals, Pesticides, and Personal Care Products		
Keerthanan et al. 2020	Biochar (Waste tea residue)	Caffeine	15.4 mg/g	Tea processing wastewater
Tomul et al. 2020	Biochar (Peanut shells)	Naproxen	324 mg/g	Simulated wastewater
Gurav et al. 2020	a. Fe_3O_4/Biochar (Banana pseudostem biomass)	Furazolidone antibiotics	96.81%	Simulated wastewater
Tam et al. 2019	Fe/Biochar (Pitch pine sawdust)	Diclofenac non-steroidal anti-inflammatory drugs	123.5 mg/g	Simulated wastewater
Ashiq et al. 2019	Montmorillonite and biochar composite (Municipal solid waste)	Ciprofloxacin antibiotic	167.36 mg/g	Simulated wastewater

(*Continued*)

TABLE 11.2 (Continued)
Applications of Biochar for Wastewater Treatment

Reference	Adsorbent/Catalyst/Filter and Source	Contaminants	Performance (Adsorption Capacity in mg/gor Efficiency in %)	Wastewater Type
Wu et al. 2020b	β cyclodextrin functionalized biochar (Rice straw)	Antibiotic resistance genes a. tetW b. tetM c. sul-1 d. sul-2 e. bla_{TEM} f. oxa-1 g. qnr-S h. intI-1	a. 68.52% b. 55.07% c. 64.56% d. 90.86% e. 87.49% f. 72.79% g. 77.43% h. 49.21%	Seed sludge with simulated wastewater
Huang et al. 2020b	Biochar (Bamboo)	a. Sulfa-methoxazole b. Sulfapyridine	a. 83.3% b. 89.6%	Simulated wastewater
Zhang et al. 2020a	Biochar (Bamboo sawdust)	Tetracycline antibiotic	4.98 mg/g	Simulated wastewater
Luo et al. 2019a	Humic acid/biochar (Sewage sludge)	Ciprofloxacin antibiotic	66.70%	Simulated wastewater
Cheng et al. 2020	KOH-activated biochar (Pomelo peel)	a. Tetracycline b. Oxytetracycline c. Chlortetracycline	a. 476.19 mg/g b. 407.5 mg/g c. 555.56 mg/g	Synthetic swine wastewater
Mandal et al. 2017	H_3PO_4/Biochar (Rice straw)	a. Atrazine Pesticide b. Imidacloprid Pesticide	a. 89.8% b. 89.5%	Simulated wastewater
Wang et al. 2020c	a. Acid-washed biochar (Peanut shell)	a. Atrazine (pesticide) b. Nicosulfuron (pesticide)	a. 1223 mg/kg b. 709 mg/kg	Simulated wastewater
Li et al. 2020a	Self-functionalized biochar (Corncobs)	a. Levofloxacin b. Amoxicillin c. Tetracycline	a. 99.93% b. 87.59% c. 98.89%	Simulated wastewater

(Continued)

TABLE 11.2 (Continued)
Applications of Biochar for Wastewater Treatment

Reference	Adsorbent/Catalyst/Filter and Source	Contaminants	Performance (Adsorption Capacity in mg/gor Efficiency in %)	Wastewater Type
Huang et al. 2018	Biochar (Rabbit feces)	Ciprofloxacin	31.88 mg/g	Simulated wastewater
Jang et al. 2019a	NaOH/Biochar (Alfalfa Hays)	Tetracycline	302.37 mg/g	Simulated wastewater
Ferreira et al. 2017	Biochar (Biological paper mill sludge)	Tricainemethane sulfonate Anesthetics	70 mg/g	Aquaculture wastewater
Jang et al. 2019b	a. Biochar (Alfalfa) b. Biochar (Bermuda grass)	Tetracycline	a. 372 mg/g b. 44.24 mg/g	Simulated wastewater
Liu et al. 2017	Cu(II)/Biochar (Peanut shells)	Doxycycline hydrochloride antibiotic	93.22%	Simulated wastewater
		Industrial Dyes		
Zubair et al. 2020	Biochar (Date palm frond)	a. Methyl Orange b. Eriochrome Black T c. Methylene Blue d. Crystal violet	a. 163.13 mg/g b. 309.59 mg/g c. 206.61 mg/g d. 934.57 mg/g	Simulated wastewater
Wu et al. 2020a	Biochar (Litchi peels)	a. Congo Red b. Malachite Green	a. 404.4 mg/g b. 2468 mg/g	Simulated wastewater
Hoslett et al. 2020	Biochar (Mixed municipal discarded materials)	Methylene Blue	7.2 mg/g	Simulated wastewater
Yang et al. 2018a	S-Fe/Biochar (Astragalus mongholicus)	Methyl Orange	76.09%	Simulated wastewater
Zhang et al. 2018b	TiO_2/biochar (Coconut shells)	Reactive Brilliant Blue KN-R dye	81.09%	Simulated wastewater
Sewu et al. 2017	a. Biochar (Korean cabbage) b. Biochar (Rice straw) c. Biochar (Wood chip)	Crystal violet	a. 1304 mg/g b. 620.3 mg/g c. 195.6 mg/g	Simulated wastewater

(*Continued*)

TABLE 11.2 (Continued)
Applications of Biochar for Wastewater Treatment

Reference	Adsorbent/Catalyst/Filter and Source	Contaminants	Performance (Adsorption Capacity in mg/gor Efficiency in %)	Wastewater Type
		Plasticizer and Organic Materials		
Wang et al. 2020a	Fe_2O_3/biochar (Grape fruit peel)	Bisphenol A organic	229.19 mg/g	Simulated wastewater
Zhang et al. 2020b	Sulfidated Fe/biochar (Seaweed biomass)	Tetrabromo bisphenol A	88.2%	Simulated wastewater
Hairuddin et al. 2019	Fe/biochar (Palm kernels)	Phenol	10.84 mg/g	Simulated wastewater
Liu et al. 2019a	Fe-Ni/biochar (*Eupatorium adenophorum*)	2,4,6 Trichlorophenol	90%	Simulated wastewater
Thang et al. 2019	Biochar (Chicken manure)	a. Phenol b. 2,4-dinitrophenol	a. 78.5 % b. 83.4%	Simulated wastewater
Lee et al. 2019	Biochar (food waste)	Phenol	14.61 mg/g	Simulated wastewater
Zhang et al. 2020c	Ca-KOH/Biochar (Bagasse)	Oxidation degradation of Phenol	90%	River water
		Inorganic Material and Microbes		
Yang et al. 2017	a. Biochar (Pine sawdust) b. Biochar (Wheat straw)	NH_4^+	a. 5.38 mg/g b. 2.08 mg/g	Simulated wastewater
Hu et al. 2020	a. Biochar (Orange peel) b. Biochar (Pineapple peel) c. Biochar (Pitaya peel)	NH_4^+	a. 4.71 mg/g b. 5.60 mg/g c. 2.65 mg/g	Simulated wastewater
Viglašová et al. 2018	Montmorillonite/biochar (Bamboo biomass)	Nitrates (NO_3^-)	9 mg/g	Simulated wastewater
Song et al. 2019	a. Biochar (pig manure) b. Biochar (rice straw) c. Biochar (cedar wood)	Ammonium nitrogen	a. 0.2 mg/g b. 0.09 mg/g c. 0.08 mg/g	Simulated wastewater
Liu et al. 2019b	Ca/biochar (Rice straw)	Phosphate	197 mg/g	Simulated wastewater

(Continued)

TABLE 11.2 (Continued)
Applications of Biochar for Wastewater Treatment

Reference	Adsorbent/Catalyst/Filter and Source	Contaminants	Performance (Adsorption Capacity in mg/gor Efficiency in %)	Wastewater Type
Jiang et al. 2019	Zn-Al-double hydroxide layer/biochar (Banana straw)	Phosphate	185.19 mg/g	Simulated wastewater
Li et al. 2019c	Polyethyleneimine modified biochar (Bamboo biomass)	Phosphorus	9.25 mg/g	Simulated wastewater
Novais et al. 2018	a. Mg/biochar (Sugarcane straw) b. Mg/biochar (Poultry manure)	Phosphorus	a. 17.7 mg/g b. 250.8 mg/g	Simulated wastewater
Li et al. 2020b	La/Biochar (Sewage sludge)	Phosphate	93.91 mg/g	Simulated wastewater
Yang et al. 2018b	Fe/biochar (Waste-activated sludge)	Phosphate	111 mg/g	Simulated wastewater
Li et al. 2019d	Dolomite/biochar (Sewage sludge)	Phosphate	29.18 mg/g	Simulated wastewater
Lau et al. 2017	H_2SO_4/biochar as a filter media (Forestry wood waste)	*Escherichia coli*	98% removal	Storm water
		Miscellaneous Applications		
Choudhary et al. 2020	Biochar (Cladodes of *Opuntia ficus-indica* (OFI) cactus)	a. Malachite Green dye b. Cu^{+2} c. Ni^{+2}	a. 1341 mg/g b. 49 mg/g c. 44 mg/g	Simulated wastewater
Liu et al. 2018	Fe^0/Biochar (Kenaf Bar)	a. Cu^{2+} b. Bisphenol A	a. 96% b. 98%	Simulated wastewater
Caprariis et al. 2017	NaOH/biochar (Poplar biomass)	Total organic carbon	840 mg/g	Pyrolysis wastewater
Zhu et al. 2020	Fe/biochar (Crop wheat straw residue)	a. Cd (II) b. As (III) c. Roxarsone antibiotic d. Phosphorus	a. 31.9 mg/g b. 65.3 mg/g c. 66.5 mg/g d. 42.7 mg/g	Simulated wastewater
Aghababaei et al. 2017	H_3PO_4/biochar (Forest residue)[1] NaOH/biochar (Forest residue)[2]	a. Oxytetracycline antibiotic b. Cd	a. 263.8 mg/g[1] b. 79.3 mg/g[2]	Simulated wastewater

11.4 REMOVAL MECHANISMS OF BIOCHAR

11.4.1 Heavy Metal Removal Mechanism by Biochar

The mechanisms of heavy metals removal such as cadmium (Cd), copper (Cu), uranium (U), strontium (Sr), nickel (Ni), mercury (Hg), barium (Ba), lead (Pb), arsenic (As), chromium (Cr), zinc (Zn) and cobalt (Co), either by biochar or surface modified biochar (modified by different minerals, metal oxides, oxidants, acids and bases) are mainly of four types;

1. **π-π surface complexation:** mainly π electron donor-accepter interaction where carboxyl, hydroxyl or aromatic ring (C–H, C=C, –OH, –COOH) present in biochar creates π electron rich system and serve as π donor, whereas the heavy metals act as a π acceptor, as a result π-π surface interaction occurs between adsorbent and adsorbate),
2. **Electrostatic interaction:** here, adsorbates (i.e., heavy metals) are attracted by oppositely charged surface functional groups of adsorbents, which means positively charged heavy metals interact with negatively charged surface functional groups such as hydroxyl, carboxyl or negatively charged heavy metals interact with positively charged surface of metal- or mineral-rich biochar,
3. **Co-precipitation mechanism:** surface functional groups of biochar such as CO_3^{2-}, PO_4^{3-}, or SiO_3^{2-} interact with heavy metal ions and form precipitation of carbonate, phosphate or silicate compounds,
4. **Ion exchange mechanism**: the metal ions such as K^+, Ca^{2+}, Na^+ or Mg^{2+} or hydroxyl, carboxyl groups present in biochar exchange their positive or negative charge(s) with heavy metal ions [Inyang et al. 2015; Li et al. 2019a].

Bandara et al. [2020] investigated that the mechanism of Cu(II) and Cd(II) adsorption on biochar depends on formation of carbonate and phosphate precipitation from interaction between metal and CO_3^{2-}, PO_4^{3-}, electrostatic interaction, π-π surface interaction between negatively charged surface functional groups (-O^-, -COO^-) of biochar and positively charged heavy metals (Cd^{2+}, Cu^{2+}). Choudhary et al. [2020] reported same mechanism for Cu^{2+} and Ni^{2+} adsorption on biochar, but at the same time, interaction of both those metal ions with Ca^{2+}, Mg^{2+} ions present in biochar through ion exchange mechanism is also proposed. Younis et al. [2020] explained that the adsorption of Ba(II) and Sr(II) on biochar was possible due to Lewis acid base interaction between metal [Lewis acid] and oxygen functional groups (–OH, –COOH, –CHO, Si-OH) [Lewis base] of biochar and ion-exchange between Na^+ ion present in biochar and Ba(II)/Sr(II). Ali et al. [2020] reported that As(III) formed the negatively charged ions $HAsO_4^{2-}$, AsO_4^{3-} in aqueous solution which interact with the positively charged K^+, Ca^{2+}, Na^+ metal ions and surface functional groups such as lactone, carboxyl, hydroxyl or aromatic ring (O=C–O–, C–H, C=C, –OH, –COOH) of biochar through electrostatic, π-π and ion exchange mechanism. Huang et al. [2020a] showed that the Cd^{2+} adsorption on biochar was done through ion-exchange with K^+, Ca^{2+}, Na^+ and π-π interaction with aromatic ring (C-H, C=C) of biochar. Zhang et al. [2019b] reported that Pb^{2+} adsorbed on biochar surface through ion exchange with Ca^{2+}, Mg^{2+} and π-π surface complexation with hydroxyl, aromatic groups of adsorbent. Another researcher showed that the cause of better adsorption of Pb and Cu was the formation of precipitation of $PbCO_3$ and CuO through co-precipitation mechanism with abandoned hydroxyl and carbonate groups of biochar [Zhang et al. 2019a]. Ho et al. [2017] showed that the presence of negatively charged surface functional groups (–OH, –COOH), Pb^{2+} was attracted through electrostatic interaction. The adsorption of Cr(VI) on biochar was the result of π-π interaction between metal (Cr) and carboxyl, hydroxyl groups, ion exchange with Ca^{2+}, Mg^{2+}, formation of chromium phosphate/carbonate precipitation due to interaction between Cr(VI) and CO_3^{2-}/PO_4^{3-} [Shakya et al. 2019]. The mechanism of uranium [U(VI)] adsorption by biochar was explained by Mishra et al. [2017] and they found that π-π interaction between uranium and surface functional groups such as hydroxyl, carboxyl groups of adsorbent was the cause of good adsorption of radioactive element. Jang et al. [2018] explored that positively charged strontium (Sr^{2+}) ion interacted with negatively charged carbonyl, carboxylate groups of biochar through electrostatic interaction for adsorption. Hanandeh et al. [2016] reported that electrostatic interaction between positively charged Hg^{2+} and negatively charged surface functional groups (–OH) of biochar played the key role in adsorption of mercury by biochar. Fig. 11.1 shows different removal mechanisms of pollutants by biochar.

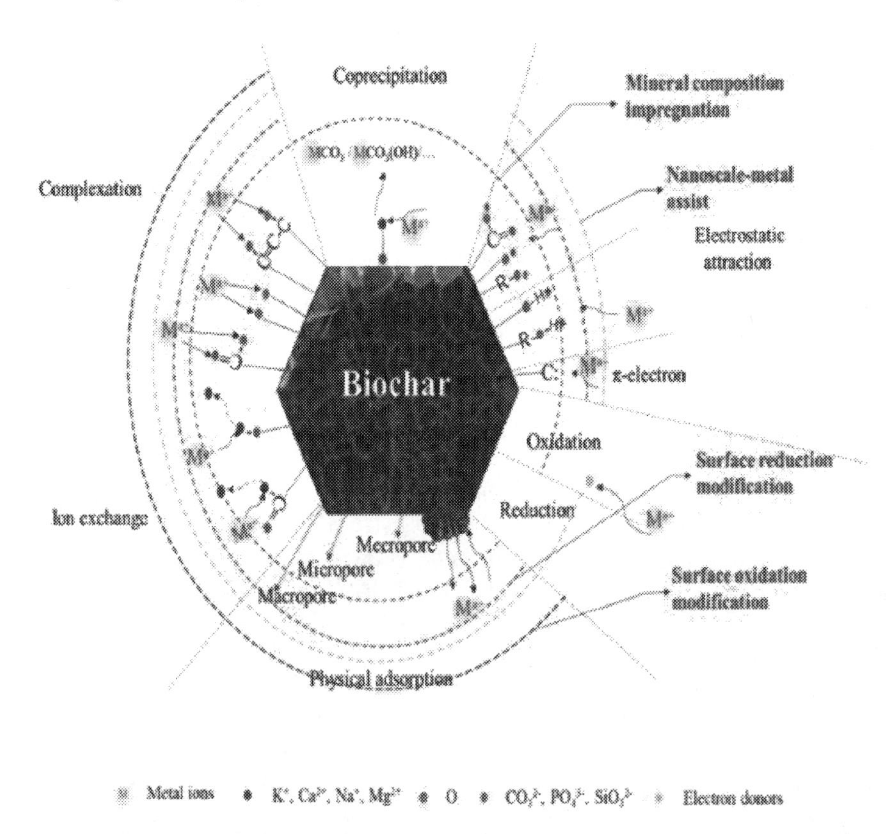

FIGURE 11.1 Mechanisms of metal removal by biochar, adapted from reference Wang et al. [2019] with permission from Elsevier.

11.4.2 Heavy Metal Removal Mechanism by Surface Modified Biochar

It was found that the adsorption capacity of biochar toward heavy metals can be improved with some folds by surface modification and this modification causes better selectivity towards heavy metals [Wang et al. 2019]. Fe impregnation in biochar is one of the most common methods for surface modification because most of the toxic heavy metals such as As, Cr, Cu, Mo, Co, Pb, Zn, Cd, Se can easily be removed by Fe incorporated biochar through either of the electrostatic interaction, ion exchange, coprecipitation and π-π surface complexation [Alqadami et al. 2018; Imran et al. 2020; Johansson et al. 2016; Zhu et al. 2020]. Except Fe, different other metals (Mn, Zn, Al, La, Ti), oxidants ($KMnO_4$, H_2O_2), acids (H_3PO_4, H_2SO_4, HNO_3), bases (NaOH, KOH) and inorganic materials (sulfur, nitrogen) are also used for surface modification of biochar for enhancement of metal active sites and oxygenated surface functional groups of adsorbent (biochar) to adsorb the toxic heavy metals [Wang et al. 2019]. Table 11.3 shows a compilation of different mechanisms of heavy metal removal by surface modified biochar.

11.4.3 Inorganic and Organic Compounds Removal Mechanism by Biochar and Surface Modified Biochar

In the previous section, it was observed that the heavy metal removal is mainly based on five interactions such as π-π surface complexation, electrostatic interaction, co-precipitation mechanism, ion exchange mechanism and acid base interaction. Removal of inorganic and organic compounds by biochar not only depends on the above mentioned mechanisms but involves other four mechanisms also, which are,

1. **Pore filling:** (The smaller molecules of adsorbate easily fills the micro and mesopores of the adsorbent due to strong affinity),
2. **Diffusion:** (Organic volatile matter diffuses from high concentration region to adsorbent pore structure.
3. **Hydrophobic interaction:** (The hydrophobic surface of biochar easily interacts with inorganic or organic compounds which are hydrophobic in nature),
4. **Hydrogen bonding:** (Electron rich polar functional groups of biochar make hydrogen bonding with inorganic or organic molecules which contain electronegative atom) [Inyang et al. 2015; Li et al. 2019 a].

Table 11.4 shows the summary of removal mechanisms of inorganic, organic (pesticides, pharmaceutical, plasticizer, dye, microbial strain and volatile organic compound) compounds by biochar and surface modified biochar.

11.5 FACTORS AFFECTING THE REMOVAL OF CONTAMINANTS BY BIOCHAR

11.5.1 Source of Feedstock for Biochar Preparation

Selection of feedstock for biochar is a critical step for adsorptive removal process. Biochar obtained from different sources show different surface morphology and elemental compositions, which have direct effect on effective surface area and surface

TABLE 11.3
Mechanisms of Heavy Metal Removal by Surface Modified Biochar (BC)

Reference	Adsorbent	Heavy Metal	Mechanism
Niu et al. 2020	Mn-Zn-Fe/BC	Pb^{2+}, Cd^{2+}	Electrostatic interaction, ion exchange and π-π surface complex formation
Zhou et al. 2017	H_3PO_4/BC	Pb(II)	π-π surface complexation
Aghababaei et al. 2017	NaOH/BC	Cd^{2+}	Electrostatic interaction
Imran et al. 2020	HNO_3/BC	Cr(VI)	π-π surface complexation
Jin et al. 2018	HNO_3/BC	U(VI)	π-π surface complexation
Biswas et al. 2019	H_2SO_4/BC	Pb(II)	π-π surface complex and electrostatic interaction
Jia et al. 2020	Mn/BC	Sb(V)	π-π surface complexation
Jeon et al. 2020	Sulfur/BC	Hg(II)	Lewis soft acid-base interaction
Chaukura et al. 2016	Sulfur/BC	Zn^{2+}	Electrostatic interaction
Wang et al. 2018b	Mn-Al-LDH/BC	Cu(II)	Isomorphic substitution
Shi et al. 2018	Fe_3O_4-SiO_2-NH_2/BC	Cr(VI)	Electrostatic interaction and Cr(III) forms chelation with amine group
Wu et al. 2017	Red mud/BC	As(V)	Electrostatic interaction
Paranavithana et al. 2016	Soil/BC	Cd^{2+}, Pb^{2+}	Electrostatic interaction
Gao et al. 2020	$KMnO_4$/BC	Pb^{2+}	Ion exchange and Coprecipitation
Li et al. 2019b	$KMnO_4$/BC	U (VI)	Lewis acid-base interaction
Lin et al. 2019	Fe-Mn-La/BC	As(III)	Electrostatic interaction and π-π surface complexation
Luo et al. 2019b	TiO_2/BC	Cd(II), As(V)	Ion exchange technique
Tang et al. 2019	NH_2/BC	Cu(II)	π-π surface complexation
An et al. 2019	$KMnO_4$-KOH/BC	Ni(II)	Ion exchange mechanism
Mian et al. 2018	N-Fe/BC	Cr(VI)	Electrostatic interaction
Gao et al. 2018	S-Fe/BC	Cr(VI)	Lewis acid-base interaction and electrostatic interaction
Ding et al. 2018	Al/BC	As(V)	Electrostatic interaction
Yu et al. 2018	ZnO/BC	Cr(VI)	Coprecipitation mechanism
Wang et al. 2020b	Mg-Al-LDH/BC	Pb^{2+}, CrO_4^{2-}	Ion exchange mechanism and electrostatic interaction

oxygen functional groups [hydroxyl, carboxyl, pyrone, aromatic, lactone, phenolic groups] (active sites of biochar), which in turn improve the removal efficiency of contaminants by biochar [Inyang et al. 2015]. Table 11.5 shows an overview of diversified characteristics of biochar obtained from different feedstock.

11.5.2 Solution pH

The electrostatic interaction between adsorbates and adsorbents mainly depends on both of their surface charges. Therefore, the solution pH has a great impact

TABLE 11.4
Summary of Removal Mechanisms of Inorganic and Organic Compounds

Contaminant	Adsorbent	Proposed Mechanism	Reference
Phosphate	Fe/BC (Biochar)	Electrostatic interaction, Ion exchange and diffusion	Yang et al. 2018b
Caffeine	BC	Hydrogen bonding and π-π surface complexation.	Keerthanan et al. 2020
Naproxen (Pharmaceutical)	Biochar	Pore filling and π-π interaction mechanism	Tomul et al. 2020
Deoxycycline hydrochloride	Cu/BC	Electrostatic interaction and π-π surface complexation	Liu et al. 2017
Tetracycline	BC	π-π interaction, hydrophobic interaction and hydrogen bonding	Zhang et al. 2020a
a. Sulfamethoxazole b. Sulfapyridine	BC	π-π interaction, hydrogen bonding and hydrophobic interaction	Huang et al. 2020b
Diclofenac	Fe/BC	Hydrophobic interaction, π-π interaction and hydrogen bonding	Tam et al. 2019
Furazolidone	Fe/BC	Hydrogen bonding	Gurav et al. 2020
Ciprofloxacin	BC and montmorillonite composite	π-π interactionand electrostatic interaction	Ashiq et al. 2019
a. Levofloxacin b. Amoxicillin	BC	π-π interaction	Li et al. 2020a
Atrazine (Pesticides)	BC	Pore filling and π-π interaction	Wang et al. 2020d
E. coli (Microbes)	Sulfuric acid-treated biochar as a filter	Surface attachment	Lau et al. 2017
a. Methyl Orange (Anion Dye) b. Eriochrome Black T (Anion Dye) c. Methylene Blue (Cation Dye) d. Crystal violet (Cation Dye)	BC	Electrostatic, ion exchange and π-π interaction	Zubair et al. 2020
a. Congo Red (CR) b. Malachite Green (MG)	BC	Pore-filling mechanism, electrostatic attraction, hydrogen bonding, and π-π interaction	Wu et al. 2020a
Phosphate	La/BC	Electrostatic interaction and π-π interaction	Li et al. 2020b
Phosphate	Polyethyleneimine/BC	Electrostatic interaction	Li et al. 2019c
Phosphate	Ca/BC	Coprecipitate mechanism	Liu et al. 2019b
Phosphate	Dolomite/BC	π-π interaction and Electrostatic interaction	Li et al. 2019d
Phosphate	Zn-Al-OH/BC	Hydrogen bonding, electrostatic interaction and coprecipitation	Jiang et al. 2019

(Continued)

TABLE 11.4 (Continued)
Summary of Removal Mechanisms of Inorganic and Organic Compounds

Contaminant	Adsorbent	Proposed Mechanism	Reference
Bisphenol A (Plasticizers)	Fe/BC	Pore filling, Hydrophobic interaction, π-π interaction and hydrogen bonding	Wang et al. 2020a
Phenol, 2-4 dinitrophenol (volatile organic compound)	BC	π-π interaction and hydrogen bonding	Thang et al. 2019
Ammonium ion (NH_4^+) Fertilizer or pesticide	BC	Electrostatic interaction	Yang et al. 2017
Nitrate	BC and montmorillonite composite	Electrostatic interaction	Viglašová et al. 2018

on adsorption efficiency of adsorbent. If the wastewater contaminant is positively charged, then the negative charge on surface of adsorbent (biochar) significantly increases with increasing solution pH; as a result, due to electrostatic interaction, the adsorption efficiency of biochar is found to be enhanced. On the other hand, if the wastewater contaminant is negatively charged, then increments of solution pH have

TABLE 11.5
Overview of Biochar Produced from Different Feedstock [Qambrani et al. 2017]

Feedstock	BET Surface Area (m^2/g)	Elemental Composition (wt%)
Anaerobic digestion residue	7.6	C:- 63.5; H:- 5.3; O:- 18.1; N:-0.9
Bamboo	10.2	C:- 76.9; H:- 3.6; O:- 18.1; N:-0.2
Brazillian pepper wood	234.7	C:- 77.0; H:- 2.2; O:- 17.7; N:-0.1
Cattle manure	1.4-58.6	C:- 52.2–65.9; N:-1.9–2.2
Corn straw	13.1	C:- 35.9; H:- 1.6; O:- 1.9; N:-0.4
Dairy manure	1.6	C:- 55.8; H:- 4.3; O:- 18.7; N:-2.6
Eucalyptus	10.4	C:- 77.8; H:- 5.4; O:- 18.3; N:-0.4
Goat manure	3.3-93.5	C:- 42.7-43.6; H:-1.7–0.8; O:- 30.1–21.7; N:-2.1–1.1
Grass	3.3	C:- 47.2-89.0; H:-7.1-2.5; O:- 45.1-7.6; N:-0.6-1.0
Orange peels	428-110.2	C:- 68.4-74.8; H:4.8-1.6; O:- 19.8–13.4; N:-2.0–1.7
Peanut shells	3.1	C:- 68.3; H:-3.9; O:- 25.9; N:-1.9
Pine needles	390.4	C:- 93.7; H:- 0.6; O:- 2.1; N:-3.6
Poultry litter	3.9	C:- 51.1; H:- 3.8; O:- 15.6; N:-4.5
Rice husks	27.8	C:- 38.6; N:-0.4
Sludge	126.4	C:- 8.5; H:- 1.0; O:- 6.4; N:-0.3
Soybean stover	420.3	C:- 82.0; H:- 1.3; O:- 15.5; N:-1.3
Sugarcane bagasse	388.3	C:- 77.9; H:- 2.4; O:- 17.8; N:-0.4
Wood	196–392	C:- 81.9-89; H:-3.5-3.0; O:- 14.5–8.0; N:-0.1

negative impact on absorptive removal of contaminant due to electrostatic repulsion. Hong et al. [2020] reported that in a system of selenium (Se) as adsorbate and Fe impregnated biochar (Fe/BC) as adsorbent, Se creates the anion SeO_4^{2-} in aqueous solution; as a result, when the solution pH was increased, the negative charge accumulation on surface of Fe/BC caused electrostatic repulsion between adsorbate (Se) and adsorbent (Fe/BC), and the adsorption efficiency dropped down. The same trend was observed by Tomul et al. [2020] and Huang et al. [2018] for adsorption of antibiotics, naproxen, and ciprofloxacin by biochar due to electrostatic repulsion between anionic antibiotics and negatively charged surface of biochar due to increments of pH. Zhu et al. [2020] reported that because of the anionic nature of arsenic and phosphate ions in aqueous solution, the removal efficiency by biochar significantly decreased by 11% and 66% during increase of solution pH from 5.2 to 9 and 2 to 5.2, respectively. But Huang et al. [2020a] reported that cadmium (Cd^{2+}) adsorption on Fe-impregnated biochar (Fe/BC) was favorable during an increase in solution pH because cadmium, a positively charged ion, was attracted to negatively charged Fe/BC by electrostatic interaction; as a result, the adsorption capacity reached 51.37 mg/g when the solution pH reached 6 (pH 3, 20 mg/g; at pH 4, 43 mg/g). Cheng et al. [2020] found that a greater number of cationic tetracycline molecules were attracted through electrostatic interaction with biochar when its negative surface charge increased significantly due to increase in pH of solution. The same trend was observed by Hu et al. [2020], Jalayeri et al. [2019], and Liu et al. [2017] for positively charged ammonium (NH_4^+), cupric (Cu^{2+}) ions and doxycycline adsorption respectively on biochar.

11.5.3 Adsorbent Dose

The amount or dose of adsorbent used in the process is an important parameter that determines the economic feasibility of adsorbent. Zhang et al. [2020b] reported that the adsorption efficiency of biochar was increased from 43% to 88.2% with an increase in the adsorbent dose from 0.2 to 0.6 g/L due to enhancement of the number of active sites and interaction between adsorbate and adsorbent. Choudhary et al. [2020] explained that an increase in the amount of adsorbent from 1 to 60 mg per 100 mL solution enhanced the adsorption efficiency of biochar from 93% to 99% due to an increase in available surface area. The same result was reported by Ali et al. [2020] and Shakya et al. [2019] for adsorption of arsenic and chromium by biochar respectively, where the adsorption efficiency was increased with an increase in the adsorbent amount. Hu et al. [2020] reported a massive increase in adsorption efficiency of biochar from 32% to 91% by increasing the adsorbent amount from 0.1 to 2 g due to the same reason of increase in surface area. Beyond 2 g, no significant change in adsorption efficiency was observed, which indicated the chances of overlap of adsorbent layers one upon another resulting the shielding of available active sites.

11.5.4 Initial Concentration of Contaminants of Wastewater

The adsorption efficiency of adsorbent (biochar) depends on the initial concentration of adsorbate. If the initial concentration of adsorbate is increased without increasing the adsorbent amount, the adsorption remarkably decreased due to the obvious reason of non-availability of sufficient number of active sites of adsorbent to interact with adsorbate. Zhang et al. [2020b] reported that, when the initial concentration of tetrabromo bisphenol A was increased from 0.2 to 20 mg/L with a fixed amount of adsorbent, then the adsorption efficiency of biochar decreased from 97.6% to 82.1% due to the same reason mentioned before. One other similar observation was by Thang et al. [2019], where with the increase in initial concentration of phenol from 10 to 200 mg/L, the adsorption efficiency was found to decrease from 98.8% to 80.5%. Shakya et al. [2019] and Imran et al. [2020] reported the same trend for adsorption of chromium using biochar and Fe-impregnated biochar (Fe/BC) as adsorbent respectively.

11.5.5 Effect of Contact Time Between Adsorbate and Adsorbent

The role of operating time for removal of contaminates from wastewater by biochar is critical. Increase in the contact time between adsorbate and adsorbent enhances the removal efficiency of adsorbent as this allows longer interaction between adsorbent active sites and adsorbate molecules by a higher degree of the mass transfer between them. But when exceeding the contact time beyond a certain limit, minor or no change in removal efficiencies of biochar are observed, which indicates that equilibrium exists and available active sites are saturated [Choudhary et al. 2020]. A study [Choudhary et al. 2020] reported that with an increase in operating time from 10 to 120 minutes, an appreciable increase of 50%–99.07% in the removal efficiency of biochar was observed. Another study of chromium removal by biochar [Shakya et al. 2019] showed an increase in the removal efficiency to 50% in first 12 hours of operation, with a higher pace than in the next phase of operation, where the removal efficiency increased very slowly with time to reach 99%. The reason may be the repulsion between surface sorbed metal ion of adsorbent and remaining metal ions present in simulated wastewater. Thang et al. [2019] reported a similar trend of result where the adsorption efficiency of biochar increased up to 90 minutes of contact time and reached to 83.5%, but after that no significant change in removal efficiency of biochar was observed due to set-in of adsorption equilibrium.

11.5.6 Effect of Pyrolysis Temperature and Operating Temperature

Variation in pyrolysis temperature during the preparation stage of biochar is responsible for diversification of its surface morphology. It was found that high pyrolysis temperature enhances the surface area of biochar, which is a determining factor for enhancement of the removal of contaminants. Li et al. [2020b] reported that when the pyrolysis temperature of biochar was increased from 400 to 600°C, the removal efficiency of the resultant biochar for phosphate removal was found to increase from 58% to 65%. A similar study [Park et al. 2019] showed that the increment of pyrolysis temperature from 300 to 600°C resulted in enhanced adsorption capacity of cadmium by biochar from 15 to 75 mg/g due to higher available surface area for adsorption to occur.

The effect of operating temperature indicated that the adsorption process is either exothermic or endothermic in nature. If the process is endothermic, then the adsorption efficiency of the adsorbent increases with increase in temperature, but if the process is exothermic, then increments of temperature have a negative effect on adsorption efficiency [Cheng et al. 2020]. Cheng et al. [2020] reported that the adsorption capacity of tetracycline by biochar increased from 401.61 to 476.19 mg/g when the operating temperature was enhanced from 20.5 to 40.5°C.

11.6 REGENERATION OF BIOCHAR

Regeneration of adsorbents is one of the key factors for assessing the economic feasibility of adsorptive removal of contaminants from industrial point of view. Study of the research done so far says that biochar is a useful, and in many cases a better, adsorbent compared to other conventional adsorbents such as activated carbon, clay material, ion exchange resin, and carbon nanotubes in the field of wastewater treatment. The feed-stocks of biochar are easily available and low cost; therefore most research concentrates on the improvement of its efficiency for wastewater treatment. Reports on the regeneration of biochar are scarce. For regaining adsorption capability of biochar, acid or base wash is a common procedure. Table 11.6 compiles there generation efficiency of the solvent washing of adsorbent (biochar).

TABLE 11.6
Regeneration Performance of Solvent Treatments on Biochar (BC)

Contaminants of Wastewater	Adsorbent	Regeneration Technique	Regeneration Efficiency	Reference
Pb	Ca/BC	Calcium disodium edentate dihydrate wash	39% of adsorption capacity retained after 5th cycle	Zhang et al. 2019b
Pb	Mn-Zn-Fe/BC	HCl wash	Adsorption efficiency decreased only 7% after 6th cycle	Niu et al. 2020
Malachite Green	BC	NaOH wash	Adsorption efficiency decreased only 15% after 6th cycle	Choudhary et al. 2020
Cr	Fe/BC	HCl wash	Adsorption capacity decreased only 11.21% after 5th cycle	Imran et al. 2020
Cr	BC	HCl wash	Adsorption efficiency decreased 34% after 5th cycle	Shakya et al. 2019
Cr	Fe_3O_4-SiO_2-NH_2/BC	NaOH wash	85% of adsorption capacity recovered after 6th cycle	Shi et al. 2018
Phenol	BC	Ethanol wash	Adsorption efficiency decreased 31.3% after 5th cycle	Thang et al. 2019
As	Fe/BC	NaOH wash	Adsorption efficiency decreased 12.5% after 3rd cycle	He et al. 2018
U	BC	HNO_3 wash	91.5% uranium recovered from used adsorbent to regain its power	Mishra et al. 2017

11.7 FUTURE PERSPECTIVES

Current investigation already proves that biochar is an environmentally sustainable product that easily removes various contaminants such as heavy metals, inorganic compounds, organic compounds, industrial dyes, antibiotics, and microbes from wastewater to a satisfactory degree. But still there are several research gaps that require more attention by researchers to establish biochar as a unique material for wastewater treatment plants.

- Biochar is produced by slow pyrolysis, fast pyrolysis, and gasification techniques. Investigation is needed to determine a cost-effective and high-yield biochar production method.
- More research is required to know how to minimize the emission of toxic compounds (dioxins) and greenhouse gases and to extract energy from biochar during pyrolysis process.
- Research is needed for the selection of best feedstock for biochar production for particular contaminants.
- In the gasification technique, a large number of hazardous inorganic minerals evolve out during water or acid washing. A proper study is required on how to minimize the release of unwanted minerals in the gasification process.
- A low yield of biochar from gasification is observed. An economical and upgraded process is required for increasing the production rate of biochar from this process.
- Some researches show various surface modifications of biochar in the aim to improve the selectivity of adsorption of contaminants from wastewater. There is much scope of further research is here that can direct the way to modify the adsorption capacity of different surface modified biochar.
- The application of biochar should be diversified. Hence, research is needed to check the performance of biochar and surface modified biochar in adsorption or filtration or disinfection processes for drinking water plants, column pilot plants, industrial wastewater treatment plants, and municipal wastewater treatment plants.
- More investigation is required on regeneration studies of biochar and finding efficient ways of regeneration.
- Further studies are required to determine the effect of competitive adsorption of different co-solutes present with contaminants in wastewater.
- An important and obvious step in research on adsorption of pollutants is the disposal problem, and that is no different with biochar. It is highly desired to determine how to convert them into useful material, such as fertilizer to improve soil fertility or other important applications.

11.8 CONCLUSION

Biochar is widely used as an efficient and low-cost adsorbent for wastewater treatment. It can easily replace traditional adsorbents such as ion exchange resin, activated carbon, and carbon nanotubes for wastewater treatment. The feedstock of biochar, such as pine sawdust, waste tea residue, food waste, peanut shells, rapeseed straw, seaweed biomass, rice straw, bamboo sawdust, pomelo peels, rabbit feces, date palm fronds, litchi peels, wheat straw residue, chicken manure, fruit peals, and municipal solid waste, etc., are very cheap as they are naturally obtained and easily available. Biochar can be prepared by slow or fast pyrolysis or gasification techniques, but recent research has shown that the production of biochar by slow pyrolysis attracts more attention due to the creation of a large number of surface oxygen functional groups on biochar surface. Biochar is an effective adsorbent for removal of heavy metals, such as cadmium, copper, uranium, nickel, mercury, arsenic, chromium, cobalt, zinc, etc., as well as alarming concentrations of other contaminants and pollutants such as various antibiotics such as chloramphenicol, erythromycin, sulfadiazine, tetracycline, norfloxacin, penicillin, carbadox etc.; different harmful dye chemicals, such as malachite green, methylene blue, eriochrome black T, congo red etc.; disease-causing microbes such as *E. Coli*, *Staphylococcus aureus*, and *Pseudomonas aeruginosa,* etc. It was found that surface modification of biochar by different oxidants, acids, metal, and bases magnificently enhanced the selectivity of contaminants of wastewater toward biochar. The adsorption mechanism followed π-π surface complexation, electrostatic interaction, coprecipitation mechanism, ion exchange mechanism, pore filling, diffusion, hydrophobic interaction, and hydrogen bonding. Finally, it can be said that there are different research gaps on biochar that require more attention to make it a universal material for wastewater treatment.

REFERENCES

Aghababaei, A., Ncibi, M. C., and Sillanpää, M. 2017. Optimized removal of oxytetracycline and cadmium from contaminated waters using chemically-activated and pyrolyzed biochars from forest and wood-processing residues. Bioresour. Technol. 239: 28–36.

Ahmed, M. B., Zhou, J. L., Ngo, H. H., and Guo, W. 2015. Adsorptive removal of antibiotics from water and waste water: Progress and challenges. Sci. Total Environ. 532: 112–126.

Ali, S., Rizwan, M., Shakoor, M. B., Jilani, A., and Anjum, R. 2020. High sorption efficiency for As(III) and As(V) from aqueous solutions using novel almond shell biochar. Chemosphere. 243: 125330.

Alqadami, A. A., Khan, M. A., Otero, M., Siddiqui, M. R., Jeon, B., and Batoo, K. M. 2018. A magnetic nanocomposite produced from camel bones for an efficient adsorption of toxic metals from water. J. Cleaner Prod. 178: 293–304.

An, Q., Jiang, Y., Nan, H., Yu, Y., and Jiang, J. 2019. Unraveling sorption of nickel from aqueous solution by $KMnO_4$ and KOH-modified peanut shell biochar: Implicit mechanism. Chemosphere. 214: 846–854.

Ashiq, A., Sarkar, B., Adassooriya, N., Walpita, J., Rajapaksha, A. U., Ok, Y. S., and Vithanage, M. 2019. Sorption process of municipal solid waste biochar-montmorillonite composite for ciprofloxacin removal in aqueous media. Chemosphere. 236: 124384.

Bandara, T., Xu, J., Potter, I. D., Franks, A., Chathurika, J., and Tang, C. 2020. Mechanisms for the removal of Cd(II) and Cu(II) from aqueous solution and mine water by biochars derived from agricultural wastes. Chemosphere. 254: 126745.

Banerjee, S., Laminka-Ot, A., Joshi, S. R., Mandal, T., and Halder, G. 2017. Optimization of Fe^{2+} removal from coal mine wastewater using activated biochar of colocasia esculenta. Water Environ. Res. 89(9): 774–782.

Biswas, S., Siddiqi, H., Meikap, B. C., Sen, T. K., and Khiadani, M. 2019. Preparation and characterization of raw and inorganic acid-activated pine cone biochar and its application in the removal of aqueous-phase Pb^{2+} metal ions by adsorption. Water Air Soil Pollut. 231(1): doi:10.1007/s11270-019-4375-7

Caprariis, B. D., Filippis, P. D., Hernandez, A. D., Petrucci, E., Petrullo, A., Scarsella, M., and Turchi, M. 2017. Pyrolysis wastewater treatment by adsorption on biochars produced by poplar biomass. J. Environ. Manage. 197: 231–238.

Chaukura, N., Gwenzi, W., Mupatsi, N., Ruziwa, D. T., and Chimuka, C. 2016. Comparative adsorption of Zn^{2+} from aqueous solution using hydroxylated and sulphonated biochars derived from pulp and paper sludge. Water Air Soil Pollut. 228(1): doi:10.1007/s11270-016-3191-6

Cheng, D., Ngo, H. H., Guo, W., Chang, S. W., Nguyen, D. D., Zhang, X., Varjani, S., and Liu, Y. 2020. Feasibility study on a new pomelo peel derived biochar for tetracycline antibiotics removal in swine wastewater. Sci.Total Environ. 720: 137662.

Chen, L., Cheng, P., Ye, L., Chen, H., Xu, X., and Zhu, L. 2020. Biological performance and fouling mitigation in the biochar amended anaerobic membrane bioreactor treating pharmaceutical waste water. Bioresour. Technol. 302: 122805.

Choudhary, M., Kumar, R., and Neogi, S. 2020. Activated biochar derived from Opuntia ficus-indica for the efficient adsorption of malachite green dye, Cu^2 and Ni^2 from water. J. Hazard. Mater. 392: 122441.

Deng, Y., Zhang, T., and Wang, Q. 2017. Biochar adsorption treatment for typical pollutants removal in livestock wastewater: A review, engineering applications of biochar. Intech Open. doi:10.5772/intechopen.68253

Ding, Z., Xu, X., Phan, T., Hu, X., and Nie, G. 2018. High adsorption performance for As(III) and As(V) onto novel aluminum-enriched biochar derived from abandoned Tetra Paks. Chemosphere 208: 800–807.

Fan, S., Li, H., Wang, Y., Wang, Z., Tang, J., Tang, J., and Li, X. 2017. Cadmium removal from aqueous solution by biochar obtained by co-pyrolysis of sewage sludge with tea waste. Res. Chem. Intermed. 44(1): 135–154.

Ferreira, C. I., Calisto, V., Otero, M., Nadais, H., and Esteves, V. I. 2017. Removal of tricaine methanesulfonate from aquaculture wastewater by adsorption onto pyrolysed paper mill sludge. Chemosphere. 168: 139–146.

Gao, J., Yang, L., Liu, Y., Shao, F., Liao, Q., and Shang, J. 2018. Scavenging of Cr(VI) from aqueous solutions by sulfide-modified nanoscale zero-valent iron supported by biochar. J. Taiwan Inst. Chem. Eng. 91: 449–456.

Gao, R., Xiang, L., Hu, H., Fu, Q., Zhu, J., Liu, Y., and Huang, G. 2020. High efficiency removal capacities and quantitative sorption mechanism of Pb by oxidized rape straw biochars. Sci. Total Environ. 699: 134262.

Gurav, R., Bhatia, S. K., Choi, T., Park, Y., Park, J. Y., Han, Y., Vyavahare, G., Jadhav, J., Song, H. S., Yang, P., Yoon, J. J., Bhatnagar, A., Choi, Y. K., and Yang, Y. 2020. Treatment of furazolidone contaminated water using banana pseudostem biochar engineered with facile synthesized magnetic nanocomposites. Bioresour. Technol. 297: 122472.

Gwenzi, W., Chaukura, N., Mukome, F.N.D., Machado, S., Nyamasoka, B. 2015. Biochar production and applications in sub-Saharan Africa: Opportunities, constraints, risks and uncertainties. J. Environ. Manage. 150: 250–261.

Gwenzi, W., Chaukura, N., Noubactep, C., and Mukome, F. N. D. 2017. Biochar-based water treatment systems as a potential low-cost and sustainable technology for clean water provision. J. Environ. Manage. 197: 732–749

Hairuddin, M. N., Mubarak, N. M., Khalid, M., Abdullah, E. C., Walvekar, R., and Karri, R. R. 2019. Magnetic palm kernel biochar potential route for phenol removal from wastewater. Environ. Sci. Pollut. Res. 26(34): 35183–35197.

Hanandeh, A. E., Abu-Zurayk, R. A., Hamadneh, I., and Al-Dujaili, A. H. 2016. Characterization of biochar prepared from slow pyrolysis of Jordanian olive oil processing solid waste and adsorption efficiency of Hg^{2+} ions in aqueous solutions. Water Sci. Technol. 74(8): 1899–1910.

Hassan, M., Naidu, R., Du, J., Liu, Y., and Qi, F. 2020. Critical review of magnetic biosorbents: Their preparation, application, and regeneration for wastewater treatment. Sci. Total Environ. 702: 134893.

He, R., Peng, Z., Lyu, H., Huang, H., Nan, Q., and Tang, J. 2018. Synthesis and characterization of an iron-impregnated biochar for aqueous arsenic removal. Sci. Total Environ. 612: 1177–1186.

Ho, S., Chen, Y., Yang, Z., Nagarajan, D., Chang, J., and Ren, N. 2017. High-efficiency removal of lead from wastewater by biochar derived from anaerobic digestion sludge. Bioresour. Technol. 246: 142–149.

Hong, S., Lyonga, F. N., Kang, J., Seo, E., Lee, C., Jeong, S., Hong, S. G., and Park, S. 2020. Synthesis of Fe-impregnated biochar from food waste for Selenium(VI) removal from aqueous solution through adsorption: Process optimization and assessment. Chemosphere. 252: 126475.

Hong, Y., Xu, Z., Feng, C., Xu, D., and Wu, F. 2019. The preparation of biochar particles from sludge and corncobs and its Pb^{2+} adsorption properties. Bull. Environ. Contam. Toxicol. 103(6): 848–853.

Hoslett, J., Ghazal, H., Mohamad, N., and Jouhara, H. 2020. Removal of methylene blue from aqueous solutions by biochar prepared from the pyrolysis of mixed municipal discarded material. Sci. Total Environ. 714: 136832.

Hu, X., Zhang, X., Ngo, H. H., Guo, W., Wen, H., Li, C., Zhang, Y., and Ma, C. 2020. Comparison study on the ammonium adsorption of the biochars derived from different kinds of fruit peel. Sci. Total Environ. 707: 135544.

Huang, F., Zhang, L., Wu, R., Zhang, S., and Xiao, R. 2020a. Adsorption behaviour and relative distribution of Cd^{2+} adsorption mechanisms by the magnetic and nonmagnetic biochars derived from chicken manure. Int. J. Environ. Res. Public Health. 17(5): 1602.

Huang, J., Zimmerman, A. R., Chen, H., and Gao, B. 2020b. Ball milled biochar effectively removes sulfamethoxazole and sulfapyridine antibiotics from water and wastewater. Environ. Pollut. 258: 113809.

Huang, W., Chen, J., and Zhang, J. 2018. Removal of ciprofloxacin from aqueous solution by rabbit manure biochar. Environ. Technol. 41(11): 1380–1390.

Imran, M., Khan, Z. U., Iqbal, M. M., Iqbal, J., Shah, N. S., Munawar, S., Ali, S., Murtaza, B., Naeem, M.A., and Rizwan, M. 2020. Effect of biochar modified with magnetite nanoparticles and HNO_3 for efficient removal of Cr(VI) from contaminated water: A batch and column scale study. Environ. Pollut. 261: 114231.

Inyang, M., and Dickenson, E. 2015. The potential role of biochar in the removal of organic and microbial contaminants from potable and reuse water: A review. Chemosphere. 134: 232–240.

Jalayeri, H., and Pepe, F. 2019. Novel and high-performance biochar derived from pistachio green hull biomass: Production, characterization, and application to Cu(II) removal from aqueous solutions. Ecotoxicol. Environ. Saf. 168: 64–71.

Jang, H. M., and Kan, E. 2019a. Engineered biochar from agricultural waste for removal of tetracycline in water. Bioresour. Technol. 284: 437–447.

Jang, H. M., and Kan, E. 2019b. A novel hay-derived biochar for removal of tetracyclines in water. Bioresour. Technol. 274: 162–172.

Jang, J., Miran, W., Divine, S. D., Nawaz, M., Shahzad, A., Woo, S. H., and Lee, D. S. 2018. Rice straw-based biochar beads for the removal of radioactive strontium from aqueous solution. Sci. Total Environ. 615: 698–707.

Jeon, C., Solis, K. L., An, H., Hong, Y., Igalavithana, A. D., and Ok, Y. S. 2020. Sustainable removal of Hg(II) by sulfur-modified pine-needle biochar. J. Hazard. Mater. 388: 122048.

Jia, X., Zhou, J., Liu, J., Liu, P., Yu, L., Wen, B., and Feng, Y. 2020. The antimony sorption and transport mechanisms in removal experiment by Mn-coated biochar. Sci. Total Environ. 724: 138158.

Jiang, Y., Li, A., Deng, H., Ye, C., and Li, Y. 2019. Phosphate adsorption from wastewater using ZnAl-LDO-loaded modified banana straw biochar. Environ. Sci. Pollut. Res. 26(18): 18343–18353.

Jin, J., Li, S., Peng, X., Liu, W., Zhang, C., Yang, Y., Han, L., Du, Z., Sun, K., and Wang, X. 2018. HNO_3 modified biochars for uranium (VI) removal from aqueous solution. Bioresour. Technol. 256: 247–253.

Johansson, C. L., Paul, N. A., Nys, R. D., and Roberts, D. A. 2016. Simultaneous biosorption of selenium, arsenic and molybdenum with modified algal-based biochars. J. Environ. Manage. 165: 117–123.

Keerthanan, S., Bhatnagar, A., Mahatantila, K., Jayasinghe, C., Ok, Y. S., and Vithanage, M. 2020. Engineered tea-waste biochar for the removal of caffeine, a model compound in pharmaceuticals and personal care products (PPCPs), from aqueous media. Environ. Technol. Innovation. 19: 100847.

Lau, A. Y., Tsang, D. C., Graham, N. J., Ok, Y. S., Yang, X., and Li, X. 2017. Surface-modified biochar in a bioretention system for *Escherichia coli* removal from stormwater. Chemosphere. 169: 89–98.

Lee, C., Hong, S., Hong, S., Choi, J., and Park, S. 2019. Production of biochar from food waste and its application for phenol removal from aqueous solution. Water Air Soil Pollut. 230(3): doi:10.1007/s11270-019-4125-x

Lellis, B., Favaro-Polonio, C.Z., Pamphile, J.A., and Polonio, J.C. 2019. Effects of textile dyes on health and the environment and bioremediation potential of living organisms. Biotechnol. Res. Innovation. 3: 275–290.

Li, H., Hu, J., Yao, L., Shen, Q., An, L., and Wang, X. 2020a. Ultrahigh adsorbability towards different antibiotic residues on fore-modified self-functionalized biochar: Competitive adsorption and mechanism studies. J. Hazard. Mater. 390: 122127.

Li, J., Li, B., Huang, H., Lv, X., Zhao, N., Guo, G., and Zhang, D. 2019d. Removal of phosphate from aqueous solution by dolomite-modified biochar derived from urban dewatered sewage sludge. Sci. Total Environ. 687: 460–469.

Li, J., Li, B., Huang, H., Zhao, N., Zhang, M., and Cao, L. 2020b. Investigation into lanthanum-coated biochar obtained from urban dewatered sewage sludge for enhanced phosphate adsorption. Sci. Total Environ. 714: 136839.

Li, L., Zou, D., Xiao, Z., Zeng, X., Zhang, L., Jiang, L., Wang, A., Ge, D., Zhang, G., and Liu, F. 2019a. Biochar as a sorbent for emerging contaminants enables improvements in waste management and sustainable resource use. J. Cleaner Prod. 210: 1324–1342

Li, N., Yin, M., Tsang, D. C., Yang, S., Liu, J., Li, X., Song, G., and Wang, J. 2019b. Mechanisms of U(VI) removal by biochar derived from Ficus microcarpa aerial root: A comparison between raw and modified biochar. Sci. Total Environ. 697: 134115.

Li, T., Tong, Z., Gao, B., Li, Y. C., Smyth, A., and Bayabil, H. K. 2019c. Polyethyleneimine-modified biochar for enhanced phosphate adsorption. Environ. Sci. Pollut. Res. 27(7): 7420–7429.

Lin, L., Song, Z., Khan, Z. H., Liu, X., and Qiu, W. 2019. Enhanced As(III) removal from aqueous solution by Fe-Mn-La-impregnated biochar composites. Sci. Total Environ. 686: 1185–1193.

Liu, C., Diao, Z., Huo, W., Kong, L., and Du, J. 2018. Simultaneous removal of Cu^{2+} and bisphenol A by a novel biochar-supported zero valent iron from aqueous solution: Synthesis, reactivity and mechanism. Environ. Pollut. 239: 698–705.

Liu, G., Tang, H., Fan, J., Xie, Z., He, T., Shi, R., and Liao, B. 2019a. Removal of 2,4,6-trichlorophenol from water by Eupatorium adenophorum biochar-loaded nano-iron/nickel. Bioresour. Technol. 289: 121734.

Liu, J., Jiang, J., Meng, Y., Aikelaimu, A., Xu, Y., Xiang, H., Gao, Y., and Chen, X. 2020. Preparation, environmental application and prospect of biochar-supported metal nanoparticles: A review. J. Hazard. Mater. 388: 122026.

Liu, S., Xu, W., Liu, Y., Tan, X., Zeng, G., Li, X., Liang, J., Zhou, Z., Yan, Z., and Cai, X. 2017. Facile synthesis of Cu(II) impregnated biochar with enhanced adsorption activity for the removal of doxycycline hydrochloride from water. Sci. Total Environ. 592: 546–553.

Liu, X., Shen, F., Smith, R. L., and Qi, X. 2019b. Black liquor-derived calcium-activated biochar for recovery of phosphate from aqueous solutions. Bioresour. Technol. 294: 122198.

Luo, K., Pang, Y., Yang, Q., Wang, D., Li, X., Wang, L., Lei, M., and Liu, J. 2019a. Enhanced ciprofloxacin removal by sludge-derived biochar: Effect of humic acid. Chemosphere. 231: 495–501.

Luo, M., Lin, H., He, Y., Li, B., Dong, Y., and Wang, L. 2019b. Efficient simultaneous removal of cadmium and arsenic in aqueous solution by titanium-modified ultrasonic biochar. Bioresour. Technol. 284: 333–339.

Mandal, A., Singh, N., and Purakayastha, T. 2017. Characterization of pesticide sorption behaviour of slow pyrolysis biochars as low cost adsorbent for atrazine and imidacloprid removal. Sci. Total Environ. 577: 376–385.

Mercado, L.F.P., Lalander, C., Joel, A., Ottoson, J., Dalahmeh, S., and Vinneras, B. 2019. Biochar filters as an on-farm treatment to reduce pathogens when irrigating with wastewater polluted sources. J. Environ. Manage. 248: 109295.

Mian, M. M., Liu, G., Yousaf, B., Fu, B., Ullah, H., Ali, M. U., Abbas, Q., Munir, M.A.M., and Ruijia, L. 2018. Simultaneous functionalization and magnetization of biochar via NH_3 ambiance pyrolysis for efficient removal of Cr (VI). Chemosphere. 208: 712–721.

Mishra, V., Sureshkumar, M. K., Gupta, N., and Kaushik, C. P. 2017. Study on sorption characteristics of uranium onto biochar derived from eucalyptus wood. Water Air Soil Pollut. 228(8): doi:10.1007/s11270-017-3480-8

Niu, Z., Feng, W., Huang, H., Wang, B., Chen, L., Miao, Y., and Su, S. 2020. Green synthesis of a novel Mn–Zn ferrite/biochar composite from waste batteries and pine sawdust for Pb^{2+} removal. Chemosphere. 252: 126529.

Novais, S. V., Zenero, M. D., Tronto, J., Conz, R. F., and Cerri, C. E. 2018. Poultry manure and sugarcane straw biochars modified with $MgCl_2$ for phosphorus adsorption. J. Environ. Manage. 214: 36–44.

Pap, S., Bezanovic, V., Radonic, J., Babic, A., Saric, S., Adamovic, D., and Sekulic, M. T. 2018. Synthesis of highly-efficient functionalized biochars from fruit industry waste biomass for the removal of chromium and lead. J. Mol. Liq. 268: 315–325.

Paranavithana, G. N., Kawamoto, K., Inoue, Y., Saito, T., Vithanage, M., Kalpage, C. S., and Herath, G. B. 2016. Adsorption of Cd^{2+} and Pb^{2+} onto coconut shell biochar and biochar-mixed soil. Environ. Earth Sci. 75(6): doi: 10.1007/s12665-015-5167-z

Park, J., Wang, J. J., Kim, S., Kang, S., Jeong, C. Y., Jeon, J., Park, K. H., Cho, J. S., Delaune, R. D., and Seo, D. 2019. Cadmium adsorption characteristics of biochars derived using various pine tree residues and pyrolysis temperatures. J. Colloid Interface Sci. 553: 298–307.

Qambrani, N. A., Rahman, M. M., Won, S., Shim, S., and Ra, C. 2017. Biochar properties and eco-friendly applications for climate change mitigation, waste management, and wastewater treatment: A review. Renewable Sustainable Energy Rev. 79: 255–273.

Sewu, D. D., Boakye, P., and Woo, S. H. 2017. Highly efficient adsorption of cationic dye by biochar produced with Korean cabbage waste. Bioresour. Technol. 224: 206–213.

Shakya, A., and Agarwal, T. 2019. Removal of Cr(VI) from water using pineapple peel derived biochars: Adsorption potential and re-usability assessment. J. Mol. Liq. 293: 111497.

Shi, S., Yang, J., Liang, S., Li, M., Gan, Q., Xiao, K., and Hu, J. 2018. Enhanced Cr(VI) removal from acidic solutions using biochar modified by Fe_3O_4@SiO_2-NH_2 particles. Sci. Total Environ. 628-629: 499–508.

Song, H., Wang, J., Garg, A., Lin, X., Zheng, Q., and Sharma, S. 2019. Potential of novel biochars produced from invasive aquatic species outside food chain in removing ammonium nitrogen: comparison with conventional biochars and clinoptilolite. Sustainability. 11(24): 7136.

Son, E., Poo, K., Chang, J., and Chae, K. 2018. Heavy metal removal from aqueous solutions using engineered magnetic biochars derived from waste marine macro-algal biomass. Sci. Total Environ. 615: 161–168.

Tam, N. T., Liu, Y., Bashir, H., Yin, Z., He, Y., and Zhou, X. 2019. Efficient removal of diclofenac from aqueous solution by potassium ferrate-activated porous graphitic biochar: Ambient condition influences and adsorption mechanism. Int. J. Environ. Res. Public Health. 17(1): 291.

Tan, X. F., Liu, Y. G., Gu, Y. L., Xu, Y., Zeng, G. M., Hu, X. J., Liu, S. B., Wang, X., Liu, S. M., and Li, J. 2016. Biochar-based nanocomposite for the decontamination of waste water: A review. Bioresour. Technol. 212: 318–333.

Tang, S., Shao, N., Zheng, C., Yan, F., and Zhang, Z. 2019. Amino-functionalized sewage sludge-derived biochar as sustainable efficient adsorbent for Cu(II) removal. Waste Manage. 90: 17–28.

Tatarchuk, T., Bououdina, M., Al-Najar, B., and Bitra, R. B. 2019. Green and eco-friendly materials for the remediation of inorganic and organic pollutants in water. A New Generation Material Graphene: Applications in Water Technology. Springer, Cham, https://doi.org/10.1007/978-3-319-75484-0_4

Thang, P. Q., Jitae, K., Giang, B. L., Viet, N., and Huong, P. T. 2019. Potential application of chicken manure biochar towards toxic phenol and 2,4-dinitrophenol in wastewaters. J. Environ. Manage. 251: 109556.

Tomul, F., Arslan, Y., Kabak, B., Trak, D., Kendüzler, E., Lima, E. C., and Tran, H. N. 2020. Peanut shells-derived biochars prepared from different carbonization processes: Comparison of characterization and mechanism of naproxen adsorption in water. Sci. Total Environ. 726: 137828.

Viglašová, E., Galamboš, M., Danková, Z., Krivosudský, L., Lengauer, C. L., Hood-Nowotny, R., Soja, G., Rompel, A., Matik, M., and Briančin, J. 2018. Production, characterization and adsorption studies of bamboo-based biochar/montmorillonite composite for nitrate removal. Waste Manage. 79: 385–394.

Water for Life 2005–2015, United Nations Department of Economic and Social Affairs, https://www.un.org/waterforlifedecade/scarcity.shtml.

Wang, C., and Wang, H. 2018a. Pb(II) sorption from aqueous solution by novel biochar loaded with nano-particles. Chemosphere. 192: 1–4.

Wang, H., Wang, S., Chen, Z., Zhou, X., Wang, J., and Chen, Z. 2020b. Engineered biochar with anisotropic layered double hydroxide nanosheets to simultaneously and efficiently capture Pb^{2+} and CrO_4^{2-} from electroplating wastewater. Bioresour. Technol. 306: 123118.

Wang, J., and Zhang, M. 2020a. Adsorption characteristics and mechanism of Bisphenol A by magnetic biochar. Int. J. Environ. Res. Public Health 17(3): 1075.

Wang, L., Wang, Y., Ma, F., Tankpa, V., Bai, S., Guo, X., and Wang, X. 2019. Mechanism and reutilization of modified biochar used for removal of heavy metals from wastewater: A review. Sci. Total Environ. 668: 1298–1309.

Wang, P., Liu, X., Yu, B., Wu, X., Xu, J., Dong, F., and Zheng, Y. 2020c. Characterization of peanut-shell biochar and the mechanisms underlying its sorption for atrazine and nicosulfuron in aqueous solution. Sci. Total Environ. 702: 134767.

Wang, T., Li, C., Wang, C., and Wang, H. 2018b. Biochar/MnAl-LDH composites for Cu (II) removal from aqueous solution. Colloids Surf. A. 538: 443–450.

Wu, C., Huang, L., Xue, S., Huang, Y., Hartley, W., Cui, M., and Wong, M. 2017. Arsenic sorption by red mud-modified biochar produced from rice straw. Environ. Sci. Pollut. Res. 24(22): 18168–18178.

Wu, J., Yang, J., Feng, P., Huang, G., Xu, C., and Lin, B. 2020a. High-efficiency removal of dyes from wastewater by fully recycling litchi peel biochar. Chemosphere. 246: 125734.

Wu, J., Wu, C., Zhou, C., Dong, L., Liu, B., Xing, D., Yang, S., Fan, J., Feng, L., Cao, G., and You, S. 2020b. Fate and removal of antibiotic resistance genes in heavy metals and dye co-contaminated wastewater treatment system amended with β-cyclodextrin functionalized biochar. Sci. Total Environ. 723: 137991.

Xiang, W., Zhang, X., Chen, J., Zou, W., He, F., Hu, X., Tsang, D. C. W., Ok, Y. S., and Gao, B. 2020. Biochar technology in wastewater treatment: A critical review. Chemosphere. 252:126539.

Yang, H. I., Lou, K., Rajapaksha, A. U., Ok, Y. S., Anyia, A. O., and Chang, S. X. 2017. Adsorption of ammonium in aqueous solutions by pine sawdust and wheat straw biochars. Environ. Sci. Pollut. Res. 25(26): 25638–25647.

Yang, L., Gao, J., Liu, Y., Zhang, Z., Zou, M., Liao, Q., and Shang, J. 2018a. Removal of methyl orange from water using sulfur-modified nzvi supported on biochar composite. Water Air Soil Pollut. 229(11): doi:10.1007/s11270-018-3992-x

Yang, Q., Wang, X., Luo, W., Sun, J., Xu, Q., Chen, F., Zhao, J., Wang, S., Yao, F., Wang, D., Li, X., and Zeng, G. 2018b. Effectiveness and mechanisms of phosphate adsorption on iron-modified biochars derived from waste activated sludge. Bioresour. Technol. 247: 537–544.

Younis, S. A., El-Salamony, R. A., Tsang, Y. F., and Kim, K. 2020. Use of rice straw-based biochar for batch sorption of barium/strontium from saline water: Protection against scale formation in petroleum/desalination industries. J. Cleaner Prod. 250: 119442.

Yu, J., Jiang, C., Guan, Q., Ning, P., Gu, J., Chen, Q., Zhang, J., and Miao, R. 2018. Enhanced removal of Cr(VI) from aqueous solution by supported ZnO nanoparticles on biochar derived from waste water hyacinth. Chemosphere. 195: 632–640.

Zhang, H., Lu, T., Wang, M., Jin, R., Song, Y., Zhou, Y., Qi, Z., and Chen, W. 2020a. Inhibitory role of citric acid in the adsorption of tetracycline onto biochars: Effects of solution pH and Cu^2. Colloids Surf. A. 595: 124731.

Zhang, C., Lu, J., and Wu, J. 2020b. One-step green preparation of magnetic seaweed biochar/sulfidated Fe^0 composite with strengthen adsorptive removal of tetrabromobisphenol A through in situ reduction. Bioresour. Technol. 307: 123170.

Zhang, J., Hu, X., Zhang, K., and Xue, Y. 2019b. Desorption of calcium-rich crayfish shell biochar for the removal of lead from aqueous solutions. J. Colloid Interface Sci. 554: 417–423.

Zhang, L., Li, W., Cao, H., Hu, D., Chen, X., Guan, Y., Tang, J., and Gao, H. 2019a. Ultra-efficient sorption of Cu^2 and Pb^2 ions by light biochar derived from Medulla etrapanacis. Bioresour. Technol. 291: 121818.

Zhang, X., Fu, W., Yin, Y., Chen, Z., Qiu, R., Simonnot, M., and Wang, X. 2018a. Adsorption-reduction removal of Cr(VI) by tobacco petiole pyrolytic biochar: Batch experiment, kinetic and mechanism studies. Bioresour. Technol. 268: 149–157.

Zhang, S., and Lu, X. 2018b. Treatment of wastewater containing Reactive Brilliant Blue KN-R using TiO_2/BC composite as heterogeneous photocatalyst and adsorbent. Chemosphere. 206: 777–783.

Zhang, H., Tang, L., Wang, J., Yu, J., Feng, H., Lu, Y., Chen, Y., Liu, Y., Wang, J., and Xie, Q. 2020 c. Enhanced surface activation process of persulfate by modified bagasse biochar for degradation of phenol in water and soil: Active sites and electron transfer mechanism. Colloids Surf. A 599: 124904.

Zhou, N., Chen, H., Xi, J., Yao, D., Zhou, Z., Tian, Y., and Lu, X. 2017. Biochars with excellent Pb(II) adsorption property produced from fresh and dehydrated banana peels via hydrothermal carbonization. Bioresour. Technol. 232: 204–210.

Zhu, S., Zhao, J., Zhao, N., Yang, X., Chen, C., and Shang, J. 2020. Goethite modified biochar as a multifunctional amendment for cationic Cd(II), anionic As(III), roxarsone, and phosphorus in soil and water. J. Cleaner Prod. 247: 119579.

Zubair, M., Mu'Azu, N. D., Jarrah, N., Blaisi, N. I., Aziz, H. A., and Al-Harthi, M. A. 2020. Adsorption behavior and mechanism of methylene blue, crystal violet, eriochrome black T, and methyl orange dyes onto biochar-derived date palm fronds waste produced at different pyrolysis conditions. Water Air Soil Pollut. 231(5): doi:10.1007/s11270-020-04595-x

12 Microalgae for Removing Pharmaceutical Compounds from Wastewater

Eliana M. Jiménez-Bambague, Aura C. Ortiz-Escobar and Carlos A. Madera-Parra
Escuela de Ingeniería de los Recursos Naturales y del Ambiente-EIDENAR, Universidad del Valle, Cali, Colombia

Fiderman Machuca-Martinez
GAOX, CENM, Escuela de Ingeniería Química, Universidad del Valle, Ciudad Universitaria Meléndez, 760032, Cali, Colombia

CONTENTS

DOI: 10.1201/9781003138303-12

12.1 INTRODUCTION

Wastewater treatment is an environmental concern, particularly in developing countries. Wastewater can contain numerous diverse contaminants; some of them are in reduced concentrations such as micropollutants, which are chemical substances found in the environment in concentrations of μgL^{-1}, or ngL^{-1}. Nowadays, the presence of micropollutants in aquatic environments has become more relevant due to their ecotoxicological effects (Verlicchi et al. 2012). These micropollutants can include pesticides, pharmaceutical compounds, personal care products, plasticizers, polyaromatic hydrocarbons, hormones, and illicit drugs (Toro-Vélez et al. 2016).

Pharmaceutical compounds are substances used by people and animals to prevent, cure, or alleviate diseases or correct the aftermath of them. Among those pharmaceuticals that are most consumed by humans are analgesics/anti-inflammatories, antibiotics, antihypertensives, barbiturates, beta-blockers, diuretics, lipid-lowering agents, psychiatric drugs, hormones, antineoplastic drugs, topical products, and contrast agents (Verlicchi et al. 2012). They are designed to resist inactivation before their desired effect. They can come into contact with the environment without having completed their metabolic process in the organism, due to being excreted through urine and feces. Therefore, some could have xenobiotic, persistent, bioaccumulative, and toxic characteristics (Fent et al. 2006).

The presence of pharmaceutical compounds in the environment is related to the discharge of domestic, hospital, and pharmaceutical industry wastewater. Many wastewater treatment systems have been evaluated, including advanced oxidation processes (AOPs), which have shown the most optimal results, despite being expensive at times and having somewhat challenging operational conditions in developing countries. Taking this into account, biological treatments are an appropriate alternative for wastewater treatment as complex and toxic chemical compounds are decomposed in simple and less toxic products through microorganisms or plants (Tiwari et al. 2017). Microalgae-based wastewater treatments are a promising alternative that have demonstrated good removal efficiency in contaminants such as nutrients, pathogens, and heavy metals (Muñoz and Guieysse 2006; Abargues et al. 2018).

Microalgae-based systems have many advantages for the treatment of hazardous contaminants and for the challenges associated with degradation, where different physicochemical and biologic processes take place. The objective of this chapter is to offer a description of the use of microalgae-based wastewater treatment systems to remove pharmaceutical compounds. The main characteristics, biotic and abiotic processes, removal efficiencies, and the main removal mechanisms will be described.

12.2 MICROALGAE-BASED WASTEWATER TREATMENTS

Microalgae-based wastewater systems can be used as a tertiary treatment process with the potential use of biomass such as fertilizers, biofuels, pigments, and animal feed. Microalgae-based treatments are usually used as tertiary treatments for removing inorganic nitrogen, phosphorus, and pathogens. Currently many studies have been developed to find the correlation between microalgae and the removal of micropollutants such as pharmaceuticals, hormones, and personal care products, among others.

Microalgae-based wastewater treatments are classified according to the design perspective and the configuration in open and closed systems. The selection of the bioreactor will depend on the safety conditions associated to the process, the land cost and biomass use (Muñoz and Guieysse 2006). The main advantages and disadvantages of open and closed bioreactors are presented in Table 12.1.

12.2.1 Open Bioreactors

Open bioreactors are most common due to their low cost and lower energy requirements. They are in contact with the atmosphere, meaning a large surface area is required to capture a greater amount of solar energy. In these bioreactors it is more difficult to control environmental conditions, whereas the natural introduction of numerous and diverse organisms such as microalgae, bacteria, predators, and other unwanted organisms is highly likely to be occur.

The natural introduction of different species of microalgae such as *Chlorella sp.*, *Desmodesmus sp.*, *Euglena sp.*, *Chilomonas sp.*, *Actinastrum sp.*, *Dictyosphaerium sp.*, *Coelastrum sp.*, *Scenedesmus sp.*, *Microactinium sp.*, and *Pediastrum sp.*, commonly occurs in open bioreactors (Craggs et al. 2011; García et al. 2000). These organisms form vast colonies that settle in diameters of 50–200 μm, which enables a profitable and simple harvest of biomass by gravity (Craggs et al. 2011; García et al. 2000; Park et al. 2011) if the biomass use is taken into account.

12.2.1.1 Open Ponds

Stabilization ponds are the simplest form of wastewater treatments. They are typically operated as plug-flow systems (Muñoz and Guieysse 2006) and can be rectangular or squared structures where physic-chemical and biological processes converge. According to their associated depth, they can be facultative, anaerobic, or maturation ponds. Facultative ponds combine aerobic, anaerobic, and anoxic conditions and have a depth of between 1.2 and 2.4 m. Anaerobic ponds are deeper than 8m and are usually used for wastewater treatment with high organic loads. Moreover, they are often used as a first treatment barrier in lagoon systems. Maturation ponds are also called polishing ponds because they are located as tertiary treatment methods for removing pathogens. They can be less than 1 m deep. In general, the aerobic

TABLE 12.1
Advantages and Disadvantages of Open and Closed Bioreactors

	Open	Closed
Advantages	• Low cost of operation. • Ease of scaling up. • Lower energy consumption. • Surface area and HRT may reduce with HRAP. • No accumulation of dissolved oxygen. • Deployment is extensive in small and medium sized communities and in industrial sectors. • High removal efficiencies demonstrated organic load, nutrients, pathogens, heavy metals and other micropollutants. • Large experience in industrial microalgal production in several applications. • Can be operated in continuous flow.	• Can be operated at controlled conditions. • High biomass production (20–40 g m^{-2} d^{-1}) that can be used in the production of other goods. • Low area requirements. • Flexible technical designs. • The introduction of unwanted organisms is less likely than in open systems. • High removal efficiencies demonstrated by organic load, nutrients, pathogens, heavy metals and other micropollutants. • Allow for the growth of monocultures. • Can distribute the sunlight over a larger surface area. • Evaporation can be avoided.
Disadvantages	• High volume-to-surface ratio, particularly with open ponds. • Low liquid velocity. • High CO_2 loss rate • Introduction of unwanted organisms. • Low biomass production (10–20 g $m^{-2}d^{-1}$) • HRAP requires access to electricity for paddle-wheel operation. • Recuperation of biomass is more difficult than with closed bioreactors. • Poor mixing, light and CO_2 use. • Temperature and evaporation cannot be controlled	• Overheating. • Cleaning issues due to bio-fouling and benthic algae growth. • High build-up of dissolved oxygen resulting in growth limitation. • High capital costs for designing and operating. • High energy requirements. • Difficult to scale up. • Removal of O_2 is necessary. • Dissolved oxygen may accumulate due to toxic levels.

Source: Cai et al. (2013); Chisti (2008); García-Galán et al. (2018); Li et al. (2019); Molina Grima et al. (1999); Pulz (2001); Shen et al. (2009); Shilton et al. (2008); and Shoener et al. (2014).

zone is located next to the pond's surface. It is in this area that the greatest oxygen transfer occurs due to the established contact with the atmosphere and the presence of algae, which contribute to increase the dissolved oxygen concentration through the process of photosynthesis (Mihelcic 2011).

12.2.1.2 High Rate Algal Pond

High rate algal ponds (HRAPs) were designed by Oswald (1988) in order to match algal growth and O_2 production with wastewater (Fig. 12.1a). They are designed 2–3 m

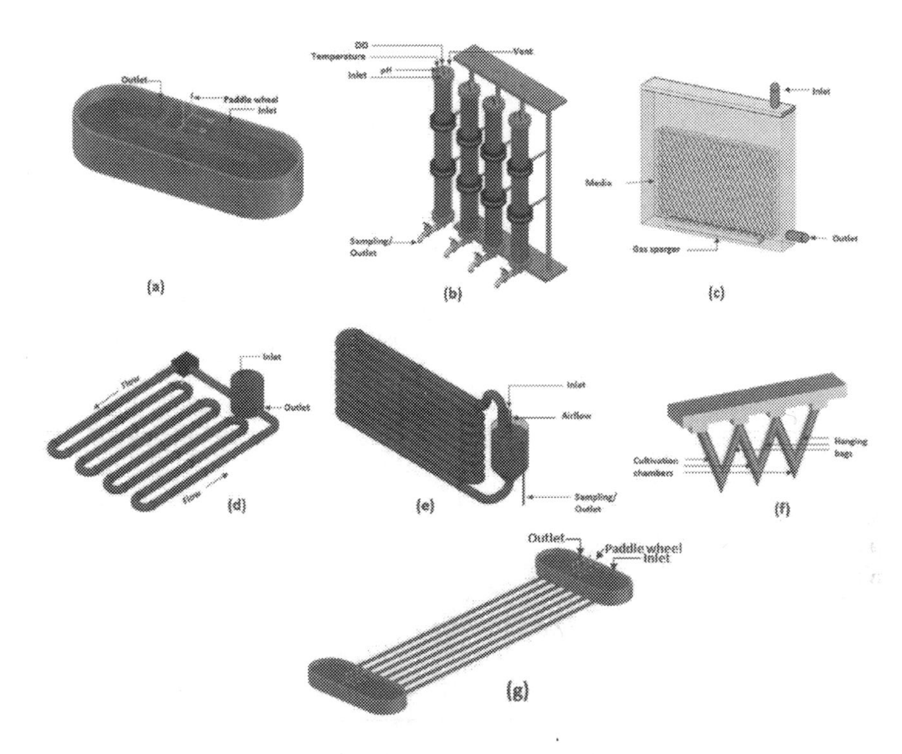

FIGURE 12.1 Examples of photobioreactors (PBRs). (a) HRAP (b) Column PBR (c) Flat-plate PBR (d) Horizontal tubular PBR (e) Vertical tubular PBR (f) Soft-frame PBR (g) Hybrid PBR. Adapted from Huang et al. (2017); Naira et al. (2020); Sierra et al. (2008); Solimeno et al. (2017); and Xu et al. (2021).

wide and 0.1–0.5 m deep (Muñoz and Guieysse 2006; Sutherland et al. 2017). The shallow reactor has a raceway pond reactor (RPR) configuration where the microalgae and bacteria are grown in symbiosis. Additionally, it has paddle wheels that are used for the mechanical agitation of mixed liquid to increase biomass production and improve the removal of contaminants (Solimeno and García 2017). Paddle wheels have velocities of between 0.15 and 0.3 $m.s^{-1}$ (Sutherland et al. 2015) and contribute to homogenous conditions and limit the formation of anaerobic zones (Grobbelaar 2000).

In this type of reactor, organic matter is also degraded by heterotrophic bacteria that consume oxygen that is generated by photosynthetic activity of microalgae (Passos et al. 2014). In addition, HRAPs have low superficial area requirements when compared to other biological treatments such as constructed wetlands and stabilization ponds and have low operational costs (Craggs et al. 2014). The design of HRAPs allows a continuous operation with a typical hydraulic retention time (HRT) between 4 and 10 days (Picot et al. 1992), which is lower than other open ponds that require 10 to 180 days (Mihelcic 2011).

Regarding open bioreactors, HRAPs are the most frequently used systems (Vassalle et al. 2020) due to their lower energy requirements and low operation and maintenance costs. However, cultures in this technology can be easily contaminated, and the control of the different growth and environmental parameters such as temperature and sunlight is worse than that of closed systems (Park and Craggs 2010).

12.2.2 Closed Bioreactors

This type of configuration can be operated at controlled conditions with less risk of evaporation, contamination, and the introduction of predators. Closed bioreactors are designed to ensure optimal light availability, gas exchange improvement, greater algal productivity, and a reduced space requirement as they can be built vertically (Muñoz and Guieysse 2006). These characteristics make closed bioreactors more efficient for wastewater treatment as they offer the advantage of algal biomass recovery and recycling. However, their installation and maintenance costs can be higher than those of open ponds (Cai et al. 2013; Li et al. 2019). Among closed photobioreactors (PBRs), the tubular, flat-plate and soft-frame systems are included.

12.2.2.1 Tubular Photobioreactors

Tubular PBRs are an arrangement of connected and closed transparent tubes within which the algae grow; they can be built in materials such as glass, PVC, or polycarbonate. They work by circulating the algae via a pump. The system also has a gas exchange unit where CO_2 can be added and photosynthetically produced O_2 is stripped from the medium. The configuration can be vertical to minimize the space requirement (Fig. 12.1e), in a horizontal configuration (Fig. 12.1d), inclined, or spirally wound around a central support (Muñoz and Guieysse 2006).

The light captured by the solar collector in the tubular PBRs is dependent on their configuration, including the tube diameter, which can be approximately 0.1 m to ensure light penetration, and the tube's spatial distribution. The tube diameter influences the light gradient inside, and therefore the light regime to which the cells are exposed in the bioreactor (Díaz et al. 2019).

The challenge with the spatial distribution of the tubes is in achieving optimal designs to maximize the amount of solar radiation intercepted and in distributing it over a larger surface to avoid light excess that can cause photoinhibition (Díaz et al. 2019).

They can be scaled-up by increasing the length and the number of tubes, as well as by the connection of several units via manifolds. These arrays are connected, and the mixing is done through pumps or airlift systems. Therefore, tubular PBRs have enough contact time for mass transfer (Muñoz and Guieysse 2006). Due to the larger reactor surface area per unit of occupied land, tubular PBRs show higher light-use efficiencies than flat-plate PBRs (Tredici and Zlttelli 1998).

Tubular PBRs involve the adjustment of operational parameters such as tube arrangement, dissolved oxygen, nutrient distribution, energy, and illumination (Vo et al. 2019). Slegers et al. (2013) examined the contribution of the reactor dimension, sunlight condition and microalgae characteristics in horizontal and vertical tubular PBRs. Vertical PBRs have a much higher productivity rate than horizontal ones.

12.2.2.2 Flat-Plate Reactors

Flat-plate PBRs consist of flat wide illuminating surface media to which microalgae can attach and develop (Fig. 12.1c). When compared to other PBRs, their high surface-to-volume ratio leads to the most optimal photosynthetic efficiencies that have been observed (Sun et al. 2016). They are often placed vertically, with the light reactor source incident on one side or tilted at an angle that allows optimal exposure to solar irradiation (Carvalho et al. 2006).

Moreover, these technologies are characterized by their stability and scalability (Sierra et al. 2008), by having minimal mechanical requirements, and by their low energy consumption (Guo et al. 2015; Sierra et al. 2008). Moreover, increased biomass yield is 5–20 times higher than with other photobioreactor types (Vo et al. 2019), and the reduction of problems related to biomass clogging, sedimentation, or leakage (Huang et al. 2017) is also evidenced.

Some of the commonly identified restrictions of flat-plate reactors are the scaling-up requirements including the need for many compartments and support materials, the possibility of algal cell clustering on the reactor's wall, the difficulty in controlling culture temperature, and the incompatibility with certain algal strains (Carvalho et al. 2006; Sierra et al. 2008). The height and width of flat-plate reactors are the two dimensions available for scaling-up with a practical limit of 2–3 m (Posten 2009).

12.2.2.3 Column Photobioreactors

Column PBRs consist in a vertical bubble column into which the algae culture is aerated by air or CO_2 to provide a constant mixture and to generate the suspension process (Fig. 12.1b). The configuration of these prototypes is based on connected modules that are combined with other techniques such as membrane processes (Chang et al. 2016). This type of PBR has sufficient gas–liquid mass transfer, biomass yield, and light/dark cycle control features (Pham et al. 2017; Sierra et al. 2008). The optimal dimensions of vertical column PBRs are about 0.2m diameter and 4 m column height (Mirón et al. 1999).

The scaling up of this PBR is difficult because the diameter increment impedes the light path along the radial direction, while the increment of column height decreases the axial mixing rate and calls for high construction costs (Janssen et al. 2003). For these reasons, it is considered a laboratory-scale device or inoculum production system (Huang et al. 2017).

12.2.2.4 Soft-Frame Photobioreactors

The soft-frame configuration, unlike most PBRs, which are made of arduous, permanently fixed materials, is recognized for its flexible, space-efficient, foldable, and mobile characteristics (Fig. 12.1f). These PBRs have different designs: they can have a floating in water configuration; they can be hanging from a shelf or lying on the ground. The manufacturing material options include ethylene vinyl acetate/low density polyethylene and polyethylene to polytetrafluoroethylene. Compared to other PBRs, the applications of soft-frame PBRs are modest, but they are becoming increasingly attractive, most notably in recent years (Vo et al. 2019).

The disadvantage of this PBR is that the material used for its construction has a substantial probability of being damaged due to the high pressure inside. In addition, there is a large number of dead zones, and the cost of agitation for industrial application could increase (Vo et al. 2019).

12.2.2.5 Hybrid Photobioreactors

Hybrid PBRs refer to the combination and/or integration of the above-mentioned PBR types with other technologies (Fig. 12.1.g). Throughout these incorporations, the aim is to exploit the prominent characteristics of traditional PBRs, thus achieving, for instance, a higher biomass yield, a volume reduction, and increasing the pollutants removal efficiency (de Jesus and Maciel Filho 2017; Vo et al. 2019).

12.3 BIOTIC- AND ABIOTIC-FACTOR–LINKED MICROALGAE-BASED WASTEWATER TREATMENTS

Microalgae are eukaryotic organisms (micrometer-sized) as plants are; they can use solar energy and transform carbon dioxide through photosynthetic activity. They prefer aquatic environments, but they can grow in terrestrial habitats. The omnipresence of microalgae in these habitats facilitate their proliferation (Xiong et al. 2018) Microalgae activity is influenced by biotic and abiotic factors such as pH levels, dissolved oxygen, light supply, temperature, reduction-oxidation processes, and interaction with other organisms.

12.3.1 pH

pH is one of the main factors in the behavior of microalgae. It has fluctuations during the day associated with the light/dark cycles that interfere in the photosynthesis and respiration process. During the day, photosynthesis and use of CO_2 raise pH levels, achieving maximum values when photosynthetic activity is greater. Dark respiration is an inverse photosynthesis process, where pH levels are reduced again. The pH levels increase due to the consumption of CO_2 and HCO_3, resulting in a basic medium with high pH values of up to 11 (Craggs et al. 2011; Park and Craggs 2010). High pH levels influence N and P removal via NH_3 volatilization, and orthophosphate precipitation at a high pH of between 9 and 11 (Craggs et al. 1996; García et al. 2000). However, the increase of pH above 10 units can generate a complete bacterial inhibition (Mara and Pearson 1986; Oswald 1988), and although it is beneficial for the inactivation of pathogen organisms, the biodegradability mechanism can be reduced due to the synergetic effect of microalgae and bacteria in the mineralization of pollutants.

The variations of pH in the media have effects on the dissociation and solubility of some components, which could have toxic or inhibitory effects on microalgae growth. According to Moheimani (2013), the optimum pH for biomass and lipid production of *Tetraselmis suecica* and *Chlorella sp.* is 7.5 and 7, respectively. Bartley et al. (2014) found that pH has an influence on *Nannochloropsis salina* growth and on lipid accumulation, with an optimum pH of between 8 and 9 (pH values that minimize the abundance of undesired organisms).

12.3.2 Dissolved Oxygen

Similarly, at high pH levels the dissolved oxygen concentration (DOC) is typically high due to the photosynthetic activity (Craggs et al. 2011; Park and Craggs 2010). The dynamic in microalgae-systems depends on the light/dark cycles with a change of aerobic to anaerobic conditions. Moreover, in open reactors, the DOC dynamics and variations can occur at different depths, with the aerobic conditions occurring often in the superficial zone and anaerobic conditions in the deeper zones (Dušek et al. 2008; Sun et al. 2019).

DOC in open ponds can achieve values of up to 36 mg/L (Jiménez-Bambague et al. 2020), and O_2 super-saturation in enclosed PBRs designed for mass algal cultivation of up to 400 percent, which severely inhibits microalgal growth due to photo-oxidative damage on microalgal cells (Oswald 1988; Suh and Lee 2003). However, O_2 super-saturation may be positive for biodegradation processes due to the continuous O_2 consumption by heterotrophic bacteria (Muñoz and Guieysse 2006).

According to the aforementioned, a high DOC is a good indicator of complete pollutant depletion in continuous processes (Muñoz et al. 2004) through mechanisms such as biodegradation and photo-oxidation. The oxygen generated by the microalgae and introduced by air–water transfer in open reactors can generate in situ radicals through the introduction of natural or artificial light. These radicals are often formed during photochemical reactions in which light-stimulating chemical species such as nitrate ion, hydrogen peroxide, or dissolved organic matter generate free radicals such as hydroxyl (•OH), carbonate (•CO_3), and peroxide (•OOH) (Abargues et al. 2018).

12.3.2.1 Hydroxyl Radical (•OH) Generation

The hydroxyl radical is generated with either irradiation of nitrate ion, hydrogen peroxide, or organic matter (Cuerda-Correa et al. 2020). Ion nitrate is stimulated by light (with a wavelength below 302 nm) and is transformed into radical •O^- (Reaction 1), which is rapidly protonated to the hydroxyl radical (Reaction 2) (Abargues et al. 2018). The cleavage of one hydrogen peroxide molecule produces two molecules of hydroxyl radical (Reaction 3) (Zuo 2003). The hydroxyl radical is also produced through irradiation of dissolved organic matter (DOM) in the presence of oxygen (Reaction 4). Hydrogen peroxide is produced after irradiation of DOM, which produces a hydroxyl radical in the presence of light (Abargues et al. 2018), according to Reaction 3.

- Nitrate ion irradiation

$$NO_3^- \xrightarrow{hv} NO_3^{-\bullet} \rightarrow NO_2 + \bullet O^- \text{ Reaction 1}$$

$$\cdot O^- + H_2O \rightarrow \bullet OH + OH^- \text{ Reaction 2}$$

- Hydrogen peroxide cleavage

$$H_2O_2 \xrightarrow{hv} 2 \bullet OH \text{ Reaction 3}$$

- Dissolved organic matter (DOM) irradiation

$$DOM \xrightarrow{hv} {}^1DOM^\bullet + {}^3DOM^\bullet \rightarrow \bullet DOM$$

$$\bullet DOM + O_2 \rightarrow \bullet DOM^+ + O_2^- \text{ Reaction 4}$$

$$2O_2^- + 2H^+ \rightarrow H_2O_2 + O_2$$

The carbonate radical ($\bullet CO_3^-$) results from the reaction between carbonate or bicarbonate and a hydroxyl radical that takes place without light (Abargues et al. 2018), according to Reaction 5.

12.3.2.2 Carbonate Radical ($\bullet CO_3^-$) Generation

$$\bullet OH + CO_3^{2-} \rightarrow OH^- + \bullet CO_3^-$$

$$\bullet OH + HCO_3^- \rightarrow OH^- + \bullet HCO_3 \text{ Reaction 5}$$

$$\bullet HCO_3 \leftrightharpoons \bullet CO_3^- + H^+$$

Peroxide radicals (•OOH) are formed from the reaction between the hydroxyl radical and DOM with unsaturated carbon-carbon bonds; reaction with molecular oxygen; and cleavage of the peroxide radical group (•OOH), (Pocostales et al. 2010).

12.3.2.3 Peroxide Radical (•OOH) Generation

The peroxide and carbonate radicals react selectively with sulphur-containing compounds, while the hydroxyl radical is a strong and non-selective oxidant that rapidly reacts with organic chemicals (Borowska et al. 2016).

12.3.2.4 Oxido-reduction Potential

This parameter may be found in nature with values between - 400 to + 800 mV (Dušek et al. 2008; Sgroi et al. 2018). The negative values correspond to a reducing environment and anaerobic conditions that are characterized by the absence of O_2, with the presence of oxidizing agents that attract electrons. Positive values are characterized by aerobic conditions and an oxidizing environment with the presence of reducing agents.

The dynamic of all micro-organisms in algal systems, such as open ponds, changes due to different depths and light/dark cycles of aerobic to anaerobic conditions, allowing for variation of the oxido-reduction potential (ORP) from negative values to positive values during the day. Moreover, as depth increases, it may vary from the highest values, measured at the surface, to lower values. The ORP gives an indication of the type of organisms present in the natural environment (Table 12.2). However, if microbial processes are intense, the ORP can temporarily decrease to values that are lower than those predicted (Dušek et al. 2008; Mahapatra et al. 2013).

According to Mahapatra et al. (2013), the vertical movement of algae varies with the time of the day and wind movement. These variations could be due to algal, bacterial, and physicochemical stratification in the water column of a facultative pond (i.e., algae dynamics at different depths). Moreover, following the progressive increase of dissolved oxygen from morning to midday, and then the drop to a

TABLE 12.2
ORP According to Micro-Organism Activity

ORP (mV)	Activity
< 400	Presence of denitrifying bacteria.
< 100	Reduction of Fe^{+3} ions.
< -100	Reduction of sulfates and organic substances (fermentation).
< -200	Activity of methanogenic bacteria

Source: Taken from Dušek et al. (2008).

minimum at night, the ORP in open PBRs such as facultative lagoons, presents oxidizing conditions at the surface during the day and negative values at night.

12.3.2.5 Temperature

According to Park et al. (2011), algal productivity rises with increasing temperature until reaching the optimum temperature, at which there is a rise of algae respiration and photorespiration. Once optimum temperature is surpassed, productivity then decreases. Optimum value of growth depends on the culture medium and the strain, but an estimated general track of between 20 and 27°C is estimated (Cerón Hernández et al. 2015). Algal growth is difficult to obtain at temperature values above 34°C, which may be deadly to the algae (Soeder et al. 1985), and the efficiency of microalgae-based treatments normally decrease at low temperatures (Muñoz et al. 2004).

Gonzalez-Camejo et al. (2019) evaluated the effect of temperature on biomass productivity of *chlorella vulgaris* in a PBR in semi-continuous operations under the same nutrient loading rate, air sparging flow rate, and 6-day hydraulic retention time (HRT). The results showed that the optimum range for *C. vulgaris* is 15–30°C, while at temperatures of 35°C the cell viability dropped due to heat stress. In this study, it was observed that microalgae growth drops much more abruptly at high temperatures than at low temperatures because most microalgae strains can tolerate temperatures of around 15°C below the optimum level, but when the optimum temperature is exceeded by only 2–4°C, this can be detrimental to algae growth (Subhash et al. 2014).

Microalgae may be affected more by high temperatures in closed PBRs than by high temperatures in open PBRs, due to the fact that there is no evaporation loss that can regulate temperature in closed PBRs (Gonzalez-Camejo et al. 2019). The temperature inside a closed PBR can be about 10–30 °C higher than ambient temperature (Wang et al. 2012).

Temperatures lower than optimal values can limit the microalgae growth rate by affecting the kinetics of cell enzymatic processes. However, when the strains have been previously cold-adapted, wastewater treatment is possible, although biological activity can be lower (Gonzalez-Camejo et al. 2019; Muñoz and Guieysse 2006). Ferro et al. (2018) found that with adapted microalgae strains of *C. vulgaris, Scenedesmus* sp., and *Desmodesmus* sp., the biomass production at 5°C was similar to standard conditions (25°C), albeit with reduced growth rates.

Vassalle (2020) reported that high temperature within tubular PBRs may have affected the microalgae and led to faster biodegradation processes and removal routes. According to studies done in wastewater treatment plants, seasonal variations in temperature do influence the removal efficiency for different contaminants (Matamoros et al. 2015).

12.3.2.6 Light Supply

Light is very important in the photosynthesis process and the production of new biomass. It is a limiting factor of microalgae development and effective behavior of microalgae-based systems. In photosynthesis, artificial light or sunlight is converted in chemical energy through pigments such as chlorophylls (Chlorophyll-a being the main pigment) and carotenoids that are absorbed in specific wavelengths of the visible spectrum. The absorption range, known as photosynthetically active radiation (PAR), is the amount of useful energy for the photosynthesis process. PAR is located within the 400 to 700 nm solar spectrum wavelength (Righini and Grossi Gallegos 2005).

Algal activity increases with light intensity of up to 200 μmol.m^{-2}s^{-1}, which can become saturated (Masojídek et al. 2003). Higher light intensities can cause photoinhibition, thereby affecting the photosynthetic activity of the microalgae, particularly in the central hours of a sunny day, when the irradiance can reach up to 4000 mE. m^{-2}s^{-1} (Muñoz and Guieysse 2006), equivalent to 4000 μmol.m^{-2}s^{-1}.

Absence of light causes a halt in photosynthesis, which leads to the occurrence of anaerobic conditions in the PBR. However, photosynthesis normally resumes once light is again available (Muñoz and Guieysse 2006) which maintains the aerobic–anaerobic cycles that support the complex dynamic of all microorganisms in the PBR.

Biomass concentration determines light-use efficiency and therefore controls the oxygenation and pollutant removal in the PBR. When light intensity is low, biomass concentration should be optimized to avoid mutual shading and dark respiration, while at a high light intensity the biomass helps to protect the microalgae from photoinhibition (Muñoz and Guieysse 2006).

12.3.2.7 Unwanted Organisms

In microalgae-based systems, the presence of unwanted organisms is probable, and this can be a risk to effective microalgae development. Included in this group of organisms are unwanted algae, yeast, molds, fungi, bacteria (Bartley et al. 2014), and predators such as herbivorous zooplankton and insect larvae, particularly in open bioreactors, where the environmental conditions cannot be easily controlled.

Van Den Hende et al. (2014) have also reported the presence of predators such as *Chironomidae* and *Tubifex* larvae, which caused a significant loss of diversity and biomass in a microalgae/bacteria consortium. However, according to Lietz (1987), the presence of *Tubifex* sp. in the biomass can be beneficial for its valorization, because appetites and the palatability of fish are increased. The *Trichoptera* genus has also marked its importance in the food chains due to the variety of niches that the larvae occupy in different predatory groups. (Guzmán-Soto and Tamarís-Turizo 2014).

Herbivorous zooplankton that could be in open PBRs include ciliates, rotifers, cladocerans, copepods and ostracod (Montemezzani et al. 2015). Among these, rotifers have the highest growth and reproduction rates, and can therefore consume the largest amount of biomass (Pagano 2008). The presence of these organisms could cause the death of all algal biomass, affecting the microalgae-based system performance and possibly requiring a new start-up. Therefore, different options for zooplankton control have been developed. Among these options are physical methods, such as filtration, hydrodynamic cavitation, shear, and bead mills; chemical methods, such as increasing night-time CO_2 concentration in a HRAP, the promotion of the lethal un-ionized ammonia toxicity, the use of biocides, and the chitinase inhibitor chitosan; and biocontrol methods using competitor and predatory organisms (Montemezzani et al. 2015).

Although the physical control methods are generally effective, the installation and operational costs are often high, and they can also remove or disrupt the microalgae–bacterial flocs, as well as damage the algae fragile cells decreasing the harvest efficiency (Montemezzani et al. 2015).

Among the chemical methods, the daily addition of O_2 and/or CO_2 for a short period of time (1 h) might help suppress the growth of higher aerobic organisms and control the Zooplankton populations, because they are able to tolerate very low dissolved oxygen conditions (Schlüter and Groeneweg 1981).

12.3.2.8 Microalgae-Bacteria Consortium

The microalgae-bacteria consortium is a determining factor when considering open reactors. In this consortium, microalgae produce O_2 to heterotrophic aerobic bacteria while bacteria release CO_2 that is captured by microalgae during photosynthesis, allowing the mineralization of biodegradable compounds (Muñoz and Guieysse 2006). Additionally, the capture of CO_2 by algae contributes to the reduction of greenhouse gas emission (Subashchandrabose et al. 2011). Therefore, algae-based technologies are more sustainable compared with conventional wastewater treatment technologies.

Moreover, this relationship of symbiosis goes further: microalgae produce sheaths (of a tri-laminar structure) that are composed of carbohydrate, protein, and metal cations that are related to the formation of algal cell aggregation, with which bacteria are associated (Croft et al. 2006). This association occurs when bacterial symbionts bind indirectly to the sheath and directly to the cell wall of the algae, a process during which substrates are exchanged (Park et al. 2008). Beyond the photosynthetic process, algae require (i) vitamins such as biotin, thiamine, and cobalamine as growth factors due to the fact that cobalamine auxotrophs are widespread (Croft et al. 2006) and (ii) bacterial siderophores for growth under iron-deficient conditions (Butler 1998).

Similarly, bacterial growth can enhance microalgal metabolism by releasing growth promoting factors such as nutrient regeneration and trace elements and releasing phytohormones (Gonzalez and Bashan 2000) that can reduce the presence of organic recalcitrant and toxic compounds that can affect microalgae. In fact, bacteria are more resistant than microalgae to changes in environmental conditions and are better equipped in the presence of hazardous compounds, and their growth rate is faster than microalgae due to their reduced size (Lee 2001; Semple et al. 1999). Therefore, this interrelation provides stability to the parts, resistance, and resilience to environmental fluctuations.

Among the various advantages of using alga-bacteria consortium-based technologies in wastewater treatment is that valuable products such as biofuel, vitamins,

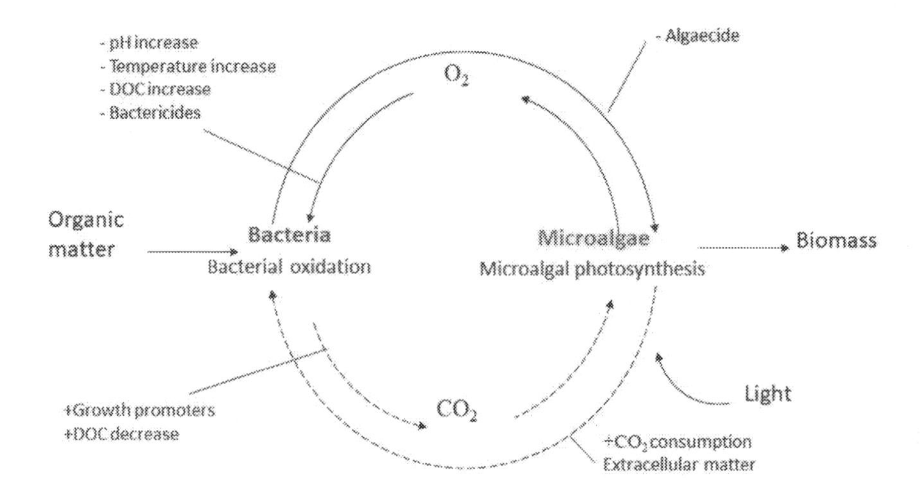

FIGURE 12.2 Symbiosis microalgae-bacteria process, modified from Muñoz and Guieysse (2006).

carbohydrates, pigments, and proteins are derived from the process. Furthermore, in comparison with conventional treatment systems, algae-bacteria technologies provide energy savings in O_2 supply, CO_2 mitigation, efficient recycling of nutrients, and revalorization of algal biomass (Subashchandrabose et al. 2011).

However, the microalgae–bacteria interaction might not only be cooperative but could also be competitive (Fig. 12.2). Algal growth can also inhibit bacterial activity by releasing toxic metabolites, increasing the temperature and keeping high O_2 levels (Skulberg 2000). A negative interaction can relate to the competing of limiting nutrients and spaces (Liu et al. 2016; Scott et al. 2008). Microalgae can affect bacteria with the increase of pH, the DOC, and temperature or by excreting inhibitory metabolites (Oswald 2003; Schumacher et al. 2003). Moreover, there might be an inhibition of microalgae when algicidal extracellular metabolites are produced or when they change the microenvironment (Fukami et al. 1997; Subashchandrabose et al. 2011; Wang et al. 2015). Likewise, Gonçalves et al. (2017) reported that some microalgal metabolites present a bactericidal effect, as in the case of chlorellin where a bactericidal effect on Gram positive and Gram negative bacteria is exerted.

The type of microorganisms present in a consortium is also determined by the composition of the effluent and the stationary environmental conditions (Fukami et al. 1997, Mara and Pearson 1986;). According to Sancho et al. (1993), Euglena and Chlamydomonas dominate at high organic loads in sewage treatment, while Chlorella and Scenedesmus are the most abundant species at medium loads. Regarding the irradiance, an increase below saturation level can generally enhance the photosynthetic activity and growth of microalgae. Similarly, microalgae and bacteria growth escalate with increasing temperatures that reach their optimum growth temperature point to rapidly decrease beyond this value. The optimal growth temperature depends on the species and typically ranges from 10 to 35°C (Breuer et al. 2013; Markou and Georgakakis 2011). For example, the Euglena and Scenedesmus species predominate over Chlorella below 15°C due to their higher tolerance to low temperatures (Mara and Pearson 1986).

According to Yao et al. (2013), bacteria are sensitive to dissolved oxygen levels, especially those that have strict requirements for aerobic or anaerobic conditions, which are also relevant to denitrification, ammonia oxidation, and nitrification. A study conducted by Abargues (2018), reported that high dissolved oxygen concentrations produced by algae enhanced pollutant degradation rates. According to the conducted experiments, there was lower degradation efficiency under non-aerated conditions.

12.4 REMOVAL OF PHARMACEUTICAL COMPOUNDS

Table 12.3 shows the removal efficiencies of pharmaceutical compounds by different microalgae-based wastewater treatment systems. The use of an open system for wastewater treatment is most common, while the closed systems are more common

TABLE 12.3
Removal Efficiencies of Pharmaceutical Compounds in Microalgae-Based Systems

Pharmaceutical Compounds	Photobioreactor Configuration	Removal Efficiencies	Microalgae
	Analgesics/Anti-inflamatories		
1-OH-ibuprofen	HRAP without sedimentation (HRT: 4.5 d)[g]	78	Microalgae biomass
	HRAP with sedimentation (HRT: 4.5 d)[g]	95	
2-OH-ibuprofen	HRAP without sedimentation (HRT: 4.5 d)[g]	85	
	HRAP with sedimentation (HRT: 4.5 d)[g]	80	
4-OH-Diclofenac	HRAP without sedimentation (HRT: 4.5 d)[g]	100	
	HRAP with sedimentation (HRT: 4.5 d)[g]	100	
5-OH-Diclofenac	HRAP without sedimentation (HRT: 4.5 d)[g]	100	
	HRAP with sedimentation (HRT: 4.5 d)[g]	100	
Acetaminophen/ Paracetamol	HRAP Warm season (HRT: 4 d)[c]	99	*Stigeoclonium sp., Diatoms, Chlorella sp., Monoraphidium*
	HRAP Warm season (HRT: 8 d)[c]	99	
	HRAP Cold season (HRT: 4 d)[c]	99	
	HRAP Cold season (HRT: 8 d)[c]	99	
	Bioreactor batch (0-31 d)[e]	100	*C. sorokiniana*
	PBR tubular (HRT: 8 d)[f]	99.2	Phormidium and cyanobacteria
	PBR tubular (HRT: 12 d)[f]	<MQL	
	HRAP American tropical conditions (HRT: 3 d)[h]	76.2-100.0	Phylum *Chlorophyta*
	HRAP After UASB (HRT: 3-9 d)[a]	99.0	*Coelastrum sp*
	HRAP Direct (HRT: 3-9 d)[a]	95.3	
	HRAP without sedimentation (HRT: 4.5 d)[g]	100	Microalgae biomass
	HRAP with sedimentation (HRT: 4.5 d)[g]	100	
	HRAP without sedimentation (HRT: 4.5 d)[g]	<0	
	HRAP with sedimentation (HRT: 4.5 d)[g]	<0	

(*Continued*)

TABLE 12.3 (Continued)
Removal Efficiencies of Pharmaceutical Compounds in Microalgae-Based Systems

Pharmaceutical Compounds	Photobioreactor Configuration	Removal Efficiencies	Microalgae
Acetylsalicylic acid	HRAP without sedimentation (HRT: 4.5 d)[g]	<0 - 52	Microalgae biomass
	HRAP with sedimentation (HRT: 4.5 d)[g]	<0 - 53	
Carboxy-ibuprofen	HRAP without sedimentation (HRT: 4.5 d)[g]	100	Microalgae biomass
	HRAP with sedimentation (HRT: 4.5 d)[g]	100	
Codeine	Open pond (HRT: 7d)[d]	-11	Green algal species
Diclofenac	HRAP After UASB (HRT: 3-9 d)[a]	58.0	*Coelastrum sp*
	HRAP Direct (HRT: 3-9 d)[a]	56.9	
	HRAP Warm season (HRT: 4 d)[c]	82	*Stigeoclonium sp., Diatoms, Chlorella sp., Monoraphidium*
	HRAP Warm season (HRT: 8 d)[c]	92	
	HRAP Cold season (HRT: 4 d)[c]	21	
	HRAP Cold season (HRT: 8 d)[c]	29	
	bioreactor batch (0-31 d)[e]	40-60	*C. sorokiniana*
	HRAP American tropical conditions (HRT: 3 d)[h]	42.3-89.4	Phylum *Chlorophyta*
	HRAP without sedimentation (HRT: 4.5 d)[g]	51.3	Microalgae biomass
	HRAP with sedimentation (HRT: 4.5 d)[g]	54.8	
Fenoprofen	HRAP Direct (HRT: 3-9 d)[a]	47.7	*Coelastrum sp*
Ibuprofen	HRAP After UASB (HRT: 3-9 d)[a]	98.5	*Coelastrum sp*
	HRAP Direct (HRT: 3-9 d)[a]	83.3	
	HRAP Warm season (HRT: 4 d)[c]	99	*Stigeoclonium sp., Diatoms, Chlorella sp., Monoraphidium*
	HRAP Warm season (HRT: 8 d)[c]	99	
	HRAP Cold season (HRT: 4 d)[c]	86	
	HRAP Cold season (HRT: 8 d)[c]	93	
	Bioreactor batch (0-31 d)[e]	100	*C. sorokiniana*
	PBR tubular (HRT: 8 d)[f]	98.5	Phormidium and cyanobacteria
	PBR tubular (HRT: 12 d)[f]	<MQL	
	HRAP American tropical conditions (HRT: 3 d)[h]	17.6-98.6	Phylum *Chlorophyta*
	HRAP without sedimentation (HRT: 4.5 d)[g]	79	Microalgae biomass
	HRAP with sedimentation (HRT: 4.5 d)[g]	78	
Indomethacine	HRAP After UASB (HRT: 3-9 d)[a]	30.9	*Coelastrum sp*
	HRAP Direct (HRT: 3-9 d)[a]	79.9	

(*Continued*)

TABLE 12.3 (Continued)
Removal Efficiencies of Pharmaceutical Compounds in Microalgae-Based Systems

Pharmaceutical Compounds	Photobioreactor Configuration	Removal Efficiencies	Microalgae
Ketoprofen	HRAP After UASB (HRT: 3-9 d)[a]	90.9	*Coelastrum sp*
	HRAP Direct (HRT: 3-9 d)[a]	60.7	
	HRAP Warm season (HRT: 4 d)[c]	87	*Stigeoclonium sp.,*
	HRAP Warm season (HRT: 8 d)[c]	95	*Diatoms,*
	HRAP Cold season (HRT: 4 d)[c]	50	*Chlorella sp.,*
	HRAP Cold season (HRT: 8 d)[c]	75	*Monoraphidium*
	PBR tubular (HRT: 8 d)[f]	36.2	Phormidium and
	PBR tubular (HRT: 12 d)[f]	84.6	cyanobacteria
	HRAP American tropical conditions (HRT: 3 d)[h]	54.5-58.3	Phylum *Chlorophyta*
	Hybrid: semi-closed horizontal tubular PBR (HRT: 5 d)[i]	94	*Cyanobacteria Synechocystis sp* (most abundant)
Menfenamic Acid	HRAP After UASB (HRT: 3-9 d)	37.5	*Coelastrum sp*
	HRAP Direct (HRT: 3-9 d)[a]	-73.7	
Naproxen	HRAP After UASB (HRT: 3-9 d)[a]	71.3	*Coelastrum sp*
	HRAP Direct (HRT: 3-9 d)[a]	58.0	
	HRAP Warm season (HRT: 4 d)[c]	83	*Stigeoclonium sp.,*
	HRAP Warm season (HRT: 8 d)[c]	89	*Diatoms,*
	HRAP Cold season (HRT: 4 d)[c]	48	*Chlorella sp.,*
	HRAP Cold season (HRT: 8 d)[c]	60	*Monoraphidium*
	PBR tubular (HRT: 8 d)[f]	10.2	Phormidium and
	PBR tubular (HRT: 12 d)[f]	69.2	cyanobacteria
	HRAP American tropical conditions (HRT: 3 d)[h]	15.6-83.1	Phylum *Chlorophyta*
Phenazone	HRAP After UASB (HRT: 3-9 d)[a]	-389.2	*Coelastrum sp*
	HRAP Direct (HRT: 3-9 d)[a]	-1466.7	
Pizotifen	Open pond (HRT: 7d)[d]	80	Green algal species
Salicylic Acid	HRAPAfter UASB (HRT: 3-9 d)[a]	99.7	*Coelastrum sp*
	HRAP Direct (HRT: 3-9 d)[a]	99.7	
	PBR tubular (HRT: 8 d)[f]	<MQL	Phormidium and
	PBR tubular (HRT: 12 d)[f]	33.4	cyanobacteria
Tramadol	Open pond (HRT: 7d)[d]	57	Green algal species
Caffeine	HRAP Warm season (HRT: 4 d)[c]	97	*Stigeoclonium sp.,*
	HRAP Warm season (HRT: 8 d)[c]	98	*Diatoms,*
	HRAP Cold season (HRT: 4 d)[c]	85	*Chlorella sp.,*
	HRAP Cold season (HRT: 8 d)[c]	91	*Monoraphidium*
	Hybrid: semi-closed horizontal tubular PBR (HRT: 5 d)[i]	100	*Cyanobacteria, Synechocystis sp* (most abundant)

(*Continued*)

TABLE 12.3 (Continued)

Removal Efficiencies of Pharmaceutical Compounds in Microalgae-Based Systems

Pharmaceutical Compounds	Photobioreactor Configuration	Removal Efficiencies	Microalgae
	Antibiotics		
Azithromycin	PBR tubular (HRT: 8 d)[f]	88.8	Phormidium and cyanobacteria
	HRAP After UASB (HRT: 3-9 d)[a]	-35.5	*Coelastrum sp*
	HRAP Direct (HRT: 3-9 d)[a]	84.2	
Chloramphenicol	HRAP Direct (HRT: 3-9 d)[a]	-150.0	*Coelastrum sp*
Ciprofloxacin	Open pond (HRT: 7d)[d]	11	Green algal species
	PBR tubular (HRT: 8 d)[f]	47.6	Phormidium and
	PBR tubular (HRT: 12 d)[f]	<MDL	cyanobacteria
	HRAP After UASB (HRT: 3-9 d)[a]	26.9	*Coelastrum sp*
	HRAP Direct (HRT: 3-9 d)[a]	-208.4	
Clarithromycin	HRAP Direct (HRT: 3-9 d)[a]	93.4	*Coelastrum sp*
	Open pond (HRT: 7d)[d]	90	Green algal species
Clindamycin	Open pond (HRT: 7d)[d]	45	Green algal species
	HRAP After UASB (HRT: 3-9 d)[a]	63.6	*Coelastrum sp*
	HRAP Direct (HRT: 3-9 d)[a]	-49.2	
Erythromycin	PBR tubular (HRT: 12 d)[f]	84.9	Phormidium and cyanobacteria
	HRAP After UASB (HRT: 3-9 d)[a]	88.0	*Coelastrum sp*
	HRAP Direct (HRT: 3-9 d)[a]	72.2	
Flumequine	HRAP Direct (HRT: 3-9 d)[a]	-150.0	*Coelastrum sp*
Lincomycin	HRAP After UASB (HRT: 3-9 d)[a]	50.0	*Coelastrum sp*
	HRAP Direct (HRT: 3-9 d)[a]	-64.9	
Metronidazole	HRAP without sedimentation (HRT: 4.5 d)[g]	> 91	Microalgae
	HRAP with sedimentation (HRT: 4.5 d)[g]	> 89	biomass
	HRAP Direct (HRT: 3-9 d)[a]	-65.2	*Coelastrum sp*
N4 –acetyl sulfamethoxazole	HRAP without sedimentation (HRT: 4.5 d)[g]	89.2	Microalgae
	HRAP with sedimentation (HRT: 4.5 d)[g]	82.4	biomass
Norfloxacin	HRAP After UASB (HRT: 3-9 d)[a]	74.0	*Coelastrum sp*
	HRAP Direct (HRT: 3-9 d)[a]	44.2	
Ofloxacin	HRAPAfter UASB (HRT: 3-9 d)[a]	64.8	*Coelastrum sp*
	HRAP Direct (HRT: 3-9 d)[a]	52.4	
	PBR tubular (HRT: 8 d)[f]	67.7	Phormidium and
	PBR tubular (HRT: 12 d)[f]	66.8	cyanobacteria
Ornidazole	HRAP Direct (HRT: 3-9 d)[a]	-100.0	*Coelastrum sp*
Roxithromycin	HRAP After UASB (HRT: 3-9 d)[a]	0.0	*Coelastrum sp*
	HRAP Direct (HRT: 3-9 d)[a]	84.2	
	Open pond (HRT: 7d)[d]	44	Green algal species
Spiramycin	HRAP Direct (HRT: 3-9 d)[a]	-39.2	*Coelastrum sp*

(Continued)

TABLE 12.3 (Continued)
Removal Efficiencies of Pharmaceutical Compounds in Microalgae-Based Systems

Pharmaceutical Compounds	Photobioreactor Configuration	Removal Efficiencies	Microalgae
Sulfadiazine	HRAP After UASB (HRT: 3-9 d)[a]	82.2	*Coelastrum sp*
	HRAP Direct (HRT: 3-9 d)[a]	53.7	
Sulfamethizole	HRAP Direct (HRT: 3-9 d)[a]	-179.5	*Coelastrum sp*
Sulfamethoxazole	HRAP After UASB (HRT: 3-9 d)[a]	52.0	*Coelastrum sp*
	HRAP Direct (HRT: 3-9 d)[a]	13.3	
	Open pond (HRT: 7d)[d]	6	Green algal species
	HRAP without sedimentation (HRT: 4.5 d)[g]	85.3	Microalgae biomass
	HRAP with sedimentation (HRT: 4.5 d)[g]	50.5	
	Hybrid: semi-closed horizontal tubular PBR (HRT: 5 d)[i]	46	Cyanobacteria *Synechocystis sp* (most abundant)
Sulfapyridine	Hybrid: semi-closed horizontal tubular PBR (HRT: 5 d)[i]	-32	Cyanobacteria *Synechocystis sp* (most abundant)
Sulfathiazole	HRAP After UASB (HRT: 3-9 d)[a]	71.4	Coelastrum sp
	HRAP Direct (HRT: 3-9 d)[a]	-27.5	
	Hybrid: semi-closed horizontal tubular PBR (HRT: 5 d)[i]	100	Cyanobacteria *Synechocystis sp* (most abundant)
Tetracycline	HRAP Direct (HRT: 3-9 d)[a]	1.0	Coelastrum sp
	HRAP (HRT: 7d)[b]	69	*C. vulgaris*
Tiamulin	HRAP Direct (HRT: 3-9 d)[a]	-70.0	*Coelastrum sp*
Trimethoprim	HRAP Direct (HRT: 3-9 d)[a]	24.2	*Coelastrum sp*
	Open pond (HRT: 7d)[d]	3.7	Green algal species
	Bioreactor batch (0-31 d)[e]	<40	*C. sorokiniana*
	Hybrid: semi-closed horizontal tubular PBR (HRT: 5 d)[i]	78	Cyanobacteria *Synechocystis sp* (most abundant)
	Antihistamine		
Azelastine	Open pond (HRT: 7d)[d]	27	Green algal species
Clemastine	Open pond (HRT: 7d)[d]	40	
Cyproheptadine	Open pond (HRT: 7d)[d]	-450	
Desloratidin	Open pond (HRT: 7d)[d]	-45	
Diphenhydramin	Open pond (HRT: 7d)[d]	89	
Fexofenadine	Open pond (HRT: 7d)[d]	-5.2	
Hydroxyzine	Open pond (HRT: 7d)[d]	87	
Orphenadrin	Open pond (HRT: 7d)[d]	-3.8	
Cilazapril	Open pond (HRT: 7d)[d]	61	
Eprosartan	Open pond (HRT: 7d)[d]	80	
Ibersartan	Open pond (HRT: 7d)[d]	6.4	
Verapamil	Open pond (HRT: 7d)[d]	-13	

(*Continued*)

TABLE 12.3 (Continued)

Removal Efficiencies of Pharmaceutical Compounds in Microalgae-Based Systems

Pharmaceutical Compounds	Photobioreactor Configuration	Removal Efficiencies	Microalgae
	Lipid Regulators		
Bezafibrate	HRAP After UASB (HRT: 3-9 d)[a]	91.3	Coelastrum sp
	HRAP Direct (HRT: 3-9 d)[a]	85.8	
Clofibric Acid	HRAP After UASB (HRT: 3-9 d)[a]	97.5	Coelastrum sp
	HRAP Direct (HRT: 3-9 d)[a]	46.0	
Fenofibrate	HRAP American tropical conditions (HRT: 3 d)[h]	54.5	Phylum *Chlorophyta*
Fenofibric acid	HRAP American tropical conditions (HRT: 3 d)[h]	68.8-84.4	Phylum *Chlorophyta*
Gemfibrozil	HRAP After UASB (HRT: 3-9 d)[a]	53.1	Coelastrum sp
	HRAP Direct (HRT: 3-9 d)[a]	1.1	
	HRAP American tropical conditions (HRT: 3 d)[h]	0-84.3	Phylum *Chlorophyta*
	Hybrid: semi-closed horizontal tubular PBR (HRT: 5 d)[i]	75	Cyanobacteria Synechocystis sp (most abundant)
	Others		
Albuterol	HRAP After UASB (HRT: 3-9 d)[a]	42.0	Coelastrum sp
	HRAP Direct (HRT: 3-9 d)[a]	6.0	
Alfuzosin	Open pond (HRT: 7d)[d]	64	Green algal species
Atracurium	Open pond (HRT: 7d)[d]	97	
Biperiden	Open pond (HRT: 7d)[d]	-490	
Clotrimazol	Open pond (HRT: 7d)[d]	19	
Dicycloverin	Open pond (HRT: 7d)[d]	71	
Diltiazem	Open pond (HRT: 7d)[d]	94	
	PBR tubular (HRT: 8 d)[f]	77.3	Phormidium and cyanobacteria
	PBR tubular (HRT: 12 d)[f]	72.6	
Famotidine	HRAP After UASB (HRT: 3-9 d)[a]	65.0	*Coelastrum sp*
	HRAP Direct (HRT: 3-9 d)[a]	39.0	
Flecainide	Open pond (HRT: 7d)[d]	58	Green algal species
Fluconazole	Open pond (HRT: 7d)[d]	-17	Green algal species
Furosemide	HRAP After UASB (HRT: 3-9 d)[a]	82.1	*Coelastrum sp*
	HRAP Direct (HRT: 3-9 d)[a]	67.9	
	PBR tubular (HRT: 12 d)[f]	<MDL	Phormidium and cyanobacteria
Glyburide	HRAP After UASB (HRT: 3-9 d)[a]	90.0	*Coelastrum sp*
	HRAP Direct (HRT: 3-9 d)[a]	-305.0	

(*Continued*)

TABLE 12.3 (Continued)
Removal Efficiencies of Pharmaceutical Compounds in Microalgae-Based Systems

Pharmaceutical Compounds	Photobioreactor Configuration	Removal Efficiencies	Microalgae
Hydrochlorothiazide	HRAP After UASB (HRT: 3-9 d)[a]	24.5	*Coelastrum sp*
	HRAP Direct (HRT: 3-9 d)[a]	66.4	
	PBR tubular (HRT: 8 d)[f]	44.3	Phormidium and cyanobacteria
	PBR tubular (HRT: 12 d)[f]	84.4	
Loperamide	Open pond (HRT: 7d)[d]	41	Green algal species
Memantin	Open pond (HRT: 7d)[d]	81	
Miconazole	Open pond (HRT: 7d)[d]	65	
Pentoxifylline	HRAP American tropical conditions (HRT: 3 d)[h]	30.0-54.5	Phylum *Chlorophyta*
Ranitidine	HRAP After UASB (HRT: 3-9 d)[a]	22.7	*Coelastrum sp*
	HRAP Direct (HRT: 3-9 d)[a]	38.1	
	Open pond (HRT: 7d)[d]	75	Green algal species
Trihexyphenidyl	Open pond (HRT: 7d)[d]	49	Green algal species
	Psychiatric Drugs		
2-Hydroxycarbamazepine	HRAP without sedimentation (HRT: 4.5 d)[g]	38-54	Microalgae biomass
	HRAP with sedimentation (HRT: 4.5 d)[g]	>50	
Alprazolam	Open pond (HRT: 7d)[d]	-49	Green algal species
	PBR tubular (HRT: 12 d)[f]	87.4	Phormidium and cyanobacteria
Amitriptiline	HRAP After UASB (HRT: 3-9 d)[a]	96.2	*Coelastrum sp*
Bupropion	Open pond (HRT: 7d)[d]	93	Green algal species
Carbamazepine	HRAP Direct (HRT: 3-9 d)[a]	18.6	*Coelastrum sp*
	HRAP Direct (HRT: 3-9 d)[a]	-6.3	
	HRAP Warm season (HRT: 4 d)[c]	46	*Stigeoclonium sp., Diatoms, Chlorella sp., Monoraphidium*
	HRAP Warm season (HRT: 8 d)[c]	62	
	HRAP Cold season (HRT: 4 d)[c]	15	
	HRAP Cold season (HRT: 8 d)[c]	34	
	Open pond (HRT: 7d)[d]	-14	Green algal species
	Bioreactor batch (0-31 d)[e]	<30	*C. sorokiniana*
	HRAP without sedimentation (HRT: 4.5 d)[g]	0	Microalgae biomass
	HRAP with sedimentation (HRT: 4.5 d)[g]	0	
	HRAP American tropical conditions (HRT: 3 d)[h]	0-48.6	Phylum *Chlorophyta*
10,11-Dihidro-10,11-dihidroxi carbamazepine	HRAP American tropical conditions (HRT: 3 d)[h]	0-10	Phylum *Chlorophyta*
Citalopram	Open pond (HRT: 7d)[d]	98	Green algal species
Clonazepam	Open pond (HRT: 7d)[d]	88	Green algal species
Desmethyldiazepam	HRAP without sedimentation (HRT: 4.5 d)[g]	100	Microalgae biomass
	HRAP with sedimentation (HRT: 4.5 d)[g]	100	

(*Continued*)

TABLE 12.3 (Continued)
Removal Efficiencies of Pharmaceutical Compounds in Microalgae-Based Systems

Pharmaceutical Compounds	Photobioreactor Configuration	Removal Efficiencies	Microalgae
Flupetixol	Open pond (HRT: 7d)[d]	-75	Green algal species
Gabapentine	HRAP American tropical conditions (HRT: 3 d)[h]	0-86.3	Phylum *Chlorophyta*
Haloperidol	Open pond (HRT: 7d)[d]	-5000	Green algal species
Lamotrigine	HRAP American tropical conditions (HRT: 3 d)[h]	47-48	Phylum *Chlorophyta*
Lorazepam	PBR tubular (HRT: 8 d)[f]	57.2	Phormidium and cyanobacteria
	PBR tubular (HRT: 12 d)[f]	30	
Mirtazapine	Open pond (HRT: 7d)[d]	39	Green algal species
N – Desmethylvenlafaxine	HRAP without sedimentation (HRT: 4.5 d)[g]	27	Microalgae biomass
	HRAP with sedimentation (HRT: 4.5 d)[g]	42	
N,N-Didesmethylvenlafaxine	HRAP without sedimentation (HRT: 4.5 d)[g]	88.6	Microalgae biomass
	HRAP with sedimentation (HRT: 4.5 d)[g]	85	
N,N-Didesmethy-Odesmethyllvenlafaxine	HRAP without sedimentation (HRT: 4.5 d)[g]	0	Microalgae biomass
	HRAP with sedimentation (HRT: 4.5 d)[g]	65	
Nefazodon	Open pond (HRT: 7d)[d]	-630	Green algal species
O – Desmethyl venlafaxine	HRAP without sedimentation (HRT: 4.5 d)[g]	13	Microalgae biomass
	HRAP with sedimentation (HRT: 4.5 d)[g]	39	
Oxazepam	Open pond (HRT: 7d)[d]	-13	Green algal species
Paroxetine	PBR tubular (HRT: 12 d)[f]	93.8	Phormidium and cyanobacteria
Risperidone	Open pond (HRT: 7d)[d]	-3.2	Green algal species
Sertraline	Open pond (HRT: 7d)[d]	17	Green algal species
Venlafaxine	Open pond (HRT: 7d)[d]	57	Green algal species
Venlafaxine	HRAP without sedimentation (HRT: 4.5 d)[g]	17-53	Microalgae biomass
	HRAP with sedimentation (HRT: 4.5 d)[g]	49-67	
	β-Blocking Agents		
Atenolol	HRAP After UASB (HRT: 3-9 d)[a]	90.4	*Coelastrum sp*
	HRAP Direct (HRT: 3-9 d)[a]	35.1	
	Open pond (HRT: 7d)[d]	99	Green algal species
	PBR tubular (HRT: 8 d)[f]	98.5	Phormidium and cyanobacteria
	PBR tubular (HRT: 12 d)[f]	85.6	
Bisoprolol	Open pond (HRT: 7d)[d]	97	Green algal species
Metoprolol	Open pond (HRT: 7d)[d]	99	Green algal species
	Bioreactor batch (0-31 d)[e]	70	*C. sorokiniana*
	HRAP without sedimentation (HRT: 4.5 d)[g]	< 0	Microalgae biomass
	HRAP with sedimentation (HRT: 4.5 d)[g]	0-36	
	HRAP Direct (HRT: 3-9 d)[a]	42.7	*Coelastrum sp*
Metoprolol acid	HRAP without sedimentation (HRT: 4.5 d)[g]	32.3	Microalgae biomass
	HRAP with sedimentation (HRT: 4.5 d)[g]	10	

(Continued)

TABLE 12.3 (Continued)
Removal Efficiencies of Pharmaceutical Compounds in Microalgae-Based Systems

Pharmaceutical Compounds	Photobioreactor Configuration	Removal Efficiencies	Microalgae
Nadolol	HRAP After UASB (HRT: 3-9 d)[a]	6.6	*Coelastrum sp*
	HRAP Direct (HRT: 3-9 d)[a]	-1.4	
Propanolol	HRAP After UASB (HRT: 3-9 d)[a]	42.9	*Coelastrum sp*
	HRAP Direct (HRT: 3-9 d)[a]	53.6	
Sotalol	Open pond (HRT: 7d)[d]	43	Green algal species
Terbutalin	Open pond (HRT: 7d)[d]	98	Green algal species
Timolol	HRAP After UASB (HRT: 3-9 d)[a]	16.7	*Coelastrum sp*
	HRAP Direct (HRT: 3-9 d)[a]	-124.1	

Notes:
[a] Villar-Navarro et al. (2018).
[b] de Godos et al. (2012).
[c] Matamoros et al. (2015).
[d] Gentili and Fick (2017).
[e] de Wilt et al. (2016).
[f] Hom-Diaz et al. (2017).
[g] García-Galán et al. (2020).
[h] Jiménez-Bambague et al. (2020).
[i] Vassalle et al. (2020).

in the harvest of microalgae and in achieving products such as biofuels, fertilizers, etc. (Vo et al. 2019).

HRAPs have been evaluated in different configurations, different climate conditions (cold, warm, and tropical conditions), and in HRTs of 3 to 9 days. The results showed that the removal efficiencies were highest when the HRAP was located after a preliminary treatment such as a settling tank or an upflow anaerobic sludge blanket (UASB) because the organic load and solids were reduced. This in turn contributed to the removal of pharmaceutical compounds and to the performance optimization of microalgae and other removal mechanisms as the competition with other pollutants is reduced. Villar-Navarro et al. (2018) compared the removal efficiency of pharmaceutical compounds in a HRAP with conventional WWTPs, and they recommend it as an alternative to tertiary treatment, after the implementation of UASB. Furthermore, García-Galán et al. (2020) indicated that the HRAP, as a secondary treatment, is a feasible alternative to activated sludge systems for wastewater and micropollutant treatment. Overall, the location of a microalgae-based system and its removal efficiency depend on the type of wastewater, the operational conditions, and environmental conditions.

Other authors have found that summer-related conditions improve the removal of pharmaceutical compounds and other micropollutants (Matamoros et al. 2015;

Tolboom et al. 2019; Villar-Navarro et al. 2018) due to temperature and solar radiation being greater and therefore microalgae activity and photo-degradation being increased. Matamoros et al. (2015) reported that the increase of HRT (of 4 to 8 days) has an important influence over biodegradation, photodegradation, sorption, and volatilization removal mechanisms in the cold season. While Jiménez-Bambague et al. (2020) found that removal efficiencies could be improved with shorter HRT (e.g., 3 days), but under tropical conditions. The results showed that the American tropics provide ideal conditions for the effective development of microalgae as well as the removal efficiency of pharmaceutical compounds such as ibuprofen, paracetamol (acetaminophen), diclofenac, and naproxen. These results were very similar to other studies that involved the highest HRT.

An open PBR constructed of thin fiberglass to allow light penetration from all sides was also evaluated by Gentili and Fick (2016) for the removal of pharmaceuticals in urban wastewater with a 7-day HRT. The highest removal rate of above 90 percent was found with compounds such as clindamycin, atracurium, diltiazem, bupropion, citalopram, atenolol, bisoprolol, and metoprolol. The removal of compounds such as ciprofloxacin, clarithromycin, clindamycin, venlafaxine, atenolol, and metoprolol were close to or higher than the HRAPs that were evaluated without any preliminary treatment, while sulfamethoxazole and carbamazepine were considerably less in the open pond than the HRAPs. Overall, not many differences were observed between the open pond and the HRAPs performance.

Hom-Diaz et al. (2017) evaluated a pilot-scale tubular photobioreactor to remove pharmaceutically active compounds of toilet wastewater. This photobioreactor was operated with an 8- and 12-day HRT. The main results showed that the system could remove 30 percent 80 percent of pharmaceutical compounds. Paracetamol, ibuprofen, paroxetine, and atenolol had a removal efficiency of above 90 percent at a 12-day HRT, despite the temperature and light irradiation being lower than when the system was evaluated with an 8-day HRT. According to Hom-Diaz et al. (2017), HRT is a key design parameter for the removal efficiency of micropollutants in this microalgal system.

Two-hybrid PBRs were evaluated by Vassalle et al. (2020). These reactors were designed as semi-closed horizontal tubular PBRs with a 5-day HRT. Eight-blade paddle wheels were installed at the center to ensure homogeneous distribution, the mixing of the liquor, and the release of the excess oxygen that was dissolved. The main results showed effective removal of the majority of analyzed compounds (of between 46 percent and 100 percent), only sulfapyridine had a negative removal (of –32 percent). The high removals were associated with the size of the reactors, the specific mixed cultures that were developed, and the high temperatures and pH levels in the closed system. Overall, for the compounds studied, this system had more optimal results than the open PBRs.

According to the results presented in Table 12.3, the highest removal efficiencies (>90 percent) in the microalgae-based systems were obtained at least once for the group of analgesics/anti-inflammatories such as paracetamol, diclofenac, and its metabolites (4-OH-diclofenac and 5-OH-diclofenac), ibuprofen and its metabolite carboxy-ibuprofen, ketoprofen, and salicylic acid; antibiotics such as clarithromycin, metronidazole, and sulfathiazole; lipid regulators such as bezafibrate and acid clofibric; psychiatric drugs such as amitriptyline, bupropion, citalopram, desmethyldiazepam, and paroxetine; β-blocking agents such as atenolol, bisoprolol, metoprolol, and terbutaline; and other compounds such as atracurium, caffeine, diltiazem, and glyburide. Other compounds, particularly antibiotics, had adverse removal efficiencies which were less than 30 percent including negative values. Despite this, the results show the great potential that these types of systems have for the removal of many regularly used micropollutants that have a high presence in wastewater. Compounds with low removal efficiency suggest the need to implement additional treatment.

12.5 REMOVAL MECHANISMS OF PHARMACEUTICAL COMPOUNDS

Microalgae-based wastewater treatment systems are a dynamic ecosystem in which different biotic and abiotic processes occur. The main removal mechanisms of micropollutants identified are sorption in the biomass and/or sediments present in the bioreactor, volatilization, photodegradation, and biodegradation. Sorption is a physical mechanism by which the contaminants transfer from the liquid matrix to the solid fraction. Volatilization is the change of a substance into a gaseous form. This mechanism is more common in compounds with moderately high Henry's Law constant values (11–12 Pam^{-3} mol^{-1}), such as musk fragrances (Matamoros et al. 2015), while in pharmaceutical compounds it is less likely, thus explaining why it will not be included in this item.

Photodegradation and biodegradation allow those complex molecules to be transformed into simpler ones. These processes have been suggested as the main removal mechanisms of pharmaceutical compounds (de Wilt et al. 2016; Jiménez-Bambague et al. 2020; Matamoros et al. 2015). The occurrence of one or more processes depends on the physicochemical characteristics of compounds and the environmental and operational conditions.

12.5.1 Sorption

Sorption is not a degradation mechanism of pollutants but a transfer of the compound from wastewater to the solid fraction (biomass, soil, sediments, sludge, or another solid surface in the bioreactor) without altering it chemically. The significance of sorption in contaminant removal is determined by the clamping force between the contaminants and the sorbents (Wang et al. 2017). In algae-based treatment systems, biosorption is highly likely. This is defined as the removal of substances from solution by biological material such as living and dead biomass, just like their excreted and derivated products (Gadd 2009), including microalgae and bacteria.

The sorption process is divided mainly into absorption and adsorption. Absorption occurs when the substance is transferred into the microalgae due to the hydrophobic interactions of the aliphatic and aromatic groups of a compound with the lipophilic cell membrane of the microorganisms (Ternes et al. 2004). Adsorption is the joining onto the surface of biomass and/or solid fraction due to electrostatic interactions of positively charged groups of chemicals with the negatively charged surfaces of the microorganisms (Ternes et al. 2004).

The fixation of micropollutants depends mainly on their physicochemical properties such as the octanol-water partition coefficient (log Kow), solid–water distribution coefficient (Kd), and acid dissociation constant (PKa) (Carballa et al. 2005). Log Kow is the affinity of a substance on the lipid fraction and determines the hydrophobicity (log Kow ≥ 3) and hydrophilicity (log Kow <3) of this substance. (The hydrophobic compounds are most effectively removed through sorption.) Pharmaceutical compounds that are more effective in removing their log Kow are above 4 (Matamoros et al. 2015). Non-polar compounds have a strong affinity to biosorbents due to hydrophobic interactions that can cross cell membranes and be absorbed into the organic matrix (Gadd 2009). Log Kd is determined by the ratio between the contaminant concentration and the amount of water in the soil. PKa defines the ionization level, the acceptance or the donation of protons under a pH value. Compounds with acid characteristics have better sorption.

The hydrophobicity and slightly acidic nature of some compounds such as diclofenac (log Kow: 4.51; PKa: 4.15), indomethacin (log Kow: 4.27; PKa: 4.5), and naproxen (log Kow: 3.18; PKa: 4.15) could suggest a similar sorption to that of the removal mechanism. However, these properties are not enough to determine sorption as the main removal mechanism in microalgae-based wastewater treatments, or any wastewater treatment for that matter. Besides, this process is not selective; the obtaining of target contaminants could be impeded by other contaminants that may occupy the adsorption surface (Sutherland and Ralph 2019; Xiong et al. 2018).

Microalgae cell walls contain polysaccharides (such as cellulose, chitin, alginate, and glycan), proteins and lipids. They contain functional groups such as amino, hydroxyl, carboxyl, and sulfate. In addition, microalgae produce peptides that also bind to micropollutants. These functional groups act as the binding site and are used to capture different contaminants through adsorption or ion exchange processes (Hom-Diaz et al. 2017; Xiong et al. 2018). Bacteria can also capture the contaminants through adsorption due to the peptidoglycan, teichoic acids, and lipo-teichoic acids that are present at the surface (Wang et al. 2017).

Adsorption of pharmaceutical compounds by microalgae has been estimated to be between 0 and 16.7 percent due to the contaminants with cationic groups being attracted to the microalgae through electrostatic interaction, resulting in an effective bio-adsorption (de Wilt et al. 2016; Xiong et al. 2017). A study carried out by de Wilt et al. (2016) with the microalgae *Chlorella Sorokiniana* reported that the sorption on algal biomass accounted for less than 20 percent of the micropollutant removal.

Sorption is not the main removal mechanism in the microalgae-based system, because when the pH level of the wastewater is close to neutral, the adsorption tends to be reduced due to some compounds being negatively charged and in turn producing a repulsion effect (Park et al. 2018; Suarez et al. 2009). This situation is highly likely due to the microalgae-based system increasing the pH values when the photosynthesis activity occurs, though it could possibly be at night when the pH level decreases. Nevertheless, Santos et al. (2019) found that the anti-depressant venlafaxine is positively charged at an experimental pH of 8.1, which can facilitate sorption in the biomass through electrostatic interactions. García-Galán et al. (2020) detected diclofenac in biomass samples in an HRAP and attributed the bioadsorption/bioaccumulation as a removal mechanism of wastewater.

Due to the sorption process transferring from wastewater to microalgae-bacteria and solid matrix without any breakdown, the biomass and sediments with these compounds could be hazardous. Therefore, an appropriate disposition to avoid the releasing of the sorbed pharmaceutical compounds is required (Wang et al. 2017).

12.5.2 Biodegradation

Biodegradation is the breakdown of organic compounds through biological processes in which the complex parent compounds are converted in simpler-form molecules by catalytic degradation (Xiong et al. 2018). Biodegradation may occur through contaminant degradation in which compounds are used as sole carbon and energy sources for microalgae and bacteria (Tiwari et al. 2017) or by extracellular degradation through the excretion of extracellular polymeric substances (EPS) including polysaccharides, protein, enzymes, substituents (polysaccharide-link methyl, and acetyl groups) and lipids (Tiwari et al. 2017; Xiong et al. 2018).

Microalgae biodegradation begins with an attack by the enzyme phase I (cytochrome 450). Cytochrome 450 is closely linked to heme-thiolate enzymes that allow the oxidative metabolism of hydrophobic compounds (Zangar et al. 2004). This process makes these compounds more hydrophilic due to the addition or unmasking of the hydroxyl group. Moreover, this process can promote enzymatic reactions such as hydroxylation, carboxylation, oxidation, hydrogenation, glycosylation, demethylation, ring breaking, decarboxylation, dehydroxylation, and bromination (Matamoros et al. 2016; Peng et al. 2014). The enzymes phase II (glutathione-S-transferase) catalyze the conjugation reaction between electrophilic compounds and glutathione. The conjugation of enzyme phase II produces the opening of the epoxide ring for protection against the oxidative damage (Ding et al. 2017).

In extracellular degradation, the EPS can form a hydrated biofilm matrix that acts as an external digestive system by keeping extracellular enzymes close to the cell. This allows for the metabolizing of organic compounds in a dissolved, colloidal, or solid form. The EPS includes poly-saccharides, proteins, nucleic acids, and lipids that allow the accumulation of nutrients and other compounds from the environment (Xiong et al. 2018). A microalga-based system, EPS may act as an anti-coagulant allowing the cell protection of contaminants, as well as contributing to extra-cellular degradation and influencing intra-cellular degradation through the formed byproducts (Xiao and Zheng 2016; Xiong et al. 2018).

Overall, pharmaceutical compounds have been designed to withstand biological degradation, and hence the majority are in the environment in their original form. However, early studies have demonstrated that some compounds can be mineralized by microalgae and bacteria (biodegradation being one of the main removal mechanisms in microalgae-based systems). This mineralization is influenced by the pollutant structure and environmental conditions, particularly pH levels, dissolved oxygen (Norvill et al. 2016), and temperature because the warm climate provides suitable conditions for algal productivity (Park et al. 2011; Soeder et al. 1985). It has also been demonstrated that a longer HRT produces better micropollutant removal efficiency due to processes such as biodegradation and photodegradation being improved with the increased time spent on them in the bioreactor (Hom-Diaz et al. 2017; Matamoros et al. 2015; Norvill et al. 2016).

Open bioreactors like HRAPs are characterized by having a long HRT, therefore eliminating contaminants with high half-lives and slow biodegradation kinetics (García-Galán et al. 2020), as well as by supporting a high microbial diversity and the interaction between them (like microalgae-bacteria consortium). According to Matamoros et al. (2016), the enhancing biodegradation effect provided by microalgae and carried out by bacteria may be at least as high as the effect from aquatic plants. In these processes the bacteria breakdown of organic material and secretion of extracellular metabolites such as auxins and vitamin B12, are important supplements for the microalgal growth. In this respect, the microalgae-bacteria consortium is more efficient in the removal of organic compounds than the individual microorganisms (Salim et al. 2014).

Previous studies have shown that analgesics/anti-inflammatories are generally biodegradable. In particular, paracetamol, salicylic acid, ibuprofen, and caffeine have achieved removal efficiencies of up to 100 percent in microalgal systems (de Wilt et al. 2016; Hom-Diaz et al. 2017; García-Galán et al. 2020; Jiménez-Bambague et al. 2020; Matamoros et al. 2015; Villar-Navarro et al. 2018). Escapa et al. (2017a; 2017b) evaluate the removal of paracetamol and salicylic acid by C. *sorokiniana*, *Chlorella vulgaris,* and *Tetradesmus obliquus.* The removal efficiencies were mainly associated with microalgae *C. sorokiniana,* with corresponding values of 41 percent to 69 percent and 93 percent to 98 percent; *C. vulgaris* >21 percent and >25 percent; and *T. obliquus* >40 percent and >93 percent for paracetamol and salicylic acid, respectively. Similarly, the removal of paracetamol has also been associated with photodegradation (Sutherland and Ralph 2019).

Matamoros et al. (2016) showed that biodegradation was the main factor for the removal of caffeine and ibuprofen in wastewater, achieving efficiencies of 99 percent and 95 percent, respectively. This study also demonstrated that the removal of caffeine increased with the presence of microalgae, and the biodegradation process was favored by the release of exudates and the degradation of bacteria (Matamoros et al. 2016).

Most compounds are more effectively biodegraded under aerobic conditions than under anaerobic conditions (Koumaki et al. 2017; Norvill et al. 2016). This is an advantage for these types of PBRs, in which the aerobic conditions prevail due to photosynthetic activity. Jiménez-Bambague et al. (2020) found a positive correlation between the removal of ibuprofen with temperature; pH levels; and dissolved oxygen that supports aerobic biodegradation as its main removal mechanism, which is associated with the enantioselective degradation (Matamoros et al. 2016). In another study, Henning et al. (2018) evaluated the biotransformation of gabapentin in different ORP conditions, and the results showed that this compound was degraded under aerobic conditions, with a half-life of two to seven days.

Other studies carried out with microalgal systems have reported biodegradation as the main removal mechanism. De Wilt et al. (2016) reported that 70 percent of the beta-blocker metoprolol was mainly removed through biotransformation by *C. sorokiniana.* A study developed by Kiki et al. (2020) found that the microalgae *Haematococcus pluvialis* had the highest removal efficiency of all antibiotics (42 percent to 100 percent) in comparison with strains of *Selenastrum capricornutum*, *C. vulgaris,* and *Scenedesmus quadricauda. H. pluvialis. These* antibiotics obtained half-lives of 2.4 to 4.2 days for sulfonamides, 3.7 to 9.4 days for macrolides, 2.1 to 12.2 days for fluoroquinolones, and 8.0 to 14.8 days for pyrimidine inhibitor trimethoprim.

Gemfibrozil has low biodegradability (Fabbri et al. 2017; Grenni et al. 2013). Nonetheless, Grenni et al. (2018) found that the copresence of naproxen had a negative influence over gemfibrozil biodegradation because the bacteria found naproxen was easier to degrade. This situation can be similar to those found in other types of organisms such as microalgae, which explains why one can infer that compounds with slight biodegradation will be scarcely removed from microalgal systems where more biodegradables compounds are present.

Carbamazepine is highly hydrophilic (log Kow 2.45), and its removal efficiency in microalgal systems is lower than the majority of pharmaceutical compounds. This is recalcitrant, very stable, and resistant to aerobic biodegradation and photodegradation (Matamoros et al. 2015; Villar-Navarro et al. 2018). Some studies have reported removal efficiencies of up to 37 percent by microalgae such as *Chlamydomonas Mexicana* and *Scenedesmus obliquus* (Xiong et al. 2016), between 10 percent and 30 percent by *C. sorokiniana* (de Wilt et al. 2016), and 20 percent in a culture mainly dominated by *Chlorella* sp. and *Scenedesmus* sp. Moreover, the removal may also be associated with anaerobic biodegradation through the hydrolysis of the amide group (Schwarzenbach et al. 2006). This process may occur at night or in dark periods when anaerobic conditions prevail.

12.5.3 Photodegradation

Photodegradation involves both direct photolysis and indirect photolysis. Direct photolysis is the breakdown of pollutants when the incident light strikes directly over them (Wang et al. 2017). Direct photolysis is caused by the presence of aromatic rings, conjugated Π systems, heteroatoms, and other functional groups of pharmaceutical compounds that can facilitate the absorption of the incident sun irradiation (Challis et al. 2014). Indirect photolysis or photo-oxidative degradation occurs due to the interaction between the contaminant with free radicals such as hydroxyl radicals (OH^-), peroxyl radicals (ROO^-), and singlet oxygen ($^1O_2$), which are produced under sunlight illumination (Wang et al. 2017). These free radicals are generated by the presence of photosensitizers such as chromophoric dissolved organic matters (CDOMs) which are released into the surrounding algal system (Norvill et al. 2016; de Wilt et al. 2016). These photosensitizers could be molecules such as hydrophilic organic acids, hemicellulose, humic acids, and fulvic acids (Sutherland and Ralph 2019). This indicates that greater algal density can contribute to indirect photolysis, thereby promoting the removal of pharmaceutical compounds which dominate over direct photolysis (Norvill et al. 2016). Moreover, the presence of microalgae can also attenuate the supply light and radiation inside a bioreactor, impeding direct photolysis.

Photodegradation is one of the main removal mechanisms in microalgae-based wastewater treatment systems, together with biodegradation. This mechanism also depends on pharmaceutical compounds and wastewater physicochemical properties, as well as environmental, design, and operational conditions because it is expected that a greater surface area allows for improving this process.

The incidence of sunlight may be related to the latitude and weather seasons. A more rapid photodegradation can be expected during summer in a low latitude area (Wang et al. 2017). A lot of research has shown that the removal efficiency of pharmaceutical compounds and other micropollutants was higher in a warm climate because the temperature and solar radiation conditions were considerably more favorable for both the photodegradation and biodegradation processes (Matamoros et al. 2015; Tolboom et al. 2019; Villar-Navarro et al. 2018).

Some authors have agreed that photodegradation and biodegradation are the main removal mechanisms of micropollutants in microalgae-based wastewater treatment systems (de Wilt et al. 2016; Jiménez-Bambague et al. 2020; Matamoros et al. 2015). However, photodegradation can play a major role in the elimination of organic contaminants such as pharmaceutical compounds (Villar-Navarro et al. 2018).

Lin and Reinhard (2005) evaluated the photodegradation of gemfibrozil, ibuprofen, ketoprofen, naproxen, and propranolol using both purified and river water under an irradiation of 765 W/m2 (wavelength between 290 and 700 nm). This study determined that direct photolysis had more influence over the removal of ketoprofen with a half-life of 2.5 minutes. Indirect photolysis was the main photodegradation mechanism of propranolol and naproxen with a half-life of 1.1 minutes and 1.4 hours, respectively, and of gemfibrozil and ibuprofen with a half-life of 15 hours.

Other analgesics/anti-inflammatories have shown elimination through photodegradation. Diclofenac has demonstrated elimination by direct photolysis forming less toxic hydrophilic products, although no significant mineralization was achieved during a 30-minute treatment time (Ardila et al. 2019). Similarly, direct photolysis has been included as another removal mechanism of paracetamol, with a half-life that varies from one day to two weeks, depending on the illumination surface (De Laurentiis et al. 2014).

Photodegradation has been reported as the main removal mechanism of antibiotics. De Godos et al. (2012) found that tetracycline was mainly removed by photodegradation and biosorption. This study was carried out under a UV irradiation of 0.8–0.9 Wm^{-2}, and in dark periods. The removal efficiencies were 92 percent to 98 percent after 43 hours, with a pseudo-first-order kinetic rate. Ryan et al. (2011) evaluated the photolysis of sulfamethoxazole and trimethoprim in wastewater effluent treatment plants. The main results reported indirect photolysis to trimethoprim (62 percent), and direct photolysis to sulfamethoxazole (48 percent). However, Bai and Acharya (2017) found that in lake water, both trimethoprim and sulfamethoxazole are unaffected by exposure to light, while ciprofloxacin and triclosan were degraded through direct photolysis with half-lives of 12.7 and 31.2 hours, respectively.

It is important to note that most of the experiments mentioned above were carried out under controlled conditions and that removal under real conditions can vary as a greater amount of interference can occur.

ACKNOWLEDGMENTS

Eliana M. Jiménez-Bambague thanks the Ministerio de Ciencia Tecnología e Innovación for the Ph.D. Scholarship Plan Bicentenario. The authors thanks Universidad del Valle for the financial support to produce this work through the project CI. 1126. Machuca-Martínez thanks to Sistema General de Regalías Colombia, Grant BPIN 2018000100096.

REFERENCES

Abargues, M.R., Giménez, J.B., Ferrer, J., Bouzas, A. and Seco, A. (2018) 'Endocrine disrupter compounds removal in wastewater using microalgae: Degradation kinetics assessment,' Chemical Engineering Journal, 334, 313–321.

Ardila, P.L.K., da Silva, B.F., Spadoto, M., Rispoli, B.C.M. and Azevedo, E.B. (2019) 'Which route to take for diclofenac removal from water: Hydroxylation or direct photolysis?', Journal of Photochemistry and Photobiology A: Chemistry, 382, 111879.

Bai, X. and Acharya, K. (2017) 'Algae-mediated removal of selected pharmaceutical and personal care products (PPCPs) from Lake Mead water', Science of the Total Environment, 581, 734–740.

Bartley, M.L., Boeing, W.J., Dungan, B.N., Holguin, F.O. and Schaub, T. (2014) 'pH effects on growth and lipid accumulation of the biofuel microalgae Nannochloropsis salina and invading organisms', Journal of Applied Phycology, 26(3), 1431–1437.

Borowska, E., Felis, E. and Kalka, J. (2016) 'Oxidation of benzotriazole and benzothiazole in photochemical processes: Kinetics and formation of transformation products', Chemical Engineering Journal, 304, 852–863.

Breuer, G., Lamers, P.P., Martens, D.E., Draaisma, R.B. and Wijffels, R.H. (2013) 'Effect of light intensity, pH, and temperature on triacylglycerol (TAG) accumulation induced by nitrogen starvation in Scenedesmus obliquus', Bioresource Technology, 143, 1–9.

Butler, A. (1998) 'Acquisition and utilization of transition metal ions by marine organisms', Science, 281(5374), 207–209.

Cai, T., Park, S.Y. and Li, Y. (2013) 'Nutrient recovery from wastewater streams by microalgae: Status and prospects', Renewable and Sustainable Energy Reviews, 19, 360–369.

Carballa, M., Omil, F. and Lema, J.M. (2005) 'Removal of cosmetic ingredients and pharmaceuticals in sewage primary treatment', Water Research, 39(19), 4790–4796.

Carvalho, A.P., Meireles, L.A. and Malcata, F.X. (2006) 'Microalgal reactors: a review of enclosed system designs and performances', Biotechnology progress, 22(6), 1490–1506.

Cerón Hernández, V.A., Madera Parra, C.A. and Peña Varón, M. (2015) 'Uso de lagunas algales de alta tasa para tratamiento de aguas residuales', Ingeniería y Desarrollo, 33, 98–125.

Challis, J.K., Hanson, M.L., Friesen, K.J. and Wong, C.S. (2014) 'A critical assessment of the photodegradation of pharmaceuticals in aquatic environments: Defining our current understanding and identifying knowledge gaps', Environmental Science: Processes & Impacts, 16(4), 672–696.

Chang, H.-X., Fu, Q., Huang, Y., Xia, A., Liao, Q., Zhu, X., Zheng, Y.-P. and Sun, C.-H. (2016) 'An annular photobioreactor with ion-exchange-membrane for non-touch microalgae cultivation with wastewater', Bioresource Technology, 219, 668–676.

Chisti, Y. (2008) 'Biodiesel from microalgae beats bioethanol', Trends in biotechnology, 26(3), 126–131.

Craggs, R., Park, J., Heubeck, S. and Sutherland, D. (2014) 'High rate algal pond systems for low-energy wastewater treatment, nutrient recovery and energy production', New Zealand Journal of Botany, 52(1), 60–73.

Craggs, R.J., Adey, W.H., Jenson, K.R., St. John, M.S., Green, F.B. and Oswald, W.J. (1996) 'Phosphorus removal from wastewater using an algal turf scrubber', Water Science and Technology, 33(7), 191–198.

Craggs, R.J., Heubeck, S., Lundquist, T.J. and Benemann, J.R. (2011) 'Algal biofuels from wastewater treatment high rate algal ponds', Water Science and Technology, 63(4), 660–665.

Croft, M.T., Warren, M.J. and Smith, A.G. (2006) 'Algae need their vitamins', Eukaryotic Cell, 5(8), 1175–1183.

Cuerda-Correa, E.M., Alexandre-Franco, M.F. and Fernández-González, C. (2020) 'Advanced oxidation processes for the removal of antibiotics from water. An overview', Water, 12(1), 102.

de Godos, I., Muñoz, R. and Guieysse, B. (2012) 'Tetracycline removal during wastewater treatment in high-rate algal ponds', Journal of hazardous materials, 229, 446–449.

de Jesus, S.S. and Maciel Filho, R. (2017) 'Potential of algal biofuel production in a hybrid photobioreactor', Chemical Engineering Science, 171, 282–292.

De Laurentiis, E., Prasse, C., Ternes, T.A., Minella, M., Maurino, V., Minero, C., Sarakha, M., Brigante, M. and Vione, D. (2014) 'Assessing the photochemical transformation pathways of acetaminophen relevant to surface waters: transformation kinetics, intermediates, and modelling', Water Research, 53, 235–248.

de Wilt, A., Butkovskyi, A., Tuantet, K., Leal, L.H., Fernandes, T.V., Langenhoff, A. and Zeeman, G. (2016) 'Micropollutant removal in an algal treatment system fed with source separated wastewater streams', Journal of Hazardous Materials, 304, 84–92.

Ding, T., Yang, M., Zhang, J., Yang, B., Lin, K., Li, J. and Gan, J. (2017) 'Toxicity, degradation and metabolic fate of ibuprofen on freshwater diatom Navicula sp', Journal of hazardous materials, 330, 127–134.

Dušek, J., Picek, T. and Čížková, H. (2008) 'Redox potential dynamics in a horizontal subsurface flow constructed wetland for wastewater treatment: Diel, seasonal and spatial fluctuations', Ecological Engineering, 34(3), 223–232.

Díaz, J.P., Inostroza, C. and Acién Fernández, F.G. (2019) 'Fibonacci-type tubular photobioreactor for the production of microalgae', Process Biochemistry, 86, 1–8.

Escapa, C., Coimbra, R., Paniagua, S., García, A. and Otero, M. (2017a) 'Comparison of the culture and harvesting of Chlorella vulgaris and Tetradesmus obliquus for the removal of pharmaceuticals from water', Journal of Applied Phycology, 29(3), 1179–1193.

Escapa, C., Coimbra, R., Paniagua, S., García, A. and Otero, M. (2017b) 'Paracetamol and salicylic acid removal from contaminated water by microalgae', Journal of environmental management, 203, 799–806.

Fabbri, D., Maurino, V., Minella, M., Minero, C. and Vione, D. (2017) 'Modelling the photochemical attenuation pathways of the fibrate drug gemfibrozil in surface waters', Chemosphere, 170, 124–133.

Fent, K., Weston, A.A. and Caminada, D. (2006) 'Ecotoxicology of human pharmaceuticals', Aquatic Toxicology, 76(2), 122–159.

Ferro, L., Gorzsás, A., Gentili, F.G. and Funk, C. (2018) 'Subarctic microalgal strains treat wastewater and produce biomass at low temperature and short photoperiod', Algal Research, 35, 160–167.

Fukami, K., Nishijima, T. and Ishida, Y. (1997) 'Stimulative and inhibitory effects of bacteria on the growth of microalgae', Hydrobiologia, 358(1-3), 185–191.

Gadd, G.M. (2009) 'Biosorption: Critical review of scientific rationale, environmental importance and significance for pollution treatment', Journal of Chemical Technology & Biotechnology: International Research in Process, Environmental & Clean Technology, 84(1), 13–28.

García, J., Hernández-Mariné, M. and Mujeriego, R. (2000) 'Influence of Phytoplankton Composition on Biomass Removal from High-Rate Oxidation Lagoons by Means of Sedimentation and Spontaneous Flocculation', Water Environment Research, 72(2), 230–237.

García-Galán, M.J., Arashiro, L., Santos, L.H., Insa, S., Rodríguez-Mozaz, S., Barceló, D., Ferrer, I. and Garfí, M. (2020) 'Fate of priority pharmaceuticals and their main metabolites and transformation products in microalgae-based wastewater treatment systems', Journal of Hazardous Materials, 390, 121771.

García-Galán, M.J., Gutiérrez, R., Uggetti, E., Matamoros, V., García, J. and Ferrer, I. (2018) 'Use of full-scale hybrid horizontal tubular photobioreactors to process agricultural runoff', Biosystems Engineering, 166, 138–149.

Gentili, F. G. and Fick, J. (2017) 'Algal cultivation in urban wastewater: an efficient way to reduce pharmaceutical pollutants', Journal of Applied Phycology, 29(1), 255–262.

Gonzalez, L.E. and Bashan, Y. (2000) 'Increased growth of the microalga chlorella vulgaris when coimmobilized and cocultured in alginate beads with the plant-growth-promoting bacterium Azospirillum brasilense', Applied and Environmental Microbiology, 66(4), 1527–1531.

Gonzalez-Camejo, J., Aparicio, S., Ruano, M., Borrás, L., Barat, R. and Ferrer, J. (2019) 'Effect of ambient temperature variations on an indigenous microalgae-nitrifying bacteria culture dominated by Chlorella', Bioresource technology, 290, 121788.

Gonçalves, A.L., Pires, J.C. and Simões, M. (2017) 'A review on the use of microalgal consortia for wastewater treatment', Algal Research, 24, 403–415.

Grenni, P., Patrolecco, L., Ademollo, N., Di Lenola, M. and Caracciolo, A.B. (2018) 'Assessment of gemfibrozil persistence in river water alone and in co-presence of naproxen', Microchemical Journal, 136, 49–55.

Grenni, P., Patrolecco, L., Ademollo, N., Tolomei, A. and Caracciolo, A.B. (2013) 'Degradation of gemfibrozil and naproxen in a river water ecosystem', Microchemical Journal, 107, 158–164.

Grobbelaar, J.U. (2000) 'Physiological and technological considerations for optimising mass algal cultures', Journal of Applied Phycology, 12(3-5), 201–206.

Guo, X., Yao, L. and Huang, Q. (2015) 'Aeration and mass transfer optimization in a rectangular airlift loop photobioreactor for the production of microalgae', Bioresource Technology, 190, 189–195.

Guzmán-Soto, C.J. and Tamarís-Turizo, C.E. (2014) 'Hábitos alimentarios de individuos inmaduros de Ephemeroptera, Plecoptera y Trichoptera en la parte media de un río tropical de montaña', Revista de Biología Tropical, 62, 169–178.

Henning, N., Kunkel, U., Wick, A. and Ternes, T.A. (2018) 'Biotransformation of gabapentin in surface water matrices under different redox conditions and the occurrence of one major TP in the aquatic environment', Water Research, 137, 290–300.

Hom-Diaz, A., Jaén-Gil, A., Bello-Laserna, I., Rodríguez-Mozaz, S., Vicent, T., Barceló, D. and Blánquez, P. (2017) 'Performance of a microalgal photobioreactor treating toilet wastewater: Pharmaceutically active compound removal and biomass harvesting', Science of the Total Environment, 592, 1–11.

Huang, Q., Jiang, F., Wang, L. and Yang, C. (2017) 'Design of Photobioreactors for Mass Cultivation of Photosynthetic Organisms', Engineering, 3(3), 318–329.

Janssen, M., Tramper, J., Mur, L.R. and Wijffels, R.H. (2003) 'Enclosed outdoor photobioreactors: Light regime, photosynthetic efficiency, scale-up, and future prospects', Biotechnology and bioengineering, 81(2), 193–210.

Jiménez-Bambague, E.M., Madera-Parra, C.A., Ortiz-Escobar, A.C., Morales-Acosta, P.A., Peña-Salamanca, E.J. and Machuca-Martínez, F. (2020) 'High-rate algal pond for removal of pharmaceutical compounds from urban domestic wastewater under tropical conditions. Case study: Santiago de cali-Colombia', Water Science and Technology.

Kiki, C., Rashid, A., Wang, Y., Li, Y., Zeng, Q., Yu, C.-P. and Sun, Q. (2020) 'Dissipation of antibiotics by microalgae: Kinetics, identification of transformation products and pathways', Journal of Hazardous Materials, 387, 121985.

Koumaki, E., Mamais, D. and Noutsopoulos, C. (2017) 'Environmental fate of non-steroidal anti-inflammatory drugs in river water/sediment systems', Journal of Hazardous Materials, 323, 233–241.

Lee, Y.-K. (2001) 'Microalgal mass culture systems and methods: their limitation and potential', Journal of Applied Phycology, 13(4), 307–315.

Li, K., Liu, Q., Fang, F., Luo, R., Lu, Q., Zhou, W., Huo, S., Cheng, P., Liu, J. and Addy, M. (2019) 'Microalgae-based wastewater treatment for nutrients recovery: A review', Bioresource technology, 121934.

Lietz, D.M. (1987) 'Potential for aquatic oligochaetes as live food in commercial aquaculture' Aquatic Oligochaeta Springer, 309–310.

Lin, A.Y.C. and Reinhard, M. (2005) 'Photodegradation of common environmental pharmaceuticals and estrogens in river water', Environmental Toxicology and Chemistry: An International Journal, 24(6), 1303–1309.
Liu, J., Wang, F., Liu, W., Tang, C., Wu, C. and Wu, Y. (2016) 'Nutrient removal by up-scaling a hybrid floating treatment bed (HFTB) using plant and periphyton: From laboratory tank to polluted river', Bioresource Technology, 207, 142–149.
Mahapatra, D.M., Chanakya, H.N. and Ramachandra, T.V. (2013) 'Treatment efficacy of algae-based sewage treatment plants', Environmental Monitoring and Assessment, 185(9), 7145–7164.
Mara, D. and Pearson, H. (1986) 'Artificial freshwater environment: Waste stabilization ponds', Biotechnology, 8, 177–206.
Markou, G. and Georgakakis, D. (2011) 'Cultivation of filamentous cyanobacteria (blue-green algae) in agro-industrial wastes and wastewaters: A review', Applied Energy, 88(10), 3389–3401.
Masojídek, J., Papáček, Š., Sergejevova, M., Jirka, V., Červený, J., Kunc, J., Korečko, J., Verbovikova, O., Kopecký, J. and Štys, D. (2003) 'A closed solar photobioreactor for cultivation of microalgae under supra-high irradiance: Basic design and performance', Journal of Applied Phycology, 15(2-3), 239–248.
Matamoros, V., Gutiérrez, R., Ferrer, I., García, J. and Bayona, J.M. (2015) 'Capability of microalgae-based wastewater treatment systems to remove emerging organic contaminants: A pilot-scale study', Journal of Hazardous Materials, 288, 34–42.
Matamoros, V., Uggetti, E., García, J. and Bayona, J.M. (2016) 'Assessment of the mechanisms involved in the removal of emerging contaminants by microalgae from wastewater: a laboratory scale study', Journal of Hazardous Materials, 301, 197–205.
Mihelcic J.R. and Zimmerman, J.B. (2011) 'Ingeniería ambiental: fundamentos, sustentabilidad, diseño', 192.
Mirón, A.S., Gómez, A.C., Camacho, F.G., Grima, E.M. and Chisti, Y. (1999) 'Comparative evaluation of compact photobioreactors for large-scale monoculture of microalgae' in Osinga, R., Tramper, J., Burgess, J. G. and Wijffels, R. H., eds., Progress in Industrial Microbiology Elsevier, 249–270.
Moheimani, N.R. (2013) 'Inorganic carbon and pH effect on growth and lipid productivity of Tetraselmis suecica and Chlorella sp (Chlorophyta) grown outdoors in bag photobioreactors', Journal of Applied Phycology, 25(2), 387–398.
Molina Grima, E., Fernández, F.G.A., García Camacho, F. and Chisti, Y. (1999) 'Photobioreactors: light regime, mass transfer, and scaleup', Journal of Biotechnology, 70(1), 231–247.
Montemezzani, V., Duggan, I.C., Hogg, I.D. and Craggs, R.J. (2015) 'A review of potential methods for zooplankton control in wastewater treatment high rate algal ponds and algal production raceways', Algal Research, 11, 211–226.
Muñoz, R. and Guieysse, B. (2006) 'Algal–bacterial processes for the treatment of hazardous contaminants: a review', Water research, 40(15), 2799–2815.
Muñoz, R., Köllner, C., Guieysse, B. and Mattiasson, B. (2004) 'Photosynthetically oxygenated salicylate biodegradation in a continuous stirred tank photobioreactor', Biotechnology and Bioengineering, 87(6), 797–803.
Naira, V.R., Das, D. and Maiti, S.K. (2020) 'A novel bubble-driven internal mixer for improving productivities of algal biomass and biodiesel in a bubble-column photobioreactor under natural sunlight', Renewable Energy, 157, 605–615.
Norvill, Z.N., Shilton, A. and Guieysse, B. (2016) 'Emerging contaminant degradation and removal in algal wastewater treatment ponds: Identifying the research gaps', Journal of Hazardous Materials, 313, 291–309.
Oswald, W.J. (1988) 'Micro-algae and waste-water treatment', in M.B.L. Borowitzka, ed., Micro-algal Biotechnology, Cambridge University Press, Cambridge, 305–328.

Oswald, W.J. (2003) 'My sixty years in applied algology', Journal of Applied Phycology, 15(2-3), 99–106.

Pagano, M. (2008) 'Feeding of tropical cladocerans (Moina micrura, Diaphanosoma excisum) and rotifer (Brachionus calyciflorus) on natural phytoplankton: Effect of phytoplankton size–structure', Journal of Plankton Research, 30(4), 401–414.

Park, J., Cho, K.H., Lee, E., Lee, S. and Cho, J. (2018) 'Sorption of pharmaceuticals to soil organic matter in a constructed wetland by electrostatic interaction', Science of The Total Environment, 635, 1345–1350.

Park, J. and Craggs, R. (2010) 'Wastewater treatment and algal production in high rate algal ponds with carbon dioxide addition', Water Science and Technology, 633–639.

Park, J., Craggs, R. and Shilton, A. (2011) 'Wastewater treatment high rate algal ponds for biofuel production', Bioresource technology, 102(1), 35–42.

Park, J.B. and Craggs, R.J. (2010) 'Wastewater treatment and algal production in high rate algal ponds with carbon dioxide addition', Water Sci Technol, 61(3), 633–9.

Park, Y., Je, K.-W., Lee, K., Jung, S.-E. and Choi, T.-J. (2008) 'Growth promotion of Chlorella ellipsoidea by co-inoculation with Brevundimonas sp. isolated from the microalga', Hydrobiologia, 598(1), 219–228.

Passos, F., Hernandez-Marine, M., Garcia, J. and Ferrer, I. (2014) 'Long-term anaerobic digestion of microalgae grown in HRAP for wastewater treatment. Effect of microwave pretreatment', Water Research, 49, 351–359.

Peng, F.-Q., Ying, G.-G., Yang, B., Liu, S., Lai, H.-J., Liu, Y.-S., Chen, Z.-F. and Zhou, G.-J. (2014) 'Biotransformation of progesterone and norgestrel by two freshwater microalgae (Scenedesmus obliquus and Chlorella pyrenoidosa): Transformation kinetics and products identification', Chemosphere, 95, 581–588.

Pham, H.-M., Kwak, H.S., Hong, M.-E., Lee, J., Chang, W.S. and Sim, S.J. (2017) 'Development of an X-Shape airlift photobioreactor for increasing algal biomass and biodiesel production', Bioresource Technology, 239, 211–218.

Picot, B., Bahlaoui, A., Moersidik, S., Baleux, B. and Bontoux, J. (1992) 'Comparison of the purifying efficiency of high rate algal pond with stabilization pond', Water Science and Technology, 25(12), 197–206.

Pocostales, J.P., Sein, M.M., Knolle, W., von Sonntag, C. and Schmidt, T.C. (2010) 'Degradation of ozone-refractory organic phosphates in wastewater by ozone and ozone/hydrogen peroxide (peroxone): The role of ozone consumption by dissolved organic matter', Environmental Science & Technology, 44(21), 8248–8253.

Posten, C. (2009) 'Design principles of photo-bioreactors for cultivation of microalgae', Engineering in Life Sciences, 9(3), 165–177.

Pulz, O. (2001) 'Photobioreactors: Production systems for phototrophic microorganisms', Applied Microbiology and Biotechnology, 57(3), 287–293.

Righini, R. and Grossi Gallegos, H. (2005) 'Análisis de la correlación entre la radiación fotosintéticamente activa y la radiación solar global en San Miguel, provincia de Buenos Aires', Avances en Energías Renovables y Medio Ambiente, 9, 11.01–11.04.

Ryan, C.C., Tan, D.T. and Arnold, W.A. (2011) 'Direct and indirect photolysis of sulfamethoxazole and trimethoprim in wastewater treatment plant effluent', Water Research, 45(3), 1280–1286.

Salim, S., Kosterink, N., Wacka, N.T., Vermuë, M. and Wijffels, R. (2014) 'Mechanism behind autoflocculation of unicellular green microalgae Ettlia texensis', Journal of Biotechnology, 174, 34–38.

Sancho, M.E.M., Castillo, J.J., Lozano, J.E. and El Yousfi, F. (1993) 'Sistemas algas-bacterias para tratamiento de residuos liquidos', Ingeniería química, (291), 131–135.

Santos, L.H., Freixa, A., Insa, S., Acuña, V., Sanchís, J., Farré, M., Sabater, S., Barceló, D. and Rodríguez-Mozaz, S. (2019) 'Impact of fullerenes in the bioaccumulation and biotransformation of venlafaxine, diuron and triclosan in river biofilms', Environmental Research, 169, 377–386.

Schlüter, M. and Groeneweg, J. (1981) 'Mass production of freshwater rotifers on liquid wastes: I. The influence of some environmental factors on population growth of Brachionus rubens Ehrenberg 1838', Aquaculture, 25(1), 17–24.

Schumacher, G., Blume, T. and Sekoulov, I. (2003) 'Bacteria reduction and nutrient removal in small wastewater treatment plants by an algal biofilm', Water Science and Technology, 47(11), 195–202.

Schwarzenbach, R.P., Escher, B.I., Fenner, K., Hofstetter, T.B., Johnson, C.A., Von Gunten, U. and Wehrli, B. (2006) 'The challenge of micropollutants in aquatic systems', Science, 313(5790), 1072–1077.

Scott, J.T., Back, J.A., Taylor, J.M. and King, R.S. (2008) 'Does nutrient enrichment decouple algal–bacterial production in periphyton?', Journal of the North American Benthological Society, 27(2), 332–344.

Semple, K.T., Cain, R.B. and Schmidt, S. (1999) 'Biodegradation of aromatic compounds by microalgae', FEMS Microbiology Letters, 170(2), 291–300.

Sgroi, M., Pelissari, C., Roccaro, P., Sezerino, P.H., García, J., Vagliasindi, F.G.A. and Ávila, C. (2018) 'Removal of organic carbon, nitrogen, emerging contaminants and fluorescing organic matter in different constructed wetland configurations', Chemical Engineering Journal, 332, 619–627.

Shen, Y., Yuan, W., Pei, Z., Wu, Q. and Mao, E. (2009) 'Microalgae mass production methods', Transactions of the ASABE, 52(4), 1275–1287.

Shilton, A., Mara, D., Craggs, R. and Powell, N. (2008) 'Solar-powered aeration and disinfection, anaerobic co-digestion, biological CO2 scrubbing and biofuel production: The energy and carbon management opportunities of waste stabilisation ponds', Water Science and Technology, 58(1), 253–258.

Shoener, B., Bradley, I., Cusick, R. and Guest, J. (2014) 'Energy positive domestic wastewater treatment: The roles of anaerobic and phototrophic technologies', Environmental Science: Processes & Impacts, 16(6), 1204–1222.

Sierra, E., Acién, F.G., Fernández, J.M., García, J.L., González, C. and Molina, E. (2008) 'Characterization of a flat plate photobioreactor for the production of microalgae' Chemical Engineering Journal, 138(1), 136–147.

Skulberg, O.M. (2000) 'Microalgae as a source of bioactive molecules–experience from cyanophyte research', Journal of Applied Phycology, 12(3-5), 341–348.

Slegers, P.M., van Beveren, P.J.M., Wijffels, R.H., van Straten, G. and van Boxtel, A.J.B. (2013) 'Scenario analysis of large scale algae production in tubular photobioreactors', Applied Energy, 105, 395–406.

Soeder, C., Hegewald, E., Fiolitakis, E. and Grobbelaar, J. (1985) 'Temperature dependence of population growth in a green microalga: Thermodynamic characteristics of growth intensity and the influence of cell concentration', Zeitschrift für Naturforschung C, 40(3-4), 227–233.

Solimeno, A., Acíen, F.G. and García, J. (2017) 'Mechanistic model for design, analysis, operation and control of microalgae cultures: Calibration and application to tubular photobioreactors', Algal Research, 21, 236–246.

Solimeno, A. and García, J. (2017) 'Microalgae-bacteria models evolution: From microalgae steady-state to integrated microalgae-bacteria wastewater treatment models–A comparative review', Science of The Total Environment, 607, 1136–1150.

Suarez, S., Lema, J.M. and Omil, F. (2009) 'Pre-treatment of hospital wastewater by coagulation–flocculation and flotation', Bioresource technology, 100(7), 2138–2146.

Subashchandrabose, S.R., Ramakrishnan, B., Megharaj, M., Venkateswarlu, K. and Naidu, R. (2011) 'Consortia of cyanobacteria/microalgae and bacteria: biotechnological potential', Biotechnology Advances, 29(6), 896–907.

Subhash, G.V., Rohit, M., Devi, M.P., Swamy, Y. and Mohan, S.V. (2014) 'Temperature induced stress influence on biodiesel productivity during mixotrophic microalgae cultivation with wastewater', Bioresource Technology, 169, 789–793.

Suh, I.S. and Lee, C.-G. (2003) 'Photobioreactor engineering: design and performance', Biotechnology and Bioprocess Engineering, 8(6), 313.

Sun, J., Xu, W., Cai, B., Huang, G., Zhang, H., Zhang, Y., Yuan, Y., Chang, K., Chen, K., Peng, Y. and Chen, K. (2019) 'High-concentration nitrogen removal coupling with bioelectric power generation by a self-sustaining algal-bacterial biocathode photo-bioelectrochemical system under daily light/dark cycle', Chemosphere, 222, 797–809.

Sun, Y., Huang, Y., Liao, Q., Fu, Q. and Zhu, X. (2016) 'Enhancement of microalgae production by embedding hollow light guides to a flat-plate photobioreactor', Bioresource Technology, 207, 31–38.

Sutherland, D.L., Howard-Williams, C., Turnbull, M.H., Broady, P.A. and Craggs, R.J. (2015) 'Enhancing microalgal photosynthesis and productivity in wastewater treatment high rate algal ponds for biofuel production', Bioresource Technology, 184, 222–229.

Sutherland, D.L. and Ralph, P.J. (2019) 'Microalgal bioremediation of emerging contaminants-Opportunities and challenges', Water research, 114921.

Sutherland, D.L., Turnbull, M.H. and Craggs, R.J. (2017) 'Environmental drivers that influence microalgal species in fullscale wastewater treatment high rate algal ponds', Water Research, 124, 504–512.

Ternes, T.A., Herrmann, N., Bonerz, M., Knacker, T., Siegrist, H. and Joss, A. (2004) 'A rapid method to measure the solid–water distribution coefficient (Kd) for pharmaceuticals and musk fragrances in sewage sludge', Water research, 38(19), 4075–4084.

Tiwari, B., Sellamuthu, B., Ouarda, Y., Drogui, P., Tyagi, R.D. and Buelna, G. (2017) 'Review on fate and mechanism of removal of pharmaceutical pollutants from wastewater using biological approach', Bioresource Technology, 224, 1–12.

Tolboom, S.N., Carrillo-Nieves, D., de Jesús Rostro-Alanis, M., de la Cruz Quiroz, R., Barceló, D., Iqbal, H.M. and Parra-Saldivar, R. (2019) 'Algal-based removal strategies for hazardous contaminants from the environment–a review', Science of the Total Environment, 665, 358–366.

Toro-Vélez, A., Madera-Parra, C., Peña-Varón, M., Lee, W., Bezares-Cruz, J., Walker, W., Cárdenas-Henao, H., Quesada-Calderón, S., García-Hernández, H. and Lens, P. (2016) 'BPA and NP removal from municipal wastewater by tropical horizontal subsurface constructed wetlands', Science of the Total Environment, 542, 93–101.

Tredici, M.R. and Zlttelli, G.C. (1998) 'Efficiency of sunlight utilization: Tubular versus flat photobioreactors', Biotechnology and Bioengineering, 57(2), 187–197.

Van Den Hende, S., Beelen, V., Bore, G., Boon, N. and Vervaeren, H. (2014) 'Up-scaling aquaculture wastewater treatment by microalgal bacterial flocs: From lab reactors to an outdoor raceway pond', Bioresource Technology, 159, 342–354.

Vassalle, L., Sunyer-Caldú, A., Uggetti, E., Díez-Montero, R., Díaz-Cruz, M.S., García, J. and García-Galán, M.J. (2020) 'Bioremediation of emerging micropollutants in irrigation water. The alternative of microalgae-based treatments', Journal of Environmental Management, 274, 111081.

Verlicchi, P., Al Aukidy, M. and Zambello, E. (2012) 'Occurrence of pharmaceutical compounds in urban wastewater: removal, mass load and environmental risk after a secondary treatment: A review', Science of the Total Environment, 429, 123–155.

Villar-Navarro, E., Baena-Nogueras, R.M., Paniw, M., Perales, J.A. and Lara-Martín, P.A. (2018) 'Removal of pharmaceuticals in urban wastewater: High rate algae pond (HRAP) based technologies as an alternative to activated sludge based processes', Water research, 139, 19–29.

Vo, H.N.P., Ngo, H.H., Guo, W., Nguyen, T.M.H., Liu, Y., Liu, Y., Nguyen, D.D. and Chang, S.W. (2019) 'A critical review on designs and applications of microalgae-based photobioreactors for pollutants treatment', Science of the Total Environment, 651, 1549–1568.

Wang, B., Lan, C.Q. and Horsman, M. (2012) 'Closed photobioreactors for production of microalgal biomasses', Biotechnology Advances, 30(4), 904–912.

Wang, Y., Guo, W., Yen, H.-W., Ho, S.-H., Lo, Y.-C., Cheng, C.-L., Ren, N. and Chang, J.-S. (2015) 'Cultivation of Chlorella vulgaris JSC-6 with swine wastewater for simultaneous nutrient/COD removal and carbohydrate production', Bioresource Technology, 198, 619–625.

Wang, Y., Liu, J., Kang, D., Wu, C. and Wu, Y. (2017) 'Removal of pharmaceuticals and personal care products from wastewater using algae-based technologies: A review', Reviews in Environmental Science and Bio/Technology, 16(4), 717–735.

Xiao, R. and Zheng, Y. (2016) 'Overview of microalgal extracellular polymeric substances (EPS) and their applications', Biotechnology Advances, 34(7), 1225–1244.

Xiong, J.-Q., Kurade, M.B., Abou-Shanab, R.A., Ji, M.-K., Choi, J., Kim, J.O. and Jeon, B.-H. (2016) 'Biodegradation of carbamazepine using freshwater microalgae Chlamydomonas mexicana and Scenedesmus obliquus and the determination of its metabolic fate', Bioresource Technology, 205, 183–190.

Xiong, J.-Q., Kurade, M.B. and Jeon, B.-H. (2018) 'Can microalgae remove pharmaceutical contaminants from water?', Trends in biotechnology, 36(1), 30–44.

Xu, J., Cheng, J., Wang, Y., Yang, W., Park, J.-Y., Kim, H. and Xu, L. (2021) 'Strengthening CO2 dissolution with zeolitic imidazolate framework-8 nanoparticles to improve microalgal growth in a horizontal tubular photobioreactor', Chemical Engineering Journal, 405, 126062.

Yao, S., Ni, J., Ma, T. and Li, C. (2013) 'Heterotrophic nitrification and aerobic denitrification at low temperature by a newly isolated bacterium, Acinetobacter sp. HA2', Bioresource Technology, 139, 80–86.

Zangar, R.C., Davydov, D.R. and Verma, S. (2004) 'Mechanisms that regulate production of reactive oxygen species by cytochrome P450', Toxicology and Applied Pharmacology, 199(3), 316–331.

Zuo, Y. (2003) 'Light-induced formation of hydroxyl radicals in fog waters determined by an authentic fog constituent, hydroxymethanesulfonate', Chemosphere, 51(3), 175–179.

13 Processes for the Treatment of Industrial Wastewater

Nimish Shah, Ankur H. Dwivedi, and Shibu G. Pillai
Department of Chemical Engineering, School of Engineering, Institute of Technology, Nirma University, Ahmedabad, Gujarat, India

CONTENTS

DOI: 10.1201/9781003138303-13

13.1 INTRODUCTION

Wastewater has a significant impact on the natural world, and it is important to treat it efficaciously. By treating wastewater, you don't just preserve the creatures thriving on it but protect the planet holistically. Wastewater treatment facilities are an essential part of any industrialized nation's infrastructure. Also known as sewage treatment facilities, they provide a critical life-sustaining resource, water, to be reused without continually damaging the ecosystems in which we live. Initially wastewater discharge was carried out in natural water bodies. But with increased industrialization, the threat to public health through spread of disease and arose, and treatment began. As population rose in cities, the quantity of wastewater increased and quality deteriorated. Available land for wastewater treatment became limited. A wide variety of onsite and offsite treatment systems were developed by various scientists. For any discharge of wastewater clean-up is compulsory, so industry needs a cost-efficient technology. For remediation of wastewater from 1900 to the early 1970s techniques were developed for removal of floating and suspended solids and reduction of biodegradables and harmful bacteria. From 1970 to 1990, additional esthetic and environmental concerns arose. Existing processes for biodegradables and solids were scaled up. Later, in some streams and lakes, treatments started to remove/reduce nutrients such as nitrogen and phosphorous. (Rajasulochana & Preethy, 2016)

Because all industries have different quality of wastewater, treatment is complicated. Not only quality but fluctuations in contaminant concentration and flow rates make the treatment more complicated (Environmental Protection Agency, Ireland, 1997). Toxic metals that are present in effluent influence the choice of the abstraction method to be used in effluent treatment plants. Fig. 13.1 shows a typical layout of a simple wastewater treatment plant.

The normal way of quantifying contamination is generally in terms of dissolved oxygen (DO), chemical oxygen demand (COD), biological oxygen demand (BOD), etc. These terms are explained below:

I. **Chemical oxygen demand (COD):** "the amplitude of oxygen required to oxidize the polluting chemicals to CO_2 and H_2O" (Environmental Protection Agency, Ireland, 1997). The local authority may declare the permissible limits for the discharge of an effluent.
II. **Biological oxygen demand (BOD):** "It is the magnitude of oxygen required by the biological microbial mass during the effluent treatment to oxidize the

biologically oxidizable pollutants and for their own sustenance. It is quantified by the oxygen consumption of a pre-inoculated sample at 20–25°C in tenebrosity over an incubation period of five days" (Environmental Protection Agency, Ireland, 1997).

III. **Dissolved oxygen (DO):** "It is the optically canvassed quantity of O_2 dissolved in the water" (Environmental Protection Agency, Ireland, 1997). Treatment of effluent will always raise the oxygen level in effluent.

IV. **Suspended solids (TSS):** "As a result of surface charge on minuscule particles in the effluent, more sizably voluminous solid particulate remains in a suspended manner in it. This is additionally termed as MLSS (Commixed Liquor Suspended Solids)" (Environmental Protection Agency, Ireland, 1997).

V. **Total Dissolved Solids (TDS):** "Many inorganic salts, which are soluble in water are arduous to abstract since they are consummately dissolved and show high solubility in water" (Environmental Protection Agency, Ireland, 1997).

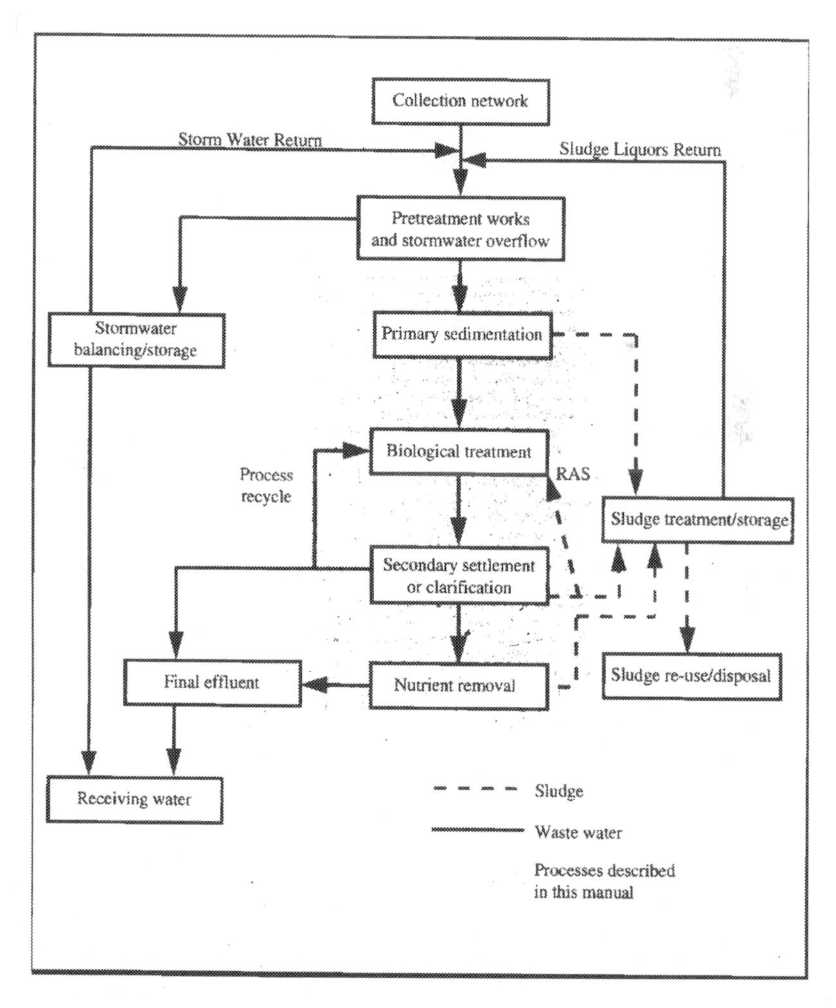

FIGURE 13.1 Typical layout of a simple waste water treatment plant.

Traditional methods for wastewater treatment can additionally be relegated on substructure of the operations involved; typical traditional wastewater treatment is accomplished by four rudimentary systems: physical, mechanical, biological, and chemical.

The relegation of techniques for the abstraction or the abbreviation of the contaminants predicated on the desideratum. Some of them are:

1. **Physical (thermal) treatment:** incineration (consummate combustion), supercritical fluid oxidation, and wet air oxidation of toxic contaminant (Kulkarni et al., 2000)
2. **Biological treatment:** oxidation (aerobic and anaerobic digestion) of the effluent to minimize waste
3. **Chemical treatment:** chemical oxidation of components with photo-processes with and without catalysts, direct reaction, and eradication by sonication

Factors such as effluent concentration, type, heterogeneity, local regulations, and economic viability will be considered during the choice of a particular effluent treatment technique. With increasing industrialization, waste contained more toxic components and precious metals; treatment of such waste involved the use of microorganisms for reduction or separation because of improved performance, low cost, and availability of raw materials. From external environment, microorganisms including yeast, fungi, algae, and bacteria can accumulate heavy metal efficiently (Ahluwalia & Goyal, 2007; Ansari & Malik, 2007; Benaïssa & Elouchdi, 2007; Bunluesin et al., 2007; Dursun, 2003; Ghimire et al., 2007; Mallick, 2003; Pan et al., 2007; Rajasulochana & Preethy, 2016; Ziagova et al., 2007). Fig. 13.2 shows the different paths of wastewater treatments.

13.1.1 Conventional Treatments of Effluents

There are many conventional physicochemical methods, such as screening, coagulation–flocculation, ozonation, biological treatment etc., for removal of metals, organic matters, nitrogenous compounds, and phosphorous in effluent from various industries (Adin & Asano, 1998; Rajasulochana & Preethy, 2016).

These conventional physicochemical and biological methods have certain limitations, such as large quantity of sludge generation, large space requirement, toxic metals that disturb the magnification of microorganisms, large non-biodegradable organic molecules such as dyes, and time-consuming biodegradation process, etc. Generally, there are three stages of these traditional process: primary, secondary, and tertiary treatments.

13.1.1.1 Primary Treatment

This is the initial stage of the traditional treatment of wastewater. Excessive amounts of oil, grease, and gritty matter and suspended settleable solids (which may inhibit anaerobic treatment later) are allowed to settle in a tank as simple as a septic tank. Floating materials are also separated by mechanical scrapping and screening so they will not enter into the secondary treatment to hinder the process. The liquid is neutralized to bring its pH in range of 5–9, which is ideal for further treatment.

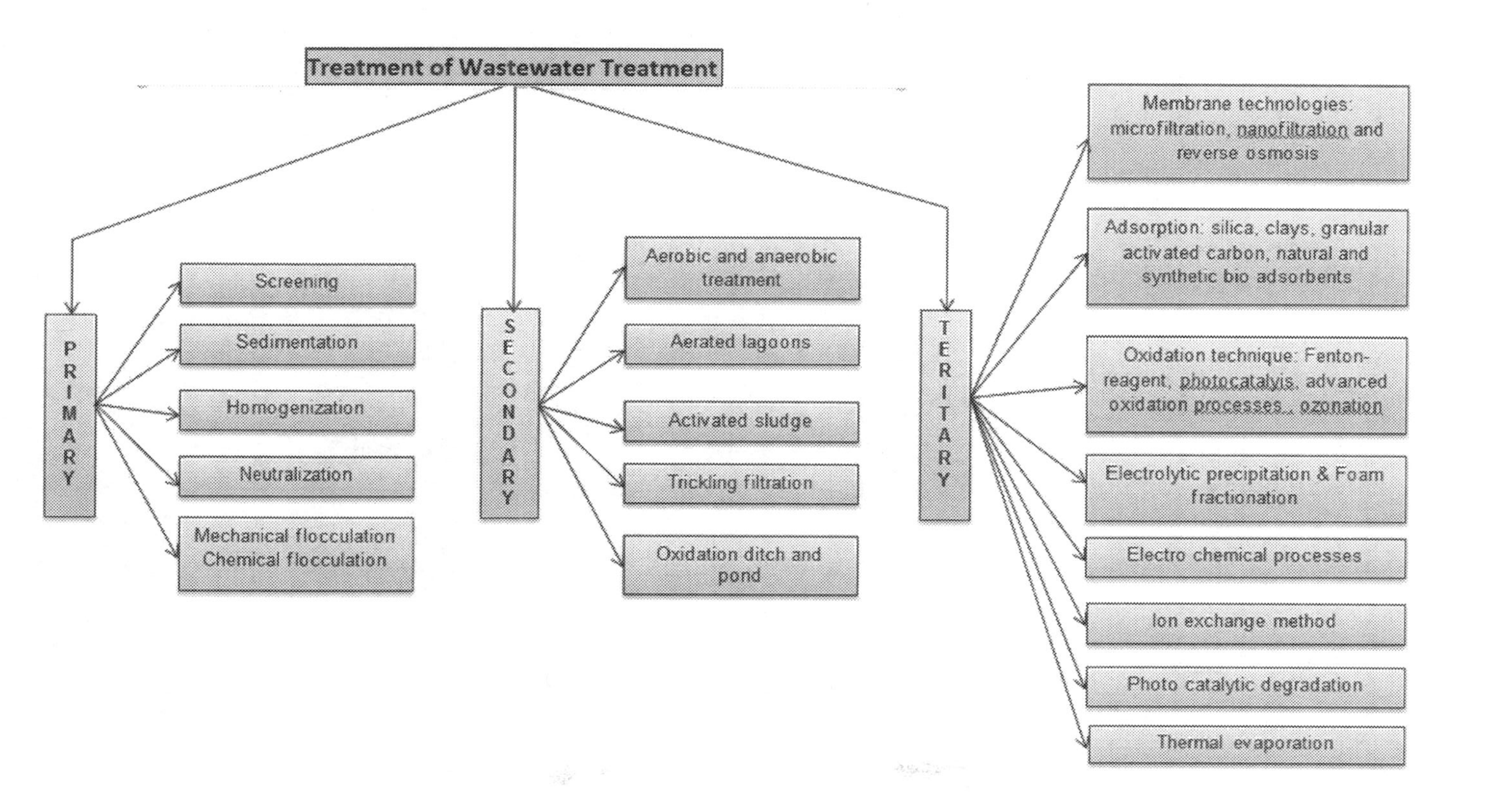

FIGURE 13.2 Different paths of waste water treatments.

There are two stages of this step: screening and settling. Screening is done in two stages: course screening and fine screening (Babu et al., 2007; Eswaramoorthi et al., 2008; Rajasulochana & Preethy, 2016). Coarse screens have openings of 0.6 cm or more to avert damage from plastics, metals, paper, and rags. After that comes a fine screen, with openings of 0.015–0.6 cm, and then sometimes very fine screen (0.02– 0.015 cm openings) too. These screens, particularly the fine one, reduce suspended solids in the effluent. The screened effluent enters the sedimentation zone, which allows gravitational settling of suspended heavy particles such as clay, silt, etc. along with liquid effluent. According to Das (2000; Rajasulochana & Preethy, 2016) simple sedimentation is not enough, as it leaves colloidal particles in the effluents. These colloidal particles can be settled by coagulation. These colloidal particles have charges on their surfaces. Through integration of chemicals, clumps of colloids will be formed. As they are heavy, they will settle. Alum and lime are the most commonly used chemicals for this step. Sometimes ferric chloride, ferric sulfate, and ferrous sulfate are also used (Das, 2000; EPA, 2000; Rajasulochana & Preethy, 2016; Chen et al., 2005).

In flocculation, the small particles are brought together by slow mixing with paddles to compose heavier particles. These coagulated heavier particles cannot remain suspended and will settle and accumulate as sludge. Sludge disposal is a major challenge of treatment plants. During flocculation care is taken to avoid a risk of short-circuiting, and the flocs in the system are obstinate. Sludge will be removed from bottom of the tank. Due care should be taken to prevent resuspension of the solids in the liquid (Heukelekian, 1941; McEwen, 1998; Rajasulochana & Preethy, 2016; Tridib & Bhudeb R., 2006). Albeit economical and requiring little maintenance, flocculation may be prone to failure due to many unknown factors. Even when operating effectively it will not be able to remove pathogens from the waste stream.

13.1.1.2 Secondary Treatment

For the removal of oil particles and phenol and reducing BOD and the color of effluent, a secondary treatment process is used. Sand filters can be used for abstraction of pathogens if there is deep permeable sand. They would not work with highly permeable soil. So that should be checked. Secondary treatment performance is based on the competency of microorganisms to use diverse wastewater constituents to provide energy for the microbial metabolism. Aerobic bacteria produce CO_2, water, and ammonia by oxidizing dissolved organic matter including nitrogenous material to get energy and nutrients. For biological treatment, activated sludge, trickling filters, or aerated lagoons systems are used in secondary treatment (Rajasulochana & Preethy, 2016).

Aerated lagoons are large tanks with rubber or polythene lining with aeration ability. Effluent that has been through primary treatment will be aerated for three to six days and sludge will be removed from it. This can reduce the BOD content by up to 99 percent and phosphorous up to 15 to 20 percent. Ammonia nitrification is also carried out. With the use of algae, TSS can also be reduced. The only limitation of lagoons is space. Care should be taken to avoid bacterial contamination. (Das, 2000; Rajasulochana & Preethy, 2016; U.S. EPA, 2000).

Trickling filters are also used in secondary treatment. These are a beds of synthetic resins, broken stones, poly vinyl chloride (PVC), gravel, coal, etc., in rectangular or circular shape. Microorganisms are kept in a gelatinous film on the surface of the filter to oxidize organic matter to CO_2 and water. Less space requirement makes this filter advantageous compared to aerated lagoons. Limitations of this filter include unwanted odor and very high investment (Etter et al., 2011; Rajasulochana & Preethy, 2016). In activated sludge processes, aerobic bacteria metabolize the suspended and soluble organic solids with aeration inside an effluent tank to generate CO_2. The sludge separated by sedimentation will be reused as a source of microbes. This process takes a long time to reduce BOD up to 90 to 95 percent (National Small Flows – Clearinghouse, 1996; Rajasulochana & Preethy, 2016; Yasui et al., 1996).

Primary and secondary treatment both produce sludge, disposal of which is a big issue. It cannot be disposed of untreated due to its composition; it causes an environmental quandary as it contains microbes and organic substances (Mittal, 2011; Rajasulochana & Preethy, 2016). Biological treatment of wastewater may also be carried out by using constructed wetlands, rotating biological contactors, biofiltration, or alternatives of these processes in fixed-media or suspended-growth reactors.

13.1.1.3 Tertiary Treatment

Technologies such as ion exchange, reverse osmosis, electrodialysis, electrolytic precipitation, incineration, etc., are used as tertiary treatments of effluent. In the ion exchange method, effluent passes through beds of cationic or anionic ion exchange resins. When effluent passes through a cationic resins bed, the cations present in the effluent will be exchanged by resins and superseded with hydrogen ions, making it acidic. In the same way, anionic ions are exchanged by resins to become alkaline (Ananthashankar, 2013; Neumann & Fatula, 2009; Rajasulochana & Preethy, 2016).

Membranes are used in reverse osmosis to separate large dissolved solid (TDS) particles, ions, and other components from the waste. Above 90 percent efficiency has been reported for this technique (Babu et al., 2007; Rajasulochana & Preethy, 2016). Electrodialysis uses semi-permeable membranes to reduce dissolved salts. Electricity is used to generate electrical potential across water to convey ions. These used membranes are charge specific. The negatively charged particles will be allowed to pass through anion-selective membrane, and positively charged particles will be trapped, and vice versa. When many membranes are placed, it will obstruct the flow of effluent. The ions will trapped or separated and neutralized. Membrane fouling may occur. (Marcucci et al., 2002; Rajasulochana & Preethy, 2016).

Electrolytic precipitation is a method using electric current for removal of dissolved metal particles. In this process electric current passed through wastewater by using electrodes. With electro-chemical reactions, the finely dispersed particles and dissolved metal ions coalesce with each other and get precipitated. Time required for this reaction is longer (Das, 2000; *Process Water Treatment by Oxidative and Electrolytic Processes*, n.d.; Rajasulochana & Preethy, 2016). Incineration or thermal evaporation is an eco-convivial technic. This high-energy–consuming method uses sodium persulfate as oxidizing agent and evaporate liquid. This method does not generate heavy sludge or toxic fumes. (Ananthashankar, 2013; Das, 2000; Rajasulochana & Preethy, 2016).

At the industrial level, due to very heavy contamination, a single method may not be effective to the desired level. So a combination of two or more techniques is employed to get better results. The pace of incipient chemical entities discovered is much faster than bio adaptability. So conventional methods may not work to generate expected results. The recently developed concept of removal of nutrients from wastewater before discharge is considered extended tertiary treatment. The term "nutrients" is used for nitrogen and phosphorous compounds present in treated wastewater. These compounds will adversely impact water resources by enriching water bodies and causing eutrophication. These compounds enrich water bodies and in turn promote algae and other photosynthetic aquatic plant development. This can be prevented by treatment with microorganisms to remove these nutrients (Environmental Protection Agency, Ireland, 1997).

13.1.2 Extended Tertiary Treatment: Nutrient Removal Processes

Extended tertiary treatment is carried out with the aim of abstracting nutrients such as nitrogen (N) and phosphorus (P) from wastewater before discharging it outside the treatment plant. There are several methods available for abstraction of N and P. Some are discussed here.

13.1.2.1 Mechanism and Application of Nitrogen Abstraction

Physicochemical methods for nitrogen abstraction: These methods are not used as part of common treatments because of cost, quality water poor, and larger quantities of sludge.

1. **Air divesting of ammonia:** If the ammonical contents are present, it can be stripped with biological filters or by scrubbing with water.
2. **Ion exchange:** Ammonical compounds can be removed by ion exchange with materials such as zeolites. Zeolites can be selected with ammonical ions such as zeolite clinoptilolite. A bed of zeolite can remove 90 to 97 percent of ammonical compounds.
3. **Chlorination reaction:** A step reaction with chlorine can convert ammonium in wastewater to nitrogen gas.

Biological methods for nitrogen abstraction: Nitrogen compounds can be removed by nitrification and denitrification steps in biological treatments such as percolating filter plants and activated sludge. Overall, the mechanisms produce nitrate and nitrite by oxidation of ammonia (nitrification) and then conversion to nitrogen, nitrous oxide, nitric acid, and nitrite (denitrification).

13.1.2.2 Mechanism and Application of Phosphorus Abstraction

1. **Chemical method:** Generally phosphorus found in wastewater has three forms: organic phosphate, polyphosphate, and orthophosphate. The first two can be converted to orthophosphate during aerobic treatment. This orthophosphate can be removed by precipitating with chemicals.
2. **Biological method:** The bacterium Acinerobacter spp. is capable of absorbing biological phosphorus and relinquishing phosphate under anaerobic conditions.
3. **Cumulated Biological N and P Abstraction:** Some research groups developed methods for removal of both nitrogenous and phosphorous compounds in biological treatments such as activated sludge and submerged filters. There are a number of patented technologies for combined removal of N and P. Both aerobic and anaerobic bacteria are used in these methods.

Over and above the mentioned traditional treatments, some other treatment processes are carried out on treated wastewater before discharge.

13.1.3 Disinfection of Wastewater

Urban wastewater may contain some pathogens or micro-organisms that may cause disease and should be removed or destroyed. Still it is not sterilization, which would remove all microbes.

Proximity to wastewater treatment plant presents some common issues common such as unpleasant odor, noise, visual impact, etc. Plants near residential area may have to address these issues too (Environmental Protection Agency, Ireland, 1997).

13.1.4 Pretreatment of Industrial Waste

The performance of any effluent treatment plant depends on chemical and biological treatments, which in turn depend on the performance of microbes. Microorganisms can work on typical compositions of toxic materials in effluent. If industries can carry out some pretreatment of their waste, it will improve performance of common facility. The Environmental Protection Agency Act, 1992 (Urban Waste Water Treatment) Regulations, fourth schedule 1994 (Sr. No 419 of 1994), states that "industrial effluent entering common collecting facility and effluent treatment plants be subjected to such pretreatment as is required in order to protect plant facility, damage to equipment and health of working staff and maintain performance of the plant to achieve targeted limits of COD, BOD, Colour, etc." Common effluent treatment plants faces difficulties such as extreme temperature, hydraulic overloads, and sometimes excess quantity of waste such as suspended solids, organic or inorganic wastes, extreme alkaline or acidic waste, oil and grease, flammable and explosive materials, odorous or corrosive volatiles.

13.2 TEXTILE EFFLUENT TREATMENT

One of the most exhaustive water-consuming industries is the textile industry. Water is used in almost each step of manufacturing such as dyeing of fabrics and cleaning the products as well as raw materials (Vineta et al., n.d.). Fundamentally, manufacturing activities in textile industries can be divided into three main activities. Yarns and threads converted from fibers. Fabrics prepared by weaving and knitting of yarns and threads. Finally as per market demand dyeing and finishing of the fabrics (Kharat, 2015). Fig. 13.3 depicts the general process flow diagram for textile industries.

In general, a typical textile industry manufacturing process include many mechanical processing and wet processing. Mechanical processes include weaving, knitting, and spinning. Wet processes include printing and finishing, dyeing, mercerizing, bleaching, scouring, sizing, and desizing.

A wide variety of pollutants are produced by textile industries during the processing of fibers, fabrics, and garments. The major concern is the chemical load and the magnitude of wastewater produced. In textile industries, water is use at many points in the process, including cleaning of raw material, flushing during process steps, etc. Some of the major applications are applying chemicals on materials and rinsing manufactured materials, etc. Vineta et al. studied and concluded that almost 200 L

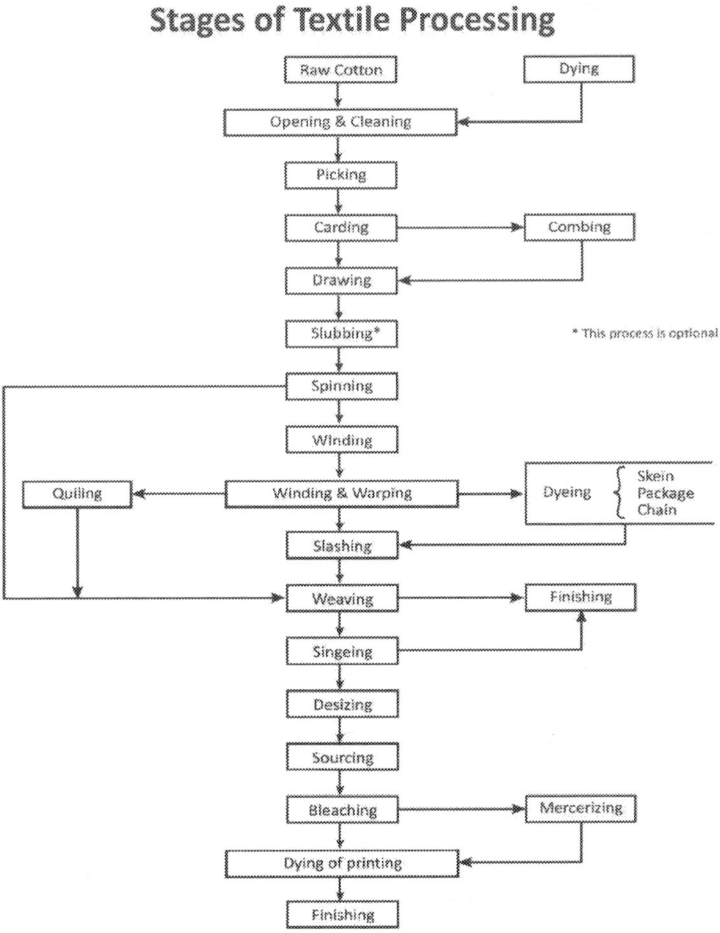

FIGURE 13.3 Textile industry general process flow diagram.

water is used per kg of textile. They reported that the bleaching operation alone accounts for almost 38 percent of the water used in a typical textile industry, while dying uses about 16 percent, printing applications about 8 percent, boilers about 14 percent and other uses almost 24 percent (from textile wet finishing operations, 2009; Vineta et al., n.d.). Characterization of effluent

13.2.1 Why the Chemical Load of the Textile Industry is High?

It is well known that textile industries have a higher chemical load than other chemical industries. The chemicals used in textile industries contribute to increase this load. Some of the chemicals are detergents (poisonous to fish), stain removers (ozone depletion), oxalic acid (very high COD), sequestering agents (ostracized chemicals), printing gums (damages organs such as kidney, liver, etc.), fine-tuning agents (the chemical uses are banned internationally), bleaches (causes itching and is harmful), and dyes (carcinogens).

Textile industries use a variety of chemicals. Some are biodegradable (e.g., starch), and some are non-biodegradable (e.g., dyes). So the wastewater could have lower DO, meaning higher COD/BOD. In textile wastewater, fibrous solids enter from raw materials and processing, which in turn perturbs aquatic life. After production, the wastewater generated contains large amounts of particles such as dyes, which make the water well beyond the standards and raise the chemical load on the treatment plant (Vineta et al., n.d.).

The large variety of chemicals used in textile industries makes effluent treatment more complex. There are organic/inorganic non-biodegradable compounds such as surfactants, pesticides, phenols, metals, dyes, phosphates. These contribute to increased suspended solids and dissolved solids, very low dissolved oxygen, higher chemical oxygen demand, and biological oxygen demand (*Water Online*, n.d.).

13.2.2 Conventional Treatments of Textile Effluents

Treatment of effluent from textile industries can also be done with traditional processes. But looking at chemical loading, conventional processes that are applicable to other chemical industrial wastewater are not enough. There are different processes can be identified on the basis of generating pollution load. Dyeing and finishing processes generate highly concentrated liquid effluents. Washing and rinsing bath produces medium polluted effluent and water from cooling tower, would not having any pollution load.

13.2.2.1 Primary Treatment

All steps are similar to conventional primary treatment method. Here the coarse suspended materials (such as rags, fibers, yarns, pieces of fabrics, lint, etc.) present in effluent are first removed use bars and fine screens (Chipasa, 2001; Rajasulochana & Preethy, 2016). Then rest of the steps have been explained in the following sections.

13.2.2.2 Secondary Treatment

Similar to traditional processes, biodegradable materials are treated with bacteria, which are then decomposed and removed as sludge.

13.2.2.3 Tertiary Treatment

Similar to traditional processes all steps will be carried out to reduce color, odor, etc. in the textile effluent. During cotton dyeing processes, a very high concentration of NaCl is used as electrolytes. For removal of such high concentrations of salts reverse osmosis is applied (Ananthashankar, 2013; Rajasulochana & Preethy, 2016; Subramanian Senthil Kannan et al., 2006).

In addition to the mentioned processes, other processes have been developed that can treat textile effluent. A wide range of dyes can be decolorized with photocatalytic degradation.(Rajasulochana & Preethy, 2016; Vautier et al., 2001). Adsorption of dyes molecules on various absorbers can also remove color effectively. Dyes waste can be oxidized with anodes coated with catalysts by anodic oxidation process. It is also known as electrochemical treatment. Ion exchange of different cations such as Cu (II), Fe (III), Fe (II) with zeolite is also reported as an efficient process. No sludge would be produced in this method of degradation of dyes.

One more method that is capable of handling textile industry effluent is electrocoagulation. It removes contaminants in a more efficient and optimized way. It works on the principles of electrochemistry, coagulation, and flocculation.

From thorough study of wastewater treatment of the textile industry, it is concluded that, compared to the end-of-the-pipe approach, a systematic approach of reducing wastewater generation during each step could be more effective. If the industry can implement an approach of reducing waste generation at source, a majority of issues related to effluent treatment can be resolved. (Vineta et al., n.d.) If the textile industry can set and achieve a sustainable goal to reuse water, it will not only solve the pollution problem but lead to cost savings too. Treatment and reuse of textile wastewater is a promising answer to water resources preservation and enhancement (*Water Online*, n.d.).

13.3 EFFLUENT FROM THE OIL AND GAS INDUSTRY

The oil and gas industry makes up about 3.8 percent of the global economy. Oil and gas companies are continuously expanding and investing millions annually to increase production. The oil and gas industry also provides feedstock such as dyes, fine chemicals, fertilizers, pesticides, textiles, plastics, synthetic rubber, and other industrial goods to a variety of other industries. It is hard to imagine the world without this major industrial sector; but the oil industry is also a major source of environmental pollution. The waste from this global industry affects the air, water, and soil extensively, leading to environmental degradation and harm to human population.

The petrochemical industry is faced with several challenges due to the highly volatile and fluctuating gas prices, the constant burden to improve and expand its base capacity, and the need to improve its environmental footprint. From offshore facilities to the refinery, a constant and crucial component in the production process is water. Advanced oil recovery techniques have led to more use of water in this industry. Hence the management of wastewater is a key component for successful operations. The most common method of disposing of wastewater is underground injection, where water can no longer be retrieved or used.

Wastewater from the petrochemical industry consists of a wide variety of pollutants such as hydrocarbons, mercaptans, grease, phenols, ammonia and ammonium salts, sulfides, and other organic molecules (Varjani et al., 2020). These compounds form a complex mixture difficult to separate and toxic to the environment (Raza et al., 2019). Hence, suitable management or treatment is needed for the discharge, treatment, and final disposal of this water. The complex composition and the rigorous environmental regulations regarding discharge limits necessitate the requirement of a combination of treatment methods (Wei et al., 2019).

Wastewater treatment plants in the oil and gas industry involve different processes such as coagulation, aerobic and anaerobic treatment, reverse osmosis, adsorption, ultrafiltration, chemical destabilization, flocculation, dissolved air flotation (DAF), membrane process, etc. (Bennett, 1988; Kriipsalu et al., 2008; Li et al., 2014; Sonune & Ghate, 2004; Tejero et al., 2017; Usman et al., 2013; Varjani et al., 2020). To decrease the toxic level of pollutants present in the effluent of oil and gas, many processes are used that depend mainly on the source and concentration of contamination present in the effluent. The treatment process can be divided into three categories: primary treatment, secondary treatment, and tertiary treatment (Varjani et al., 2020).

13.3.1 Primary Treatment

Primary treatment is a physical treatment process wherein removal or separation of suspended and particulate solids and immiscible liquids such as oil and grease, etc., is achieved. In this treatment no chemical or biological treatment is carried out on the petroleum wastewater (Renault et al., 2009).

Primary treatment is an important step because it allows waste for the secondary treatment unit. The removal of oil from wastewater involves the gravity separation followed by skimming (Al-Shamrani et al., 2002). The cost effective and most efficient method is the American Petroleum Institute (API) oil–water separator, which works on the principle of difference in specific gravity. But the disadvantage API separator is that it is not efficient for removing smaller oil droplets and emulsions present in the effluent (Umar et al., 2018). The other significant method, dissolved air flotation (DAF), promotes coagulation to reduce the suspended colloidal solids or dispersion and immiscible liquids from the effluent (Al-Shamrani et al., 2002). A coagulation process is used to remove turbidity and reduce organic load from the effluent.

The induced air flotation (IAF) method used in the primary treatment method is cost effective, has a compact unit size, and effectively separates the suspended solids, oil and grease, and other insoluble impurities from effluents. In this process, air is dissolved in water under high pressure, and micro-bubbles are formed when the pressure is released. The particles get trapped in these micro-bubbles and float on the surface of the water, where they are skimmed and separated (Bennett, 1988).

The other important method to remove the hydrocarbons from the effluent is the adsorption method, in which activated carbon, zeolites, etc. are used (Fakhru'l-Razi et al., 2009a). Evaporation is also suggested to remove oil residuals in brine wastewater (Pichtel, 2016).

Currently, sedimentation is one of the primary treatment methods used before biological operation methods to remove suspended solids. However, literature also revealed that physical or primary treatment processes were comparatively fruitless for the treatment of petroleum wastewater because of its complexity, and therefore other practices might be used for pretreatment (Aljuboury et al., 2017).

13.3.2 Secondary Treatment

The various oil and gas effluents are characterized by the presence of large amounts of crude oil products, polyaromatic hydrocarbons (PAHs), ammonia, phenolic derivatives, metals in various forms, suspended solids, sulfides etc. (Diya'uddeen et al., 2011; Mustapha et al., 2018). The main focus of this treatment stage is to decrease the pollutant levels of effluent to an acceptable limit before discharge into the environment.

The secondary treatment stage mainly consists of three methods: flocculation, coagulation, and biological treatment, which integrates the action of various kind of microbes to eradicate the stabilized organic pollutants in the effluent (Changmai et al., 2019; Varjani et al., 2020). In modern effluent treatment, the coagulation and flocculation methods are significant for overall effluent treatment processes. In this process, coagulants or flocculants, such as ferric chloride, aluminum chloride, aluminum hydroxide chloride, aluminum sulfate, chitosan, carboxy methyl cellulose, etc., are added to accelerate the agglomeration of solid particles, and the sedimentation process occurs in the clarification tank (Farajnezhad & Gharbani, 2012; Hosny et al., 2016; Kriipsalu et al., 2008). Research studies have also reported a high success rate of the electrocoagulation process for separation of oil from oily effluent; hence it plays a significant role in the treatment of oil and gas effluent (Chen et al., 2002). Hybrid electrocoagulation methods with reverse osmosis, nanofiltration, and microfiltration can also be used for treatment of oily effluents (Changmai et al., 2019).

Effluent from oil and gas contains a high load of organic contents, oil, and grease, etc. These high organic loads are mainly treated by biological treatment methods. The biological treatment can be achieved by aerobic or anaerobic treatment methods.

The anaerobic digestion method requires less space (i.e., it is compact) and also generates low sludge compared to the aerobic method. The efficiency of anaerobic digestion depends on various parameters, such as the constituent material, type of the reactor, temperature conditions, loading rate, source of effluent, and other operational conditions. In anaerobic digestion, the effluent is commonly treated in various reactors, such as anaerobic baffled reactor (ABR), up-flow sludge blanket reactor (UASB), etc. (Ji et al., 2009). Compared to anaerobic processes, aerobic processes are typically characterized as less sensitive to toxic effects and have greater growth of organisms and better ease of operations. Conventional aerobic reactors, such as traditional active sludge (TAS), sequence batch reactor (SBR), membrane bioreactor (MBR), moving bed biofilm reactor (MBBR), biological aerated filter (BAF), aerobic submerged fixed-bed reactor (ASFBR) etc., have been established to treat various petrochemical wastes. The efficiency of aerobic treatment depends on important factors such as temperature, feed-to-organism ratio, sludge retention time, dissolved oxygen level etc. (Ebrahimi et al., 2016; Ghimire & Wang, 2018; Thakur et al., 2014).

13.3.3 Tertiary Treatment

After the secondary treatment process, effluent may contain contaminants such as TSS, COD, dissolved metal ions, and organic impurities such as PAHs. Such contaminants can be removed by tertiary treatment methods such as sand filtration, activated carbon process, and chemical oxidation methods (Ghimire & Wang, 2018; Li et al., 2014).

Effluent from secondary treatment, especially a biological treatment system usually contains suspended solids in clarifiers. These effluents are passed through a filter bed made of a layer of sand below a layer of anthracite; the larger particles are trapped by the anthracite, and the finer solids are held up in the sand. This arrangement is known as a sand filtration system.

Removal of dissolved organic components from the oil and gas effluent can be done by the carbon adsorption method. In this method the effluent is passed through a granular activated carbon (GAC) bed, where the dissolved organic constituents present in the effluent are adsorbed by the activated carbon (Altmann et al., 2016).

The chemical oxidation method is commonly used for removing residual COD and trace organic compounds such as PAHs. In general, hydrogen peroxide, chlorine dioxide, and ozone are used as oxidizing reagents. In some cases, a photo-oxidation method is also used where ultraviolet (UV) light acts as a catalyst.

13.3.4 Treatment of Flow Backwater and Produced Water

The natural gas trapped below Earth's surface is released when hydraulic fracturing is used; two types of wastewater is produced: flowback wastewater and produced wastewater. Although these two types of wastewater are from different sources, their compositions are very similar and therefore challenging to differentiate without conducting a decisive chemical analysis (Gregory et al., 2011; Mantell, 2011).

Flowback wastewater is the water-based solution that is injected into the ground during drilling and contains the chemical additives necessary to fracture the shale basin and release natural gas. Hydrofracking flowback contains a mixture of chemical additives, dissolved metal ions, and total dissolved solids. The hydrofracking fluid is ultimately returned to the surface as flowback water and will require treatment (Balashov et al., 2015).

Produced water is different from flowback water because produced water is naturally occurring water that has already formed in the shale. Produced water is "mainly salty water that is trapped in reservoir rock and brought up along with oil and gas during production. The produced water leaches minerals such as barium, calcium, iron and magnesium dissolved hydrocarbons, dissolved inorganic salts and dispersed oil droplets, naturally-occurring organic compounds, including benzene, toluene, ethylbenzene, oil & grease, radioactive materials, bacteria and other living organisms, under the earth's surface. In certain cases, the produced water from these oil exploration operations can either be reused or it can be used for other applications, including agriculture or irrigation or other non-potable applications based on monetary feasibility" (U.S. Environmental Protection Agency, Washington, 2016).

Flowback water from hydraulic fracturing operations and produced water from the extraction process of oil and gas reserves must be properly managed during oil drilling or fracturing in order to mitigate any environmental effects to existing water bodies. In general, these waters are composed of a brackish solution containing relatively high levels of dissolved minerals, certain heavy metals, salts and organic compounds, and potential sulfur-reducing bacteria (SRB). The composition of these waters is variable and depends on the geographical area.

Decreased availability of water and increased environmental regulations relating to produced water from hydraulic fracking and conventional oil and gas field operations across the world look for innovative treatment solutions to meet the environmental and regulatory issues. The scientific world is aware that flowback and produced water are usually mixed together by the time they reach the surface; hence currently there is no single treatment method for hydrofracking wastewater. Therefore, any technologies or solutions that have been included and are specific to the treatment of one type of wastewater will be considered as only a part of a treatment plan for treating the entire mix of flowback and produced water (Igunnu & Chen, 2014).

There are three potential wastewater treatment and disposal issues that the Environmental Protection Agency (EPA) has recognized. First, the EPA is alarmed with the hydrofracking fluid chemicals being discharged. Second, it is concerned about the possible quantities of wastewater not being fully treated. Lastly, it is concerned about the accidents that could occur during the transportation process of wastewater.

13.3.4.1 Injection Wells Process

Injection wells processes are cheap, energy efficient, safe, and effective, but the energy requirements for treating some fluids make these techniques economically unviable and environmentally disagreeable. This method is very popular in the United States; the government estimated that about 90 percent of produced waters are disposed of using deep well injection. Although this process is widely used in the oil and natural gas industry, there are still issues that could potentially threaten the drinking water supply.

13.3.4.2 Membrane Filtration Treatment Method

The membrane filtration method is widely used for treating produced water and others. Basically, membranes are microporous films with specific pore grades, which selectively separate a fluid from its components. This filtration comprises microfiltration (MF), ultrafiltration (UF), reverse osmosis (RO), and nanofiltration (NF) (Xu & Drewes, 2006). Membrane systems can compete with more complex treatment technologies for treating water with high oil content, low mean particle size, and flow rates higher than 150 m^3/h. Membrane technology operates two types of filtration processes, cross-flow filtration or dead-end filtration, which can be a pressure (or vacuum)-driven system (Igunnu & Chen, 2014).

(i) Microfilteration/Ultrafiltration

Microfilters have a large pore size of 0.1–3 μm. These filters are used to remove suspended solids and turbidity from the wastewater. Microfilteration can operate either in cross-flow or dead-end filtration. MF operates at a low transmembrane pressure of 1–30 psi and can serve as a pretreatment to desalination but cannot remove salt from water (Madaeni, 1999). Utrafilters pore sizes are between 0.01 and 0.1 μm. The ultrafiltration (method is the most effective method for removing oil from produced water in comparison with traditional separation methods, and it also removes odor, color, living organisms, hydrocarbons, suspended solids, colloidal organic matter, etc. (Ghimire & Wang, 2018; Han et al., 2010). UF also operates at low trans membrane pressure.

(ii) Ceramic/Polymeric Membranes

Polymeric membranes have become a modest technology for wastewater treatment for produced water. These membranes are extensively used in membrane separation processes because of properties such as hardiness, flexibility, economic value, and excellent separation. This method can be termed a green method because of less sludge formation and effective removal of various chemicals (Mondal, 2016). Ceramic membranes are capable of removing metal oxides, organic matter, particulates, oil, and grease. They are better than polymeric membranes due to increased chemical resistivity, mechanical and thermal stability, and higher productivity. While the capital costs of ceramic membranes are higher than the polymeric membrane method, they have a longer life span. Hence that means the overall cost for long term usage will be educed drastically (Fakhru'l-Razi et al., 2009b).

(iii) Bentonite Clay and Zeolite Membranes

In membrane separation, various materials are used for edifice the membranes in numerous operational conditions, including sensitivity to variation in flow rate of produced water, etc. These membranes must possess stable thermal, mechanical and chemical properties (Zaidi et al., 1992). Zeolite membranes are stable in high temperatures and pressure conditions and also can be used in strong chemical environments. Researchers have used RO-zeolite membrane for separating the alkali and alkaline metal ions from produced water. Silicate zeolite membranes are also studied for desalination of brine in produced water of oilfield (Li et al., 2008). Bentonite clay membranes are also studied for treatment of produced water because of their low cost, lack of requirement of chemicals, and excellent capability to adsorb ions. These clay membranes effectively reduce the mineral content but are not suitable for high TDS content present in produced water (Liangxiong et al., 2003).

(iv) Reverse Osmosis and Nanofiltration

Reverse osmosis (RO) and nanofiltration (NF) are pressure-driven membrane processes. In general, an RO process is used to separate ionic and dissolved constituents present in water bodies (Madaeni, 1999), while NF membranes are selectively designed for separation of multivalent ions. The technology of RO membrane also shows an outstanding separation technique along with necessary pretreatment for oilfield-produced water treatment. NF membranes can also be employed for produced water treatment on both bench and pilot scales (Mondal & Wickramasinghe, 2008).

13.3.4.3 Thermal Remediation

Thermal remediation is a process that can be applied in the treatment of flowback water and produced water. The treatment can be done by using multiple effect distillations, vapor compression distillation, steam-enhanced extraction, electrical-resistance heating, and thermal conductive heating. These methods are simple and highly effective for high concentration of TDS, organic constituents, suspended solids, etc. The main disadvantage of these methods is high cost due to high energy consumption. According to current research, thermal process engineering make thermal process more attractive and competitive in treating highly contaminated water (Igunnu & Chen, 2014).

13.3.4.4 Biological Treatments

Biological treatment can be achieved by aerobic or anaerobic treatment methods. In aerobic treatment, various methods are used, such as activated sludge method, trickling filters, chemostate reactors, biological aerated filters (BAF), sequencing batch reactors (SBRs), and lagoons. The activated sludge method is used to remove total petroleum hydrocarbons (TPH) with efficiency of 98 to 99 percent. Total influent organic carbon (TOC) can be removed more efficiently by the SBR method compared to trickling filters and chemostate reactors. COD can also be removed by the SBR method with efficiency of up to 50 percent. Biological aerated filter (BAF) enables the biochemical oxidation and removal of organic constituents in wastewater. BAF effectively removes COD, BOD, oil, suspended solids, ammonia, hydrogen sulfide, and heavy metal ions present in the produced water. This method does not required any chemical treatment during its operation, and its life span is very long.

Anaerobic degradation will be a cost-effective alternative method if the treated water is in concentrated form (Fakhru'l-Razi et al., 2009b).

13.3.4.5 Ion Exchange Technology

Applied technologies such as ion exchange methods are used in industrial operations for various purposes. This method is widely used for the removal of ions and metals by ion exchange resins from polluted water. This technology has a good life span but requires pretreatment for removal of solids and also requires chemicals for regeneration of resins (Brown & Sheedy, 2002).

13.3.4.6 Chemical Oxidation Method

Another useful method is chemical oxidation, a reliable, applied technological method for removing color, odor, COD, organic constituents, BOD from produced water. In this method, oxidants such as oxygen, peroxide, ozone, chlorine dioxide, chlorine, permanganate, etc., are used for removing the pollutants from wastewater. The main disadvantage of this method is maintenance of the pump and other equipment (Renou et al., 2008).

13.4 MINING EFFLUENTS

A lot of wastewater (e.g., stone cutting water, wash water, scrubber water) is generated during the processes of mining of gold, nickel, copper, uranium, coal, and other minerals. These processes involve cooling equipment, dewatering from mines, and storm water runoff at mines and processing plants. Minerals are obtained from underground mines, surface mines, and quarries with the help of different types of machinery and explosives. A large quantity of water is used in extraction processes. Many processes are performed even after the extraction processes, such as crushing, milling, washing, and other chemical treatments in which water is used for different purposes and ultimately discarded into the environment.

The process of mining seriously affects freshwater sources through the use of large quantities of water in the ore processing. Acid mine drainage, contamination of heavy metals, and increase in sediment levels in streams are the results of the water pollution caused by the mining processes. Water pollution is caused from the discharged mining effluent and waste from investigations and waste rock internments (Ali et al., 2016; Marcus, 1997).

Continuously increasing the human activities of mining is dangerous for the water sources on which the whole world depends. Water is considered "mining's most common casualty" (Lyon, n.d.). Even though taking extreme care and making improvements in mining practices, significant environmental risks remain during the mining processes. We should manage water pollution from mine waste rock and tailings. The impacts of mining depend on a variety of factors, such as the local terrain, the composition of minerals; the type of technology used for the mining process; the expertise, knowledge, and environmental commitment of the company responsible for the mining contract; and finally, the monitoring and enforcement of compliance with government regulations.

Due to the development of new technologies, the extent of mining has been increased and waste generation has been multiplied enormously. As mining technologies are developed for profit making, more and more waste will be generated in future, which is certainly harmful for animals and human beings. After the removal of waste rocks often containing sulfides, heavy metals, and other contaminants, they are generally stored above the ground in large free-draining piles. These waste rocks and the exposed substrata walls are the major source of the heavy metals pollution. Hence care should be taken for the implementation of systematic water treatment of the mining wastewater to protect our environment, which will indirectly protect the life on Earth. This treatment will help in processing wastewater from mines, in the extraction of minerals from water streams, in the treatment of mine sludge, and in reusing process water. The processes should be able to remove the pollutants such as suspended solids of different heavy metals such as lead, arsenic, mercury, cadmium, chromium, copper, nickel, iron, selenium, cobalt, molybdenum, manganese, antimony, zinc, and many more to give zero liquid discharge. Different types of technologies can be implemented for mine water treatment, such as filtration, evaporation, crystallization, desalination, ion exchange, membrane separation and other various other technologies.

13.4.1 Heavy Metals: Toxic Effects and Remediation

13.4.1.1 Toxic Effects of Heavy Metals

Heavy metals are defined as metallic or metalloid materials that have higher density than water and have toxic effects on living beings (Olías et al., 2004). Some of these heavy metals, such as Co, Cr, Cu, Mn, Fe, Mo, Se, Ni, and Zn, are also essential nutrients required for various biological functions and their deficiency can cause various diseases (WHO, 1996).

Heavy metal contamination is currently the biggest issue for the water environments and has gained considerable attention all over the globe due to the toxicity, abundance, persistence, and bio-exaggeration of heavy metals in the environment and their subsequent accumulation in aquatic life (Förstner & Wittman, 1983; Meng et al., 2016). Heavy metals in water environments enter from numerous sources, including natural processes such as bedrock weathering, volcanism, erosion, and anthropogenic activities such as agriculture fertilization, industry and drainage, and especially metal smelting, mining, and refining. The pathways of heavy metal intoxication in living organism are explained in Fig. 13.4 (Azeh Engwa et al., 2019).

Researchers in this area are continuously reporting that mining activity is the most critical contributor of heavy metal pollution in river basins (Förstner & Wittman, 1983; Kumar et al., 2017; Meng et al., 2016; Shi et al., 2013; Zhou et al., 2007). These heavy metals can cause considerable harm to the environment when they are above certain permissible concentrations. These elements can leach into the surface water and groundwater, which is taken up by the plants. They can bond with soil components such as clay, sand, or some organic matter, which later affects human health (Gómez-Alvarez et al., 2009; Kabata-Pendias & Szteke, 2015; Sierra et al., 2017; Zhang et al., 2016). The Surface sediments have significantly higher amount of heavy metals compared with those in bodies of water, and hence it is essential to determine the heavy metal contents in the surface sediments (Sundaray et al., 2011; Wang et al., 2014).

Once heavy metals enter a water body, they can harm aquatic life. Through the processes of chemical and physical adsorption followed by the precipitation, they

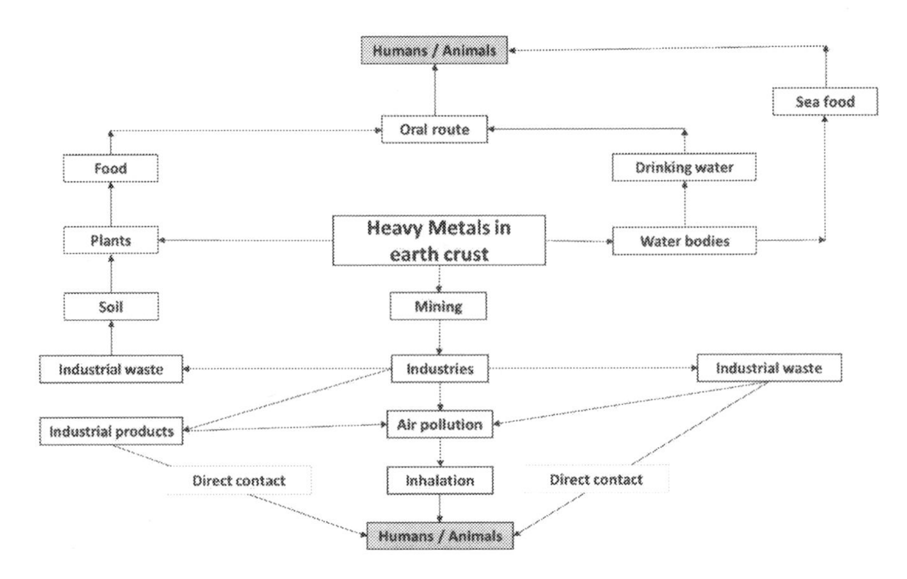

FIGURE 13.4 Pathways of heavy metals intoxication in living organisms.

accumulate in the sediments of the water environment (Gao et al., 2014). Both natural weathering and mining cause heavy metal contamination of water. There are various toxic effects due to the mining water waste in which substantial amount of heavy metals are disposed of into local water sources. It is our responsibility to maintain the ecosystems of the Earth. There are two pathways applied for the release of heavy metals, depending on the mining technique involved. First is the open-pit mining, which primarily contributes the heavy metals via enhanced weathering; and second is underground mining, which contributes heavy metals to acid mine discharge due to the weak weathering processes. Very little attention has been given to the deterioration of water quality in response to human activities such as mining (Wei et al., 2018).

Various adverse effects have been observed due to the bioaccumulation of heavy metals in the human body as well as other living organisms. These heavy metals make complexes with proteins, nucleic acids, and other bio-macromolecules and damage them thus causing toxicity. It can affect the neurological system to cause mental disorder and can damage blood constituents such as red and white blood cells, etc. It can also affect the respiratory system, renal system, and other vital organs. Moreover, studies by Jaishankar et al., and Järup et al., show that longer to heavy metals leads to muscular dystrophy, multiple sclerosis, Parkinson's disease, Alzheimer's, and cancer, etc. (Jaishankar et al., 2014; Järup, 2003). The pathways of heavy metals into a living organism and their toxic effects are depicted in Fig. 13.5 (Azeh Engwa et al., 2019).

13.4.1.2 Remediation of Heavy Metals

Today, many researchers are working on remediation technology for heavy metal removal from soil and water as it is the demand of the day due to high toxicity and harmful effects on living organisms (Ertan Akün, 2020; González-Martínez et al., 2019). Various processes are used for remediation of heavy metals. Currently physicochemical remediation (precipitation, ion exchange, and reverse osmosis),

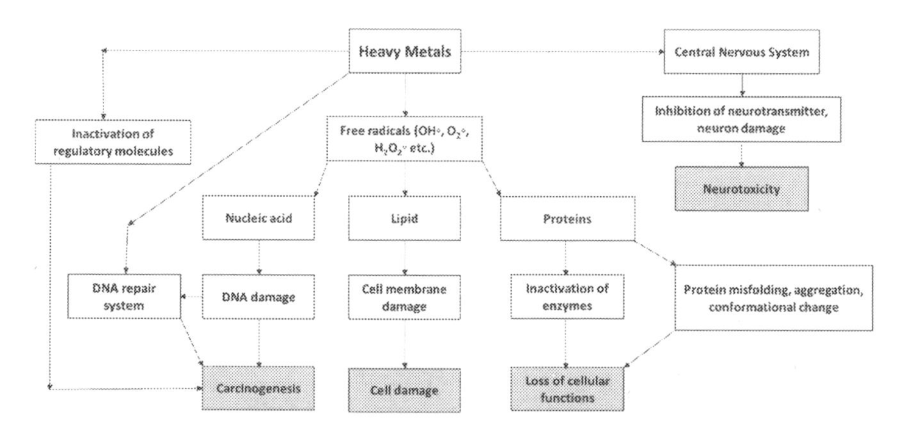

FIGURE 13.5 Pathways of heavy metals entry into living organism and their toxic effects.

phytoremediation, or microbial remediation are the most used technologies for heavy metal removal, but all these techniques have their own advantages and disadvantages. Hence there is a need for safe and economical processes to be implemented.

There are chemical and biological remediation methodologies available, but chemical remediation should be avoided because contamination of heavy metal is a chemical process itself; hence biological processes are preferred. Phytoremediation is a biological process that is listed and concisely explained here (Akpor & Muchie, 2010; Ertan Akün, 2020).

Phytoremediation is a bioremediation process in which various types of plants are used to remove, transfer, stabilize, and/or destroy contaminants in the soil, sediment, sludge, surface water, and groundwater.

- Phytoaccumulation
- Phytoextraction
- Phytostimulation
- Phytostabilization
- Phytodegradation
- Phytovolatilization
- Hydraulic control
- Microbial remediation

Phytoaccumulation

This method is also known as phytoextracion, in which, at a specific region, the heavy metals are accumulated in soil and water. These heavy metals are removed under field conditions by plants or algae. This process is a hybrid chemical bio-concentration and bio-magnification.

Phytoextraction

In this method, plant-root contaminants can be absorbed, translocated, and stored along with other nutrients and water. For wastes containing heavy metals, this method is very useful. Different plants can be used according to their capability to take up and withstand high levels of pollutants. This can either be performed continuously using hyper accumulators or induced through the addition of chelates to increase bioavailability.

Phytostimulation

In this method, natural substances are released through their roots, thereby supplying nutrients to microorganisms, which in turn enhances biological degradation. By this the microbial activity increases, which will stimulate the natural degradation of organic contaminants.

Phytostabilization

In this method, mobility of heavy metals is reduced in the soil by various methods such as adding organic matter, controlling soil erosion, providing an aerobic environment conditions, etc.

Phytodegradation

The other name of this method is phytotransformation. In this method, plants break down organic contaminants into simple and biodegradable molecular form in the soil through metabolic processes with the help of enzymes.

Phytovolatilization

This process involves the absorption of water and organic contaminants from the soil by the plants, which discharge them in the volatile form into the environment via transpiration.

Hydraulic Control

In this process, the flow direction of a contaminated aqueous medium is altered, and contaminated flow is oriented. With this method, stopping or reversing contaminant migration is possible.

Microbial Bioremediation

In this method, microorganisms are stimulated to rapidly degrade hazardous organic contaminants to environmentally safe compounds in soils, subsurface materials, water, sludge, and residues. This method plays a key role in detoxification of water. Researchers are taking a keen interest in the studies of the interactions of microorganisms with heavy metals. Some of the research is established on elucidation of different metal resistance mechanisms and their interactions and processes, especially those used by bacteria, protozoa, and fungi.

In addition to these phytoremediation processes, various physicochemical methods such as precipitation, ion exchange, reverse osmosis, photocatalytic and electrocatalytic methods, etc. are also used for the purification of water contaminated with heavy metals, and the selection of the method depends on the nature and extent of contamination.

13.5 AGROCHEMICALS

For human beings and animals, a pollution-free environment is essential for a healthy life. Water and food are the most important part of human life, and that must be pollution free. One-sixth of the world population is facing the facing the problem

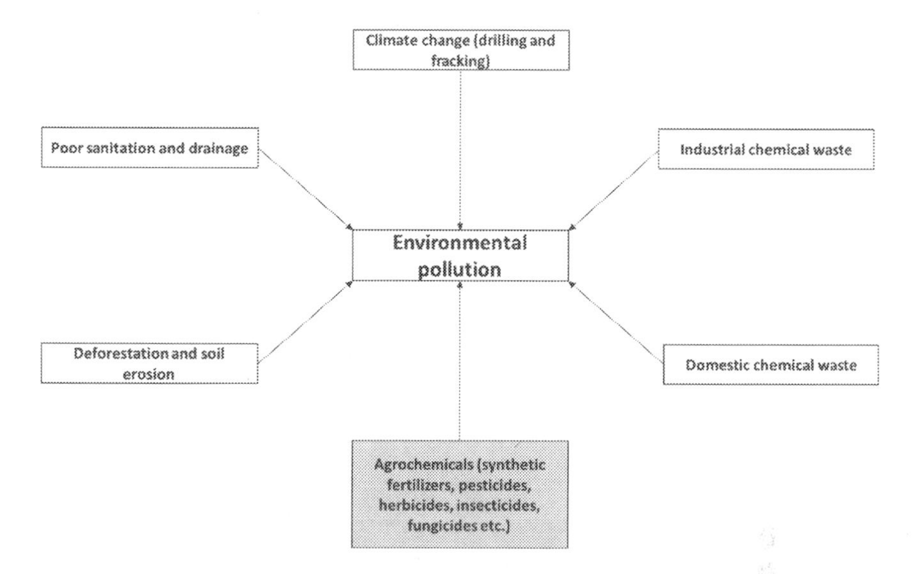

FIGURE 13.6 Pathways of environmental pollution.

of lack of access to fresh drinking water (Elimelech, 2006). Agrochemicals include fertilizers rich in potassium, nitrogen, and phosphorus; insecticides; herbicides; and fungicides used to protect the crops from insects and pests and to increase the productivity of crops. Due to lax environmental laws, excessive use of agrochemicals has been found to be a major source of water and soil contamination (Pelig-Ba, 2011). Besides pesticides, herbicides, insecticides, and fungicides, synthetic fertilizers are also major contributors to water pollution. Moreover, atmospheric deposition and animals' excreted waste in open land and water are also intensive donors in water and soil pollution (Tanko et al., 2012).

Today, a global problem is the contamination of water and soil through human and industrial activities and agrochemical waste in water sources due to the agrochemical runoff. A key cause of pollution of rivers, ponds, lakes, oceans, reservoirs, and watersheds is misuse or mishandling of agrochemicals due to the irresponsible behavior of farmers, agrochemical industries, and sellers. Such pollution can be indeed fully preventable by educating farmers and enforcing strict environmental laws on agrochemical industries, distributors, and sellers by the government. The different pathways of environmental pollution are shown in Fig. 13.6 (Wimalawansa & Wimalawansa, 2014). Hence, the increasing usage of agrochemicals in agricultural activities suggests that agrochemicals are responsible for the surface, groundwater, and soil contamination worldwide, and if no major steps are taken to stop fertilizer runoff into water bodies, severe environmental problems related to the quality of water will arise (Bowyer-Bower & Drakakis-Smith, 1996).

13.5.1 Hazardous Effects of Agrochemicals

Agrochemicals are very useful to achieve enhanced quality and quantity of crops; however, they are also capable of causing occupational health diseases, especially in

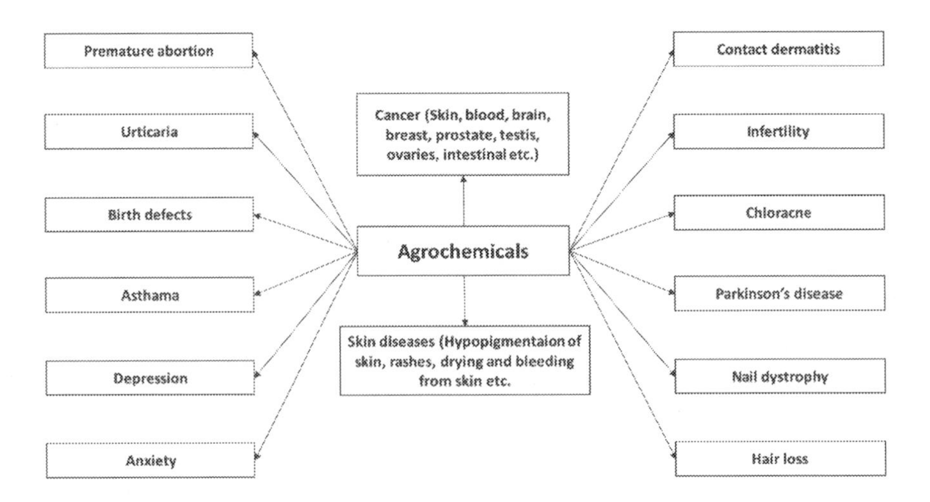

FIGURE 13.7 Different types of diseases caused by the agrochemicals.

farmers as they are the major direct users. Farmers are largely exposed to pesticides and other agrochemical while mixing and loading them and cleaning the equipment and disposing of waste after usage. Moreover, they are also exposed to agrochemicals while sowing pesticide-treated seeds, weeding, cutting, and harvesting sprayed crops. Prolonged exposure increases the risk of irritation, allergies, and carcinogenic effects on skin. Different types of diseases caused by the agrochemicals are shown in Fig. 13.7 (Śpiewak, 2001).

People affected by agrochemicals may suffer from chronic kidney disease and diseases of the heart, liver, skin, birth, and brain. Agrochemical pollution can also cause mental health and behavioral problems; and the intelligence quotient is lowered due to fat-soluble toxic agrochemicals. Children and pregnant women are the most vulnerable (Śpiewak, 2001). Therefore, affordable and easily available water purification technology should be developed to reduce the death rate due to the agrochemical pollution in water and soil. Some important and widely used processes are listed here (Sharma & Bhattacharya, 2017).

- Precipitation and coagulation
- Adsorption
- Distillation
- Purification with activated carbon, activated alumina, and silica gel
- Ion exchange
- Electrodialysis
- Reverse osmosis
- Electrocatalytic oxidation
- Photochemical degradation
- Bioremediation

Continued efforts in the research area of water purification technology will result in some processes/technologies to eliminate the contaminations from polluted water (Sharma & Bhattacharya, 2017).

13.6 FINE CHEMICALS

Fine chemicals are an important class of chemicals that are produced by batch processes or by using biotechnological methods in small amounts for specific purposes. These chemicals may be simple, complex, or high-purity substances. Fine chemicals comprise an extensive range of chemical products that serve many important industries such as agrochemicals, pharmaceuticals, food, flavorings, life sciences, agrochemicals, dyes, etc. Fine chemical industries are diversified and their processes always depend on the nature of the different products and their process synthesis; hence the wastewater composition will vary, and for this reason there is not a common wastewater treatment for all the effluents from these fine chemical industries.

For wastewater treatment there are many chemical, biological, physical, and mechanical methods available (Gitis & Hankins, 2018). Additionally, researchers are constantly inventing new and efficient technologies, including nanotechnology for wastewater treatment and reusable ability (Yaqoob et al., 2020). Two of the most important methods for wastewater treatment are nanosorption and degradation.

13.6.1 Nanosorption and Degradation

In nanosorption, nanosorbents are used for wastewater treatment due to their unique physicochemical properties. The most common nanosorbents are based on carbon and its analogues, such as graphite, carbon black, graphene oxide, and carbon nanotubes (CNT) (Salim & Ho, 2015). Likewise, polymeric nanoadsorbents are also used to remove organic impurities, dissolve metal ions etc. from wastewater. The important advantage of these adsorbents is their reusability by regeneration, biocompatibility, biodegradability, etc. (Shen et al., 2014). Another important class of adsorbent is metal ion-impregnated zeolites, used to remove metal ions and microorganisms (Giwa et al., 2017). Magnetic nanoadsorbents are also an important class of nanoadsorbents that effectively remove toxic metal ions from wastewater (Perez et al., 2019).

An additional method in wastewater water treatment using nanotechnology is the nanophotocatalytic degradation method. The efficiency of degradation of pollutants is mainly dependent on the preparation of photocatalysts. In this method contaminants, especially organic moieties, are degraded by nanocatalyst such as TiO_2, SiO_2, ZnS, WO_3, ZnO, Fe_2O_3, $SrTiO_3$, CdS, etc., in the presence of light. Among these nanocatalyst, TiO_2, TiO_2-based composites, and ZnO have been widely applied as photocatalysts because of their nontoxicity, stability, activity, lower cost, eco-friendly properties, etc. (Gnanaprakasam et al., 2015).

REFERENCES

Adin, A., & Asano, T. (1998). The role of physical-chemical treatment in wastewater reclamation and reuse. *Water Science and Technology*. https://doi.org/10.2166/wst.1998.0381

Ahluwalia, S. S., & Goyal, D. (2007). Microbial and plant derived biomass for removal of heavy metals from wastewater. *Bioresource Technology*. https://doi.org/10.1016/j.biortech.2005.12.006

Akpor, O. B., & Muchie, M. (2010). Remediation of heavy metals in drinking water and wastewater treatment systems: Processes and applications. In *International Journal of Physical Sciences,* p. vol. 5, no. 12, 1807–1817.

Ali, M. M., Ali, M. L., Islam, M. S., & Rahman, M. Z. (2016). Preliminary assessment of heavy metals in water and sediment of Karnaphuli River, Bangladesh. *Environmental Nanotechnology, Monitoring and Management, 5*, 27–35. https://doi.org/10.1016/j.enmm.2016.01.002

Aljuboury, D. A. D. A., Palaniandy, P., Abdul Aziz, H. B., & Feroz, S. (2017). Treatment of petroleum wastewater by conventional and new technologies: A review. *Global Nest Journal, 19*(3), 439–452. https://doi.org/10.30955/gnj.002239

Al-Shamrani, A. A., James, A., & Xiao, H. (2002). Separation of oil from water by dissolved air flotation. *Colloids and Surfaces A: Physicochemical and Engineering Aspects, 209*, 15–26. https://doi.org/10.1016/S0927-7757(02)00208-X

Altmann, J., Rehfeld, D., Träder, K., Sperlich, A., & Jekel, M. (2016). Combination of granular activated carbon adsorption and deep-bed filtration as a single advanced wastewater treatment step for organic micropollutant and phosphorus removal. *Water Research.* https://doi.org/10.1016/j.watres.2016.01.051

Ananthashankar, R. A. G. (2013). Production, characterization and treatment of textile effluents: A critical review. *Journal of Chemical Engineering & Process Technology, 05*(01). https://doi.org/10.4172/2157-7048.1000182

Ansari, M. I., & Malik, A. (2007). Biosorption of nickel and cadmium by metal resistant bacterial isolates from agricultural soil irrigated with industrial wastewater. *Bioresource Technology.* https://doi.org/10.1016/j.biortech.2006.10.008

Azeh Engwa, G., Udoka Ferdinand, P., Nweke Nwalo, F., & N. Unachukwu, M. (2019). Mechanism and Health effects of heavy metal toxicity in humans. In *Poisoning in the Modern World: New Tricks for an Old Dog?* https://doi.org/10.5772/intechopen.82511

Babu, B. R., Parande, A. K., Raghu, S., & Prem Kumar, T. (2007). Cotton textile processing: Waste generation and effluent treatment. *Journal of Cotton Science, 11*(1), 110–122.

Balashov, V. N., Engelder, T., Gu, X., Fantle, M. S., & Brantley, S. L. (2015). A model describing flowback chemistry changes with time after Marcellus Shale hydraulic fracturing. *AAPG Bulletin, 99*(1), 143–154. https://doi.org/10.1306/06041413119

Benaïssa, H., & Elouchdi, M. A. (2007). Removal of copper ions from aqueous solutions by dried sunflower leaves. *Chemical Engineering and Processing: Process Intensification.* https://doi.org/10.1016/j.cep.2006.08.006

Bennett, G. F. (1988). The removal of oil from wastewater by air flotation: A review. *Critical Reviews in Environmental Control, 18*(3), 189–253. https://doi.org/10.1080/10643388809388348

Bowyer-Bower, T., & Drakakis-Smith, D. W. (1996). *The needs of the urban poor versus environmental conservation: Conflict in urban agriculture.* Authors. https://books.google.co.in/books?id=aAc7QwAACAAJ

Brown, C., & Sheedy, M. (2002). A New ion exchange process for softening high TDS produced water. In *SPE International Thermal Operations and Heavy Oil Symposium and International Horizontal Well Technology Conference* (p. 9). Society of Petroleum Engineers. https://doi.org/10.2118/78941-MS

Bunluesin, S., Kruatrachue, M., Pokethitiyook, P., Upatham, S., & Lanza, G. R. (2007). Batch and continuous packed column studies of cadmium biosorption by Hydrilla verticillata biomass. *Journal of Bioscience and Bioengineering.* https://doi.org/10.1263/jbb.103.509

Changmai, M., Pasawan, M., & Purkait, M. K. (2019). Treatment of oily wastewater from drilling site using electrocoagulation followed by microfiltration. *Separation and Purification Technology, 210*, 463–472. https://doi.org/10.1016/j.seppur.2018.08.007

Characterization of Effluent from Textile Wet Finishing Operations. (2009). *Lecture Notes in Engineering and Computer Science.*

Chen, Xuejun, Shen, Z., Zhu, X., Fan, Y., & Wang, W. (2005). Advanced treatment of textile wastewater for reuse using electrochemical oxidation and membrane filtration. *Water SA*. https://doi.org/10.4314/wsa.v31i1.5129

Chen, Xueming, Chen, G., & Yue, P. L. (2002). Novel electrode system for electroflotation of wastewater. *Environmental Science and Technology*, *36*, 778–783. https://doi.org/10.1021/es011003u

Chipasa, K. B. (2001). Limits of physicochemical treatment of wastewater in the vegetable oil refining industry. *Polish Journal of Environmental Studies*.

Das, S. (2000). Textile effluent treatment: A solution to the environmental pollution. *Fibre2Fashion: Ahmedabad, India*.

Diya'uddeen, B. H., Daud, W. M. A. W., & Abdul Aziz, A. R. (2011). Treatment technologies for petroleum refinery effluents: A review. *Process Safety and Environmental Protection*, *89*(2), 95–105. https://doi.org/10.1016/j.psep.2010.11.003

Dursun, A. (2003). The effect of pH on the equilibrium of heavy metal biosorption by Aspergillus niger. *Fresenius Environmental Bulletin*, *12*, 1315–1322.

Ebrahimi, M., Kazemi, H., Mirbagheri, S. A., & Rockaway, T. D. (2016). An optimized biological approach for treatment of petroleum refinery wastewater. *Journal of Environmental Chemical Engineering*, *4*(3), 3401–3408. https://doi.org/10.1016/j.jece.2016.06.030

Elimelech, M. (2006). The global challenge for adequate and safe water. *Journal of Water Supply: Research and Technology - AQUA*, 3–10. https://doi.org/10.2166/aqua.2005.064

Environmental Protection Agency, Ireland. (1997). *Waste water treatment manuals: Primary, secondary and tertiary treatment*. EPA.

EPA. (2000). Wastewater Technology Fact Sheet (Screening and Grit Removal). *Environmental Protection*. https://doi.org/EPA 832-F-99-062

Ertan Akün, M. (2020). Heavy metal contamination and remediation of water and soil with case studies from Cyprus. In *Heavy Metal Toxicity in Public Health*. https://doi.org/10.5772/intechopen.90060

Eswaramoorthi, S., Dhanapal, K., & Chauhan, D. S. (2008). Advances in textile wastewater treatment: The case for UV-ozonation and membrane bioreactor for common effluent treatment plants in Tirupur, India. *Environmental Technology Awareness Series*, 1–17.

Etter, B., Tilley, E., Khadka, R., & Udert, K. M. (2011). Low-cost struvite production using source-separated urine in Nepal. *Water Research*, *45* (2011), 852–862. https://doi.org/10.1016/j.watres.2010.10.007

Fakhru'l-Razi, A., Pendashteh, A., Abdullah, L. C., Biak, D. R. A., Madaeni, S. S., & Abidin, Z. Z. (2009a). Review of technologies for oil and gas produced water treatment. *Journal of Hazardous Materials*. https://doi.org/10.1016/j.jhazmat.2009.05.044

Fakhru'l-Razi, A., Pendashteh, A., Abdullah, L. C., Biak, D. R. A., Madaeni, S. S., & Abidin, Z. Z. (2009b). Review of technologies for oil and gas produced water treatment. *Journal of Hazardous Materials*, *170*(2–3), 530–551. https://doi.org/10.1016/j.jhazmat.2009.05.044

Farajnezhad, H., & Gharbani, P. (2012). Coagulation treatment of wastewater in petroleum industry using poly aluminum chloride and ferric chloride. *IJRRAS*, *13*, 306–310.

Förstner, U. & Wittman, G. T. W. (1983). Metal pollution in the aquatic environment. In *Folia geobotanica & phytotaxonomica*. Springer. https://doi.org/10.1007/bf02857454

Gao, C., Lin, Q., Bao, K., Zhao, H., Zhang, Z., Xing, W., Lu, X., & Wang, G. (2014). Historical variation and recent ecological risk of heavy metals in wetland sediments along Wusuli River, Northeast China. *Environmental Earth Sciences*, *72*(11), 4345–4355. https://doi.org/10.1007/s12665-014-3334-2

Ghimire, K. N., Inoue, K., Ohto, K., & Hayashida, T. (2007). Adsorptive separation of metallic pollutants onto waste seaweeds, Porphyra yezoensis and Ulva japonica. *Separation Science and Technology*. https://doi.org/10.1080/15363830701313461

Ghimire, N., & Wang, S. (2018). Biological treatment of petrochemical wastewater. *Petroleum Chemicals: Recent Insight* (55–74). https://doi.org/10.5772/intechopen.79655

Gitis, V., & Hankins, N. (2018). Water treatment chemicals: Trends and challenges. *Journal of Water Process Engineering*, *25*, 34–38. https://doi.org/10.1016/j.jwpe.2018.06.003

Giwa, A., Hasan, S. W., Yousuf, A., Chakraborty, S., Johnson, D. J., & Hilal, N. (2017). Biomimetic membranes: A critical review of recent progress. *Desalination*, *420*, 403–424. https://doi.org/10.1016/j.desal.2017.06.025

Gnanaprakasam, A., Sivakumar, V. M., & Thirumarimurugan, M. (2015). Influencing parameters in the photocatalytic degradation of organic effluent via nanometal oxide catalyst: A review. *Indian Journal of Materials Science*, *2015*, 601827. https://doi.org/10.1155/2015/601827

Gómez-Alvarez, A., Meza-Figueroa, D., Villalba-Atondo, A. I., Valenzuela-García, J. L., Ramírez-Hernández, J., & Almendariz-Tapia, J. (2009). Estimation of potential pollution from mine tailings in the San Pedro River (1993-2005), Mexico-US border. *Environmental Geology*, *57*(7), 1469–1479. https://doi.org/10.1007/s00254-008-1424-8

González-Martínez, A., de Simón-Martín, M., López, R., Táboas-Fernández, R., & Bernardo-Sánchez, A. (2019). Remediation of potential toxic elements from wastes and soils: Analysis and energy prospects. *Sustainability (Switzerland)*, *11*(12), 3307. https://doi.org/10.3390/SU11123307

Gregory, K. B., Vidic, R. D., & Dzombak, D. A. (2011). Water management challenges associated with the production of shale gas by hydraulic fracturing. *Elements*, *7*(3), 181–186. https://doi.org/10.2113/gselements.7.3.181

Han, R., Zhang, S., Xing, D., & Jian, X. (2010). Desalination of dye utilizing copoly(phthalazinone biphenyl ether sulfone) ultrafiltration membrane with low molecular weight cut-off. *Journal of Membrane Science*, *358*, 1–6. https://doi.org/10.1016/j.memsci.2010.03.036

Heukelekian, H. (1941). Mechanical flocculation and bioflocculation of sewage. *Sewage Works Journal*, *13*(3), 506–522.

Hosny, R., Fathy, M., Ramzi, M., Abdel Moghny, T., Desouky, S. E. M., & Shama, S. A. (2016). Treatment of the oily produced water (OPW) using coagulant mixtures. *Egyptian Journal of Petroleum*, *25*(3), 391–396. https://doi.org/10.1016/j.ejpe.2015.09.006

Igunnu, E. T., & Chen, G. Z. (2014). Produced water treatment technologies. *International Journal of Low-Carbon Technologies*, *9*(3), 157–177. https://doi.org/10.1093/ijlct/cts049

Jaishankar, M., Tseten, T., Anbalagan, N., Mathew, B. B., & Beeregowda, K. N. (2014). Toxicity, mechanism and health effects of some heavy metals. *Interdisciplinary Toxicology*, 7(2), 60–72. https://doi.org/10.2478/intox-2014-0009

Järup, L. (2003). Hazards of heavy metal contamination. *British Medical Bulletin*, 68(1), 167–182. https://doi.org/10.1093/bmb/ldg032

Ji, G. D., Sun, T. H., Ni, J. R., & Tong, J. J. (2009). Anaerobic baffled reactor (ABR) for treating heavy oil produced water with high concentrations of salt and poor nutrient. *Bioresource Technology*, *100*, 1108–1114. https://doi.org/10.1016/j.biortech.2008.08.015

Kabata-Pendias, A., & Szteke, B. (2015). Trace elements in abiotic and biotic environments. In *Trace Elements in Abiotic and Biotic Environments*. Taylor & Francis Group. https://doi.org/10.1201/b18198

Kharat, D. S. (2015). Treatment of textile industry effluents: Limitations and scope. *Journal of Environmental Research and Development*, *9*(4), 1210–1213.

Kriipsalu, M., Marques, M., & Maastik, A. (2008). Characterization of oily sludge from a wastewater treatment plant flocculation-flotation unit in a petroleum refinery and its treatment implications. *Journal of Material Cycles and Waste Management*, *10*, 79–86. https://doi.org/10.1007/s10163-007-0188-7

Kulkarni, A. A., Deshpande, M., & Pandit, A. B. (2000). Techniques of wastewater treatment. *Resonance*, *5*(12), 64–74. https://doi.org/10.1007/bf02840396

Kumar, M., Ramanatahn, A. L., Tripathi, R., Farswan, S., Kumar, D., & Bhattacharya, P. (2017). A study of trace element contamination using multivariate statistical techniques and health risk assessment in groundwater of Chhaprola Industrial Area, Gautam Buddha Nagar, Uttar Pradesh, India. *Chemosphere*, *166*, 135–145. https://doi.org/10.1016/j.chemosphere.2016.09.086

Li, L., Liu, N., McPherson, B., & Lee, R. (2008). Influence of counter ions on the reverse osmosis through MFI zeolite membranes: implications for produced water desalination. *Desalination*, *228*, 217–225. https://doi.org/10.1016/j.desal.2007.10.010

Li, X., Cao, X., Wu, G., Temple, T., Coulon, F., & Sui, H. (2014). Ozonation of diesel-fuel contaminated sand and the implications for remediation end-points. *Chemosphere*, *109*, 71–76. https://doi.org/10.1016/j.chemosphere.2014.03.005

Liangxiong, L., Whitworth, T. M., & Lee, R. (2003). Separation of inorganic solutes from oil-field produced water using a compacted bentonite membrane. *Journal of Membrane Science*, *2017*(1–2), 215–225. https://doi.org/10.1016/S0376-7388(03)00138-8

Lyon, James. (n.d.). *Mineral Policy Center.*

Madaeni, S. S. (1999). The application of membrane technology for water disinfection. *Water Research*, *33*, 301–308. https://doi.org/10.1016/S0043-1354(98)00212-7

Mallick, N. (2003). Biotechnological potential of Chlorella vulgaris for accumulation of Cu and Ni from single and binary metal solutions. *World Journal of Microbiology and Biotechnology*. https://doi.org/10.1023/A:1025104918352

Mantell, M. (2011). Produced water reuse and recycling challenges and opportunities across major shale plays. *US Environmental Protection Agency #4 Water Resources Management March 29–30.*

Marcucci, M., Ciardelli, G., Matteucci, A., Ranieri, L., & Russo, M. (2002). Experimental campaigns on textile wastewater for reuse by means of different membrane processes. *Desalination*. https://doi.org/10.1016/S0011-9164(02)00745-2

Marcus, J. J. (1997). *Mining Environmental Handbook: Effects of Mining on the Environment and American Environmental Controls on Mining*. Imperial College Press.

McEwen, J. B. (1998). *Treatment process selection for particle removal*. American Water Works Association.

Meng, Q., Zhang, J., Zhang, Z., & Wu, T. (2016). Geochemistry of dissolved trace elements and heavy metals in the Dan River Drainage (China): Distribution, sources, and water quality assessment. *Environmental Science and Pollution Research*, *23*, 8091–8103. https://doi.org/10.1007/s11356-016-6074-x

Mittal, A. (2011). Biological wastewater treatment. *Water Today*, *2011*, 32–44.

Mondal, S., & Wickramasinghe, S. R. (2008). Produced water treatment by nanofiltration and reverse osmosis membranes. *Journal of Membrane Science*, *322*, 162–170. https://doi.org/10.1016/j.memsci.2008.05.039

Mondal, Subrata. (2016). Polymeric membranes for produced water treatment: An overview of fouling behavior and its control. *Reviews in Chemical Engineering*, *32*(6), 611–628. https://doi.org/10.1515/revce-2015-0027

Mustapha, H. I., van Bruggen, H. J. J. A., & Lens, P. N. L. (2018). Vertical subsurface flow constructed wetlands for the removal of petroleum contaminants from secondary refinery effluent at the Kaduna refining plant (Kaduna, Nigeria). *Environmental Science and Pollution Research*, *25*(30), 30451–30462. https://doi.org/10.1007/s11356-018-2996-9

National Small Flows – Clearinghouse. (1996). Home aerobic wastewater treatment: An alternative to septic systems. *Pipeline*, *7*(1), 1–8.

Neumann, S., & Fatula, P. (2009). Principles of ion exchange in wastewater treatment. *Asian Water*, *19*, 14–19.

Olías, M., Nieto, J. M., Sarmiento, A. M., Cerón, J. C., & Cánovas, C. R. (2004). Seasonal water quality variations in a river affected by acid mine drainage: The Odiel River (South West Spain). *Science of the Total Environment*, *333*(1–3), 267–281. https://doi.org/10.1016/j.scitotenv.2004.05.012

Pan, B., Zhang, Q., Du, W., Zhang, W., Pan, B., Zhang, Q., Xu, Z., & Zhang, Q. (2007). Selective heavy metals removal from waters by amorphous zirconium phosphate: Behavior and mechanism. *Water Research*. https://doi.org/10.1016/j.watres.2007.03.004

Pelig-Ba, K. B. (2011). Levels of Agricultural pesticides in sediments and irrigation water from Tono and Vea in the Upper East of Ghana. *Journal of Environmental Protection*, *2*(6), 761–768. https://doi.org/10.4236/jep.2011.26088

Perez, T., Pasquini, D., de Faria Lima, A., Rosa, E. V., Sousa, M. H., Cerqueira, D. A., & de Morais, L. C. (2019). Efficient removal of lead ions from water by magnetic nanosorbents based on manganese ferrite nanoparticles capped with thin layers of modified biopolymers. *Journal of Environmental Chemical Engineering*, *7*(1), 102892. https://doi.org/10.1016/j.jece.2019.102892

Pichtel, J. (2016). Oil and gas production wastewater: Soil contamination and pollution prevention. *Applied and Environmental Soil Science*, 1–24. https://doi.org/10.1155/2016/2707989

Rajasulochana, P., & Preethy, V. (2016). Comparison on efficiency of various techniques in treatment of waste and sewage water: A comprehensive review. *Resource-Efficient Technologies*, *2*(4), 175–184. https://doi.org/10.1016/j.reffit.2016.09.004

Raza, W., Lee, J., Raza, N., Luo, Y., Kim, K. H., & Yang, J. (2019). Removal of phenolic compounds from industrial waste water based on membrane-based technologies. *Journal of Industrial and Engineering Chemistry*, *71*, 1–18. https://doi.org/10.1016/j.jiec.2018.11.024

Renault, F., Sancey, B., Badot, P.-M., & Crini, G. (2009). Chitosan for coagulation/flocculation processes: An eco-friendly approach. *European Polymer Journal*, *45*(5), 1337–1348. https://doi.org/10.1016/j.eurpolymj.2008.12.027

Renou, S., Givaudan, J. G., Poulain, S., Dirassouyan, F., & Moulin, P. (2008). Landfill leachate treatment: Review and opportunity. *Journal of Hazardous Materials*, *150*, 468–493. https://doi.org/10.1016/j.jhazmat.2007.09.077

Salim, W., & Ho, W. W. (2015). Recent developments on nanostructured polymer-based membranes. *Current Opinion in Chemical Engineering*, *8*, 76–82. https://doi.org/10.1016/j.coche.2015.03.003

Sharma, S., & Bhattacharya, A. (2017). Drinking water contamination and treatment techniques. *Applied Water Science*, *7*, 1043–1067. https://doi.org/10.1007/s13201-016-0455-7

Shen, Y., Saboe, P. O., Sines, I. T., Erbakan, M., & Kumar, M. (2014). Biomimetic membranes: A review. *Journal of Membrane Science*, *454*, 359–381. https://doi.org/10.1016/j.memsci.2013.12.019

Shi, X., Chen, L., & Wang, J. (2013). Multivariate analysis of heavy metal pollution in street dusts of Xianyang city, *NW China. Environmental Earth Sciences*, *69*, 1973–1979. https://doi.org/10.1007/s12665-012-2032-1

Sierra, C., Ruíz-Barzola, O., Menéndez, M., Demey, J. R., & Vicente-Villardón, J. L. (2017). Geochemical interactions study in surface river sediments at an artisanal mining area by means of Canonical (MANOVA)-Biplot. *Journal of Geochemical Exploration*, *175*, 72–81. https://doi.org/10.1016/j.gexplo.2017.01.002

Sonune, A., & Ghate, R. (2004). Developments in wastewater treatment methods. *Desalination*, *167*, 55–63. https://doi.org/10.1016/j.desal.2004.06.113

Śpiewak, R. (2001). Pesticides as a cause of occupational skin diseases in farmers. In *Annals of Agricultural and Environmental Medicine* (pp. 1–5).

Subramanian Senthil Kannan, M., Gobalakrishnan, M., Kumaravel, S., Nithyanadan, R., Rajashankar, K. J., & Vadicherala, T. (2006). Influence of cationization of cotton on reactive dyeing. *Journal of Textile and Apparel, Technology and Management*, *5*(2) 1–16.

Sundaray, S. K., Nayak, B. B., Lin, S., & Bhatta, D. (2011). Geochemical speciation and risk assessment of heavy metals in the river estuarine sediments: A case study: Mahanadi basin, India. *Journal of Hazardous Materials*, *186*(2–3), 1837–1846. https://doi.org/10.1016/j.jhazmat.2010.12.081

Tanko, J. A., Oluwadamisi, E. A., & Abubakar, I. (2012). Agrochemical concentration level in Zaria Dam Reservoir and ground waters in the environs. *Journal of Environmental Protection*, *3*(2), 225–232. https://doi.org/10.4236/jep.2012.32028

Tejero, M. D. À., Jové, E., Carmona, P., Gomez, V., García-Molina, V., Villa, J., & Das, S. (2017). Treatment of oil–water emulsions by adsorption onto resin and activated carbon. *Desalination and Water Treatment*, *100*(21), 28. https://doi.org/10.5004/dwt.2017.21689

Thakur, C., Srivastava, V. C., & Mall, I. D. (2014). Aerobic degradation of petroleum refinery wastewater in sequential batch reactor. *Journal of Environmental Science and Health - Part A Toxic/Hazardous Substances and Environmental Engineering*, *49*, 1436–1444. https://doi.org/10.1080/10934529.2014.928557

Förstner, U. & Whitman, G. T. W. (1983). Metal pollution in the aquatic environment. In *Folia geobotanica & phytotaxonomica*. Springer. https://doi.org/10.1007/bf02857454

U.S. Environmental Protection Agency, Washington. (2016). *No impacts from the hydraulic fracturing water cycle on drinking water resources in the United States.*

U.S. EPA. (2000). Wastewater Technology Fact Sheet, Aerated Partial Mix Lagoon. *United States Environmental Protection Agency*. https://doi.org/EPA 832-F-99-062

Umar, A. A., Saaid, I. B. M., Sulaimon, A. A., & Pilus, R. B. M. (2018). A review of petroleum emulsions and recent progress on water-in-crude oil emulsions stabilized by natural surfactants and solids. *Journal of Petroleum Science and Engineering*, *165*, 673–690. https://doi.org/10.1016/j.petrol.2018.03.014

Usman, M., Faure, P., Lorgeoux, C., Ruby, C., & Hanna, K. (2013). Treatment of hydrocarbon contamination under flow through conditions by using magnetite catalyzed chemical oxidation. *Environmental Science and Pollution Research*, *20*(1), 22–30. https://doi.org/10.1007/s11356-012-1016-8

Varjani, S., Joshi, R., Srivastava, V. K., Ngo, H. H., & Guo, W. (2020). Treatment of wastewater from petroleum industry: current practices and perspectives. *Environmental Science and Pollution Research*, *27*, 27172–27180. https://doi.org/10.1007/s11356-019-04725-x

Vautier, M., Guillard, C., & Herrmann, J. M. (2001). Photocatalytic degradation of dyes in water: Case study of indigo and of indigo carmine. *Journal of Catalysis*. https://doi.org/10.1006/jcat.2001.3232

Vineta, S., Silvana, Z., Sanja, R., & Golomeova, S. (n.d.). Methods for waste waters treatment in textile industry. *International Scientific Conference "UNITECH 2014" – Gabrovo*, *10*, 93–125.

Wang, J., Liu, R., Zhang, P., Yu, W., Shen, Z., & Feng, C. (2014). Spatial variation, environmental assessment and source identification of heavy metals in sediments of the Yangtze River Estuary. *Marine Pollution Bulletin*, *87*(1), 364–373. https://doi.org/10.1016/j.marpolbul.2014.07.048

Wei, W., Ma, R., Sun, Z., Zhou, A., Bu, J., Long, X., & Liu, Y. (2018). Effects of mining activities on the release of heavy metals (HMs) in a typical mountain headwater region, the Qinghai-Tibet Plateau in China. *International Journal of Environmental Research and Public Health*, *17*, 588–594. https://doi.org/10.3390/ijerph15091987

Wei, X., Zhang, S., Han, Y., & Wolfe, F. A. (2019). Treatment of petrochemical wastewater and produced water from oil and gas. *Water Environment Research*, *91*, 1025–1033. https://doi.org/10.1002/wer.1172

WHO. (1996). *Trace elements in human nutrition and health World Health Organization*. World Health Organization. WHO/FAO/IAEA.

Wimalawansa, S. A., & Wimalawansa, S. J. (2014). Agrochemical-related environmental pollution: effects on human health. *Global Journal of Biology, Agriculture & Health Sciences*, *3*(3), 72–83.

Xu, P., & Drewes, J. E. (2006). Viability of nanofiltration and ultra-low pressure reverse osmosis membranes for multi-beneficial use of methane produced water. *Separation and Purification Technology*, *52*, 67–76. https://doi.org/10.1016/j.seppur.2006.03.019

Yaqoob, A. A., Parveen, T., Umar, K., & Mohamad Ibrahim, M. N. (2020). Role of nanomaterials in the treatment of wastewater: A review. *Water*, *12*(2), 495. https://doi.org/10.3390/w12020495

Yasui, H., Nakamura, K., Sakuma, S., Iwasaki, M., & Sakai, Y. (1996). A full-scale operation of a novel activated sludge process without excess sludge production. *Water Science and Technology*, *34*(3–4), 395–404.

Zaidi, A., Buisson, H., Sourirajan, S., & Wood, H. (1992). Ultra- and nano-filtration in advanced effluent treatment schemes for pollution control in the pulp and paper industry. *Water Science and Technology*. https://doi.org/10.2166/wst.1992.0254

Zhang, Z., Juying, L., & Mamat, Z. (2016). Sources identification and pollution evaluation of heavy metals in the surface sediments of Bortala River, Northwest China. *Ecotoxicology and Environmental Safety*, *126*, 94–101. https://doi.org/10.1016/j.ecoenv.2015.12.025

Zhou, J. M., Dang, Z., CAI, M. F., & Liu, C. Q. (2007). Soil Heavy metal pollution around the Dabaoshan Mine, Guangdong Province, China. *Pedosphere*, *17*, 588–594. https://doi.org/10.1016/S1002-0160(07)60069-1

Ziagova, M., Dimitriadis, G., Aslanidou, D., Papaioannou, X., Litopoulou Tzannetaki, E., & Liakopoulou-Kyriakides, M. (2007). Comparative study of Cd(II) and Cr(VI) biosorption on Staphylococcus xylosus and Pseudomonas sp. in single and binary mixtures. *Bioresource Technology*. https://doi.org/10.1016/j.biortech.2006.09.043

14 Sustainable Technologies for Wastewater Treatment

Pablo Ortiz
Chemical Engineering Department, Universidad de los Andes, Bogotá, Colombia

CONTENTS

14.1 INTRODUCTION

Water is necessary to support any form of life on the planet, from microorganisms to complex ecosystems. It makes up more than 60 percent of the weight of human beings, and some fruits and vegetables retain more than 95 percent. If humanity's survival depends on it, so does the livelihood and well-being of any social structure at any scale or economic model, as water plays a role in the creation of every single good and service we require, from basic sanitation to luxury products (Oki and Kanae 2004). In particular, freshwater has no substitute, and while it is renewable, its amount is finite and not homogenously distributed on the planet (Cosgrove and Loucks 2015).

Today more than 2 billion people are living in areas of physical or economic water scarcity, and their health, development, and security are threatened (U.N. 2018). The challenge of providing clean water to all has been recognized as a priority policy by worldwide organizations since the 1970s, when climate change and energy problems

DOI: 10.1201/9781003138303-14

began to be undeniable. Included first in the Millennium Development Goals, the water challenge is one of the Sustainable Development Goals adopted by the United Nations member states in 2015 (U.N. 2018).

The causes of water problems are strongly related to population growth, the increase in water demand per capita as a result of an increase in the demand of products and services, and the fact that water once used is returned to the sources with reduced quality. Moreover, it is estimated that globally over 80 percent of all wastewater is discharged without any treatment (Sedlak and Schnoor 2013, WWAP 2017). Eventually, the impact on the environment by wastewater, as well as by air and soil pollution, decreases its intrinsic potential to serve as a natural treatment reservoir and freshwater source and threatens the resilience of ecosystems. Consequently, the water cycle and the spatial and temporal distributions of precipitation are affected. Changes in landscape due to agriculture and energy production, as well as population concentration in cities, alter locally and globally the quality of freshwater sources (Cosgrove and Loucks 2015).

In this context, it is unquestionable that to solve the water crisis, a change in the current intensive development and productive processes is necessary. Optimization and reduction of water consumption in domestic and industrial processes are clearly fields where citizens, engineers, and researchers must contribute and participate. On the other hand, a decisive evolution to a less hazardous and biodegradable chemistry (Anastas and Zimmerman 2003) would significantly reduce the pollution discharge in any natural reservoir. If the former preventive actions are both possible and effective, it is also true that they are far from being fully implemented in the current global production model. Mitigation relies then on the generalized implementation of efficient wastewater treatment (WWT) technologies and systems that can provide reusable water or reduce contaminant concentrations to acceptable levels which can be handled by natural remediation systems (Yenkie 2019).

Wastewater treatment systems have been implemented since ancient times, mainly at the local level to manage blackwater through cesspools or to provide drinkable water thanks to coagulation, sand or gravel filtration, and heating operations. With the Industrial Revolution and the growth of cities, it became necessary to develop sewer systems, and consequently the local domestic and relatively small industrial pollution spots evolved to centralized discharges to the nearby bodies of water. In addition, the industrialization era brought a new set of dangerous pollutants, and consequently the development of treatment plants was necessary. The construction of the first centralized sewage treatment plants dates from the late 19th and early 20th centuries, principally in the United Kingdom and the United States. Since then, a wide spectrum of physical, chemical, and biologically based treatment technologies has been employed, and, at the system level, several configurations or arrangements have been implemented and studied. At the same time, the knowledge of biological sciences and the understanding of an ecosystem's equilibrium led to the establishment of quality parameters and standards for both water reuse and discharge (Lofrano and Brown 2010).

Generally, a wastewater treatment process involves a primary stage where large particles and suspended solids are removed through coagulation, precipitation, or sedimentation in order to assure the efficiency of subsequent phases. Next, a secondary treatment reduces concentration of organic and inorganic dissolved compounds, such as nutrients (i.e., N and P), pathogens, and hazardous substances. Depending on the desired quality of the effluent, a tertiary or advanced treatment could be necessary, which continues to decrease pollutant levels or alternatively looks to eliminate some specific and recalcitrant compounds in water. In both secondary and tertiary treatments, techniques based on adsorption, filtration, chemical or biological reduction or oxidation, or a combination of them, are common. As result, a liquid effluent and a contaminated sludge, which will require further treatment or specific disposal, are obtained (Galanakis and Agrafioti 2019).

Currently, multiple alternatives and mature technologies are available for the treatment of conventional domestic, agricultural, or industrial wastewaters. In recent decades, emerging and persistent contaminants related to new health-care products, cosmetics, and the pharmaceutical industry have imposed new challenges and have fostered the development of advanced treatments and hybrid technologies to reach regulatory standards (Rene et al. 2019). However, as in many other fields in modern engineering, in the science and market of wastewater treatment the technical issues need to be articulated in a holistic way in keeping with environmental protection and the economic and social background at local and regional levels. The water crisis, as well as the frequent absence of wastewater treatment across the world, is in most of cases due not to technical limitations but to a lack of sustainable alternatives or knowledge on how to design and implement them. At the same time, some of the existing systems and operating plants inefficiently consume resources, energy, and chemicals, emitting greenhouse gases and impacting the environment contrary to their main purpose. The wastewater–energy nexus shows that is possible to turn the current negative balance of WWT operations into a neutral or even positive one as the energy contained in wastewater is about 5 to 10 times greater than that which is needed to treat it (Liu et al. 2015, U.N. 2018). Research on sustainable wastewater treatment is an open field (section 14.1, p319), and the problem is a hot topic in several journals and has been analyzed by many authors. Simultaneously, comprehensive methodologies to assess wastewater treatment sustainability are required to support technological development and decision makers. As will be seen further on, many of the studies on sustainability have focused on technical and environmental dimensions, ignoring social aspects (Padilla-Rivera et al. 2016).

In this chapter, the concept of sustainability applied to wastewater treatment systems will be presented, and a review of the sustainability assessment methods will be discussed. As wastewater treatment process involves several stages, it will be necessary to analyze the individual advantages and drawbacks of the main subprocesses that comprise the whole system. The focus will be both in conventional technologies as a baseline and in the most important emerging techniques that, despite still being in development, constitute potential breakthroughs in the wastewater treatment industry. Finally, a bibliographic review of comparative sustainability studies between different wastewater technologies and systems published in the last years will be presented.

14.2 CONCEPTUAL AND PRACTICAL FRAMEWORK

14.2.1 Sustainability in Wastewater Treatment Processes

Sustainability of any service, process, or product could be defined as an integral characteristic that guarantees that obtaining this service or good does not affect or compromise, at any stage of its life cycle, the economic, social, or environmental stability and well-being of present and future generations. This concept, presented in the Brundtland Report, extends to the global economic scale and serves as a framework for the development of societies and nations (World Commission on Environment and Development 1987). In 2015, the United Nations defined 17 Sustainable Development Goals (SDGs) as an evolution of the millennial goals proposed in 2000. Among these, SDG 6, "Ensure availability and sustainable management of water and sanitation for all" depicts a global reference and a top-down approach to address the efforts of society, scientists, politicians, and stakeholders in the solution of the water crisis (U.N. 2018). Emphasis is on providing universal and equitable access to safe and affordable water and on adequate sanitation systems for all, especially those in vulnerable situations. Actions should be reinforced in order to reduce pollution, eliminate dumping, avoid hazardous chemicals and materials, develop appropriate wastewater treatments, and foster water recycling and reuse. SDG 6 points out the urgency of the implementation of integrated water management strategies at all levels, with the participation of local communities in improving water and sanitation management, and stresses the importance of expanding international cooperation and capacity-building support to developing countries in water- and sanitation-related activities and programs. No wonder SDG 6 makes a direct call to protect and restore water-related ecosystems, including mountains, forests, wetlands, rivers, aquifers, and lakes. In this context, the critical role that worldwide implementation of wastewater treatment processes has in the achievement of sustainable water management policies and global development is explicit. This also stresses the importance of feasible projects more than just efficient wastewater treatment technologies and forces the adoption of new strategies.

Selection of technologies and treatment systems depends usually on the scale of the plant, flow and characteristics of the influent, and its variability in time as well as the desired properties of the effluent. When the treatment alternatives include biological processes, the weather conditions and fluctuations are also relevant as they affect the system performance. As in any other engineering project, economical investment in both capital and operation stages balances technical performance in the design phase and final selection. In the quest for sustainability, this classical and narrow approach needs to be expanded and include the assessment of environmental and social impacts. In this sense, the following elements should be considered:

1. The net benefits of the WWT system should be assessed. A life cycle approach is necessary, and consequently, besides the positive effects of remediation processes, the social, economic, and environmental impacts related to obtaining and transporting infrastructure materials, plant construction and decommissioning, resource consumption, emissions during operation, and the end-of-life or disposal of all secondary streams must be

taken into account. The diagram in Fig. 14.1 shows the boundaries for the life- cycle analysis of WWT systems, including sludge treatment train and product recovery options.

2. Removal of pollutants in water is the main goal of any WWT facility. However, the selected technology, the designed system, and the final effluent quality must follow the fit-for-purpose approach as impacts related to overdesign tend to be high (Anastas and Zimmerman 2003).
3. Water reuse and resource recovery from wastewater are opportunities toward WWT sustainability. Depending on the influent characteristics, each treatment technology presents advantages and disadvantages for this purpose (Diaz-Elsayed et al. 2020).
4. Environmental impacts of WWTs occur spatially at local, regional, and global levels, covering different time spans. The use of comprehensive environmental models that could assess effects on human health, ecosystems, and resource depletion, including those effects produced by very specific pollutants in wastewater (e.g., pharmaceuticals and recalcitrant compounds), is necessary.
5. The social dimension must be analyzed for WWT systems as it is critical for the feasibility and long-term stability of the project. That includes operational complexity; public acceptance; reliability and safety; local employment and community participation; and local disturbance in terms of noise, odors, or visual aspects; among other elements.

Sustainability is a complex and multidimensional attribute that integrates the above-mentioned themes and several specific aspects derived from them. Frequently these aspects are interrelated or are conflicting. As an example, a WWT plant could reinforce economic and environmental dimensions of sustainability only through the

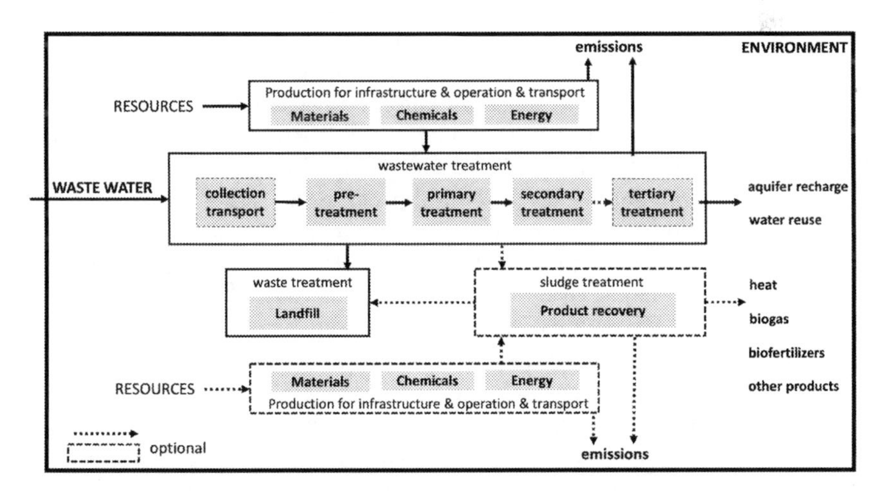

FIGURE 14.1 WWT systems boundaries. Life cycle assessment includes impacts related to the necessary materials, chemicals, and energy to allow WWT plant construction and operation. Inputs/outputs for end-of-life processes must be considered.

implementation of additional product recovery processes from sludge. However, this action also adds complexity to the system, demanding skilled workers not necessarily available locally in the short term. Finally, recognizing that sustainability is a holistic concept that cannot be represented simply as the sum of the parts, it is still necessary to propose practical assessment methods in order to support the task of designers and decision makers. In the next section, the main approaches to assess sustainability will be presented.

14.2.2 Sustainability Assessment

14.2.2.1 Life Cycle Assessment

Life cycle assessment (LCA) is a methodology to assess potential environmental impacts of any good or service and completely applicable to WWT systems. A life cycle uses the "cradle-to-grave" approach and identifies energy use, material inputs, and waste and emissions generated during the whole life cycle of the wastewater treatment system: obtaining raw materials or reactants, plant and infrastructure construction, transport of materials or streams, operation and maintenance, and the final use or disposal of water and sludge. This last phase is of paramount importance in WWT as water reuse or recycling and material or energy recovery will positively count as credits in the environmental balance, allowing researchers to reflect on and quantify the benefits of circular economy or industrial ecology concepts. Indeed, when no waste is generated and every output stream in the system is designed to be useful or to serve as raw material for other systems, then the loop is closed and the "cradle-to-cradle" status is achieved. Principles and requirements of LCA methodology are defined by ISO 14000 series standards and consist of four main activities: goal and scope, life cycle inventory (LCI), life cycle impact assessment (LCIA), and interpretation (ISO 2006 14040, ISO 2006 14044).

The goal and scope phase presents motivations behind the study and the stakeholders involved and defines the functional unit and the time during which the service will be offered. In the case of wastewater treatment, the functional unit is a certain volume of treated water (e.g., one cubic meter or influent generated by one person per day), clearly defining its quality at the entrance and at the exit of the WWT system, plant or technology. When comparing different technologies through LCA, it is also important that the scale or capacity of the systems should be similar. The LCI is an explicit list of inputs and outputs of material and energy flows required or produced by the functional unit. Here the list of infrastructure building resources or transportation systems should also be considered but adjusted according to the ratio between the functional unit and the treatment capacity of the plant during the lifetime of the facilities. LCIA converts these flows in environmental indicator metrics that express potential impacts. This is achieved through classification of emissions in different reservoirs — water, air, and soil — and through characterization of flows with equivalent units depending on the environmental indicator that they affect. The most common environmental indicators in LCIA models are related to global warming, eutrophication potential, ozone depletion, water source acidification, photochemical effect, fossil depletion, cumulative energy, ecotoxicity, and human health toxicity. These are called mid-point indicators, and they could be aggregated or grouped through normalization and weighting to obtain more general, and less precise, end-point impacts such as damage to human health, to ecosystems, or to mineral and fossil resources. Several LCIA models and methodologies have been developed that differ in the mid- and end-point indicators, the normalization process, the methods and levels of aggregation, and the determination of weighting factors. The work of Hauschild et al. presents a comprehensive review and description of the available LCIA methods, including those most employed in WWT studies (e.g., ReCiPe, DML 2000, Eco-Indicator 99, USE tox) (Hauschild et al. 2017). Finally, it must be pointed out that because normalization and weighting methodologies depend heavily on specific contexts and involve subjective decisions, the ISO standards require only the classification and characterization steps as mandatory. The upper section in Table 14.1 shows the key steps of the LCA process.

LCA is the most common approach to assessment of the environmental dimension of sustainability of WWT systems, and several comprehensive reviews have been reported (Corominas et al. 2013; Nguyen et al. 2020; Parra-Saldivar et al. 2020;

TABLE 14.1

Example of the Framework for Sustainability Assessment Where the Main Steps of Multi-Criteria Decision Analysis and Environmental LCA Methodology Are Simultaneously Indicated Column Headers Are Actions that Practitioners Should Carry Out

Grouping	Flow or Item	metric (/f.u)	Classification Indicator	Characterization (mid-point)	Normalization (e.g, score/10)	Weighting (0 to 1)	Composite Indicator	Weighting (0 to 1)
environment	CO_2	...kg						
	NO_2	...kg	I_1: Global Warming	...kg CO_2eq	sI_1	w_1		
	CH_4	...kg						
	CFCS	...kg						
	HCFCS	...kg	I_2: O_3 Depletion	...kg CFC11eq	sI_2	w_2		
	SO_X	...kg						
	NO_X	...kg	I_3: Acidification	...kg SO_2eq	sI_3	w_3	$C.I_{env} = \sum_{i=1}^{8} w_i \cdot sI_i$	w_{env}
	PO_4	...kg						
	NH_4	...kg	I_4: Eutrophication	...kg PO_4eq	sI_4	w_4		
	Nitrates	...kg						
	Toxics	...kg	I_5: Toxicity	...kg 1,4DBeq	sI_5	w_5		
	Land Occupation	...m²	I_6: Land	...m²	sI_6	w_6		
	Water	...m³	I_7: Water depletion	...m³	sI_7	w_7		
	Energy	...MJ	I_8: Energy	...MJ	sI_8	w_8		
	Etc...					$\sum_{i=1}^{8} w_i = 1$		
economic	Investment	...$	I_9: Capital		sI_9	w_9		
	Operation and Maintenance	...$	I_{10}: Operation		sI_{10}	w_{10}	$C.I_{econ} = \sum_{i=9}^{10} w_i \cdot sI_i$	w_{econ}
	Etc..					$\sum_{i=9}^{10} w_i = 1$		
social	Complexity	Qual.	I_{11}: Complexity		sI_{11}	w_{11}		
	Safety	Qual.	I_{12}: Safety		sI_{12}	w_{12}	$C.I_{soc} = \sum_{i=11}^{14} w_i \cdot sI_i$	w_{soc}
	Odors	Qual.	I_{13}: Odors		sI_{13}	w_{13}		
	Noise	Qual.	I_{14}: Noise		sI_{14}	w_{14}		
	Etc...					$\sum_{i=11}^{14} w_i = 1$		
Sustainability Index (S.I)							$S.I = w_{env}C.I_{env} + w_{econ}C.I_{econ} + w_{soc}C.I_{soc}$	

Note: In this sense it must be pointed out that in the LCA context, LCI corresponds to the flow and metric columns, while LCIA includes grouping, classification, characterization, normalization, and weighting (f.u: functional unit).

Sabeen et al. 2018). LCA has been used to evaluate the environmental performance of single WWT technologies and WWT plants (Arashiro et al. 2018, Arias et al. 2019; Fuchs et al. 2011; Saad et al. 2019) or water management strategies at city (Hadjimichael et al. 2016; Polruang et al. 2018; Tabesh et al. 2019) or country level (McNamara et al. 2016; Ng et al. 2014; Niero et al. 2014; Lu et al. 2017; Fernández-Fernández et al. 2019). Specific WWT issues have been addressed through LCA studies: sewage collection (Rashid and Liu 2020; Risch et al. 2015); plant infrastructure impacts (Morera et al. 2020); plant upgrading alternatives (Guven et al. 2018; Pasqualino et al. 2009; Remy et al. 2016); WWT systems for small communities (Machado et al. 2007) or developing countries (Awad et al. 2019; Gallego-Schmid and Tarpani 2019, Lopes et al. 2020); product recovery from sludge (Alyaseri and Zhou 2017; Fang et al. 2016; Harclerode et al. 2020; Lam et al. 2020; Pradel and Aissani 2019; Remy et al. 2016); water reuse applications (Buyukkamaci and Karaca 2017; Carre et al. 2017; Opher and Friedler 2016); definition of water effluent standards (Bai et al. 2019); treatment of pharmaceutical (Li et al. 2019), agricultural or specific industry (Chatzisymeon et al. 2013; Chen et al. 2020) wastewaters; odor control in WWT (Alfonsin et al. 2015); evaluation of appropriate plant scale or decisions on convenience of centralized or decentralized systems (De Feo and Ferrara 2017; Igos et al. 2012).

14.2.2.2 Sustainability Indices and Metrics

Capturing the complexity of sustainability and ranking different WWT alternatives is not an easy task, and a bibliography review clearly shows there is no consensus on a unique assessment methodology. However, a common approach follows the framework of multi-criteria decision making (MCDM) (Tzeng and Huang 2011). Accordingly, in a first stage, the dimensions to be analyzed (i.e., social, environmental, and economic) are defined. Next, the elements or indicators that could constitute each dimension must be listed and selected. In doing so, it is important to take into account specific issues related to WWT and to consider previous works and recommendations that have been reported in the field of sustainability (Wu and Wu 2012). Indicators should be meaningful, specific, observable and measurable characteristics of the WWT system. As the number of selected indicators used can be high, the process of weighting and aggregation is essential to allow an integral assessment and to simplify the task of designers and decision makers (Gan et al. 2017; Nardo et al. 2005). Weighting is the step where the relative importance of indicators belonging to each dimension is established. Because indicators have different units and could be quantitative or qualitative, their normalization (i.e., obtaining a unified scale for all of the indicators) is a necessary previous step. Weighting allows practitioners to obtain composite indices through aggregation or grouping. Usually, the first aggregation level produces a composite index for each of the sustainability dimensions, and further weighting of dimensions allows researchers to obtain a single sustainability index. Also, intermediate aggregation levels are possible. Metrics for each indicator are calculated, taking into account the entire life cycle of the system and being adjusted to the functional unit, (i.e., a certain amount and quality of a product or service which in WWT is the volume of treated water). The methodology is illustrated in Table 14.1, including a list of some common indicators found in WWT sustainability studies. The approach described above has been the basis of many of the studies conducted to assess sustainability or sustainability dimensions of WWT systems (Mannina et al. 2019).

There exist different sources that provide comprehensive and in-depth information about each of the sustainability dimensions. Regarding environmental aspects, the models and indicators available in the life cycle assessment (LCA) universe have had an enormous relevance in WWT sustainability studies. LCA environmental indicators are usually expressed in terms of emissions to several reservoirs (i.e., air, soil, water). However, due to the remediation nature of the WWT processes, many authors also report more treatment aimed indicators, expressed as removal efficiency of organic and nutrient contents such as chemical oxygen demand (COD), biological oxygen demand (BOD), or total nitrogen (TN) and phosphorous (TP). Regarding social and socio-economic assessment of life cycle impacts of goods and services, the UNEP and SETAC guidelines are sound and valuable references (Swarr et al. 2011; UNEP 2009;). With respect to the economic dimension, the work of Rebitzer et al. focused specifically on particularities of WWT systems (Rebitzer et al. 2003).

Several specific methodologies have been proposed to obtain sustainability indices for WWT systems. Molinos-Senante et al. developed a composite index based on 14 individual indicators belonging to the three sustainability dimensions. The approach was based on the application of the analytic hierarchy process (AHP) to assign weights to each indicator taking into account the preferences of experts. The methodology was applied to compare the global sustainability of seven different technologies for the secondary treatment of wastewater under different scenarios (Molinos-Senante et al. 2014). A similar multi-criteria decision approach but with different indicators was carried out by Plakas et al. in the sustainability assessment of tertiary wastewater treatment technologies (Plakas et al. 2016). Recently, the AHP was used to evaluate four WWT alternatives through a composite index based on total annual equivalent costs, carbon emission intensity, eutrophication, and resilience toward climate change and natural disasters (Sun et al. 2020). In the work of Padilla et al., a Sustainable Global Index (SGI) was obtained for the analysis of four WWT plants located in Mexico and USA. Environmental, economic, and social, both quantitative and qualitative, indicators were defined, and fuzzy logic was applied to aggregate them and to rank the different alternatives (Padilla-Rivera and Güereca 2019). Herein, social indicators are classified into four stakeholder categories (community and society, workers, consumers, and supply chain) following the methodology previously developed by the authors (Padilla-Rivera et al. 2016). Popovic et al. also focused on the development of quantitative social indicators and proposed a method to understand their interdependencies and to enable their prioritization according to the stakeholders' interests. In this study, the social sustainability dimension included health, safety and security, and comfort as main categories (Popovic and Kraslawski 2018). An interesting approach followed by Wang et al. focused on the importance of capturing in an explicit way the product recovery potential of WWT facilities in the sustainability assessment. For this, they developed a specific composite indicator that balances environmental burdens with nutrients and bio-energy recovery and applied it in a case study (Wang et al. 2012). Akhoundi and Nazif presented a sustainability assessment method based on AHP and evidential reasoning (ER) to rank different tertiary wastewater treatment systems for water reuse applications in Iran. The work integrated the environmental LCA results and cost analysis from previous studies with social/cultural concerns. In a context of water scarcity reuse for agricultural irrigation, industrial services or groundwater recharge were considered urgent strategies to assure a sustainable water management (Akhoundi and Nazif 2018). Kadava et al. also developed a sustainability index based on the LCA methodology for the evaluation of environmental effects. In this case, quality of life, leadership, resource allocation, natural world and climate, and risk were the main proposed categories, and LCA supported assessing impacts through them all (Kadava et al. 2018). Bolissety et al. recently reviewed tertiary wastewater treatment technologies for heavy metal removal and introduced the Ranking Efficiency Product (REP) index based on environmental and economic indicators. In this case, all indicators were considered equally important, but the authors point out the possibility to introduce weights and to include additional characteristics (Bolisetty et al. 2019).

14.3 TECHNOLOGIES AND SUSTAINABILITY

14.3.1 Activated Sludge Process

The activated sludge (AS) process, which has been used for more than a century, is the most widely applied secondary treatment method for municipal wastewaters and could be considered a reference system in comparative sustainability studies. It was developed in the early 1900s as an evolution of the irrigation fields and trickling filters, the first artificial biological treatments, when the pressure of population increase and new pollutants appeared. The introduction of AS was of paramount importance as it enabled achievement of normative standards and mitigated the effects on public health and the environment due to the uncontrolled release of wastewater into the ecosystems that was usual at this time (Lofrano and Brown 2010; Sharma and Sanghi 2012). In an activated sludge system, large particles are removed from wastewater in a preliminary stage, and suspended matter is separated in the primary clarifier. The overflow goes to the aeration tank where the biological oxidation of dissolved compounds occurs due to the action of microorganisms. To maintain high levels of oxygen in the tank, it is necessary to vigorously blow air through diffusers or to use mechanical surface aerators helping also to keep the contents stirred. Hydraulic retention time varies depending on the biological oxygen demand of the wastewater. The effluent is then treated in a secondary clarifier, and part of the settled sludge is recycled to the aeration tank in order to maintain the required concentration of microorganisms. There are several common variations of the basic activated sludge process, such as extended aeration and oxidation ditches, but the principles are similar.

AS is a mature process that has proven to be successful over decades thanks to the low capital costs, relatively simple infrastructure, and flexibility of the system to adjust its operational parameters to the treatment, especially of different municipal wastewaters. However, the AS treatment presents several shortcomings that make it unsustainable and that have stimulated researchers and the WWT market to propose technological improvements and to offer new alternatives. Possibly the main drawback is related to the high energy requirement during the AS operation due to the aeration and pumping systems (Guven et al. 2019; Nancharaiah et al. 2019). Besides, as a result of the treatment process, a high volume of concentrated sludge is generated, and further energy and activities are necessary for its adequate disposal. On the other hand, if industrial wastewater or high loads of nutrients must be treated, the efficiency of the system decays, and longer retention times or additional and complex stages could be necessary (Guven et al. 2019; Massara et al. 2017). In this sense, it is important to guarantee the microbiological equilibrium in the aeration tank demanding permanent supervision to control the system. Other important concerns are safety and health risks to operating personnel and possible disturbances due to odor and noise emissions (Mena-Ulecia and Hernández 2015; Molinos-Senante et al. 2014).

Several works analyze the problems and improvement opportunities related to conventional AS treatment. At a fundamental level, recent studies have focused on the understanding of the diversity and assembly of the microbial communities responsible for the pollutant degradation in order to increase the system efficiency and to predict its operation and stability (Xia et al. 2018; Yang et al. 2020). This basic knowledge also fosters efforts in the fields of modeling and optimization of AS systems (Dai et al. 2016; Karpinska and Bridgeman 2016). On the other hand, energy optimization has been addressed, for instance, through the improvement of aeration and agitators systems in the tank (Fureder et al. 2018; Lee et al. 2015) but maybe more importantly through an appropriate management of the generated sludge. Indeed, an extensive field of research focuses on product and energy recovery in the form of biofertilizers or biogas, among others (Agarwal et al. 2015; Guven et al. 2019; Waclawek et al. 2019).

14.3.2 Membrane Bioreactor

Membrane bioreactors (MBRs) are compact hybrid systems based on a combination of activated sludge treatment and membrane filtration for biomass retention. MBRs were first implemented in the late 1970s and are now a relatively mature technology for secondary and tertiary wastewater treatment, which technically improves the performance of conventional systems through extended removal capacity of N and P compounds and organic content. From small to large municipal plants, MBR systems have been considered the best treatment choice for municipal wastewater requirements by some authors (Dezotti 2017; Melin et al. 2006). The MBR process uses a microfiltration (MF) or ultrafiltration (UF) membrane to perform the solid–liquid separation and combines aeration tank and clarification stages into one-unit operation. Depending on the membrane module configuration, MBRs can be classified as side-stream/external circulation or submerged/immersed systems (Melin et al. 2006).

MBR technology has demonstrated important advantages over conventional wastewater treatment processes due to its small footprint and reactor requirements, high effluent quality, good disinfection capability, higher volumetric loading, and less sludge production (Assayie et al. 2017; Xiao et al. 2019). On the other hand, some impediments to further implementation are high capital costs, high energy demand, and membrane fouling problems, which affect operation stability and require chemicals for maintenance routines (Nancharaiah et al. 2019). As fouling is one of the main drawbacks of this technology, several methods have been proposed and investigated for its control (Arefi-Oskoui et al. 2019; Bagheri and Mirbagheri 2018; Huang and Lee 2015; Zheng et al. 2018).

A comparative study of environmental and economic aspects of MBR and AS systems was carried out by Pretel et al. The work analyzed LCA and life cycle costing (LCC) of simulated large-scale plants, considering several configurations and also energy and fertilizer recovery from sludge post-treatment. Direct comparison between single AS and aerobic MBR shows that membrane technology requires overall about 50 percent more electricity in the operation, mainly related to air sparging for filter scouring. This is proportionally reflected in the total cost indicator. Similar impacts were obtained in eutrophication potential for both systems, but interestingly the AS system was more efficient in global warming, abiotic depletion, and acidification indicators (Pretel et al. 2016). The better global behavior of conventional AS technology was also pointed out in the works of Wang et al., Hao et al., and Sun et al. (Hao et al. 2018; Sun et al. 2020; Wang et al. 2020). In these studies, a composite sustainability index was calculated, considering technical, economic, and management aspects of different MBR and AS facilities. The authors found that capital and operational costs and energy sub-indices negatively affect the sustainability of MBR when compared with conventional AS. These remarks were also quoted in the review by Xiao et al., pointing out, however, that MBR produces a better effluent and has an extended capacity to remove trace organic pollutants and microplastics and that costs are comparable with AS when followed by a tertiary treatment (Xiao et al. 2019). On the other hand, Kalbar et al. developed a decision support tool to evaluate sustainability characteristics of WWT technologies under different scenarios. In particular, MBR, sequencing batch reactor (SBR), and AS technologies were considered for wastewater treatment of residential townships. In the study, MBR showed higher warming potential but both less eutrophication potential and less required land. Social indicators such as acceptability, community participation, and promotion of sustainable behavior were included. Overall, both MBR and SBR sustainability indices for the required situation largely exceed those of the AS system (Kalbar et al. 2016). The performance of MBR was evaluated by Høybe et al. for the tertiary treatment of wastewater and the removal of hazardous substances, including heavy metals, endocrine disruptors, and pathogens. Comparison of environmental impacts with SF and O_3 technologies indicated that MBR plant operation is responsible for higher impacts in global warming, acidification, and nutrient enrichment categories (LCIA- EDIP model). Despite its better results in preventing freshwater eco-toxicity, the net environmental benefit of MBR alternative is not clear. Besides, the sum of investment and operational costs for MBR treatment is about two or three times that of O_3 and SF, respectively (Høibye et al. 2008).

14.3.3 High Rate Algal Ponds

High rate algal ponds (HRAPs) are an efficiently engineered evolution of natural lagoons and facultative ponds (Craggs et al. 2013). First proposed by Oswald in 1957 (Oswald et al. 1957), small–medium HRAP plants have been implemented since then for the treatment of domestic and industrial wastewater until secondary and tertiary purification levels (Mehrabadi et al. 2015; Young et al. 2017). HRAPs are shallow, gentle paddle-wheel mixed ponds where a microalgae–bacteria consortium degrades pollutants through photosynthesis and absorbs nutrients present in wastewaters. As algal biomass is mainly carbon constituted, it could be necessary to inject CO_2 to compensate the low C:N ratio of some domestic wastewaters (Craggs et al. 2013).

HRAP systems require simple infrastructure, low capital investment when inexpensive land is available, and low operating costs, and they are more efficient than AS treatment in terms of pollutant and nutrient depletion (Mehrabadi et al. 2015). Several studies have shown their capacity to degrade a wide spectrum of organic compounds, including persistent and pharmaceutical chemicals (Singh 2015; Villar-Navarro et al. 2018). The main drawbacks are the large land area demand, relatively high retention times, and performance sensitivity to weather conditions. Besides, produced biomass should be harvested, dried, and disposed. However, the harvesting process has low cost and low impacts for some algal species due to the natural and spontaneous sedimentation of large flocs formed (Park et al. 2013). HRAP systems considerably increase their sustainability potential when biomass is converted into biofuels, biofertilizers, and other valuable products (Anto et al. 2020; Brodie et al. 2017; Juneja and Murthy 2017; Kumari and Singh 2019; Murthy and Chanakya 2015). Moreover, in the case of integrating algae biofuels into WWT system boundaries, the process is optimized if the CO_2 emissions during power generation supply the necessary CO_2 to maintain biomass growth in the ponds (Craggs et al. 2013).

Comparative studies between HRAP and other WWT technologies have been performed to assess their relative sustainability. Garfí et al. analyzed the environmental impacts (LCIA- ReCiPe model) and costs during the construction, operation, and sludge/algae transportation and disposal phases for HRAP, hybrid constructed wetland (combination of vertical and horizontal flow), and AS technologies. Results show that the conventional AS system had two to five times more detrimental effects than those of natural-based systems and was two to three times more expensive. At the same time, there were just slight differences between constructed wetland and HRAP technologies. Interestingly, environmental impacts related specifically to construction and sludge handling phases are higher in constructed wetland and HRAP technologies than in AS systems (Garfí et al. 2017). Kohleb et al. evaluated environmental and cost impacts of an HRAP plant, without product recovery, against an activated sludge-based sequencing batch reactor (AS-SBR). The environmental burdens, warming potential, and energy consumption of AS-SBR are higher, but in the specific case of eutrophication potential, the system shows a slight benefit, mainly due to higher impacts for this indicator in the algae harvesting and thickening steps of HRAP systems. Overall, the cost analysis favors HRAP technology despite higher capital costs related to large surface and construction material demand (Kohlheb et al. 2020). Arashiro et al. carried out an LCA study on a small simulated HRAP plant with biogas recovery and an HRAP plant with biofertilizer recovery. Their performances were compared with that of a conventional AS plant with sludge incineration. Mid-point environmental impacts (LCIA- ReCiPe model) indicate the benefits of biogas recovery in global warning potential, ozone depletion, photochemical effect, and fossil depletion categories. On the other hand, the

HRAP plant with biofertilizer production shows less eutrophication potential than the other two alternatives. The relatively high impact due to materials required during the construction for HRAP systems is noteworthy. Meanwhile, the burdens are in resource consumption during operation in the case of the AS system. The study also shows that some environmental benefits of HRAP systems are significantly reduced in winter and cold seasons (Arashiro et al. 2018).

14.3.4 Constructed Wetlands

A constructed wetland (CW) is an artificially engineered system to treat municipal or industrial wastewater, graywater, or stormwater runoff based on controlled natural processes involving wetland vegetation, soil, and microbial consortia (Tanaka et al. 2011; Vymazal 2010; Wu et al. 2015). The first full-scale projects were developed in the 1960s, and since then CWs have been successfully implemented around the world for secondary and/or tertiary treatment (Vymazal 2011). Mechanisms involve settling of suspended particulate matter, filtration and chemical precipitation, chemical and biological transformations, adsorption and uptake of nutrients, and natural die-off of pathogens (Tanaka et al. 2011). CWs are classified depending on the hydrology (open water-surface flow and subsurface flow), type of macrophytic growth (emergent, submerged, free-floating), and flow path (HFCW: horizontal and VFCW: vertical) (Vymazal 2011).

CWs are cost-effective systems as construction and resource supply during operation, including human on-site labor, are relatively low. They have a considerable buffering capacity, so influent fluctuations do not affect reliability. They facilitate water reuse and recycling, and because they integrate well into the landscape, public acceptance is high. The main drawbacks are a large land area requirement and, as for other biological based systems, the CW performance being affected by seasonal changes. Despite their favorable visual esthetics, CWs could be a source of malodors and insects if the system is overloaded (Tanaka et al. 2011). On the other hand, the design should be carefully adjusted, especially when high levels of N or P removal are required (Shved et al. 2015; Vymazal 2011). Several current research lines have been highlighted by Wu et al. to improve the sustainable potential of CWs, notably on plant and substrate selection, plant harvesting, reclamation and recycling, and enhancing strategies such as artificial aeration, microbial augmentation and external carbon addition (Wu et al. 2015).

CWs are considered green technologies, and the assessment of their sustainability relative to other treatments has gained attention in the past decade. It should be noted that differences in CW designs could significantly affect the overall performance of the systems. For instance, an LCA study by Fuchs et al. shows the environmental advantages of VFCW relative to HFCW systems, based on mid-point and end-point indicators (LCIA ecoindicator 99 model). In particular, fewer emissions and infrastructure-related impacts, due to a more compact and reduced system, favor selection of vertical technology (Fuchs et al. 2011). Kalbar et al. carried out a comparative LCA analysis on four different secondary treatment technologies — AS, up-flow anaerobic sludge blanket reactor (UASB), sequencing batch reactor (SBR), and CW — for municipal wastewater in India, focusing only on operation and maintenance phases.

Impact assessment results (LCIA- CML 2 baseline 2000 model) indicated that CW overcomes conventional AS as well as the other alternatives in most of the environmental indicators. Important exceptions are eutrophication and human aquatic ecotoxicity potentials where SBR excels. Regarding energetic consumption, CW is 7 and 15 times lower than AS and SBR systems, respectively. As expected, the high land requirement of CW is viewed as the main obstacle to its implementation in dense areas (Kalbar et al. 2013). In another work by the same authors, CW was considered an alternative to lake rejuvenation. In this case, an up-flow anaerobic sludge blanket reactor was followed by a facultative aerobic lagoon (UASB-FAL) and AS systems were also analyzed with the support of a sustainability index. Despite a required area 7 times higher than that of the other technologies, the CW technology excels in all the environmental, economic, and social categories (Kalbar et al. 2016). Molinos-Senante et al. also analyzed the three sustainability dimensions of CW and other WWT alternatives for their implementation in small communities. The CW system obtained the highest score in the economic composite indicator and one of the highest in social aspects, while technologies such as SBR and MBR were clearly superior in the environmental dimension (Molinos-Senante et al. 2014). The work of Garfi et al., discussed earlier in Section 14.3.3, confirmed the better environmental performance of CW vs. AS and its convenience for small communities (Garfí et al. 2017). Machado et al. analyzed the impacts of construction/assembly, operation/maintenance, and dismantling/final disposal phases for a combined VFCW+HFCW, a conventional AS, and small decentralized slow-rate infiltration (SRI) plants. LCA results (LCIA- CML 2 Baseline 2000 model) show the predominance of construction impacts for CW and the operational impacts for AS. Meanwhile, SRI effects are distributed evenly along its life cycle. A final comparison indicates CW and SRI as the best choices (Machado et al. 2007). The effect of scale on environmental impacts of decentralized CW and MBR systems relative to a conventional AS-based plant was recently investigated by Kobayashi et al. Results showed that global warming, eutrophication, and human health carcinogenic potentials (LCIA- Traci 2.1 model) for both CW and MBR technologies at the household level significantly exceed those of conventional systems despite the larger collection system required. At community and neighborhood scales CW appears to be the best choice based on these three indicators. Considerations on the effects of water reuse and energy mix were presented (Kobayashi et al. 2020).

14.3.5 Membrane and Direct Filtration Technologies

Membranes have been used in several wastewater treatments systems due to their capacity to retain and remove microorganisms, organic compounds, nutrients, pathogens, and suspended and colloidal particles from polluted water. Membranes are the key element in MBR or reverse osmosis (RO) technologies, and since the 1990s their potential as a stand-alone direct-pressure–driven filtration process for water reuse and product recovery has gained increased attention (Hube et al. 2020). Membranes for microfiltration (MF), ultrafiltration (UF), or nanofiltration (NF) processes have proven to be effective at laboratory scale and pilot scale for tertiary treatment of synthetic, industrial, domestic, and agricultural wastewaters (Hube et al. 2020; Obotey Ezugbe and Rathilal 2020).

Direct filtration (DF) is a compact, easy-to-install, and simpler scale-up process. Energy and resource demand during operation are low, as no high pressures or chemical reagents are needed. Due to its modularity, replacement is simple and fast, and no specialized knowledge is required. Besides, the process performance is adaptable given the broad available spectrum of membrane materials and porosity ranges. However, the main challenges to its practical and wide implementation are membrane fouling and degradation, which increase maintenance and unit replacement frequency and could affect process performance (Sikdar and Criscuoli 2017). Finally, retained materials require disposal and landfill, although here there is an opportunity toward process sustainability improvement as valuable product recovery is possible (Hube et al. 2020). Because of the appealing advantages of DF systems, several strategies have been investigated to improve fouling issues and durability, including physical and chemical cleaning, pretreatment of feed water, and integration with coagulation/adsorption processes (Hube et al. 2020; Kimura et al. 2017).

The sustainability dimensions of direct filtration processes have been evaluated against other tertiary treatment technologies. Carre et al. carried out an LCA study on pilot-scale systems based on UF and combinations of sand filtration (SF), MF, and UF with ultraviolet (UV) post-treatments. Membrane scheme boundaries included chemical and mechanical washing cycles. Interestingly, UF alone demonstrated compliance with the required quality standards for the effluent at a lower impact in several mid-point environmental indicators (LCIA- ReciPe method), notably in accumulative energy demand and fossil fuel depletion, acidification, and ecotoxicity. In terms of eutrophication potential, UF has higher effects than SF+UV alternatives but fewer effects than UF or MF combined with UV. Overall, UF benefits from a simple infrastructure and operational requirements compared with hybrid systems (Carre et al. 2017). Buyukkamaci et al. also compared polishing processes based on membrane technology and UV treatment for water agricultural reuse. However, in this study the UF-based alternative presented largest global warming, human toxicity, ecotoxicity, and ozone depletion impacts (LCIA- CML 2001 model) due to the energy necessary for maintaining high pressure in the system. Even if there is slight improvement in eutrophication potential for UF, the authors pointed out that overall a better environmental balance is reached by the UV system (Buyukkamaci and Karaca 2017). Remy et al. focused on advanced treatment and phosphorous removal processes to upgrade WWT plants in Germany. Polymer UF and ceramic MF were compared with high rate sedimentation, microsieve, and dual media filter processes. LCA results indicated that DF technologies produce a better effluent quality and reduced eutrophication potential but at cost of higher global warming indicator (LCIA- ReCiPe 2008 model) and cumulative energy caused by pressure pumps and membrane washing requirements (Remy et al. 2014).

14.3.6 Adsorption Processes

Adsorption technology is considered one of the most interesting alternatives for tertiary treatment of wastewater due to its proven capacity to efficiently remove heavy metals, phosphorous, dyes, pesticides, pharmaceuticals, and recalcitrant micropollutants (Guillossou et al. 2019; Sharma and Sanghi 2012; Sophia and Lima 2018). Adsorption systems in the form of sorbent layers or beds involve molecular forces between the pollutant and the sorbent, allowing for surface compound retention or even degradation in the case of chemisorption. Carbon-based materials such as granular activated carbon (GAC) or conventional activated carbon (AC) are the most used sorbents thanks to their capacity to interact with multiple pollutants, their chemical resistance, high surface area, and controllable microporosity.

Adsorption is an attractive technology due to its simple installation, operation, and maintenance and its low energy and chemical demand. When required, adsorption selectivity could be designed through chemical modification and functionalization of the sorbent. However, its main shortcomings are the high cost of the conventional sorbents and the complex reactivation processes (Sharma and Sanghi 2012). In recent years, extensive research has focused on the quest of low-cost and renewable alternatives, including biosorbents, natural materials, and agricultural and industrial waste materials (Ahmad et al. 2010; Alaba et al. 2018; De Gisi et al. 2016; Zanoletti and Bontempi 2017). Among them, biochar has been considered one of the best options to develop sustainable adsorption technologies due to the ideal balance among technical performance, low cost, and simple production process from biomass waste (Gwenzi et al. 2017; Huggins et al. 2016; Vikrant et al. 2018; Xiang et al. 2020).

The sustainability dimensions of adsorption technologies have been evaluated against other tertiary methods. In the LCA comparison carried out by Igos et al., one of the scenarios studied consisted of the decentralized treatment of pharmaceutical wastewaters, first using MBR followed by AC, O_3, or UV+H_2O_2 systems. The inventory analysis (LCI) indicated in general a high pharmaceutical removal rate for all the alternatives but a negligible energy consumption for AC compared with the others. Regarding environmental performance, the single point score (LCIA- EDIP model) shows better and similar results for O_3 and AC options due to the high impact related to H_2O_2 demand in the UV hybrid system (Igos et al. 2012). More recently Rahman et al. investigated the same systems plus RO technology in the tertiary removal of emerging pollutants. For this an LCA study based on the LCIA USEtox model was carried out to evaluate ecotoxicity and human toxicity (carcinogenic and non-carcinogenic) impacts. AC and O_3 were clearly the best alternatives thanks to the relatively low chemical load and energy demand. Ecotoxicity impacts in higher energy demand techniques, UV-H_2O_2 and RO, are magnified by the fact that it was assumed that electricity was provided by the fossil-based U.S. grid (Rahman et al. 2018). In the multi-criteria analysis done by Plakas et al., the sustainability index of the AC-UF hybrid system and O_3+UV, PMR, and RO technologies was calculated. Under a specific scenario where the weights of economic, environmental, and social dimensions were 30.9 percent, 55.5 percent, and 14.3 percent, respectively, the adsorption-based system obtained the higher score due mainly to low operation and maintenance costs, low energy consumption, and high reliability (Plakas et al. 2016). Looking for alternatives to fossil-based activated carbon, Koziatnik et al. evaluated the environmental performance of low-cost sorbents from lignocellulosic sources, notably biochar (BC) and hydrochar (HC). The LCA study of conventional AC, BC, and HC materials for industrial wastewater treatment considered different end-of-life scenarios: landfill, combustion for energy recovery, and sorbent reactivation. Mid-point impacts (LCIA- ILCD v2.0 method) showed that, in general, all scenarios with CA are more environmentally friendly than those with biomass-based options. Remarkably, the production phase of these natural-based materials represented the most important burden (Kozyatnyk et al. 2020).

14.3.7 Advanced Oxidation Processes

In recent decades, advanced oxidation processes (AOPs) have been proposed as highly efficient alternatives for the tertiary treatment and removal of pathogens and recalcitrant compounds from wastewater (Garrido-Cardenas et al. 2019; Miklos et al. 2018). Though based on different strategies, all AOPs share the in-situ production of reactive oxygen species (ROS), mainly hydroxyl radicals, as the main reagent to allow complete mineralization of organic and inorganic pollutants. AOPs established at full scale are ozonation (O_3), ozonation plus hydrogen peroxide (O_3/H_2O_2), ultraviolet plus hydrogen peroxide (UV/H_2O_2), ultraviolet plus chlorine (UV/Cl_2), and Fenton (F) and photo-Fenton (pF) processes. Other systems, such as catalytic ozonation, photocatalytic ozonation, and heterogeneous photocatalysis have been implemented and investigated at laboratory or pilot scale (Miklos et al. 2018). Although several differences exist among AOPs, in general they are potentially easy to install and operate, and catalyst and reagent demand is considered low. When required, light energy is supplied by low-cost UV lamps or by direct solar radiation.

In recent years, comparative sustainability analysis among different AOPs or among AOP systems and other tertiary wastewater treatments has been carried out. Arzate et al. evaluated the life cycle environmental impacts of ozonation and solar photo-Fenton systems, both followed by a sand filtration stage, for treatment of the effluent of an AS system under a water reclamation for crop irrigation scenario. Mid-point and end-point indicators (LCIA- ReCiPe model) show that the ozone-based technology is the best alternative due to the simple ozone production system and the higher reactant demand and direct gas emissions for the solar pF system (Arzate et al. 2019). Chatzisymeon et al. investigated at laboratory scale the performance of an UV-TiO_2 photocatalytic reactor, a wet air oxidation (WAO) reactor, and an electrochemical oxidation (EO) cell in the treatment of olive mill wastewater. In this case, the energy demand of the photocatalytic system largely exceeded those of the WAO and EO systems, especially affecting global warming (LCIA- IPCC 2007 model), fossil depletion, and human toxicity (LCIA- ReCiPe model). The study was performed taking into account the Greek energy mix, which is based mainly on coal and gas combustion (Chatzisymeon et al. 2013). Foteinis et al. examined the effective removal from water of ethynylestradiol, an endocrine-disrupting chemical, by the following light-driven processes: solar pF, solar pF/H_2O_2, UVA irradiation, UVA/TiO_2, UVC irradiation, and UVC/H_2O_2. End-point indicators (LCIA- ReCiPe model) showed that the environmental footprint increases with energy consumption and treatment time. Thereby, the impacts related to the use of reactants (Fenton's reagent, H_2O_2, or TiO_2) were largely offset by the marked improvement in the removal efficiency. Overall, solar pF/H_2O_2 showed the better results, followed by UVC/H_2O_2 (Foteinis et al. 2018).

14.4 CONCLUSIONS

In this chapter, a brief review of the most important sustainability assessment methodologies and the main mature and emerging systems for wastewater treatment was presented. Individual advantages and disadvantages of WWT toward sustainability and current research areas for their improvement were highlighted. Finally, comparative sustainability studies between selected technologies and other alternatives for treating wastewater under specific scenarios were discussed. In this regard, it should be noted that WWT systems frequently involve several stages and hybrid technologies that could be arranged under different configurations. This makes it difficult and even useless to rank individual technologies instead of integral solutions. Besides, the assessment of sustainability is performed under different contexts, and realities and extracted results and conclusions of some studies must not be generalized.

Despite the above, there are several conclusions that can frame the discussion on WWT sustainability and serve as guidelines to technology selection. One important point is the relation among energy, costs, and pollution. As we can see in some of the previously discussed works, high energy requirement during operation increases costs, but even more significantly it could produce a larger environmental impact than the benefit inherent to water treatment due to the emissions during energy production. When analyzing energy-intensive methods, the sustainability assessment results could depend more on the energy mix composition than on the treatment technology itself. On the other hand, natural-based technologies such as CWs and HRAPs are highly sustainable alternatives as long as enough inexpensive land is available. Besides, because they are non-complex systems, they could represent a decentralized solution in many developing countries, small rural communities, and agricultural areas. Also, some LCA and sustainability studies showed that including product recovery in the boundaries of the system can shift both the economic and environment balance of WWTs to a profitable operation in the short term. Finally, plant size considerations in WWT design should take into account the impact of collection systems for large centralized plants but also the benefits of economy of scale.

The studies reviewed in this work confirm that the assessment of the social dimension remains limited in many discussions on WWT sustainability. One of the reasons is the logical lack of field experience on many techniques that are still under development. A relevant example is photocatalysis, which has been proposed in recent decades as a potential breakthrough in WWT but that up to this point has been tested only at the pilot scale. When no full-scale data is available, the opinion of a panel of experts could be supported with preliminary predictive models that take into account the socioeconomic context. Despite that, considerable advances have been done in the field of WWT sustainability research. In particular, a comprehensive life cycle approach on economic and environmental impacts is more frequent, contributing to a broad understanding of the real challenges behind WWT sustainability.

REFERENCES

Agarwal, M., J. Tardio and S. Venkata Mohan (2015). "Pyrolysis of activated sludge: energy analysis and its technical feasibility." *Bioresour Technol* **178**: 70–75.

Ahmad, T., M. Rafatullah, A. Ghazali, O. Sulaiman, R. Hashim and A. Ahmad (2010). "Removal of pesticides from water and wastewater by different adsorbents: A review." *J Environ Sci Health C Environ Carcinog Ecotoxicol Rev* **28**(4): 231–271.

Akhoundi, A. and S. Nazif (2018). "Sustainability assessment of wastewater reuse alternatives using the evidential reasoning approach." *J Cleaner Prod* **195**: 1350–1376.

Alaba, P. A., N. A. Oladoja, Y. M. Sani, O. B. Ayodele, I. Y. Mohammed, S. F. Olupinla and W. M. W. Daud (2018). "Insight into wastewater decontamination using polymeric adsorbents." *J Environ Chem Eng* **6**(2): 1651–1672.

Alfonsin, C., R. Lebrero, J. M. Estrada, R. Munoz, N. J. Kraakman, G. Feijoo and M. T. Moreira (2015). "Selection of odour removal technologies in wastewater treatment plants: a guideline based on Life Cycle Assessment." *J Environ Manage* **149**: 77–84.

Alyaseri, I. and J. Zhou (2017). "Towards better environmental performance of wastewater sludge treatment using endpoint approach in LCA methodology." *Heliyon* **3**(3): e00268.

Anastas, P. T., and J. B. Zimmerman (2003). "Design through the 12 principles of green engineering." *Environ Sci Technol* **37**(5): 94A–101A.

Anto, S., S. S. Mukherjee, R. Muthappa, T. Mathimani, G. Deviram, S. S. Kumar, T. N. Verma and A. Pugazhendhi (2020). "Algae as green energy reserve: Technological outlook on biofuel production." *Chemosphere* **242**: 125079.

Arashiro, L. T., N. Montero, I. Ferrer, F. G. Acién, C. Gómez and M. Garfí (2018). "Life cycle assessment of high rate algal ponds for wastewater treatment and resource recovery." *Sci Total Environ* **622-623**: 1118–1130.

Arefi-Oskoui, S., A. Khataee, M. Safarpour, Y. Orooji and V. Vatanpour (2019). "A review on the applications of ultrasonic technology in membrane bioreactors." *Ultrason Sonochem* **58**: 104633.

Arias, A., I. Vallina, Y. Lorenzo, O. T. Komesli, E. Katsou, G. Feijoo and M. T. Moreira (2019). Water footprint of a decentralised wastewater treatment strategy based on membrane technology. *Environ Footprints Eco Prod Processes*: 85–119.

Arzate, S., S. Pfister, C. Oberschelp and J. A. Sanchez-Perez (2019). "Environmental impacts of an advanced oxidation process as tertiary treatment in a wastewater treatment plant." *Sci Total Environ* **694**: 133572.

Assayie, A. A., A. Y. Gebreyohannes and L. Giorno (2017). Municipal wastewater treatment by membrane bioreactors. *Sustainable Membrane Technology for Water and Wastewater Treatment*. A. Figoli and A. Criscuoli. Singapore, Springer Singapore: 265–294.

Awad, H., M. Gar Alalm and H. K. El-Etriby (2019). "Environmental and cost life cycle assessment of different alternatives for improvement of wastewater treatment plants in developing countries." *Sci Total Environ* **660**: 57–68.

Bagheri, M. and S. A. Mirbagheri (2018). "Critical review of fouling mitigation strategies in membrane bioreactors treating water and wastewater." *Bioresour Technol* **258**: 318–334.

Bai, S., X. Zhang, Y. Xiang, X. Wang, X. Zhao and N. Ren (2019). "HIT.WATER scheme: An integrated LCA-based decision-support platform for evaluation of wastewater discharge limits." *Sci Total Environ* **655**: 1427–1438.

Bolisetty, S., M. Peydayesh and R. Mezzenga (2019). "Sustainable technologies for water purification from heavy metals: review and analysis." *Chem Soc Rev* **48**(2): 463–487.

Brodie, J., C. X. Chan, O. De Clerck, J. M. Cock, S. M. Coelho, C. Gachon, A. R. Grossman, T. Mock, J. A. Raven, A. G. Smith, H. S. Yoon and D. Bhattacharya (2017). "The Algal Revolution." *Trends Plant Sci* **22**(8): 726–738.

Buyukkamaci, N. and G. Karaca (2017). "Life cycle assessment study on polishing units for use of treated wastewater in agricultural reuse." *Water Sci Technol* **76**(11-12): 3205–3212.

Carre, E., J. Beigbeder, V. Jauzein, G. Junqua and M. Lopez-Ferber (2017). "Life cycle assessment case study: Tertiary treatment process options for wastewater reuse." *Integr Environ Assess Manag* **13**(6): 1113–1121.

Chatzisymeon, E., S. Foteinis, D. Mantzavinos and T. Tsoutsos (2013). "Life cycle assessment of advanced oxidation processes for olive mill wastewater treatment." *J Cleaner Prod* **54**: 229–234.

Chen, W., T. L. Oldfield, S. I. Patsios and N. M. Holden (2020). "Hybrid life cycle assessment of agro-industrial wastewater valorisation." *Water Res* **170**.

Corominas, L., J. Foley, J. S. Guest, A. Hospido, H. F. Larsen, S. Morera and A. Shaw (2013). "Life cycle assessment applied to wastewater treatment: state of the art." *Water Res* **47**(15): 5480–5492.

Cosgrove, W. J., and D. P. Loucks (2015). "Water management: Current and future challenges and research directions." *Water Resourc Res* **51**(6): 4823–4839.

Craggs, R., T. Lundquist and J. Benemann (2013). "Wastewater treatment and algal biofuel production." *Algae for Biofuels and Energy*. Dordrecht; New York, Springer, 153–163.

Dai, H., W. Chen and X. Lu (2016). "The application of multi-objective optimization method for activated sludge process: A review." *Water Sci Technol* **73**(2): 223–235.

De Feo, G. and C. Ferrara (2017). "Investigation of the environmental impacts of municipal wastewater treatment plants through a Life Cycle Assessment software tool." *Environ Technol* **38**(15): 1943–1948.

De Gisi, S., G. Lofrano, M. Grassi and M. Notarnicola (2016). "Characteristics and adsorption capacities of low-cost sorbents for wastewater treatment: A review." *Sustainable Mater Technol* **9**: 10–40.

Dezotti, M. (2017). *Advanced biological processes for wastewater treatment: emerging, consolidated technologies and introduction to molecular techniques*. New York, NY, Springer Science+Business Media.

Diaz-Elsayed, N., N. Rezaei, A. Ndiaye and Q. Zhang (2020). "Trends in the environmental and economic sustainability of wastewater-based resource recovery: A review." *J Cleaner Prod* **265**: 1–16.

Fang, L. L., B. Valverde-Perez, A. Damgaard, B. G. Plosz and M. Rygaard (2016). "Life cycle assessment as development and decision support tool for wastewater resource recovery technology." *Water Res* **88**: 538–549.

Fernández-Fernández, M. I., P. T. M. de la Vega and M. A. Jaramillo-Morán (2019). "Pollution and sustainability indices for small and medium Wastewater Treatment Plants in the southwest of Spain." *Water (Switzerland)* **11**(3).

Foteinis, S., A. G. L. Borthwick, Z. Frontistis, D. Mantzavinos and E. Chatzisymeon (2018). "Environmental sustainability of light-driven processes for wastewater treatment applications." *J Cleaner Prod* **182**: 8–15.

Fuchs, V. J., J. R. Mihelcic and J. S. Gierke (2011). "Life cycle assessment of vertical and horizontal flow constructed wetlands for wastewater treatment considering nitrogen and carbon greenhouse gas emissions." *Water Res* **45**(5): 2073–2081.

Fureder, K., K. Svardal, W. Frey, H. Kroiss and J. Krampe (2018). "Energy consumption of agitators in activated sludge tanks: Actual state and optimization potential." *Water Sci Technol* **77**(3-4): 800–808.

Galanakis, C. M., and E. Agrafioti (2019). *Sustainable water and wastewater processing*. eBook. Elsevier.

Gallego-Schmid, A. and R. R. Z. Tarpani (2019). "Life cycle assessment of wastewater treatment in developing countries: A review." *Water Res* **153**: 63–79.

Gan, X., I. C. Fernandez, J. Guo, M. Wilson, Y. Zhao, B. Zhou and J. Wu (2017). "When to use what: Methods for weighting and aggregating sustainability indicators." *Ecol Indic* **81**: 491–502.

Garfí, M., L. Flores and I. Ferrer (2017). "Life cycle assessment of wastewater treatment systems for small communities: Activated sludge, constructed wetlands and high rate algal ponds." *J Cleaner Prod* **161**: 211–219.

Garrido-Cardenas, J. A., B. Esteban-Garcia, A. Aguera, J. A. Sanchez-Perez and F. Manzano-Agugliaro (2019). "Wastewater treatment by advanced oxidation process and their worldwide research trends." *Int J Environ Res Public Health* **17**(1): 1–19.

Guillossou, R., J. Le Roux, R. Mailler, E. Vulliet, C. Morlay, F. Nauleau, J. Gasperi and V. Rocher (2019). "Organic micropollutants in a large wastewater treatment plant: What are the benefits of an advanced treatment by activated carbon adsorption in comparison to conventional treatment?" *Chemosphere* **218**: 1050–1060.

Guven, H., R. K. Dereli, H. Ozgun, M. E. Ersahin and I. Ozturk (2019). "Towards sustainable and energy efficient municipal wastewater treatment by up-concentration of organics." *Prog Energy Combust Sci* **70**: 145–168.

Guven, H., O. Eriksson, Z. Wang and I. Ozturk (2018). "Life cycle assessment of upgrading options of a preliminary wastewater treatment plant including food waste addition." *Water Res* **145**: 518–530.

Gwenzi, W., N. Chaukura, C. Noubactep and F. N. D. Mukome (2017). "Biochar-based water treatment systems as a potential low-cost and sustainable technology for clean water provision." *J Environ Manage* **197**: 732–749.

Hadjimichael, A., S. Morera, L. Benedetti, T. Flameling, L. Corominas, S. Weijers and J. Comas (2016). "Assessing urban wastewater system upgrades using integrated modeling, life cycle analysis, and shadow pricing." *Environ Sci Technol* **50**(23): 12548–12556.

Hao, X. D., J. Li, M. C. M. van Loosdrecht and T. Y. Li (2018). "A sustainability-based evaluation of membrane bioreactors over conventional activated sludge processes." *J Environ Chem Engin* **6**(2): 2597–2605.

Harclerode, M., A. Doody, A. Brower, P. Vila, J. Ho and P. J. Evans (2020). "Life cycle assessment and economic analysis of anaerobic membrane bioreactor whole-plant configurations for resource recovery from domestic wastewater." *J Environ Manage* **269**.

Hauschild, M., R. Rosenbaum and S. Olsen (2017). *Life cycle assessment: Theory and practice*. Cham, Springer International Publishing.

Høibye, L., J. Clauson-Kaas, H. Wenzel, H. F. Larsen, B. N. Jacobsen and O. Dalgaard (2008). Sustainability assessment of advanced wastewater treatment technologies. *Water Sci Technol* **58**: 963–968.

Huang, L. and D. J. Lee (2015). "Membrane bioreactor: A mini review on recent R&D works." *Bioresour Technol* **194**: 383–388.

Hube, S., M. Eskafi, K. F. Hrafnkelsdottir, B. Bjarnadottir, M. A. Bjarnadottir, S. Axelsdottir and B. Wu (2020). "Direct membrane filtration for wastewater treatment and resource recovery: A review." *Sci Total Environ* **710**: 136375.

Huggins, T. M., A. Haeger, J. C. Biffinger and Z. J. Ren (2016). "Granular biochar compared with activated carbon for wastewater treatment and resource recovery." *Water Res* **94**: 225–232.

Igos, E., E. Benetto, S. Venditti, C. Kohler, A. Cornelissen, R. Moeller and A. Biwer (2012). "Is it better to remove pharmaceuticals in decentralized or conventional wastewater treatment plants? A life cycle assessment comparison." *Sci Total Environ* **438**: 533–540.

ISO (2006). 14040. Environmental management: Life cycle assessment; principles and framework, ISO.

ISO (2006). 14044. Environmental management: Life cycle assessment; requirements and guidelines, ISO Geneva.

Juneja, A. and G. S. Murthy (2017). "Evaluating the potential of renewable diesel production from algae cultured on wastewater: Techno-economic analysis and life cycle assessment." *AIMS Energy* **5**(2): 239–257.

Kadava, A., S. Murthy and A. R. Shaw (2018). "Envisioning the LCA of a wastewater treatment plant." *Water Prac Technol* **13**(3): 583–588.

Kalbar, P. P., S. Karmakar and S. R. Asolekar (2013). "Assessment of wastewater treatment technologies: Life cycle approach." *Water Environ J* **27**(2): 261–268.

Kalbar, P. P., S. Karmakar and S. R. Asolekar (2016). "Life cycle-based decision support tool for selection of wastewater treatment alternatives." *J Cleaner Produc* **117**: 64–72.

Karpinska, A. M. and J. Bridgeman (2016). "CFD-aided modelling of activated sludge systems: A critical review." *Water Res* **88**: 861–879.

Kimura, K., D. Honoki and T. Sato (2017). "Effective physical cleaning and adequate membrane flux for direct membrane filtration (DMF) of municipal wastewater: Up-concentration of organic matter for efficient energy recovery." *Sep Purif Technol* **181**: 37–43.

Kobayashi, Y., N. J. Ashbolt, E. G. R. Davies and Y. Liu (2020). "Life cycle assessment of decentralized greywater treatment systems with reuse at different scales in cold regions." *Environ Int* **134**: 105215.

Kohlheb, N., M. van Afferden, E. Lara, Z. Arbib, M. Conthe, C. Poitzsch, T. Marquardt and M. Y. Becker (2020). "Assessing the life-cycle sustainability of algae and bacteria-based wastewater treatment systems: High-rate algae pond and sequencing batch reactor." *J Environ Manage* **264**.

Kozyatnyk, I., D. M. M. Yacout, J. Van Caneghem and S. Jansson (2020). "Comparative environmental assessment of end-of-life carbonaceous water treatment adsorbents." *Bioresour Technol* **302**.

Kumari, N. and R. K. Singh (2019). "Biofuel and co-products from algae solvent extraction." *J Environ Manage* **247**: 196–204.

Lam, K. L., L. Zlatanovic and J. P. van der Hoek (2020). "Life cycle assessment of nutrient recycling from wastewater: A critical review." *Water Res* **173**: 115519.

Lee, I., H. Lim, B. Jung, M. F. Colosimo and H. Kim (2015). "Evaluation of aeration energy saving in two modified activated sludge processes." *Chemosphere* **140**: 72–78.

Li, Y., S. Zhang, W. Zhang, W. Xiong, Q. Ye, X. Hou, C. Wang and P. Wang (2019). "Life cycle assessment of advanced wastewater treatment processes: Involving 126 pharmaceuticals and personal care products in life cycle inventory." *J Environ Manage* **238**: 442–450.

Liu, Y., H. Q. Yu, W. J. Ng and D. C. Stuckey (2015). "Wastewater-energy nexus." *Chemosphere* **140**: 1.

Lofrano, G., and J. Brown (2010). "Wastewater management through the ages: A history of mankind." *Sci Total Environ* **408**(22): 5254–5264.

Lopes, T. A. S., L. M. Queiroz, E. A. Torres and A. Kiperstok (2020). "Low complexity wastewater treatment process in developing countries: A LCA approach to evaluate environmental gains." *Sci Total Environ* **720**: 137593.

Lu, B., X. Du and S. Huang (2017). "The economic and environmental implications of wastewater management policy in China: From the LCA perspective." *J Cleaner Produc* **142**: 3544–3557.

Machado, A. P., L. Urbano, A. G. Brito, P. Janknecht, J. J. Salas and R. Nogueira (2007). Life cycle assessment of wastewater treatment options for small and decentralized communities. *Water Sci Technol* **56**: 15–22.

Mannina, G., T. F. Reboucas, A. Cosenza, M. Sanchez-Marre and K. Gibert (2019). "Decision support systems (DSS) for wastewater treatment plants: A review of the state of the art." *Bioresour Technol* **290**: 121814.

Massara, T. M., O. T. Komesli, O. Sozudogru, S. Komesli and E. Katsou (2017). "A Mini review of the techno-environmental sustainability of biological processes for the treatment of high organic content industrial wastewater streams." *Waste Biomass Valorization* **8**(5): 1665–1678.

McNamara, G., L. Fitzsimons, M. Horrigan, T. Phelan, Y. Delaure, B. Corcoran, E. Doherty and E. Clifford (2016). "Life cycle assessment of wastewater treatment plants in Ireland." *J Sustainable Develop Energy, Water Environ Syst* **4**(3): 216–233.

Mehrabadi, A., R. Craggs and M. M. Farid (2015). "Wastewater treatment high rate algal ponds (WWT HRAP) for low-cost biofuel production." *Bioresour Technol* **184**: 202–214.

Melin, T., B. Jefferson, D. Bixio, C. Thoeye, W. De Wilde, J. De Koning, J. van der Graaf and T. Wintgens (2006). "Membrane bioreactor technology for wastewater treatment and reuse." *Desalination* **187**(1-3): 271–282.

Mena-Ulecia, K. and H. H. Hernández (2015). "Decentralized peri-urban wastewater treatment technologies assessment integrating sustainability indicators." *Water Sci Technol* **72**(2): 214–222.

Miklos, D. B., C. Remy, M. Jekel, K. G. Linden, J. E. Drewes and U. Hubner (2018). "Evaluation of advanced oxidation processes for water and wastewater treatment: A critical review." *Water Res* **139**: 118–131.

Molinos-Senante, M., T. Gomez, M. Garrido-Baserba, R. Caballero and R. Sala-Garrido (2014). "Assessing the sustainability of small wastewater treatment systems: a composite indicator approach." *Sci Total Environ* **497-498**: 607–617.

Morera, S., M. V. E. Santana, J. Comas, M. Rigola and L. Corominas (2020). "Evaluation of different practices to estimate construction inventories for life cycle assessment of small to medium wastewater treatment plants." *J Cleaner Prod* **245**: 1–11.

Murthy, P. and H. N. Chanakya (2015). "Nutrient recovery and energy efficient algal harvest from anaerobic digestor wastewater." *Carbon* **7**(2): 109–121.

Nancharaiah, Y. V., M. Sarvajith and T. V. Krishna Mohan (2019). "Aerobic granular sludge: The future of wastewater treatment." *Curr Sci* **117**(3): 395–404.

Nardo, M., M. Saisana, A. Saltelli and S. Tarantola (2005). *Tools for Composite Indicators Building*. Report. Joint Research Center, European Commission.

Ng, B. J., J. Zhou, A. Giannis, V. W. Chang and J. Y. Wang (2014). "Environmental life cycle assessment of different domestic wastewater streams: policy effectiveness in a tropical urban environment." *J Environ Manage* **140**: 60–68.

Nguyen, T. K. L., H. H. Ngo, W. S. Guo, S. W. Chang, D. D. Nguyen, L. D. Nghiem and T. V. Nguyen (2020). "A critical review on life cycle assessment and plant-wide models towards emission control strategies for greenhouse gas from wastewater treatment plants." *J Environ Manage* **264**: 110440.

Niero, M., M. Pizzol, H. G. Bruun and M. Thomsen (2014). "Comparative life cycle assessment of wastewater treatment in Denmark including sensitivity and uncertainty analysis." *J Cleaner Produc* **68**: 25–35.

Obotey Ezugbe, E. and S. Rathilal (2020). "Membrane Technologies in Wastewater Treatment: A Review." *Membranes (Basel)* **10**(5).

Oki, T., and S. Kanae (2004). "Virtual water trade and world water resources." *Water Sci Technol* **49**(7): 203–209.

Opher, T. and E. Friedler (2016). "Comparative LCA of decentralized wastewater treatment alternatives for non-potable urban reuse." *J Environ Manage* **182**: 464–476.

Oswald, W. J., H. B. Gotaas, C. G. Golueke, W. R. Kellen, E. F. Gloyna and E. R. Hermann (1957). "Algae in Waste Treatment [with Discussion]." *Sewage Indus Wastes* **29**(4): 437–457.

Padilla-Rivera, A. and L. P. Güereca (2019). "A proposal metric for sustainability evaluations of wastewater treatment systems (SEWATS)." *Ecol Indic* **103**: 22–33.

Padilla-Rivera, A., J. M. Morgan-Sagastume, A. Noyola and L. P. Güereca (2016). "Addressing social aspects associated with wastewater treatment facilities." *Environ Impact Assess Rev* **57**: 101–113.

Park, J. B., R. J. Craggs and A. N. Shilton (2013). "Enhancing biomass energy yield from pilot-scale high rate algal ponds with recycling." *Water Res* **47**(13): 4422–4432.

Parra-Saldivar, R., M. Bilal and H. M. N. Iqbal (2020). "Life cycle assessment in wastewater treatment technology." *Curr Opin Environl Sci Health* **13**: 80–84.

Pasqualino, J. C., M. Meneses, M. Abella and F. Castells (2009). "LCA as a decision support tool for the environmental improvement of the operation of a municipal wastewater treatment plant." *Environ Sci Technol* **43**(9): 3300–3307.

Plakas, K. V., A. J. Karabelas and A. A. Georgiadis (2016). "Sustainability assessment of tertiary wastewater treatment technologies: A multi-criteria analysis." *Water Sci Technol* **73**(7): 1532–1540.

Polruang, S., S. Sirivithayapakorn and R. Prateep Na Talang (2018). "A comparative life cycle assessment of municipal wastewater treatment plants in Thailand under variable power schemes and effluent management programs." *J Cleaner Produc* **172**: 635–648.

Popovic, T. and A. Kraslawski (2018). "Quantitative indicators of social sustainability and determination of their interdependencies. Example analysis for a wastewater treatment plant." *Period Polytec Chem Engin* **62**(2): 224–235.

Pradel, M. and L. Aissani (2019). "Environmental impacts of phosphorus recovery from a 'product' life cycle assessment perspective: Allocating burdens of wastewater treatment in the production of sludge-based phosphate fertilizers." *Sci Total Environ* **656**: 55–69.

Pretel, R., A. Robles, M. V. Ruano, A. Seco and J. Ferrer (2016). "Economic and environmental sustainability of submerged anaerobic MBR-based (AnMBR-based) technology as compared to aerobic-based technologies for moderate-/high-loaded urban wastewater treatment." *J Environ Manage* **166**: 45–54.

Rahman, S. M., M. J. Eckelman, A. Onnis-Hayden and A. Z. Gu (2018). "Comparative life cycle assessment of advanced wastewater treatment processes for removal of chemicals of emerging concern." *Environ Sci Technol* **52**(19): 11346–11358.

Rashid, S. S. and Y. Q. Liu (2020). "Assessing environmental impacts of large centralized wastewater treatment plants with combined or separate sewer systems in dry/wet seasons by using LCA." *Environ Sci Pollut Res Int* **27**(13): 15674–15690.

Rebitzer, G., D. Hunkeler and O. Jolliet (2003). "LCC — the economic pillar of sustainability: Methodology and application to wastewater treatment." *Environ Prog* **22**(4): 241–249.

Remy, C., M. Boulestreau, J. Warneke, P. Jossa, C. Kabbe and B. Lesjean (2016). "Evaluating new processes and concepts for energy and resource recovery from municipal wastewater with life cycle assessment." *Water Sci Technol* **73**(5): 1074–1080.

Remy, C., U. Miehe, B. Lesjean and C. Bartholomaus (2014). "Comparing environmental impacts of tertiary wastewater treatment technologies for advanced phosphorus removal and disinfection with life cycle assessment." *Water Sci Technol* **69**(8): 1742–1750.

Rene, E. R., L. Shu and V. Jegatheesan (2019). "Editorial: Sustainable eco-technologies for water and wastewater treatment." *J Water Supply* **68**(8): 617–622.

Risch, E., O. Gutierrez, P. Roux, C. Boutin and L. Corominas (2015). "Life cycle assessment of urban wastewater systems: Quantifying the relative contribution of sewer systems." *Water Res* **77**: 35–48.

Saad, A., N. Elginoz, F. Germirli Babuna and G. Iskender (2019). "Life cycle assessment of a large water treatment plant in Turkey." *Environ Sc Poll Res* **26**(15): 14823–14834.

Sabeen, A. H., Z. Z. Noor, N. Ngadi, S. Almuraisy and A. B. Raheem (2018). "Quantification of environmental impacts of domestic wastewater treatment using life cycle assessment: A review." *J Cleaner Produc* **190**: 221–233.

Sedlak, D. L., and J. L. Schnoor (2013). "The challenge of water sustainability." *Environ Sci Technol* **47**(11): 5517.

Sharma, S. K. and R. Sanghi (2012). *Advances in water treatment and pollution prevention.* Dordrecht; New York, Springer.

Shved, O., R. Petrina, V. Chervetsova and V. Novikov (2015). "Enhancing efficiency of nitrogen removal from wastewater in constructed wetlands." *Eastern-Europ J Enterpr Technol* **3**(6): 63–38.

Sikdar, S. and A. Criscuoli (2017). "Sustainability and how membrane technologies in water treatment can be a contributor." *Sustainable membrane technology for water and wastewater treatment*. Singapore, Springer Singapore, 1–21.

Singh, B. (2015). *Algae and environmental sustainability*. New York, NY, Springer Berlin Heidelberg.

Sophia, A. C. and E. C. Lima (2018). "Removal of emerging contaminants from the environment by adsorption." *Ecotoxicol Environ Saf* **150**: 1–17.

Sun, Y., M. Garrido-Baserba, M. Molinos-Senante, N. A. Donikian, M. Poch and D. Rosso (2020). "A composite indicator approach to assess the sustainability and resilience of wastewater management alternatives." *Sci Total Environ* **725**: 1–15.

Swarr, T., D. Hunkeler, W. Klöpffer, H.-L. Pesonen, A. Ciroth, A. Brent and R. Pagan (2011). "Environmental life-cycle costing: A code of practice." *Int J Life Cycle Assess* **16**: 389–391.

Tabesh, M., M. Feizee Masooleh, B. Roghani and S. S. Motevallian (2019). "Life-cycle assessment (LCA) of wastewater treatment plants: A case study of Tehran, Iran." *Int J Civil Engin* **17**(7): 1155–1169.

Tanaka, N., W. J. Ng and K. B. S. N. Jinadasa (2011). *Wetlands for tropical applications: wastewater treatment by constructed wetlands*. London; Hackensack, NJ, Imperial College Press. Distributed by World Scientific Pub. Co.

Tzeng, G.-H. and J.-J. Huang (2011). *Multiple attribute decision making. Methods and applications*. Boca Raton, FL, CRC Press, Taylor and Francis Group.

U.N. (2018). SDG 6 Synthesis Report 2018 on Water and Sanitation.

UNEP (2009). Guidelines for social life cycle assessment of products.

Vikrant, K., K. H. Kim, Y. S. Ok, D. C. W. Tsang, Y. F. Tsang, B. S. Giri and R. S. Singh (2018). "Engineered/designer biochar for the removal of phosphate in water and wastewater." *Sci Total Environ* **616-617**: 1242–1260.

Villar-Navarro, E., R. M. Baena-Nogueras, M. Paniw, J. A. Perales and P. A. Lara-Martin (2018). "Removal of pharmaceuticals in urban wastewater: High rate algae pond (HRAP) based technologies as an alternative to activated sludge based processes." *Water Res* **139**: 19–29.

Vymazal, J. (2010). "Constructed wetlands for wastewater treatment." *Water* **25**.

Vymazal, J. (2011). "Constructed wetlands for wastewater treatment: Five decades of experience." *Environ Sci Technol* **45**(1): 61–69.

Waclawek, S., K. Grübel, D. Silvestri, V. V. T. Padil, M. Waclawek, M. Cerník and R. S. Varma (2019). "Disintegration of wastewater activated sludge (WAS) for improved biogas production." *Energies* **12**(1).

Wang, S., L. Zou, H. Li, K. Zheng, Y. Wang, G. Zheng and J. Li (2020). "Full-scale membrane bioreactor process WWTPs in East Taihu basin: Wastewater characteristics, energy consumption and sustainability." *Sci Total Environ* **723**.

Wang, X., J. Liu, N. Q. Ren, H. Q. Yu, D. J. Lee and X. Guo (2012). "Assessment of multiple sustainability demands for wastewater treatment alternatives: A refined evaluation scheme and case study." *Environ Sci Technol* **46**(10): 5542–5549.

World Commission on Environment and Development. (1987). *Our common future*. Oxford; New York, Oxford University Press.

Wu, H., J. Zhang, H. H. Ngo, W. Guo, Z. Hu, S. Liang, J. Fan and H. Liu (2015). "A review on the sustainability of constructed wetlands for wastewater treatment: Design and operation." *Bioresour Technol* **175**: 594–601.

Wu, J. and T. Wu (2012). "Sustainability indicators and indices: an overview." *Handbook of sustainable management*. London, Imperial College Press, 65–86.

WWAP (2017). *The United Nations World Water Development Report 2017: Wastewater, the Untapped Resource*. Paris, UNESCO.

Xia, Y., X. Wen, B. Zhang and Y. Yang (2018). "Diversity and assembly patterns of activated sludge microbial communities: A review." *Biotechnol Adv* **36**(4): 1038–1047.
Xiang, W., X. Zhang, J. Chen, W. Zou, F. He, X. Hu, D. C. W. Tsang, Y. S. Ok and B. Gao (2020). "Biochar technology in wastewater treatment: A critical review." *Chemosphere* **252**: 126539.
Xiao, K., S. Liang, X. Wang, C. Chen and X. Huang (2019). "Current state and challenges of full-scale membrane bioreactor applications: A critical review." *Bioresour Technol* **271**: 473–481.
Yang, Y., L. Wang, F. Xiang, L. Zhao and Z. Qiao (2020). "Activated sludge microbial community and treatment performance of wastewater treatment plants in industrial and municipal zones." *Int J Environ Res Public Health* **17**(2).
Yenkie, K. M. (2019). "Integrating the three E's in wastewater treatment: Efficient design, economic viability, and environmental sustainability." *Curr Opin Chem Engin* **26**: 131–138.
Young, P., M. Taylor and H. J. Fallowfield (2017). "Mini-review: high rate algal ponds, flexible systems for sustainable wastewater treatment." *World J Microbiol Biotechnol* **33**(6): 117.
Zanoletti, A. and E. Bontempi (2017). Case study of raw materials substitution: Activated carbon substitution for wastewater treatments. *SpringerBriefs Appl Sci Technol*: 63–77.
Zheng, Y., B. Tang, J. Ye, L. Bin, P. Li, F. Xue, S. Huang, F. Fu and Y. Xiao (2018). "A crucial factor towards a sustainable process for municipal wastewater treatment: Fouling effects of different statuses of biomass in the membrane bioreactors with no sludge discharge." *J Cleaner Produc* **192**: 877–886.

15 Challenges, Innovations, and Future Prospects in Transforming Future Wastewater Treatment Plants into Resource Recovery Facilities

Krishna Kumar Nedununi
Environmental Engineering Program, International Center for Water Resources Management, Wilberforce, OH

CONTENTS

DOI: 10.1201/9781003138303-15

15.1 INTRODUCTION

Early human settlements developed their civilizations around natural water bodies such as rivers and lakes. They used these sources for domestic water supply, agriculture, and disposal of human waste. There was no systematic understanding of the characteristics of human waste, its pathological impact on the environment, its treatment, and its safe disposal. As a consequence, water-borne diseases such as cholera and typhoid fever were widespread to an extent that they put a check on population growth. This forced many communities to start paying more attention to understanding the nature of human waste, its treatment, and the formulation of sanitation laws so that the cities and local governments could do their due diligence in addressing the serious problem of waste disposal. The evidence for the first man-made wastewater management system can be found around 1500 BC in the Indus Valley civilization. At the beginning of the modern era, there were few simple wastewater treatment plants (WWTPs) in addition to onsite septic tanks. Over the years, some small towns experimented with letting the wastewater into sewage lagoons and soon came to discover that there was no putrid odor. Water engineers found that evaporation, seepage, and biodegradation by bacteria broke down the organic wastes and eliminated bad odors. The presence of bacteria, oxygen, and sunlight facilitated the decomposition process. The stabilization ponds were found effective in the treatment of small amounts of wastewater, and they initiated the idea of biological treatment of the wastewater (Laki, Nedunuri, and Kandiah, 2019a).

Traditional biological treatment of wastewater using either an activated sludge process (AS) or a tricking filter process (TF) had been in use for over one and a half centuries (Metcalf and Eddy, 2002). The objective of these plants was the prevention of water contamination and associated health risks for the downstream users due to the presence of pathogenic organisms growing on un-stabilized human waste (Verstraete et al., 2009). Traditionally, a large-scale wastewater treatment facility has three treatment steps as shown in Fig. 15.1: (1) preliminary (pretreatment), (2) primary, and (3) secondary. Preliminary treatment involves the removal of coarse solids using

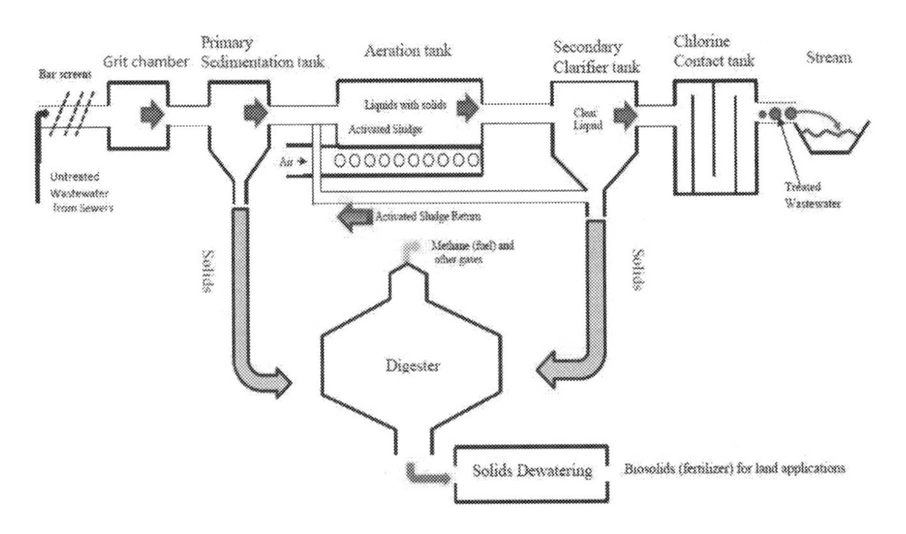

FIGURE 15.1 Schematic of a typical municipal wastewater treatment plant.

screens and a grit chamber. Primary treatment involves removing suspended (mainly inorganic) solid materials from the wastewater in sedimentation tanks. Secondary treatment removes mainly dissolved organic matter in the wastewater as sludge in aeration tanks and generates methane by sludge digestion in a digester before discharging the treated wastewater into natural water bodies after disinfection using chlorine. However, over the past 30–40 years humans have seen a significant deterioration in water quality across water bodies globally. The deterioration stems from an excessive release of nutrients N and P into rivers and oceans due to an exponential increase in agricultural, municipal, and industrial discharges. These increases in discharges are commensurate with the population explosion during the same period. Consequently, N and P removal technologies have been to added AS plants as a tertiary step using anoxic, anaerobic, and aerobic processes using bacteria have been developed (Hammer and Hammer, 2008). The AS processes have succeeded in meeting the permitted effluent quality standards, such as suspended solids, biological oxygen demand (BOD), chemical oxygen demand (COD), N, P, and pathogens; they have been criticized for consuming high energy; occupying a lot of land; and for their inability to conserve freshwater, nutrients, plastics, and other precious minerals. These limitations made AS processes unsustainable (Verstraete and Vlaeminck, 2011).

This chapter will discuss these challenges to draw readers' attention to a paradigm shift in WWTP technologies. Climate change forces our generation to minimize fossil-based fuel consumption to harness future energy; hence we can no longer continue to operate WWTPs without looking into ways of recovering the energy lost to the environment. It is clear that globally we will see an impact on water resources from climate change. For example, the eastern United States will experience significant increases in the runoff from excessive flooding and associated deterioration in water quality, whereas the western and southwestern parts of the country will suffer from severe droughts. These not so encouraging scenarios prompt water managers to look into reclaiming wastewater and recycling nutrients and precious metals (Laki, Nedunuri, and Kandiah, 2019a). The paradigm shift, therefore, involves the transformation of these historic WWTPs from serving as only contamination removal facilities into resource recovery facilities (RRF). Although, technologies in RRF are either mature or in pilot stages, issues associated with quantities, distribution, and applications of the recovered resources still remain unresolved (van der Hoek et al., 2016). Further, there have been few or no developments in developing regulatory frameworks for the RRFs, the acceptability of recovered water and other resources by the public, and the competition of the recovered resources with those manufactured in the industry (Zorpas, 2016). Public acceptance of these technologies may grow over time as the discharge of pollutants into global waters is no longer sustainable under additional stressors imposed from climate change.

15.2 GLOBAL WATER QUALITY CHALLENGES FROM CONTAMINANTS RELEASED FROM WWTPS

The Gulf of Mexico has been experiencing hypoxia due to the accumulation of nutrients from the Mississippi watershed, the world's third largest watershed. A record 7000 square miles of hypoxia was observed in 2016 in the northern Gulf of Mexico (NOAA, 2010). With dissolved oxygen levels below 2 ppm, the entire area would

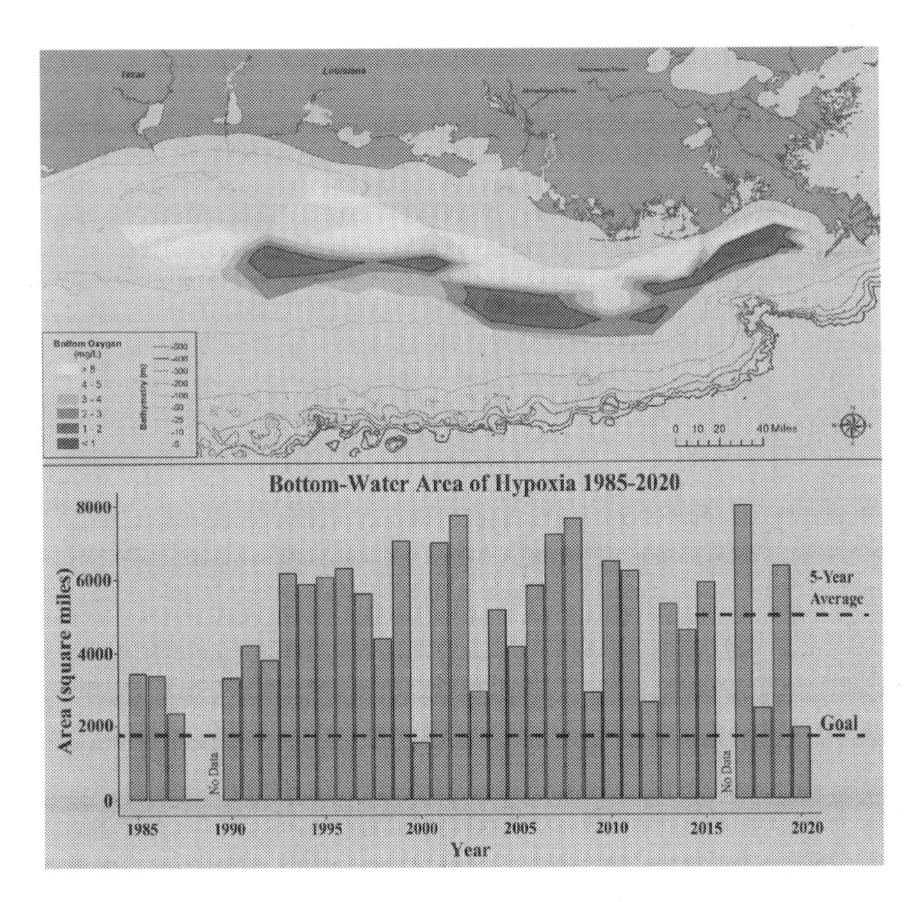

FIGURE 15.2 Historic trends of hypoxia in the Gulf of Mexico (NOAA, 2010).

be dead without marine life. It was a global disaster resulting from the discharge of excess nutrients into one of the major continental watersheds in the world. Though extensive conservation measures had been in place on these watersheds to curb the release of nutrients, the problem has not been solved as was evident from the high five-year moving average of about 5500 square miles in contrast to the target goal of reduction of the hypoxia to about 2000 square miles a shown in Fig. 15.2. The smallest value in 2020 is the result of mixing and dilution by waters dumped by Hurricane Hanna before the sampling period (U.S. EPA, 2020). Both nitrogen (nitrite/nitrate) and phosphorus (orthophosphate) are the major contributors to the nutrient loading into the Mississippi watershed. Although major sources are fertilizers and manure used on farms, about 7 percent of the nitrogen in the Mississippi watershed comes from wastewater treatment plants and 7 percent from the urban areas (Ritter and Chitikela, 2020). Recent technologies in phosphorus removal using a combination of biological and chemical processes have been able to significantly reduce phosphorus contribution from the wastewater treatment plants to watersheds. However, the cost of removal depends on the effluent requirements. When a river in an impaired watershed is required to lower its total phosphorus (TP) effluent from 2 mg/l to 0.5 mg/l,

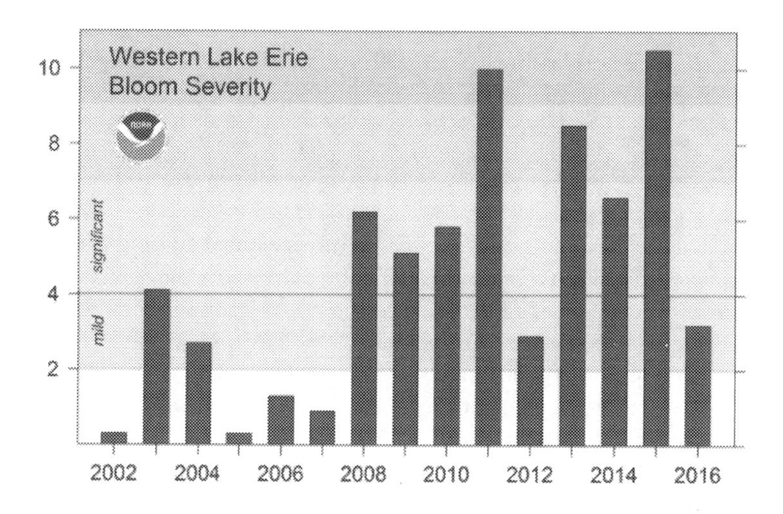

FIGURE 15.3 Historic trends of harmful algal blooms (NOAA, 2016).

the wastewater treatment plant using a combined biological and chemical treatment and discharging to the watershed would incur a three-fold increase in the capital cost of the equipment (Tetratech, 2013).

Harmful algal blooms (HABs), yet another environmental disaster, have been threatening the aquatic life and also the drinking water supply in the Great Lakes region, particularly in the Lake Erie basins; this also arises from the problem of excessive nutrient loading, in this case into freshwater lakes. The severity of algal blooms has also increased in the past decade, as shown in Fig. 15.3, although several mitigation measures, such as a 40 percent reduction in total and soluble reactive phosphorus in the western and central Lake Erie basins, were in place (U.S. EPA 2016).

While focusing on technologies in removing nitrogen and phosphorus, it is necessary to recognize that the ratio of N/P controls the growth of HAB addition in addition to individual N and P levels. It has been observed in most treatment plants that more focus is on TP removal than that on nitrogen (Qin et al., 2020). WWTPs should investigate biological nutrient removal containing anaerobic/anoxic and aerobic zones with primary and secondary clarifiers to facilitate the removal of nitrogen and phosphorus.

Long-term exposure to heavy metals in humans causes serious health concerns. Almost all of these metals are known to be either potential or known carcinogens. They are released into the water through industrial, agricultural, and technological applications. Millions of tube wells were dug beginning in the 1960s and 1970s in Bangladesh and West Bengal, India, for use in agriculture and human consumption (Mehovic and Blum, 2004). The wells became contaminated with arsenic, exposing about 75 million people and causing 200 000 to 270 000 deaths from cancer (Gilbert, 2004), considered to be one of the worst mass poisonings in history (Mehovic and Blum, 2004). The release of mercury from combustion processes in coal-based power plants and other industrial processes such as mining and smelting has been an ongoing environmental problem. Once released into the atmosphere in nano-size elemental and gaseous form (Hg0), it travels downwind and is deposited on the land as minerals in its mercuric (Hg^{2+}) form and transported by the runoff into rivers changing into organic (Hg-C) and more bioavailable forms (O'Connor et al., 2019). It is known for its biomagnification in fish. The World Health Organization lists mercury as one of the ten priority chemicals of concern due to its known effects such as Minamata disease (WHO, 2017). While U.S. EPA's Safe Drinking Water Act has a regulatory limit of 15 ppb, the lead levels in water tested in Flint, Michigan, homes reached hazardous waste levels of 5000 ppb. When water was supplied from a local river instead of Lake Huron, corrosion of old lead pipes had resulted in such elevated levels in the water. Elevated levels of lead in blood were found in more predominantly disadvantaged neighborhoods (Laki, Nedunuri, and Kandiah, 2019). Municipal wastewater treatment plants that also receive sewage from industries typically discharge higher levels of heavy metals into waterbodies. Azizullah et al. (2011) report that about 2 million tons per day of sewage sludge and effluents are being released into natural rivers and lakes globally. According to Ullah et al. (2020), about 24 percent of wastewater is treated in India and only 2 percent in Pakistan. Industrial effluents contain a variety of toxic metals such as Pb, Cd, Cr, and As (Afzal et al. 2019; Hussain et al. 2019; Rehman et al., 2018, 2019).

Per- and polyfluoroalkyl substances (PFASs) are synthetic chemicals that include compounds such as PFOA, PFOS, GenX, and many other chemicals that are highly soluble in water. Their water solubility, estimated at 680 mg/l, and negligible vapor pressure make them persistent when released into freshwater. PFASs are used in a variety of manufacturing and consumer products (food packaging, commercial household products, and workplace construction material) and are very persistent in the environment. Long-term exposure to PFASs can lead to adverse human health effects such as cancer and thyroid hormone disruption. Most nations with minimal or no treatment of sewage would experience the release of PFASs into rivers and aquifers. PFASs are being detected in drinking water (CDC, 2017; U.S. EPA, 2020).

15.3 INNOVATIONS IN RESOURCE RECOVERY TECHNOLOGIES

Several technologies were developed in the past 25 years in water reclamation, energy recovery, nutrient recovery, and recovery of other chemicals. However, most of these have been confined to either laboratory-scale experiments or pilot-scale plants well known in only scientific communities. The transfer of research-based technologies into actual implementation within WWTPs is still in its infancy. There are a few that have shown the potential to migrate into actual plants. We will discuss these technologies in four areas: water reclamation, energy recovery, nutrient recovery, and recovery of other chemicals.

15.3.1 Water Reclamation Technology

Reclaiming water from WWTPs may significantly reduce the demand for water supply in water-stressed regions of the world. A few examples where there have been long-term droughts in recent history are California in the U.S., entire sub-Saharan

Africa, and certain parts of southern India that fall under rain shadow. Californians transports water using tunnels from the Sacramento River to the south underneath the San Joaquin delta. Farmers in the south demand this water for irrigation. Residents living in the northern delta of the valley argue that it would dry up their important natural resource, leading to saltwater intrusion from the San Francisco Bay area to ruin the delta region. Monterey city in the United States supplied about 80,000 m^3 per day of nutrient-rich reclaimed municipal wastewater to irrigate crops (McCarthy et al., 2011). About 25 percent of the drinking water supply in Namibia was abstracted from wastewater (Verstraete and Vlaeminck, 2011). The city of Chennai in south India could meet 15 percent of its water demand from reusing 40 percent of its wastewater (IWA, 2018).

Water reclamation technologies use filtration and advanced oxidation processes to convert wastewater into potable water (Bruggen, 2013). Significant technologies currently available under filtration for wastewater reclamation are based on the use of cellulose-based semi-permeable membranes and activated carbon. Membrane processes can effectively remove pathogenic bacteria and even viruses without generating toxic byproducts. They are also efficient in removing toxic trace metals and certain organics to produce potable water. By consuming less space compared to other processes, such as sand filtration followed by chlorination, membranes offer the advantages of minimum land use and lower environmental footprints. However, problems associated with frequent fouling due to the formation of precipitates and high energy costs offset the benefits and often warrant the use of technology for large-scale applications involving high-volume throughputs. Ultrafiltration membranes (UF) remove colloids, proteins, polysaccharides, most bacteria, and even some viruses and produce high-quality treated effluents (Rao, 2013). Both nanofiltration (NF) and reverse osmosis (RO) are useful to separate trace metals and other ions from water containing excessive salts (Wintgens et al., 2005). NF/RO membrane technology was used to reclaim municipal wastewater as non-potable water. A fraction of reclaimed water was used in recharging local aquifers to meet Singapore's growing demand for water (Lee and Tan, 2016). An operational full-scale wastewater reuse plant and a pilot plant showed higher than 75 percent removal rates and over 99 percent rejection efficiencies for most organic compounds in the full-scale or pilot plants, including ibuprofen and nonyl phenol (Garcia et al., 2013).

Granular activated carbon (GAC) filtration may be employed in reclaiming wastewater to produce potable water when used as an advanced treatment step after it goes through the first four steps in WWTPs shown in Fig. 15.1. Locally available materials such as coal, peat, petroleum coke, and nutshells may be used to prepare GAC, which are further activated by high temperatures and chemicals to produce active adsorbents for the removal of COD, total organic carbon (TOC), chlorine, and a wide range of organic pollutants (Stephankis, 2016). GAC may be directly added to wastewater as a powder (PAC), to the activated sludge, or as granular activated carbon (GAC) in a separate pressurized filtration unit. While PAC requires disposal with the sludge, GAC can be regenerated cost-effectively on site (Trussel, 2012). Various studies have shown the effectiveness of combining GAC filtration with other advanced treatment steps for the removal of water pollutants. Biological AC in combination with ozone was investigated in simultaneous biodegradation and oxidation of certain pollutants and disinfection. BAC was found to be superior to membrane filtration using RO in a pilot study conducted at a water reclamation facility (Gerrity et al., 2011). When PAC was integrated with biological processes, it was found to be effective in the full-scale treatment of combined municipal and industrial wastewaters. The economics of GAC combined with biodegradation in wastewater treatment plants would depend on the extent of industrial components present in the raw wastewater (Oller et al., 2011).

15.3.2 Biofuel Recovery from Wastewater

Wastewater contains BOD typically in the range of 110–400 mg/l, organic carbon 80–290 mg/l, and COD 250–1000 mg/l, which suggests a possibility to recover a lot of organics, most importantly biofuels. Microorganisms play an important role in chemical transformations that convert wastewater sludge into useful biochemicals. Aerobic, anaerobic, and anoxic conditions typically maintained in the WWTPs would yield distinct products. A well-known process is the production of methane under anaerobic conditions during the digestion of wastewater sludge. Aerobic processes will lead to the production of ammonia and further conversion to nitrite and nitrate. Recovery of methane (Li et al., 2019) and nutrients (Bartkowska et al., 2020) from WWTP sludge has been well studied. We will review new technologies to recover other biofuels such as ethanol, syngas, hydrogen, and nitrous oxides.

Conversion to ethanol from food crops such as corn has increased recently as a cleaner alternative to energy consumption to reduce greenhouse gases. With increasing global demand for food, alternatives to ethanol production are being sought; among these, wastewater sludge shows promise due to its distinct advantage of the lower raw material cost compared to other sources. However, the energy involved in sludge concentration for optimal yields and the removal of impurities poses economic challenges (Chang et al., 2010; Goto et al., 1999; Manara and Zabaniotou, 2012). Syngas, hydrogen, biodiesel, and nitrous oxides may be recovered from wastewater.

Syngas, a mixture of carbon monoxide, carbon dioxide, and hydrogen is produced by catalytic conversion of hydrocarbons at high temperatures. The feedstock is coal, petroleum-based naphtha, natural gas, or biomass from agricultural and yard wastes. Steam reforming, auto-thermal forming, or partial oxidation using oxygen are typically used to produce syngas. The process, however, is energy-intensive due to the requirement of high pressure steam. The heat released from the co-generation engines that use digester gases may be used to produce steam. Energy is also released during the conversion of CO and H_2O to CO_2 and H_2 in high-temperature and low-temperature shift reactors. One of the disadvantages of using the above processes is the requirement of dewatering of the sludge. Partial oxidation can be used to convert the carbonaceous material in wastewater sludge into CO and H_2 at 15–150 bar (Sohi et al., 2009). Supercritical water gasification processes convert wastewater sludge into syngas yielding clean water, metal oxides, ad salts as byproducts (Goto et al., 2009). The residence time in the supercritical reactors where sludge is converted into energy is only a few minutes, and as a consequence these reactors can handle high throughputs of feedstock and yield high operational efficiencies. The disadvantages are corrosion of the reactors and clogging from salt precipitation under supercritical conditions (Yakaboylu et al., 2015).

Hydrogen is recovered from wastewater by controlling the anaerobic digestion of its sludge to enhance conditions favorable for hydrolysis to form sugars and fatty acids followed by *acidogenesis* fermentation to produce H_2, CO_2, formate, methanol, acetate, and methylamines. Photofermentation favors the formation of hydrogen, whereas dark fermentation favors the production of volatile fatty acids (VFA). Photo fermentation is further used to convert VFAs into H_2 (Akhlaghi and Najafpour-Darzi, 2020, Lee et al., 2014). However, the yield of H_2 in fermentative production is low (Lee at al., 2010).

Biodiesel may be produced from wastewater by harvesting lipids using heterotrophic bacteria anaerobically and by skimming the oil from the surface of primary and secondary clarifiers in WWTPs (Mueller et al., 2014). The resulting long-chain fatty acid hydrocarbons could serve as raw materials in the production of biodiesel. Biodiesel can also be produced from wastewater sludge by cultivating micro algal biomass using the nutrients from wastewater. Lipids produced from algal biomass and those present in wastewater can be used in the production of biodiesel. Ammonia oxidizing heterotrophic bacteria in wastewater enhanced the growth of microalgae and improved the production of lipid production (Zhou et al., 2020). High-rate algal ponds may be used to produce lipids; however their performance is limited by the seasonal dependence of the growth of algae. The size of the pond in biodiesel production changes the biomass production and nutrient removal (Park et al., 2011; Sutherland et al., 2020) and is often a determinant in the economic viability of these systems (Gao et al., 2014).

Nitrous oxides can be produced from wastewater by nitrification of ammonium to nitrite followed by anoxic reduction to nitrous oxide fuel (Zhang, Gu, and Liu, 2019). However, the process is energy-intensive and is seldom used as a source of energy.

15.3.3 Recovery of Nutrients from Wastewater

Nitrogen and phosphorus are the major nutrients present in wastewater. Total nitrogen in untreated domestic wastewater is typically in the range of 20–85 mg/l, existing in the forms of organic or free ammonia. Total phosphorus is in the range of 4–15 mg/l and may be found primarily as organic or inorganic forms. The sources of N and P in wastewater are sanitary sewers, storm sewers, and agricultural runoff, among which those released from the agricultural runoff could be significant in those parts of the world where conservation practices are not fully developed.

Most of the nitrogen released from farms still originates from the use of commercial fertilizer, which accounts for about 1 percent of the world's greenhouse gas emissions and 1.4 percent of energy consumption globally (Capdevila-Cortada, 2019). About 2.867 tons of CO_2 are released for every ton of ammonia produced, which accounts for about 90 percent of emissions globally. Production of ammonia is energy intensive, requiring about KWH of ammonia production (U.S. DOE 2000). Ammonia fertilizer used in agriculture and in residential lawns is removed in WWTPs as nitrogen gas by biological nitrification and denitrification, which also consumes significant energy. The recovery of nitrogen from WWTPs offers potential fertilizer and energy savings (Sheik et al., 2014). Recovery of P is also important P due to its use in manufacturing household detergents and industries. Most of the untreated P can impair water quality in streams and freshwater lakes by eutrophication, sometimes producing toxic algal blooms and hypoxia.

15.3.3.1 Biosolids

Biosolids are organic matter recovered from sewage sludge of WWTPs; they are rich in nutrients as well as some essential micronutrients such as zinc, copper, and nickel and are used as natural fertilizers. These solids also improve soil texture and water-holding capacity and are applied directly onto agricultural lands. About 51 percent of the biosolids generated in the United States are used for land application on agricultural fields (U.S. EPA, 2019). High contaminant loads have been found in bacterial biomass, leaving WWTPs as secondary sludge (Sheik et al., 2014). Hence the sewage sludge is dewatered and stabilized by adjusting the pH adding lime, through anaerobic digestion, or by composting or heat drying. Biosolids are classified as Class A or Class B depending on the degree of stabilization received. Class A biosolids have been treated to an exceptional quality and do not contain detectable pathogens. These solids also have the lowest levels of heavy metals. Class B biosolids do contain pathogens and heavy metals; however, their application would require the adoption of technologies to significantly restrict pathogens and also restrict public access to the application site, limit livestock grazing, and control crop-harvesting schedules. The land application of Class B biosolids is correlated to human health symptoms, acute and chronic diseases among residents who lived within one to five miles of the permitted fields (Czajkowski et al., 2010; Khuder et al., 2007). Also, the transport of the sludge (Kirchmann et al., 2017) and its processing requires extensive dewatering from 3 to 6 percent solids to about 30 percent water followed by pretreatment (U.S. EPA, 2010).

15.3.3.2 Struvite

A mineral that has been known to cause serious operational problems, such as fouling of effluent pipes from anaerobic digesters in WWTPs, could prove to be an asset if recovered as a resource for both nitrogen and phosphorus (le Corre et al., 2009). Struvite is magnesium ammonium phosphate (MgNH4PO4·6H2O), a mineral formed when the Mg concentrations are high enough, although this is often not the case. The formation and growth of struvite crystals in WWTPs is affected by various parameters, such as pH, temperature, mixing energy and turbulences, and the presence of other ions, most importantly calcium or carbonates (Jaffer et al., 2002). The presence of calcium ions will reduce the formation of struvite nucleation as it forms a more stable apatite by combining with PO4. The log stability constant for struvite varies from –13 to –15 (Aage et al., 1997; Babi´c-Ivanˇci´c et al., 2002; Loewenthal et al., 1994). The thermodynamic stability of this mineral allows its constituent ions N and P to dissolve gradually and serve as slow-release nutrients in soils (Xie et al., 2016).

Struvite precipitation is thermodynamically favorable when P is about 0.07 M, ammonium about 0.1 M as total N, and magnesium at higher levels of 5 mM (Snoeyink and Jenkins, 1980). Lower P concentrations reduce recovery rates, increase reaction times, and require higher pH values. Wastewater effluent low in

P may not precipitate sufficient quantities of struvite (Xie et al., 2016; Zhang et al., 2013). Nutrient enrichment is usually required prior to struvite precipitation. Sludge collected from primary and secondary stages of wastewater treatment is sent to anaerobic digesters (AD) where the digestion breaks down organic matter, producing methane while concentrating the nutrients in the solids and supernatant liquid. When the pressure in the pipes carrying the supernatant liquid from AD is reduced, struvite precipitates on the joints of the pipes on the suction side of the pumps due to a sudden increase in pH from the release of CO_2 from the solution. Precipitation of struvite will severely clog the pipes, resulting in fouling and breakdown of the pumps (Zhang et al., 2013). Instead, the supernatant can be sent to a crystallizer, where additional Mg salt is typically added to remove soluble P and N as struvite crystals (Münch and Barr, 2001; Uysal et al., 2010). Mg is added as $MgSO_4$, $MgCl_2$, $Mg(OH)_2$; however, these chemicals make the recovery expensive and also add additional impurities. Certain low-cost resources of Mg, such as seawater (Rubio-Rincón et al., 2014), sea water bittern (Siciliano and De Rosa, 2014), and combustion ash (Sakthivel et al., 2012) have been tried with limited success. The combustion ash contains a lot of heavy metals and does not comply with disposal regulations.

The overall cost of recovering struvite from AD could be economically viable (Santos Sánchez, 2020). The continuous recovery of struvite from the supernatant in AD reduces the sludge volume and decreases sludge dewatering and associated disposal costs. The savings in reduced operational costs may outweigh the costs of recovery. The market value of struvite as a relatively new fertilizer depends on the production rates, its demand regionally, and its market value compared to other sources (le Corre et al., 2009). Trace metals and organic contaminants from wastewater may be concentrated in struvite, making it unsafe for use as a fertilizer. The crystals can contain arsenic concentrations of up to 570 mg/kg (Lin et al., 2013). Thus struvite recovery without purification may be restricted by federal regulations. The legal process in allowing struvite in agriculture took considerable time in the Netherlands (van der Hoek, de Fooij and Struker, 2016). The readers may further benefit from Siciliano et al. (2020), who discussed the potential for recovering struvite from WWTP sludge under various operational conditions within the stirred tank and fluidized bed reactors. These authors reviewed the economics of struvite recovery through the use of low-cost reagents.

15.3.3.3 Humic Acids

Humus is an organic substance formed by the decomposition of dead plant and animal matter beneath the Earth's surface. It is comprised of complex aliphatic-aromatic assemblages of natural biopolymers synthesized by soil microorganisms, primarily bacteria and fungi from organic residues such as peptides, lipids, proteins, and polysaccharides. Once formed, these compounds remain in the soil for thousands of years and are highly resistant to further degradation. Humic substance is the fraction of humus available as dark black to brown colored material known as humic acid (HA) and fulvic acids (FA), which are complex groups of high-molecular-weight organic acids. HAs are large molecules that tend to remain in the solid phase of the soil, while FAs are much smaller molecules that tend to remain in the soil solution. Though HAs do not add nutrients to the soil, they serve as long-term storage reservoirs or chelates of micronutrients such as Fe, Zn, etc., due to the presence of large negatively charged surfaces (oxygen-containing carboxylate, phenolic, hydroxyl, carbonyl ligands) that can bind these cations. Due to their high cation exchange capacities and pH buffering capacities, HAs release these micronutrients slowly to the plant roots without leaching them into soils or water causing nutrient enrichment and eutrophication while retaining heavy metals and organic contaminants in soils (Réveillé et al., 2003; Sposito, 2016). HAs improve the water-holding capacity and aeration in soils through the enhancement in soil structure. The enzymes present in HAs enhance the microbial mineralization, further causing the availability of N, P, and K to plants and enhancement in plant metabolic rates.

HAs are present in sewage sludge in the range of 7.7–28 percent of volatile solids (Gonzalez et al., 2018). Further concentration of HAs occurs in AD, increasing the functional groups responsible for binding nutrients and their solubility (Liu et al., 2019). HA and FA in sewage sludge varies between 24 and 76 percent, depending on the wastewater and operational conditions of the treatment plant (Zhang et al., 2019). HAs reduced the volume of sludge in the AD and during the dewatering phase; when treated with NaOH and concentrated using membrane filtration, the supernatant yielded large amounts of dissolved organics that may be used as humic acid fertilizer (Li et al, 2009). HAs were extracted from sewage sludge using KOH and K4P2O7, analyzed, and encapsulated into alginate beads to test their potential as a soil conditioner to grow lettuce in the greenhouse. About 28 percent HAs were extracted and concentrated from the original sludge, which contained only 12.5 percent. Lettuce grown on HAs produced more biomass than the control plants and did not contain any heavy metals toxic to the plants (Cristina et al., 2020). Used alongside mechanical and thermal methods, alkaline treatment is a simple and highly efficient chemical means of disintegrating sludge. The techno-economics and the environmental impacts of HA recovery from WWTPs remain to be analyzed.

15.3.4 Recovery of Metals from Wastewater

Heavy metals such as As, Pb, Cd, Cr, Hg (specific gravities of 3.7–7.5 or above) are extremely toxic even at low concentrations. These metals are persistent and bio-accumulate in the environment, concentrating along the food chain. Arsenic found in natural waters is known to produce cancer of the bladder, kidneys, and lungs upon long-term exposure. It was estimated that between 35 and 77 million residents of Bangladesh consumed water from tube wells that tapped into groundwater containing naturally occurring inorganic arsenic (Smith et al., 2000). Cadmium is released into natural waters from smelter wastes, food wastes, air pollution, and cigarette smoke. Long-term exposure to cadmium may result in renal dysfunction. Chromium exists in trivalent and hexavalent forms in natural waters. The hexavalent form is the most toxic, and its long-term exposure has been known to produce intestinal, gastric, and liver diseases. Mercury is mostly associated with coal. It was used in the past as a fungicide. It is still released worldwide from smelters, boilers, and municipal and industrial waste incinerators. The inorganic form will be converted by bacteria in the sediment into toxic methyl mercury, which accumulates in fish and is passed along the food chain to humans (Laki, Nedunuri, and Kandiah, 2019b). The bioaccumulation in fish and human beings is found an order of magnitude greater than inorganic mercury in natural water, which endangers human health with effects such as mental disturbance, ataxia, hearing speech, and visual impairment. Lead exposure occurs through the ingestion of soil, dust, paint, and drinking water. It affects the nervous system and brain function in young children.

The sources for these metals are untreated or inadequately treated industrial effluents, urban runoff, and agricultural discharges that often mix with domestic sewage entering the municipal wastewater treatment plants. Treated wastewater will not contain high amounts because it is required to meet water quality standards before being discharged into natural water bodies. For example, the Environmental Protection Agency in the U.S. restricts their releases through the National Pollutant Discharge Elimination Standards. However, the sludge collected from the primary, secondary, and tertiary stages of WWTPs accumulate significant metals and other xenobiotics, restricting the land application of biosolids processed from the sludge. Hence the recovery of metals from wastewater sludge is not only profitable but is also desirable for the protection of environment and human health. Metal recovery from wastewater is accomplished by several processes, among which electrokinetic dialysis, supercritical fluid extraction, chemical precipitation, and bioleaching have received attention recently.

15.3.4.1 Electrokinetic Process

The electrokinetic technique is performed by inserting electrode arrays into activated sludge solution and passing a mild electric current. Metals move toward the cathode and are reduced, whereas they are converted into metal oxides at the anode. The high pH in digested sludge will inhibit the reduction process because the hydroxides reach with metals to form oxides, which are less mobile. Several modifications were done to the EK process through augmentation of the digested sludge with nitric acid, chelating agents, surfactants, and acid solution to maintain low pH conditions and increase ion mobility. EK pre-acidification using nitric acid followed by pH adjustment in the cathode chamber of anaerobically digested and dewatered sludge yielded the maximum recovery of Zn, Ni, Cr, and Cu. A detailed discussion and review of metal recovery using EK process was discussed elsewhere (Geng et al., 2020).

15.3.4.2 Supercritical Fluid Extraction

The supercritical fluid extraction (SCF) technique uses supercritical CO_2 at about 200 atm and 600C to extract metals from the sewage sludge. While SCF CO_2 molecules are neutral and nonpolar, heavy metals are positively charged and polar in nature. SCF CO_2 will not dissolve metals unless functional groups such as methanol, ethanol, or octanol are attached to create a polar CO_2 that can complex with heavy metal and remove it from the sludge. Sulfur-containing organophosphinic functional groups attached to branched hydrocarbons serve as ligands to SCF CO_2 and have been proven to be effective in the extraction of a range of metals. Chao et al. (2010) used SCF-CO_2 with sodium diethyldithiocarbamate, acetylacetone, tributyl phosphate-nitric acid mixture, 8-hydroxyquinoline, and tetrabutylammonium and modifiers methanol, ethanol, water, and acetone. These authors reported removals of 78.3 percent Cu, 73.2 percent Pb and 69.5 percent Cd. SCF-CO_2 when used with Cyanex 301 (bis(2,4-4-trimethylpentyl)-dithiophosphinic acid) extracted 82 percent Cu, 100 percent Pb, 100 percent Zn and

95 percent Cu from a cellulose matrix (Smart et al., 1997). More detailed discussions on different ligands and their ability to extract metals with SCF-CO_2 can be found in Geng et al. (2020) and Babel and de Cera (2006). SCF-CO_2 extraction has a distinct advantage over chemical methods since it does not use chemical solvents or generate byproducts that add to the environmental footprint. However, the use of chelating agents is expensive that makes the SCF-CO_2 technique economically infeasible.

15.3.4.3 Chemical Precipitation

This technique first dislodges metals from the sludge matrix using an acid or a chelate. The metals released into solution as ions or metal-ligand complexes will be separated from the sludge through precipitation using bases such as CaO, NaOH, NaHCO3 (Brooks, 1991) or sulfides such as NaS, H2S, or FeS (Veeken and Hamelers, 1999). Marchioretto et al. (2002) used nitric, hydrochloric, oxalic, and citric acids to remove Cd, Cr and Zn, Cd, and Pb from anaerobically digested sludge from a wastewater treatment plant in the Netherlands. Citric acid at pH 3 and 4 extracted 85 percent Cd, Cr and Zn; however, the extraction is more due to its chelation with metals. At pH lower than 3, HNO_3 or HCl removed 100 percent Cd and Pb. Lo and Chen (1990) used chelates EDTA and NTA to solubilize heavy metals from municipal sludge in Taiwan using 2 mol of EDTA/NTA per mole of metal using a contact time of four hours. The order of removal was Pb > Cd > Zn > Ni > Cu > Cr.

15.3.4.4 Bioleaching

Bioleaching releases metals bound to the solid organic matter in the sludge into solution. The bacteria naturally present in the sludge accelerate the solubilization of metals through redox transformations. The energy released from the transformations is used by the bacteria for growth and maintenance. The organisms also oxidize the sludge, creating acidic conditions that facilitate metal solubilization. Reduced forms of Fe and S in the sludge act as electronic acceptors that aid in the microbial synthesis of these indigenous bacteria in the sludge and also serve as the sources of energy for the bacteria. Specific bacteria known for bioleaching of metals are classified as mesophiles or thermophiles, depending on optimal temperatures selected for their growth. Mesophiles are heterotrophic sulfur-oxidizing *At. Thiooxidans* and iron-oxidizing *At. Ferrooxidans* bacteria that grow in the temperature range 20–40°C and pH between 1 and 4.5. Moderate thermophiles employed in bioleaching are *Sulfobacillus thermosulfidoxidans* and the extreme thermophiles grow at 70°C and use sulfur or thiosulfate. The species involved in metal solubilization are *Sulfolobous* viz. S. ambivalens (Kletzin, 2006), *S. brierleyi* (Konishi et al., 1998) and *Thiobacter subterraneus* (Hirayama et al., 2005).

The relevant process equations are adapted from Pathak et al; 2009 and are shown below.

- **Biosynthesis of *At. Thioxidans*:** Elemental sulfur or other reduced forms such as sulfides are oxidized to sulfuric acid by *At. Thioxidans.*

$$S^0 + H2O + 1.5\ O2 - \rightarrow H_2SO_4$$

The sulfuric acid will solubilize the metal from the organic sludge matrix.

$$H_2SO_4 + Me - Sludge- \rightarrow MeSO_4 + Sludge - 2H$$

- **Biosynthesis of *At, Ferroxidans*:** Iron (Fe^{2+}) is oxidized to Iron (Fe^{3+}) by *At. Ferroxidans*

$$2FeSO_4 + 0.5\ O_2 + H_2SO_4 - \rightarrow Fe_2(SO_4)_3 + H_2O$$

Ferric sulfate will then solubilize metal sulfide present in the sludge in the presence of oxygen.

$$Fe_2(SO_4)_3 + 2MeS + 2O_2 + 4H_2O\ - \rightarrow 2Me^{2+} + 2SO_4^{2} + 8\ FeSO_4 + 4H_2SO_4$$

Bioleaching is carried out either in batch mode or in continuous mode in laboratory conditions. The presence of organic acids is shown to inhibit metal solubilization. Growth of *At. ferrooxidans* was strongly inhibited by the organic acids present in anaerobically digested sewage sludge. Yeasts were found to utilize the organic acids as their sole carbon source and create a favorable condition for the growth of iron-oxidizing *At. Ferrooxidans* (Gu and Wong, 2007). Bioleaching dissolved 72 percent Cd, 92.5 percent Mn, and 89 percent Zn, whereas chemical leaching dissolved only 22 percent Cd, 25 percent Mn, and 14 percent Zn from an anaerobically digested municipal sewage sludge. Metal dissolution during bioleaching treatment is more efficient when mediated by microorganisms through redox transformations (Ghavidel et al., 2018). Since most of the studies on bioleaching have been successful in laboratory settings, further study on adapting the technology under field conditions and associated economics using locally available cheaper and recoverable forms of sulfur as energy sources and by activating indigenous microorganisms to the sludge and through integration of recovery of metals is required (Pathak et al., 2009).

15.3.4.5 Economics of Metal Recovery

Westerhoff et al. (2015) analyzed U.S. sewage sludge for 58 regulated and non-regulated elements for removal and recovery. The metals are significantly attached to biosolids in the sludge. The sludge contained rare-earth elements minor metals such as Y, La, Ce, Pr, Nd, Sm, Eu, Gd, Tb, Dy, Ho, Er, Tm, Yb, and Lu. However, these are derived from natural environments such as soil or dust. Most platinum group elements Ru, Rh, Pd, Pt were derived from anthropogenic sources. For a community of 1 million people, metals in biosolids were valued at up to the U.S. $13 million annually. The profitability of recovered metals depends on their adsorption to the sludge, their enrichment in industrial applications, and their marketability. The analysis revealed 14 elements (Ag, Cu, Au, P, Fe, Pd, Mn, Zn, Ir, Al, Cd, Ti, Ga, and Cr) as most valuable, with a combined value of US $280/ton of sludge (Westerhoff et al., 2015).

15.4 FUTURE PROSPECTS AND CONCLUDING REMARKS

Domestic wastewater can serve as a potential resource of reclaimed water, nutrients, biofuels, and metals. The abundance of these resources depends on the extent to which the municipal sewage sludge combines with industrial effluents and agricultural runoff. The market potentials of these products and that of energy would depend on their demand, alternative sources, and their prices, availability in wastewater, and the recovery yields. The recovery of specific products from a wastewater treatment plant must be done strategically to target those with finite supply and that are being depleted from their parent natural resources.

The innovations in the recovery of water, energy, fertilizer, and metals from wastewater have been mostly limited to laboratory scale or pilot plant studies. Kehrein et al (2020) discuss the bottlenecks for the transfer of technology as primarily economic, environmental, and social in nature. Economics is related to process costs, resource quantities and quality, market value, application, and distribution. The environmental aspects are associated with emissions, contamination, and health risks. The social issues are acceptance of recovery products by society, policy, and regulations. The perception of management should slowly transition from considering these facilities as treatment plants to resource recovery industries. Though the goal is still to meet or exceed the water quality standards, there should be a paradigm shift in modifying the traditional processes to explore the recovery of products that have a market value.

Any modifications to conventional wastewater processes with a goal of recovering resources should integrate the elements of financial feasibility into traditional economic viability assessments. The return of investment (ROI) should be an added goal in addition to the existing goal of wastewater treatment as a service to the communities and the protection of the environment. The market share for the resource, customer acceptance of a recycled resource, economies of scale, government regulations and subsidies are among several financial variables that resource recovery wastewater plants should consider. A gradual transition to assessing the performance using financial indicators would be necessary to compete with companies that produce the same product. When multiple resource recovery wastewater plants in a region plan to recover the same resource, it can increase the collective market power where economies of scale for individual companies is a challenge.

High future global temperatures will increase evaporation, snowmelt, and loss of available water for human consumption. The high temperatures in mid-continental regions such as sub-Saharan Africa will expand the deserts. The reduction in mean annual runoff will result in the intensification of droughts in Europe and South Africa. The increase in high-intensity storms will increase precipitation, the variability of the water cycle, and the areas of tropical disturbances. This will cause flooding in low-lying areas of the world. Regions closer to coastal plains that accommodate about 30 percent of the world's population will be susceptible to severe salinity and seawater intrusion. Whether it is water stress in the mid-continental regions of the world or water quality problems along the coastline, there will be a future scarcity in clean water. Therefore, water is the most precious resource in municipal wastewater. The reclaimed wastewater will be an important alternative source of freshwater in various parts of the world. As nations experience deterioration of water quality, the wastewater treatment plants would be under pressure to update their technologies to meet stricter discharge standards. As a consequence, these plants would need more financial resources and recovery of resources would be a welcome opportunity for these plants to meet their costs of treatment, and also avoid the transfer of costs to consumers.

Biofuels from municipal wastewater treatment plants will become economical only if governments can provide subsidies to compete with commercial supplies and those that produce and distribute traditional coal, natural gas, and petroleum. The effluent from the wastewater treatment has a high heat content and is currently being dissipated into the environment. Wastewater treatment plants should focus on the recovery of this waste heat in addition to the traditional capture of the heat from the co-generation using anaerobic sludge digestion. As nations plan on seriously cutting greenhouse gas emissions and limiting global temperature rise this century to well below 2°C above pre-industrial levels, a serious commitment to the reduction in the fresh consumption of fossil fuels is being sought. The recovery of energy from wastewater is more important now than ever before. Energy companies can work with wastewater treatment plants to achieve carbon neutrality as part of the economic and market-based approach of carbon trading. The same philosophy may be applied to fertilizer production facilities to consider wastewater resource recovery units as their partners in the conservation of nutrients and thus lower some of their own environmental footprints. Phosphorus removal technologies have been very expensive, and hence future efforts should focus more on technologies that recover phosphorus than its removal from wastewaters. Technologies must be further improved with a goal to reduce energy consumption in nitrogen recovery.

Future environmental regulations, climate change policies, and depletion of resources necessitate reuse and recycling of global resources and energy. The transformation of wastewater treatment plants as resource recovery facilities may contribute to this global call for conservation if we address the limitations associated with field-scale implementation of innovative technologies discussed in this chapter. Moreover, it is time to make a paradigm shift in the way we manage these treatment facilities by incorporating the marketability and sustainability of recovered resources in the design, planning, and implementation stages of these facilities.

REFERENCES

Aage, H.K., B.L. Andersen, A. Blom, I. Jensen (1997). The solubility of struvite. J. Radio Anal. Nucl. Chem. 223, 213–215.

Afzal, M., K. Rehman, G. Shabir, R. Tahseen, A. Ijaz, A.J. Hashmat, H. Brix (2019). Large scale remediation of oil-contaminated water using floating treatment wetlands. NPJ Clean Water. DOI: 10.1038/s41545-018-0025-7.

Akhlaghi, N., G. Najafpour-Darzi (2020) A comprehensive review on biological hydrogen production. Int. J. Hydrogen Energy, 45 (43), 2020, 22492–22512.

Azizullah, A., M.N.K. Khattak, P. Richter, D.P. Hader (2011). Water pollution in Pakistan and its impact on public health: A review. Environ. Int. 37, 479–497.

Babel, S., D.M. Dacera (2006). Heavy metal removal from contaminated sludge for land application: A review. Waste Manage. 26, 988–1004.

Babi´c-Ivan˘ci´c, V.J. Kontrec, D. Kralj, L. Bre˘cevi´c (2002). Precipitation diagrams of struvite and dissolution kinetics of different struvite morphologies. Croatica Chemica Acta, 75, 89–106.
Bartkowska, I., P. Biedka, I.A. Talalaj (2020). Autothermal thermophilic aerobic digestion of municipal sewage sludge in Poland. Rev. Proc. 51, 12.
Brooks, C.S. (1991). Metal recovery from industrial waste. Lews Publishers, Chelsea, MI.
Bruggen, B.V. (2013). Integrated Membrane separation processes for recycling of valuable wastewater streams: Nanofiltration, membrane distillation, and membrane crystallizers revisited. Ind. Eng. Chem. Res., 52(31), 10335–10341.
Capdevila-Cortada, M. (2019). Electrifying the Haber–Bosch. Nat Catal 2, 1055.
CDC Center for Drug Control and Prevention. (2017). Per- and Polyfluorinated Substances (PFAS) Factsheet, Chemical Factsheets, Resources, National Biomonitoring Program.
Çeçen, F., Ö Aktas (2011). Activated carbon for water and wastewater treatment: Integration of adsorption and biological treatment. John Wiley and Sons: Weinheim, Germany.
Chao, Y., W. Chan, C. Changshui (2010). Extraction of copper, lead and cadmium in green tea by supercritical carbon dioxide chelating extraction. J. Anal. Sci. 26, 211e214, 02.
Chang, H.N., N.-J. Kim, J. Kang, C. M. Jeong (2010). Biomassderived volatile fatty acid platform for fuels and chemicals. Biotechnol. Bioprocess Eng., 15(1), 1–10.
Cristina, G., Enrico Camelin, Carminna Ottone, Silvia Fraterrigo Garofalo, Lorena Jorquera, Mónica Castro, Debora Fino, María Cristina Schiappacasse, Tonia Tommasi (2020). Recovery of humic acids from anaerobic sewage sludge: Extraction, characterization and encapsulation in alginate beads. Int. J. Biol. Macromolecules, 164, 277–285.
Crutchik, D., J.M. Garrido (2016). Kinetics of the reversible reaction of struvite crystallization. Chemosphere, 154, 567–572.
Czajkowski, K.P, A. Ames, B. Alam, S. Milz, R. Vincent, W. McNulty, T. W. Ault, M. Bisesi, B. Fink, S. Khuder, T. Benko, J. Coss, D. Czajkowski, S. Sritharan, K. Nedunuri, S. Nikolov, J. Witter, A. Spongberg (2010). Application of GIS in evaluating the potential impacts of land application of biosolids on human health. Geospatial technologies and environment: Applications, policies and management, Ed. Nancy Pullen and Mark Patterson, Springer New York City, New York., Vol 3, p. 200.
Esseili, M.A., I.I. Kassem, V. Sigler, K. Czajkowski, A. Ames (2012). Genertic evidence for the offsite transport of *E. coli* associated with land application of Class B biosolids on agricultural fields. Sci. Total Envir. 433, 273–280.
Gao, H., Y. D. Scherson, G. F. Wells (2014). Towards energy neutral wastewater treatment: methodology and state of the art. Environ. Sci. 16(6), 1223–1246.
Garcia, N., J. Moreno, E. Cartmell, I. Rodriguez-Roda, S. Judd (2013). The application of microfiltration-reverse osmosis/nanofiltration to trace organics removal for municipal wastewater reuse. Environ. Technol. 34(24), 3183–3189. DOI: 10.1080/09593330.2013.808244.
Geng, H., Ying Xu, Linke Zheng, Hui Gong, Lingling Dai, Xiaohu Dai (2020). An overview of removing heavy metals from sewage sludge: Achievements and perspectives. Environ. Pollut. 266(2), 115375.
Gerrity, D., S. Gamage, J.C. Holady, D.B. Mawhinney, O. Quiñones, R.A. Trenholm et al. (2011). Pilot-scale evaluation of ozone and biological activated carbon for trace organic contaminant mitigation and disinfection. Water Res., 2011, 45(5), 2155–2165.
Ghavidel, A., S. Naji Rad, H.A. Alikhani (2018). Bioleaching of heavy metals from sewage sludge, direct action of Acidithiobacillus ferrooxidans or only the impact of pH?. J. Mater. Cycles Waste Manag. 20, 1179–1187.
Gilbert. S.G. (2004). A small dose of toxicology: The health effects of common chemicals, CRC Press, New York.
Gonzalez, A., A. Hendriks, J.B. van Lier, M. de Kreuk (2018). Pre-treatments to enhance the biodegradability of waste activated sludge: Elucidating the rate limiting step. Biotechnol. Adv. 36, 1434–1469.

Goto, M., T. Nada, A. Kodama, T. Hirose (1999). Kinetic analysis for destruction of municipal sewage sludge and alcohol distillery wastewater by supercritical water oxidation. Ind. Eng. Chem. Res. 38(5), 1863–1865.

Gu, X.-Y., J.W.C. Wong (2007). Degradation of inhibitory substances by heterotrophic microorganisms during bioleaching of heavy metals from anaerobically digested sewage sludge. Chemosphere, 69(2), 311–318.

Hammer Sr, J. Mark, Mark J. Hammer Jr. (2008). Water and wastewater technology. Pearson/Prentice Hall: New York City, New York., pp. 553.

Hirayama, H., K. Takai, F. Inagaki, K.H. Nealson, K. Horikoshi (2005). *Thiobacter subterraneus gen.* nov., sp. nov., an obligately chemolithoautotrophic, thermophilic, sulfur-oxidizing bacterium from a subsurface hot aquifer. Int. J. Syst. Evol. Microbiol. 55(1), 467–472.

Hussain, A., M. Priyadarshi, S. Dubey (2019). Experimental study on accumulation of heavy metals in vegetables irrigated with treated wastewater. Appl. Water Sci. 9, 122.

IWA (2018). Wastewater report 2018: The reuse opportunity. The International Water Association, London.

Jaffer, Y., T.A. Clark, P. Pearce, S.A. Parsons (2002). Potential phosphorus recovery by struvite formation. Water Res. 36(7), 1834–1842.

Kehrein, K., Mark van Loosdrecht, P. Osseweijer, M. Garfí, Jo Dewulf, John Posada (2020). A critical review of resource recovery from municipal wastewater treatment plants – market supply potentials, technologies and bottlenecks. Environ. Sci., 6, 877–910.

Khuder, S., S. Milz, M. Bisesi, R. Vincent, W. McNulty, K. Czajkowski (2007). Health survey of residents living near biosolids permitted farm fields. Arch. Occup. Environ. Health, 62, 5–11.

Kirchmann, H., G. Börjesson, T. Kätterer, Y. Cohen (2017). From agricultural use of sewage sludge to nutrient extraction: A soil science outlook. Ambio, 46(2), 143–154.

Kletzin, A. (2006). Metabolism of inorganic sulfur compounds in archea. Archaea: Evolution, physiology, and molecular biology, Ed. R.A. Garrett and H.P. Klenk, Blackwell Publishing, Oxford, 262–264.

Konishi, Y., H. Nishimura, S. Asai (1998). Bioleaching of sphalerite by the acidophilic thermophile Acidianus brierleyi. Hydrometallurgy, 47(2–3), 339–352.

Laki, S.L., K.V. Nedunuri, R. Kandiah (2019a). Introduction to water resources, Kendall Hunt Publishing Company: Dubuque, IA.

Laki, S.L., K.V. Nedunuri, R. Kandiah (2019b). Water quality. In Introduction to Water Resources Management, Second Edition. Kendall Hunt.

Le Corre, K.S., E. Valsami-Jones, P. Hobbs, S.A. Parsons (2009). Phosphorus recovery from wastewater by struvite crystallization: A review. Crit. Rev. Environ. Sci. Technol. 39(6), 433–477.

Lee, H., T. P. Tan (2016). Singapore's experience with reclaimed water: NEWater. Int. J. Water Resources Dev. 32(4), 611–621.

Lee, H.-S., W.F.J. Vermaas, B.E. Rittmann (2010). Biological hydrogen production: Prospects and challenges. Trends Biotechnol. 28(5), 262–271.

Lee, W.S., A.S.M. Chua, H.K. Yeoh, G.C. Ngoh (2014). A review of the production and applications of waste-derived volatile fatty acids. Chem. Eng. J. 235, 83–99.

Li, H., Y. Jin, Y. Nie (2009). Application of alkaline treatment for sludge decrement and humic acid recovery. Bioresour. Technol. 100(24), 6278–6283.

Li, Y., Yinguang Chen, Jiang Wu (2019). Enhancement of methane production in anaerobic digestion process: A review. Appl. Energ. 240, 120–137.

Lin, J., N. Chen, Y. Pan (2013). Arsenic incorporation in synthetic struvite (NH4MgPO4.6H2O): A synchrotron XAS and single-crystal EPR study. Environ. Sci. Technol. 47(22), 12728–12735.

Liu, R., Xiaodi Hao, Mark C.M. van Loosdrecht, Peng Zhou, Ji Li (2019). Dynamics of humic substance composition during anaerobic digestion of excess activated sludge. Int. Biodeterior. Biodegrad., 145, 104771.

Lo, K.S.L., Y.H. Chen (1990). Extracting heavy metals from municipal and industrial sludges. Sci. Total Environ. 90, 99–116.
Loewenthal, R.E., U.R.C. Kornmuller, E.P. Van Heerden (1994). Modelling struvite precipitation in anaerobic treatment systems. Water Sci. Technol. 30, 107–116.
Manara, P., A. Zabaniotou (2012). Towards sewage sludge based biofuels via thermochemical conversion: A review. Renewable Sustainable Energy Rev. 2012, 16(5), 2566–2582.
Marchioretto, M.M., H., Bruning, N.T.P. Loan, W.H. Rulkens (2002). Heavy metals extraction from anaerobically digested sludge. Water Sci. Technol. 46 (10), 1–8.
McCarty, P.L., J. Bae, J. Kim (2011). Domestic wastewater treatment as a net energy producer: Can this be achieved? Environ. Sci. Technol. 45(17), 7100–7106.
Mehovich, J., J. Blum (2004). Arsenic Poisoning in Bangladesh, South Asia Research Institute for Policy and Development.
Metcalf, L., and H.P. Eddy. (2002). Wastewater Engineering: Treatment and Reuse. 4th Edition. Ed. George Tchobanoglous, Franklin L. Burton, and H. David Stensel. McGraw-Hill, Inc., pp. 1848
Muller, E.E., A.R. Sheik, P. Wilmes (2014). Lipid-based biofuel production from wastewater. Curr. Opin. Biotechnol. 2014, 30, 9–16.
Münch, E.V., K. Barr (2001). Controlled struvite crystallization for removing phosphorus from anaerobic digester sidestreams. Water Res. 35(1), 151–159.
NOAA (2010) Hypoxia, https://oceanservice.noaa.gov/hazards/hypoxia/
NOAA (2016), News and Features: NOAA partners predict smaller harmful algal bloom for western Lake Erie, https://www.noaa.gov/media-release/noaa-partners-predict-smaller-harmful-algal-bloom-for-western-lake-erie.
Nouha, K., R.S. Kumar, S. Balasubramanian, R.D. Tyagi (2018). Critical review of EPS production, synthesis and composition for sludge flocculation. J. Environ. Sci. 66 (2018) 225–245.
O'Connor, D., D.Y. Hou, Y.S. Ok, J. Mulder, L. Duan, Q.R. Wu, S.X. Wang, F.M.G. Tack, J. Rinklebe (2019). Mercury speciation, transformation, and transportation in soils, atmospheric flux, and implications for risk management: A critical review. Environ. Int. 126, 747–761.
Oller, I., S. Malato, J.A. Sánchez-Pérez (2011). Combination of advanced oxidation processes and biological treatments for wastewater decontamination: A review. Sci. Total Environ. 409(20), 4141–4166.
Park, J.B.K., R. J. Craggs, A. N. Shilton (2011). Wastewater treatment high rate algal ponds for biofuel production. Bioresour. Technol. 102(1), 35–42.
Pathak, A., M.G. Dastidar, T.R. Sreekrishnan (2009). Bioleaching of heavy metals from sewage sludge: A review. J. Environ. Manage. 90(8), 2343–2353,
Qin, B., Yunlin Zhang, Guangwei Zhu, Zhijun Gong, Jianming Deng, David P. Hamilton, Guang Gao, Kun Shi, Jian Zhou, Keqiang Shao, Mengyuan Zhu, Yongqiang Zhou, Xiangming Tang, Liang Li (2020). Are nitrogen-to-phosphorus ratios of Chinese lakes actually increasing? Proc. National Acad. Sci. 117(35), 21000–21002.
Rao, D. G. (ed.) (2013). Wastewater treatment: Advanced processes and technologies, environmental engineering, CRC Press, Boca Raton, FL., p. 365.
Rehman, K., A. Ijaz, M. Arslan, M. Afzal (2019). Floating treatment wetlands as biological buoyant filters for wastewater reclamation. Int. J. Phytorem. 21 (13), 1273–1289.
Rehman, K., A. Imrana, I. Amina, M. Afzal (2018). Inoculation with bacteria in floating treatment wetlands positively modulates the phytoremediation of oil field wastewater. J. Hazard. Mater. 349, 242–251.
Réveillé, V., L. Mansuy, É. Jardé, É. Garnier-Sillam (2003). Characterisation of sewage sludge-derived organic matter: Lipids and humic acids. Org. Geochem. 34(4), 615–627.
Ritter, W. F., S.R. Chitikela (2020). The Mississippi River Basin Nitrogen problem: Past history and future challenges to solve it, ASCE Watershed Management, 109–123.

Rubio-Rincón, F.J., C.M. Lopez-Vasquez, M. Ronteltap, D. Brdjanovic (2014). Seawater for phosphorus recovery from urine. Desalination 348, 49–56.

Sakthivel, S.R., E. Tilley, K.M. Udert (2012). Wood ash as a magnesium source for phosphorus recovery from source-separated urine. Sci. Total Environ. 419, 68–75.

Santos Sánchez, A. (2020). Technical and economic feasibility of phosphorus recovery from wastewater in São Paulo's Metropolitan Region. J. Water Process Eng. 38, 101537.

Sheik, AS.R., E.E.L. Muller, P. Wilmes (2014). A hundred years of activated sludge: Time for a rethink. Front Microbiol. 5, 47.

Siciliano, A., S. De Rosa (2014). Recovery of ammonia in digestates of calf manure through a struvite precipitation process using unconventional reagents. Environ. Technol. 35, 841–850.

Siciliano, A., C. Limonti, G.M. Curcio, R. Molinari (2020). Advances in struvite precipitation technologies for nutrients removal and recovery from aqueous waste and wastewater. Sustainability 12, 7538.

Smart, N.G., T.E. Carleson, S. Elshani, S. Wang, C.M. Wai (1997). Extraction of toxic heavy metals using supercritical fluid carbon dioxide containing organophosphorus reagents. Ind. Eng. Chem. Res. 36(5), 1819–1826.

Smith, A.H., Elena O. Lingas, M. Rahman (2000). Contamination of drinking-water by arsenic in Bangladesh: A public health emergency. Bull. World Health Organ. 78 (9), 1093–1103.

Snoeyink, V.L., D. Jenkins (1980). Water chemistry, John Wiley and Sons, New York, pp. 306–310.

Sohi, S., E. Loez-Capel, E. Krull, R. Bol (2009). CSIRO Land and Water Science Report: Biochar's roles in soil and climate change: A review of research needs.

Sposito, G. (2016). The chemistry of soils. Third Edition. Oxford University Press: New York, NY, pp. 44–60.

Stefanakis, A.I. (2016). Modern water reuse technologists: Tertiary membrane and activated carbon filtration, In: Saeid Eslamian (Ed.), Urban water reuse handbook, 2016.

Sutherland, D.L., Jason Park, Stephan Heubeck, Peter J. Ralph, Rupert J. Craggs (2020). Size matters: Microalgae production and nutrient removal in wastewater treatment high rate algal ponds of three different sizes. Algal Res. 45, 101734

Tetratech (2013). Cost estimate of phosphorus removal at wastewater treatment plants: A technical support document prepared for Ohio Environmental Protection Agency.

Trussel, R. R. (2012) Water reuse: Potential for expanding the nation's water supply through reuse of municipal wastewater. National Research Council (U.S.), ed. National Research Council (U.S.), National Academies Press, Washington, D.C, p. 262.

Ullah, S., Z. Iqbal, S. Mahmood, K. Akhtar, R. Ali (2020) Phytoextraction potential of different grasses for the uptake of cadmium and lead from industrial wastewater. Soil Environ. 39(1), 77–86.

U.S. DOE (2000). Energy & Environment Profile of the US Chemical Industry, DOE Report.

U.S. EPA (2000). Biosolids Technology Fact Sheet Land Application of Biosolids, 832-F-00-064.

U.S. EPA (2016). U.S. Action Plan for Lake Erie 2018–2023.

U.S. EPA (2019). Biosolids use and disposal from major POTWS, Enforcement and compliance History Online.

U.S. EPA (2020). Mississippi River/Gulf of Mexico Hypoxia Task Force Update, https://www.epa.gov/ms-htf/northern-gulf-mexico-hypoxic-zone

Uysal A., Y.D. Yilmazel, G.N. Demirer (2010). The determination of fertilizer quality of the formed struvite from effluent of a sewage sludge anaerobic digester. J Hazard Mater. 181(2), 48–54.

van der Hoek, J.P., H. de Fooij, A. Struker (2016). Wastewater as a resource: Strategies to recover resources from Amsterdam's wastewater, Resour. Conserv. Recycl. 113, 53–64.

Verstraete, W., S. E. Vlaeminck (2011). Zero wastewater: Shortcycling of wastewater resources for sustainable cities of the future. Int. J. Sustainable Dev. World Ecol. 18(3), 253–264.

Verstraete, W., P. Van de Caveye, V. Diamantis (2009). Maximum use of resources present in domestic "used water." Bioresour. Technol. 100(23), 5537–5545.

Veeken, A.H.M., H.V.M. Hamelers (1999). Removal of heavy metals from sewage sludge by extraction with organic acids. Water Sci. Technol. 40(1), 129–136.

Westerhoff, P., S. Lee, Y. Yang, G.W. Gordon, K. Hristovski, R.U. Halden et al. (2015). Characterization, recovery opportunities, and valuation of metals in municipal sludges from U.S. wastewater treatment plants nationwide. Environ. Sci. Technol. 49(16), 9479–9488.

WHO (2017). Ten chemicals of major health concern. Retrieved from www.who.int/ipcs/assessment/public_health/chemicals_phc/en/index.html

Wintgens, T., T. Melin, A. Schäfer, S. Khan, M. Muston, D. Bixio (2005). The role of membrane processes in municipal wastewater reclamation and reuse. Desalination 178(1–3), 1–11.

Xie, M., H.K. Shon, S.R. Gray, M. Elimelech (2016). Membrane-based processes for wastewater nutrient recovery: Technology, challenges, and future direction. Water Res. 89, 210–221.

Yakaboylu, O., J. Harinck, K. Smit, W. de Jong (2015). Supercritical water gasification of biomass: A literature and technology overview. Energies 8(2), 859–894.

Yang, W., Wei Song, Ji Li, Xiaolei Zhang (2020). Bioleaching of heavy metals from wastewater sludge with the aim of land application. Chemosphere 249, 126134.

Zhang, J., Baoyi Lv, Meiyan Xing, Jian Yang (2015). Tracking the composition and transformation of humic and fulvic acids during vermicomposting of sewage sludge by elemental analysis and fluorescence excitation–emission matrix. Waste Manage. 39, 111–118.

Zhang, M., Jun Gu, Yu Liu (2019). Engineering feasibility, economic viability and environmental sustainability of energy recovery from nitrous oxide in biological wastewater treatment plant, Bioresour. Technol. 282, 514–519.

Zhang, Y., E. Desmidt, A. Van Looveren, L. Pinoy, B. Meesschaert, B. Van der Bruggen (2013). Phosphate separation and recovery from wastewater by novel electrodialysis. Environ. Sci. Technol. 2013, 47(11), 5888–5895.

Zhou, Xu et al. (2020). Enhancement of productivity of Chlorella pyrenoidosa lipids for biodiesel using co-culture with ammonia-oxidizing bacteria in municipal wastewater. Renewable Energy 151, 598–603.

Zorpas, A.A. (2016). Sustainable waste management through end-of- waste criteria development. Environ. Sci. Pollut. Res. 23(8), 7376–7389.

Index

D

E

L

M

N

O

P

Q

R

S

Printed in the United States
by Baker & Taylor Publisher Services